Engineering Materials

For further volumes:
http://www.springer.com/series/4288

Damien Alloyeau · Christine Mottet
Christian Ricolleau
Editors

Nanoalloys

Synthesis, Structure and Properties

Springer

Editors
Damien Alloyeau
Matériaux et Phénomènes Quantiques
Université Paris Diderot
Bâtiment Condorcet
75205 Paris
France

Christian Ricolleau
Matériaux et Phénomènes Quantiques
Université Paris Diderot
Bâtiment Condorcet
75205 Paris
France

Christine Mottet
Campus de Luminy
CINaM-CNRS
Case 913
13288 Marseille
France

ISSN 1612-1317
ISSN 1868-1212 (electronic)
ISBN 978-1-4471-4013-9 (Hardcover)
ISBN 978-1-4471-4014-6 (eBook)
DOI 10.1007/978-1-4471-4014-6
Springer London Heidelberg New York Dordrecht

Library of Congress Control Number: 2012937648

Printed on acid-free paper

Springer is part of Springer Science+Business Media (www.springer.com)

Preface

Since the Bronze Age, 4000 years ago, man has applied the common saying "unity is strength" to materials science by using metallic alloys for their fascinating physical properties. With the development of metallurgy since the industrial revolution, many combinations of metals have been exploited to meet the technological needs created by the world modernization. Nowadays, recent technological advances in materials science go through their size reduction. In a general manner, when the dimensions of a material raise the same order of magnitude as a characteristic length of the system (mean free path of electrons, correlation length in phase transitions, Bohr radius of the exciton), its properties may be modified from those of bulk, being then dominated by finite-size effects. The size-dependent properties of materials have generated a tremendous interest in nanoscale systems for the last 30 years. Although discovered in 1857 by Faraday, metallic nanoparticles are still at the center of this intense research effort. The development of nanoscale investigation techniques allowed studying the unusual properties of metallic nanoparticles, which are now well documented and exploited in electronics, optics, magnetism, catalysis and medicine. The idea to combine finite-size effects with the adaptability of metallic alloys has added a new dimension to the study of metallic clusters. Understanding the variability of the properties of bimetallic or multi-metallic alloy clusters—so-called nanoalloys—has emerged as one of the most exciting topics in nanoscience, fascinating both physicists and chemists. The unique potential of nanoalloys arises from the fact that their physical or chemical properties can be tuned by varying their composition, their type of atomic arrangement (segregation, solid solution, ordering), as well as their size and morphology. However, the complexity of nanoalloys requires a multidisciplinary approach, because well-controlled synthesis methods and both experimental and theoretical studies of their atomic structure are essential to understand their many technologically relevant properties.

Although the interest of the scientific community in nanoalloys is substantial, there was, so far, no book dedicated to this topic. The authors of the following 11 chapters have all been studying specific aspects of nanoalloys for many years, from fabrication (chemical and physical routes) to physical and chemical properties using various dedicated methods of characterization. This collaborative effort aims to give, from both experimental and theoretical points of view, the basis for the comprehension of such complex nanosystems. This book should provide a deeper understanding of the mechanisms involved in the growth of bimetallic nanoparticles and their essential properties (thermodynamic, electronic, optical, magnetic, and catalytic), depending on their size and chemical composition. This work is divided into three parts.

(i) *Growth and structural properties* (Chaps. 1–4). Part I aims to describe the nucleation and growth mechanisms, while taking into account the important kinetic limitations involved in nanoalloy synthesis. This part also presents a broad overview of the experimental techniques giving access to morphological, structural, and chemical information at the atomic scale (scanning probe microscopy, X-ray synchrotron experiments and transmission electron microscopy).
(ii) *Theoretical investigations of electronic, atomic structure, and thermodynamics* (Chaps. 5–8). In Part II, the electronic properties of alloys and nanoalloys are developed giving rise to their energetics and thermodynamics in order to predict the most favorable structures and chemical arrangements as a function of their composition, temperature, and size.
(iii) *Technologically relevant properties* (Chaps. 9–11). In Part III the authors describe the complex phenomena that arise from combinations of size and composition effects in the fields of magnetism, optics, and catalysis.

Damien Alloyeau
Christine Mottet
Christian Ricolleau

Contents

Nucleation and Growth of Bimetallic Nanoparticles

Christophe Petit and Vincent Repain

Abstract In the large family of nanomaterials and more specifically of inorganic nanocrystals, bimetallic alloy nanocrystals represent a particularly interesting class of materials owing their potential use in catalysis, ultra-high density magnetic recordings, and new development in sustainable energy (*i.e.* fuel cells). In order to study these specific properties, a wide variety of both chemical and physical routes have been developed for the synthesis of model nanoparticles. In particular, one of the specific challenges in the synthesis of nanoalloys is to control and to characterize the alloy composition at the nanometer scale. In this chapter, we give an overview of the main chemical and physical techniques used to synthesize bimetallic nanoparticles together with a discussion of the concepts of nucleation and growth for such objects and their consequence on their structural properties (size, shape, composition, ordering...).

1 Introduction

Metallic nanoalloys will initiate important development in nanotechnologies due to their specific chemical and physical properties (*i.e.* in catalysis, magnetism, optics, etc.). It is now well known that these properties are mainly controlled by

C. Petit (✉)
Laboratoire des Matériaux Mésoscopiques et Nanométriques (UMR CNRS 7070), Bâtiment F, Case 52, Université Pierre et Marie Curie—Paris 6, 75252, Paris cedex 05, France
e-mail: christophe.petit@upmc.fr

V. Repain
Laboratoire Matériaux et Phénomènes Quantiques (UMR CNRS 7162), Bâtiment Condorcet, Case courrier 7021, Université Paris Diderot—Paris 7, 75205, Paris cedex 13, France
e-mail: vincent.repain@univ-paris-diderot.fr

D. Alloyeau et al. (eds.), *Nanoalloys*, Engineering Materials,
DOI: 10.1007/978-1-4471-4014-6_1,

the fine tuning of structural parameters such as the size, the bimetallic composition and segregation processes. Concerning their fabrication, the bottom up approach, either physical or chemical, is ideal to design this specific class of nanomaterials due to its versatility, facility and low cost. However, the realization of well controlled bimetallic nanoparticles is not always straightforward from the know-how developed for monometallic nanoparticles. In the following, we review some of the most classical chemical and physical routes to produce nanoalloys, focusing on the growth processes and their specificities in the case of bimetallic nanoparticles. We discuss the influence of growth parameters on the size, shape, density and composition of the nanoalloys by using experimental examples in both chemical and physical routes.

In a sustainable approach, soft chemistry is well-adapted to produce such nanoalloys in large amount. However at the nanometer scale, as the properties are strongly dependent on the size and the surface state (raw or passivated), it is crucial to develop method where the polydispersity in size and composition is finely controlled. These impose to clearly separate the nucleation step from the growth process and also to control this latter to limit the size. This can be done by using colloidal assemblies, as the micellar media or the two phase system, where both the nucleation and growth process are clearly separate in space and time. More recently the organo-metallic approach has taken a growing place in this chemical route as this process allows decreasing size distribution. Nevertheless, depicted the large amount of work made on the synthesis of nanoalloys by the chemical way, there is still open questions considering the control of composition and especially the segregation process. For example, in the chemical approach, the nanoparticles are always passivated by an organic molecule. The role of this capping agent on shape and segregation control is still under discussion.

In a second part, the main physical vapor deposition techniques are described, focusing on their particular interest for the growth of nanoalloys, illustrated by few examples. We discuss in more detail the key point of the experimental control of composition and size, highlighting the crucial role of the substrate and growth parameters in the final result. Finally, we give a short overview of a mean field approach of nucleation and growth in different growth regimes and discuss the predictions of such a model on the case of supported bimetallic nanoparticles.

2 Chemical Routes

Chemical routes are directly deduced from the general framework of the chemical synthesis of inorganic nanocrystals which offers a wide variety of experimental conditions. Nanocrystals from these syntheses are usually in the form of colloids. One of the advantages of chemical methods is the ability to manipulate nanocrystals after synthesis. Indeed the post-synthesis treatments, the functionalization of surfaces or self-assembly process allow the chemist to manufacture new materials with specific properties bearing on the properties of new nanoalloys.

Two aspects of these materials have to be taken into account to elaborate the chemical process: on one hand the controls of the size and the size distribution, in order to control the physical and chemical properties and on the other hand, in the specific case of nanoalloys, the control of the composition.

2.1 General Concepts

The simplest and most often used method to produce metallic nanoparticles is to reduce the corresponding metal precursor in a solution in presence of protecting agent or in confined media to limit the growth [1]. In this last case, the size is determined by the size of confined media. More generally the size is determined by the number of atoms produced and the number of the nanocrystals formed, which are dependant on the kinetics of nucleation and growth of the nanocrystals.

As the physical and chemical properties of the metallic nanocrystals depend on their size, one of the key point, especially for application development, is to control the size distribution and, in case of the nanoalloys, the homogeneity of the composition. This can be done by the separation of nucleation and growth (as in the organometallic method) and diffusion controlled growth (as in liquid–liquid phase method or by using colloidal systems as nanoreactor). The kinetics of these competing processes can be altered by changing the experimental conditions (temperature, pressure, solvent type or nature of the metallic precursor, reducing agent or capping agent). LaMer in his pioneering work published in 1946 [2] proposed a general framework to understand qualitatively the role of the control of nucleation process.

LaMer considered the case of a homogeneous nucleation process. Due to the evolution of the system from a homogeneous phase to a heterogeneous phase (liquid + nanocrystals), there exists a high energy barrier, the LaMer plot shows how this energy barrier works to separate nucleation and growth step (Fig. 1). Three periods can be identified:

(i) *In the initiation phase* (Step 1), the concentration of monomer (the smallest subunit of the crystal) increases continuously even under supersaturated conditions because the energy barrier for spontaneous homogeneous nucleation is very high.
(ii) *In the nucleation phase* (Step 2), the degree of supersaturation is high enough to overcome the energy barrier and nucleation occurs, yielding to the formation of stable nuclei. These nuclei start to growth. As a consequence of these two processes, the monomer concentration decreases until it reaches the level at which the net nucleation rate is zero.
(iii) *In the growth phase* (Step 3), the nucleation is effectively stopped and the particles growth as long as the solution is supersaturated.

In this framework we can understand the concept of 'burst nucleation' which is at the origin of the recent development of the organo-metallic route to produce

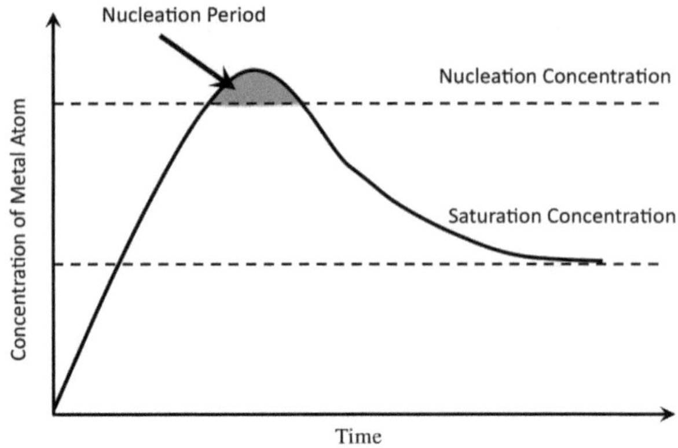

Fig. 1 The LaMer diagram illustrating variation of the monomer concentration with time during the growth process to obtain monodisperse population of nanocrystals. Redrawn from Ref. [3]

nanocrystals: in this process, nuclei are produced at the same time and then nuclei start to grow without additional nucleation. Conversely if nucleation process occurs during the formation of nanocrystals, the growth histories differ largely from one particle to another yielding to a large size distribution [4].

In the following, we will describe some of these chemical routes (mainly the reduction of metallic salt and the decomposition of organic precursor and their application to the specific case of the nanoalloys). It is not an exhaustive list but the most common process, which can be used to synthesize bimetallic nanocrystals. We will explain the role of the capping agent and the structuration of the liquid media to limit the growth, yielding to size control.

2.2 Reduction of Metallic Salts

This is the simplest method used to produce metallic nanoparticles, more often monometallic as Ag, Au, Pd, Ru, Pt…. This occurs in presence of polymer or capping agent or in confined media to control the size. Most commonly used reductants are sodium borohydride, hydrogen, alcohols… This method is illustrated by the work of Turkevich synthesizing stable solution of gold nanocrystals in water by using citrates ions both as reductant and stabilizer (stabilization occurs by electrostatic repulsion due to the charge of the citrates ions surrounding the nanocrystals) [5]. Reduction of metallic salts by sodium borohydryde has been largely used to synthesized mono or bimetallic nanocrystals [1, 4].

2.2.1 Synthesis in Reverse Micelles

In order to control the size and also to separate nucleation and growth, confined media has been largely used. This is the case of reverse micelles (water in oil nano-droplets stabilized by a surfactant). The in situ synthesis in the water-pools of reverse micelles was developed in the 1980s by Pileni et al. [6, 7]. In this method,

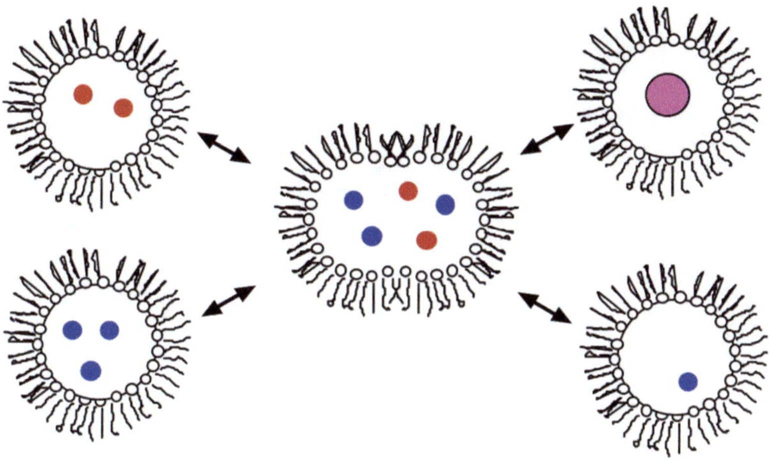

Fig. 2 Scheme of the inorganic synthesis using reverse micelle as a nanoreactor. Solution A (*blue*) and B (*red*) are mixed. Due to Brownian motion, inter-micellar exchanges are possible, yielding to reaction in confined media and formation of inorganic nanocrystals in the water-pool

the inner core of the reverse micelles can be considered as a nanoreactor and the size of the nanoparticles obtained is often approximately limited by that of the water pools. This is illustrated by the scheme presented in Fig. 2. Nucleation first takes place inside the water-pool and then growth process occurs at the minute scale due to inter-micellar collision.

This method has been used for synthesis of semiconductor materials such as CdSe [7, 8], of metallic nanoparticles such as Pt [9, 10], Cu [11, 12], Co [13], Ag [14], and Au [15] but also of nanoalloys such as CoPt [16–18], PtPd [19], FeCu [20] or AuAg [21]. In this last case control of composition is reached due to the simultaneous location at the interface of the micelle of both metallic precursors yielding to an effective control of the average composition. Nanocrystals can be extracted from the micellar media by anchoring a strong capping agent as alkane thiols or dodecanoic acid to the metallic surface of the nanocrystals. Micellar media is then broken and passivated metallic nanocrystals can be recovered as a powder easily dispersible in organic solvent [13–15].

There is, however, some limitation to the micellar way, mainly the low yield of formation of the nanocrystals and the difficulty to control the size polydispersity as the growth process is not completely separated from the nucleation step. This last difficulty can be overcome by using post-synthesis process as the size selection precipitation [14]. Figure 3 shows some example of metallic and bimetallic nanoparticles obtained by this way.

2.2.2 Two-Phase Synthesis

The second method involves transfer of the metal ion from a polar phase to a non-polar phase using a transferring agent. It has been used for synthesis of metallic

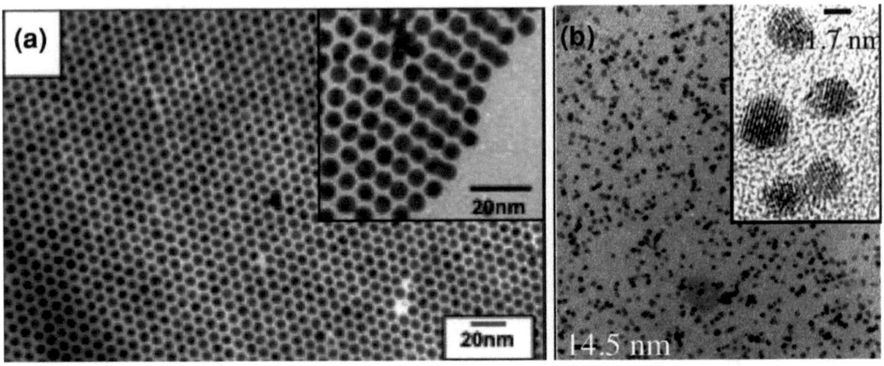

Fig. 3 **a** Gold Nanocrystals synthesized in situ in AOT reverse micelles (from Ref. [15], Copyright (2010) American Chemical Society). **b** CoPt nanocrystals synthesized in situ in AOT reverse micelles (from Ref. [16], copyright (2004), American Institut of Physics)

nanoparticles and the phase transfer method, also called two-phase synthesis, has been developed by Brust et al. in the 1990s [22–24]. It has been largely used to synthesize metallic nanoparticles as silver, gold, platinum or palladium but also bimetallic nanoalloys as AuAg [25] or CoPt [26, 27]. It typically involves the transfer of the metal precursor (metallic ions) from an aqueous solution to an organic solution containing a capping molecule as alkane thiol or amine. The transfer is assisted by a phase transfer agent such as tetradecylamonium bromide (TDAB). Reduction of metallic precursor is then carried by adding an aqueous solution of reducing agent (mainly $NaBH_4$) under vigorous stirring (Fig. 4).

The reduction takes place at the interface between the two phases. The nuclei are mainly solubilized in the organic phase due to the presence of both the capping molecule and phase transferring agent where the growth process takes place. The interest of this method is that the kinetics of nanocrystal growth is controlled by the surface coverage and thus cluster size is controlled by the reaction conditions at the interface and not by the metal-ion reduction kinetics in the homogenous aqueous phase [29]. Nanocrystals coated by capping molecules can then be recovered as a powder and dispersed in an organic solvent. Synthesis conditions such as concentrations of the metallic salt, the reducing agent and the nature of the capping agent (strongly or weakly anchored, the length of the alkyl chains… [28].) allow to control the kinetics of nucleation and growth of the nanocrystals and then the size. In most of cases, this method can produce large amount of nanocrystals with a low size distribution (around 10%).

Only few works deal on nanoalloys obtained by this two-phase synthesis due to the difficulty to control their composition. Let us illustrated this by the case of CoPt synthesis [26, 27]. Perfect control on the composition can be only achieved if the two precursors are similar in structure and in location in the liquid media. As a matter of fact the large difference in redox potential of platinum and cobalt can induce a variation in the reduction kinetics. As an example, if platinum precursor

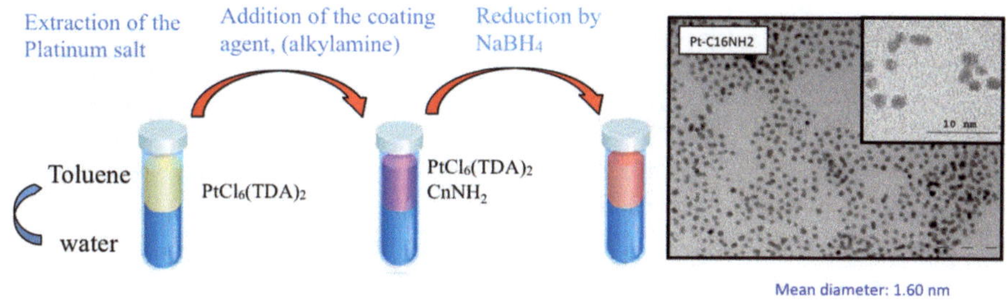

Fig. 4 On the *left*, two phase liquid synthesis of platinum nanocrystals. On the right, TEM picture of Platinum nanocrystals after extraction from the media and dispersion in toluene. From Ref. [28]

is in the organic phase (complexed by the transfer agent) and the cobalt precursor in the water phase (as an aqueous salt), the reduction takes place during the emulsification of the solution obtained by stirring when the reducing agent is added. If the reduction of platinum is dominated by the interface, the reduction of cobalt is dominated by the reduction kinetics of the salt in the aqueous phase. Thus, the co-reduction is carried out in two distinct ways. Therefore, a strong discrepancy occurs in the average content of the cobalt in the nanocrystal compared to the expected ratio. Due to the change in the characteristics of the emulsion droplets from one to another, the homogeneity in composition is low. Conversely if both cobalt and platinum salts are in the organic phase interacting with an interface, only the reaction conditions are predominant and the difference in redox potentials is no longer a problem for this interfacial reaction [29]. Thus, the best results are obtained when cobalt is in the same form as the platinum: $CoCl_2(TDA)_2$ and $PtCl_4(TDA)_2$, *i.e.* both complexed by the same agent transfer. These two molecules have a similar structure and then the composition of the interface where the reduction takes place is directly related to the initial composition of metallic precursor. Hence, the reduction yields to a precise control of the composition. Figure 5 shows typical CoPt nanoalloys obtained by this two-phase synthesis.

Another point should be mentioned considering difficulties that could occur by using two-phase synthesis for bimetallic nanoalloy. In order to increase the size of the metallic nanocrystals, the reducing agent is sometime added before the introduction of the capping agent (which limits the growth). This allows tuning the size in case of gold nanocrystals [23]. In the case of CoPt, the resulting nanocrystals obtained using this procedure are composed of pure platinum without cobalt whatever is the initial composition of the metallic salt [30]. This illustrates the importance of the complexation of the monomer by the capping agent to obtain a perfect control of the composition. In this case, it is probably due to the fact that a stable composition of the interface during the reduction could only take place in presence of the passivating agent. This could also be due to the difference between the stability of initial nuclei composed either of pure platinum or both cobalt and platinum, which can be corrected by complexation with the capping agent.

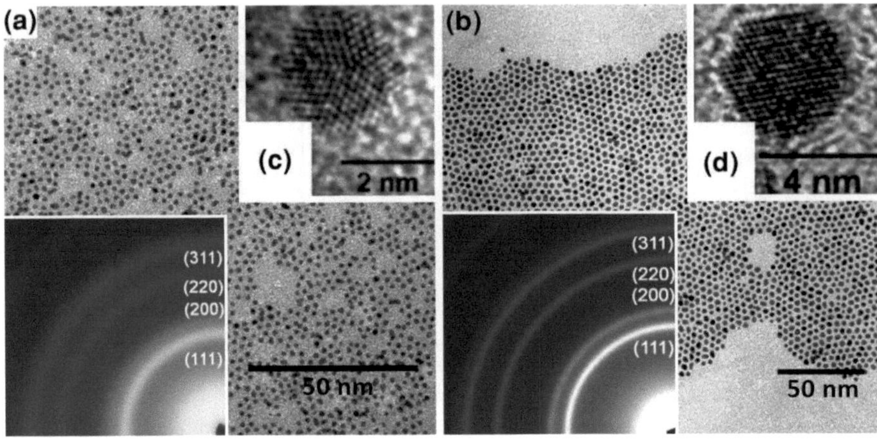

Fig. 5 TEM pictures of CoPt nanocrystals of 2 nm (**a**) and 4 nm (**b**) self-organized in local hexagonal network (inset) electronic diffraction pattern of CoPt nanocrystals shows the disordered structure and figures (**c**) and (**d**) typical HRTEM images of the CoPt nanocrystals. From Ref. [26], Copyright (2007) American Chemical Society

Thus, even if a chemical route works well in case of monometallic nanocrystals, it is not obvious that it can be directly transferred to the case of nanoalloys.

2.2.3 Single Phase Synthesis

Several groups develop single phase methods to synthesize metallic and bimetallic nanocrystals. In these methods, the metal precursor, the reducing agent and the capping agent are all dispersed in the same solvent. This can be done either using $NaBH_4$ to reduce metallic ions in a water/methanol solution in presence of hydrophilic capping agent to control the growth [31, 32] or using a strong hydrophobic reducing agent to reduce metallic complex solvated in an organic solvent in presence of hydrophobic capping agent [33, 34]. Thus the reduction, nucleation and growth occur homogeneously and not at an interface like previously. This could yield to a better control of the nanocrystals nucleation and growth.

Another method based on single phase synthesis is the "polyol process". In this case, the solvent, diol or polyalcohol (as ethylene glycol for example) acts as a reducing agent to reduce the metal salts. However, contrary to the previous one, this reaction is performed at high temperature (typically 100–200°C). This is has been largely used in case of nanoalloys as FePt [35] or NiPd [36]. As example, the use of iron acetylacetonate [$Fe(Acac)_3$] and platinum acetylcetonate [$Pt(acac)_2$] in ethylene glycol or tetraethylene glycol, generates FePt nanocrystals that show partially ordered tetragonal structures [37]. Oleic acid or oleic amine are often used as capping agent and added directly in the chemical bath to limit the growth process. Furthermore, this high temperature process often allows to reach a better crystal quality and to avoid boron contamination often observed in monometallic nanocrystals obtained by single phase borohydride reduction [4].

2.3 The Organometallic Route: Thermal Decomposition Method

In case of the nanoalloys, the derivated co-reduction method is often used. However as the two metal precursors are involved in the reduction reactions, the influence of the experimental conditions on the nucleation and growth is complex as it has been illustrated in the case of CoPt obtained by the two phase method. Thus, the size distribution and crystalline structure of the nanoalloys are difficult to control by these methods. Some problem of reproducibility could occur depending on the purity of the metallic precursor or of the reducing agent [38]. Furthermore, changes in the composition of the nanoalloys have been reported, often coupled with changes in the nanocrystals size and size distribution (C. Petit and D. Alloyeau, Juin 2010, Ecole thématique Nanoalliage, unpublished result) [39].

This is the reason of the development of the organometallic route, in which fast thermal decomposition reactions of organometallic or metal–surfactant complexes were performed at high temperature in presence of surfactant molecules acting as a capping agent. This method is now widely used to synthesize inorganic nanocrystals (not only metallic) because it is a clear example of the concept of "burst nucleation". In fact, as all the precursors have the same structure, they decompose massively at the same time and the subsequent growth by ageing takes place at slightly lower temperature in a media containing the capping agent. All the metal atoms generated from the thermal decomposition of the precursor are transformed into polynuclear clusters, which in turns lead to the nucleation and growth of the metallic nanocrystals (*cf.* Fig. 6) [40].

A better separation between the nucleation and the growth step is observed, which yields to a very narrow size distribution. Typically a size distribution between 5 and 10% could be achieved [4, 35]. This control of the monodispersity is essentially a kinetic process driven by high initial supersaturation. It requires that the precursor be reactive enough to induce high supersaturation immediately after injection of the precursor in the heated solution (burst nucleation) [41]. This allows also a better control of the crystallinity of the nanocrystals, which can be easily dispersed in organic solvents. This route of synthesis, sometimes called "hot injection method", has been developed by the pioneering group of Bawendi for the synthesis of quantum dots [42] and then extended to metal or metal oxides [4]. It has been used by the Murray group to synthesize FePt nanoalloys by thermal decomposition of $Fe(CO)_5$ in presence of oleic acid and oleylamine [35]. The composition of the nanoalloys was varied by changing the molar ratio of the two metal complexes. Size could be controlled by the concentration of initial precursors between 3 and 10 nm. This has been also used to synthesize $CoPt_3$ or CoPt or even FeCoPt [4]. One of the advantages of this technique is the high yields of formation of metallic nanocrystals, some of the variations of this process allow

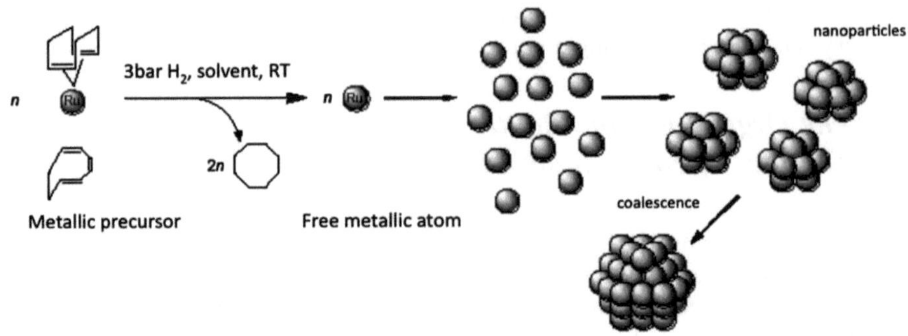

Fig. 6 Schematic of the organo-metallic route for synthesis of ruthenium nanocrystals by decomposition at low temperature under H_2 of ruthenium precursor. From LCC Toulouse (France) and Ref. [40]

gram scale synthesis of nanocrystals. However, it is more complex than the co-reduction method and some problem of reproducibility have been observed. It should be mentioned that, it is also possible to control the structure of the nanoalloys, either homogeneous or core/shell [41]. Hence, by using metallic complexes having different decomposition temperature, it is possible to synthesize first nuclei of one component and then a subsequent increase of the temperature induces the decomposition of the second complex to generate a metallic shell on the top of the initial metallic core [43]. Conversely, to obtain a perfect control of the structure of the nanoalloys, it is important to use precursors having similar temperature of decomposition. The organo-metallic route was also widely developed by Chaudret and collaborators [44]. For example, they synthesize PtRu nanocrystals by decomposition at low temperature of organometallic precursors under dihydrogen in the presence of polyvinylpyrolidone, PVP, as stabilizer [45].

2.4 Other Chemical Methods

Methods presented above are the main routes yielding to elaboration of metallic or bimetallic nanocrystals. However, it has been reported other chemical routes to synthesize nanoalloys. Hence, Byrappa et al. report the synthesis of nanoalloys by using solvothermal and hydrothermal processing [46]. In this method, sealed vessel (as bomb, autoclave…) are used to bring solvents at very high temperature above their boiling points, allowing solubilization of insoluble metallic precursors, which can then react in presence of capping agent to limit and control the size. Metallic nanocrystals and nanoalloys have been synthesized by this way. Ultrasonic fabrication of metallic nanomaterials and nanoalloys have been reported by Möhwald et al. [47], by using the cavitation energy resulting from irradiation of chemical bath by ultrasonic waves to generates the nanocrystals again in presence of surfactant or capping agent to control the growth. Mecano-chemistry (ball-milling) is also used to elaborate nanoalloys. Here, elemental metallic powders in micron size are mechanically alloyed in atomic proportion in a planetary miller running at high speed (around 500 rpm) for a long period (20 h) [48]. This yields to nanocrystalline materials however very polydisperse in size even if the adding of capping agent in the initial mixture allows reduce size and polydispersity. This method is in fact between the chemical method and the physical method as it is more relevant of the so called "top-down" approach.

3 Physical Routes

The realization of nanoparticles or thin films on surfaces under vacuum conditions is generally called physical vapour deposition (PVD) and is widely used, both for fundamental research studies and at an industrial level. The basic processes at the origin of nucleation and growth (diffusion and aggregation) are rather similar than in the chemical syntheses but it is worth noting that they generally happens on a surface whereas in a bulk solution in chemistry. Moreover, nanoalloys realized by PVD are naturally free of capping agents and can therefore be considered as the most simple model system for the comparison between experiments and theoretical predictions. Finally, in the same spirit than with chemical methods, the PVD allows a great variety of bimetallic nanostructures, from metastable states to local equilibrium. At sufficiently low temperature, *i.e.* when the exchange process between two atoms in a cluster is slow compared to the deposition time, it is possible to realize out-of-equilibrium atomic structure. Typically, a co-deposition can lead to disordered alloy whereas subsequent depositions can realize core–shell structures. When the temperature is higher, thermodynamics plays a crucial role and different driving forces can change particle structure, such as chemical hybridization, strain, magnetism... Many different recent examples have been reported in [49] or reviewed in [50].

In the following, we recall the typical experimental setups that have to be used for such studies, with a particular attention on the control of the composition. We show by different examples that the growth conditions and the nature of the substrate can sometimes lead to unexpected results. In a second part, we introduce typical mean field models that are used for the understanding of nucleation and growth and we develop the specific case of bimetallic nanoparticles in different regimes.

3.1 Experimental Techniques and Examples

3.1.1 Typical Experimental Setup

A common feature of PVD technique is the need of a vacuum environment. A typical base pressure in the range 10^{-6}–10^{-8} mbar is generally enough for the different evaporation techniques but the research on model *i.e.* free of contamination nanoparticles generally needs Ultra-High Vacuum (UHV) range, *i.e.* below 10^{-9} mbar, which is achieved by the baking of vacuum chambers. A fast entry small chamber is used for the introduction of substrates from ambient conditions and specific equipment can be used for their in situ cleaning such as heating and ion sputtering. As developed in the following, the temperature of the substrate is a crucial parameter in the final state of nanoparticles and therefore a precise temperature control of the samples is generally required, above and below room temperature.

3.1.2 Evaporation Techniques

Different techniques of PVD have been developed for specific needs. From the point of view of atomic processes leading to nanoparticles, we can distinguish evaporation techniques where the nucleation and growth of clusters happen on the substrate (molecular beam epitaxy, pulsed laser deposition, sputtering…) from those where the nucleation and growth happen in the gas phase (low energy cluster beam deposition…), before landing on the substrate, in a more similar way than what is done by chemical routes. Practically, what differentiates these different techniques is the way to obtain the vapour pressure, related to the purity and the range of accessible vapour pressure (or flux of incident atoms on the substrate). In the following, we describe briefly some of the main PVD techniques that have led to model nanoalloys.

Molecular Beam Epitaxy

The Molecular Beam Epitaxy (MBE) is the simplest method of evaporation, usually run under UHV conditions. It consists in heating the evaporant, either in a crucible or directly from a rod, close to the melting or sublimation temperature in order to generate a given vapour pressure. The deposition time on the substrate is controlled by the opening and closing of a mechanical shutter located close to the evaporant. The heating process can be either resistive or by electron bombardment (generally called Electron Beam Physical Vapor Deposition). The advantages of such techniques are their versatility, the potential great purity of deposited materials and the precise control of a regular flux over a wide range. In UHV conditions, the typical rates of deposition are low *i.e.* nm.min^{-1} and less, what is well adapted to the growth of nanoparticles in the contrary of coating processes that are generally done by sputtering techniques with higher fluxes. It is important to note that MBE gives the best crystalline layers and is therefore widely used in the microelectronic industry. The way of making a nanoalloy is simply to co-evaporate different materials from different sources on the same substrate (*cf.* Fig. 7a, b). The ratio of fluxes should fix the concentration of the alloy,

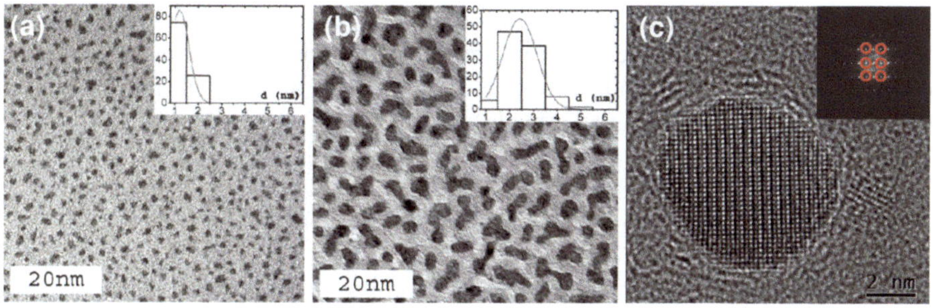

Fig. 7 Transmission Electron Microscope images of CoPt nanoparticles realized by MBE in (**a**) and (**b**) and by PLD in (**c**). **a** Room temperature co-deposition of 2.10^{15} atoms/cm^2 on amorphous carbon. The inset shows the corresponding size distribution. **b** Identical to (**a**) with a deposit of 8.10^{15} atoms/cm^2 (from Ref. [51], copyright (2008), with permission from Elsevier). **c** High resolution image of a chemically ordered $Co_{50}Pt_{50}$ particle realized by PLD at room temperature on amorphous carbon and annealed at 1,023 K. In inset, Fourier Transform of the lattice showing the spots associated to the $L1_0$ order

the deposition time and the temperature of the substrate control the cluster size, as discussed in Sect. 3.2.

Pulsed Laser Deposition

Another PVD technique is the Pulsed Laser Deposition (PLD), also called laser ablation. The vapour phase is generated by the shooting of a target (evaporant) by a short (typically ns) and energetic (typically J) laser pulse, periodically in time (typically Hz). This process generates directional plasma of evaporant that is condensed on the substrate, facing the target. In a typical setup, the amount of matter deposited on the substrate by one shoot can be as small as 10^{-3} nm, meaning that thousands of cycles can be needed to realize a nanometre scale deposit. In average, the typical fluxes are therefore rather similar to those of MBE. In order to evaporate homogeneously the target, this latter is usually rotating. A rotating mirror can also focus the laser spot alternatively on different targets. The advantages of using such a setup is the possibility to evaporate easily almost every materials among which oxides like alumina and refractory elements like tungsten. Moreover, the evaporants can be introduced and removed from the vacuum chamber in a more versatile way than MBE. One of the main drawbacks of such a technique in the realization of nanoparticles is that the huge power density generated by the laser pulse generally makes a small percentage of micron scale droplets of matter, what limits the large scale homogeneity of the sample. Concerning the realization of nanoalloys, two procedures can be used. The first one is to ablate a target of alloy, what imposes the concentration. The second one is to shoot alternatively two targets of pure elements. By changing the ratio between the numbers of shoots on the two elements, it is possible to control the concentration with a great reproducibility (*cf.* Fig. 7c) [52]. It is worth noting that this latter technique is rather different from the common MBE co-evaporation and could lead to different morphologies although the amount of pure materials deposited by each pulse is very small.

Low Energy Cluster Beam Deposition

Another original PVD technique is the production of nanoparticles by gas phase condensation. Typically a vapour of evaporants is produced by pulsed laser shots or magnetron sputtering. This vapour is thermalized through a supersonic expansion of an inert gas (He). The first stage of condensation is stopped in the so-called freezing zone when the temperature ceases to decrease in the directed gas flow (*cf.* Fig. 8a). The initial vapour pressure of evaporants and the velocity of the inert gas mainly control the size of aggregates that can be as small as few atoms. This technique has been developed and extended to the case of bimetallic compounds in the 1980s [53, 54]. It is worth noting that the very high cooling rate (typically 10^{11} K/s) generated by the supersonic expansion makes this technique extremely out-of-equilibrium with the interesting possibility to freeze some metastable configurations of thermodynamically unstable alloys. Moreover, once formed, the aggregates can be ionized and mass selected through a quadrupolar mass spectrometer, what allows unprecedented narrow size distribution (*cf.* Fig. 8b). These clusters with a rather small kinetic energy (around 0.1 eV/atom) can therefore be deposited on a solid substrate without any fragmentation, giving the name of this complex but powerful deposition technique (LECBD for Low Energy Cluster Beam Deposition).

3.1.3 Control of the Composition

The precise control of the composition of nanoalloys is an important issue in their synthesis. The first important remark is that the nucleation and growth process is inherently based on stochastic processes and therefore the composition is not perfectly fixed from particles to particles. Moreover, it often happens that the averaged composition is different from expected for different reasons. Whereas the use of a bulk alloy source as evaporant seems to be a simple way to define a precise composition, it can lead to unexpected results due to a non-stoichiometric sublimation of the different elements. The use of highly out-of-equilibrium sublimation processes like PLD or sputtering generally help to keep a stoichiometric evaporation as compared to MBE. However, the co-deposition technique with the real-time monitoring of the different fluxes is the best way of controlling the final composition x of A_xB_{1-x}. It is indeed simply given by the ratio between the flux F_A over the total flux $F_A + F_B$. The flux monitoring is generally performed using a quartz microbalance. The principle is to measure the frequency shift of a golden quartz oscillator due to the change of mass when depositing materials. The high

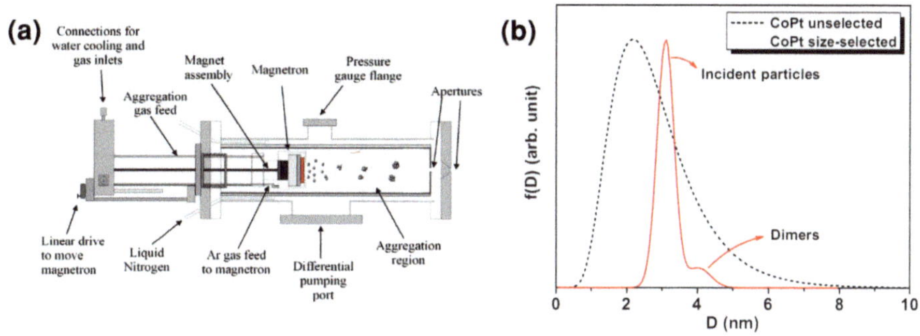

Fig. 8 **a** Schematic drawing of a gas condensation aggregates source based on magnetron sputtering (courtesy of M. Hillenkamp, LASIM, Lyon, France). **b** Size distribution of CoPt aggregates deposited on amorphous carbon with and without size selection by quadrupolar deflector (from Ref. [55], Copyright (2010) by The American Physical Society)

quality factor of such oscillators combined with their high oscillation frequency (around 5 MHz) allows single atomic layers detection. However, it is rather delicate to obtain an accurate absolute value of flux with this technique which needs several inputs as geometrical factor, material density and acoustic wave properties. A calibration of a quartz microbalance with a more quantitative measurement like Rutherford Backscattering Spectrometry is therefore highly recommended for precise determination of both amount of deposited materials and concentration of alloys. It is worth noting that the temperature of the microbalance is regulated around room temperature with a water flow. Together with its golden coating, this implies that the sticking coefficient of most of vapours from metallic elements on the balance is close to one. Depending on the nature of the substrate and on the deposition parameters (fluxes, temperature of the substrate), the sticking coefficients of the evaporated elements can be different than one on the sample, giving rise to discrepancies in the amount of material and concentration as determined by the microbalance.

3.1.4 Role of the Substrate

Substrates with Low Adsorption Energy

As discussed in the preceding section, the nature of the substrate can drastically influence the concentration of nanoalloys for given deposition parameters. Typically, insulating materials or amorphous carbon generally display low energy of adsorption for metals species. Therefore, the sticking coefficients on such samples are often lower than one for room temperature deposition and above. A direct consequence on the growth of nanoalloys is shown in Fig. 9. In this example of CuPd growth on a NaCl substrate at 280°C (*cf.* Fig. 9a), the mean concentration in nanoparticles, checked under the TEM by EDX analysis (see Chap. 4, Sect. 4.1),

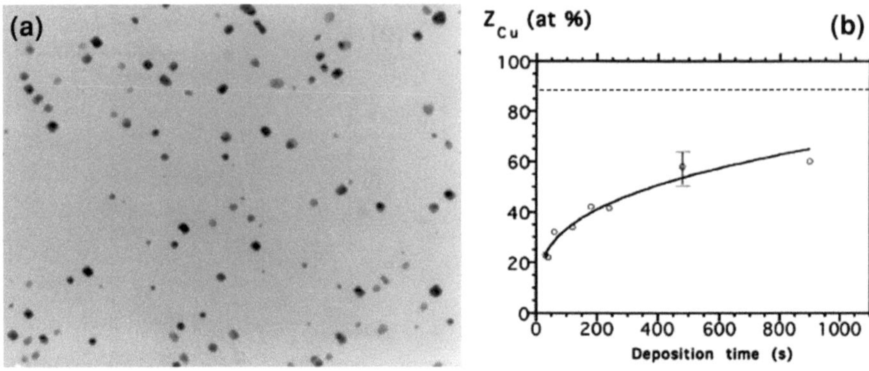

Fig. 9 **a** TEM image of a co-deposition of Cu and Pd at T = 280°C on NaCl(001). **b** Mean concentration of Cu as function of the deposition time in this sample. The dotted line indicates the nominal expected concentration. From Ref. [56]

varies with the deposition time (*cf.* Fig. 9b) with an excess of Pd as compared to the nominal composition x = 0.89 given by $F_{Pd} = 1.10^{13}$ cm^{-2} s^{-1} and $F_{Cu} = 8.10^{13}$ cm^{-2} s^{-1} [56]. This result has been interpreted by different diffusion length before evaporation for both species *i.e.* smaller for Cu than for Pd, as will be discussed in more detail in Sect. 3.2. Such results have been widely reported in the literature for different nanoalloys and can also lead to important size dependant concentration.

Self-Organized Templates

An important issue in the growth of nanoparticles is the size distribution. Indeed, monodisperse samples are generally required for the understanding of physical properties and increase the efficiency for most applications. As shown in Sect. 2, chemical routes have developed several tricks to narrow the natural size distribution. In PVD techniques, if one excludes LECBD, the size distribution is controlled by the diffusion processes on the substrate. In the last decade, patterned substrates, either artificially or naturally (self-organized surfaces), have been widely used to organize nanoparticles and reduce the size distributions.

Metallic [57] and insulating [58] templates have been studied for such purposes, and more recently graphene sheets [60]. Figure 10 shows typical examples of such ordered growth of nanoparticles arrays. Unfortunately, ordered arrays of bimetallic particles realized by this method are still rather scarce [58] whereas it typically narrows the particles size distribution by a factor more than two [61]. It is very likely that such a route can be very fruitful in a near future for the realization of model samples dedicated to a fundamental understanding of various properties of nanoalloys such as magnetic, catalytic or optical ones by means of averaging methods.

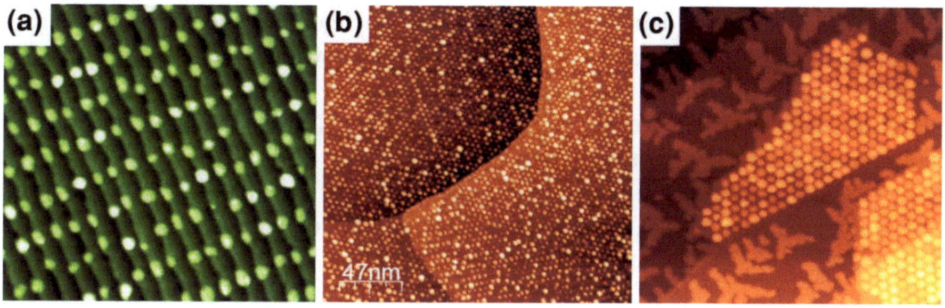

Fig. 10 **a** 60 nm STM image of Co nanoparticles organized on a Au(788) surface (from Ref. [57]). **b** Large scale STM image of AuPd nanoparticles organized on $Al_2O_3/Ni_3Al(111)$ surface (from Ref. [58]). **c** 70 nm STM image of Pt nanoparticles locally organized on graphene flakes grown on a Ir(111) surface (from Ref. [59])

3.2 Rate Equations Model for Nucleation and Growth on a Surface

The nucleation and growth on surfaces is an old topic which has been widely studied in the 1970s by Electronic Microscopy and explored in details at the atomic scale since the 1990s with the discovery of scanning probe microscopies (SPM) [62, 63]. Very similarly to what has been developed for chemical synthesis, the nucleation occurs in the presence of a supersaturation generated by a two-dimensional vapour pressure higher than the equilibrium one. The gain in chemical potential due to this supersaturation can overcome the surface energy term above a certain cluster size which is called the critical nucleus size, often given in number of atoms. In standard evaporation conditions (room temperature and nm/min fluxes) of metals, this critical cluster size is generally very small, *e.g.* dimers being stable nuclei (critical cluster size equal to one). Above this size, the cluster can grow by aggregation of diffusing atoms on the surface (adatoms).

The goal of a proper description of nucleation and growth would be to describe accurately different important experimental observations such as the nanoparticles density (*i.e.* their mean size for a given coverage), their shape and their local composition in the case of nanoalloys. The great amount of works dedicated to this topic in the 1990s has led to a good understanding of such parameters in the case of monometallic nanostructures [62]. The case of bimetallic particles is more complex due to the presence of two distinct classes of atomic processes. However, as discussed in the following, rather simple models can give some insights to this problem.

3.2.1 Introduction: Monometallic Particles

In nucleation and growth phenomena on surfaces, in a similar manner to what has been introduced for chemical synthesis, one usually distinguishes three stages versus the coverage of deposited atoms: (i) the nucleation stage where the density of stable islands is increased, (ii) the growth stage where the density is almost constant (and called the maximum cluster density n_c) but the size of islands increases, (iii) the coalescence stage where the density of islands decreases since neighbouring islands start to coalesce. In the case of adatoms moving on a homogeneous substrate (in opposite to a patterned self-organized surface), the first two stages are well described by the following rate equations treated in a mean field theory (considering stable dimers):

$$dn_1/dt = F - n_1/\tau_a - 2\sigma D n_1^2 - \sigma D n_1 n_x \tag{1}$$

$$dn_x/dt = \sigma D n_1^2 \tag{2}$$

where F is the incident flux, n_1 is the adatoms density and n_x the density of all the clusters, D is the diffusion coefficient of an adatom, σ is a capture number (in general dependant of the cluster size σ_i [63]) and τ_a is the characteristic time of re-evaporation of an adatom from the substrate to the vacuum. For example, Eq. (1) says that the growth rate of adatoms is increased by the incident flux but decreased by the re-evaporation of adatoms, the nucleation of dimers and the attachment of adatoms to clusters. Finally, one can see that growth characteristics are essentially driven by atomistic parameters for surface diffusion (diffusion energy E_d) and adsorption energy of adatoms to the surface (adsorption energy E_a). One has also to consider the binding energy to nucleated clusters in the more general case of a critical cluster size bigger than one [63] (binding energy to a cluster of i atoms E_i). Values for these parameters can be determined by the comparison between scaling predictions and experimental measurements [64]. Indeed, the measurement of the maximum cluster density versus the temperature generally gives valuable information on the dominant atomic mechanisms responsible for the nucleation stage. Such a measurement can be extracted, for example, from variable temperatures SPM experiments. The variation of n_c with temperature strongly depends from the condensation regime that gives rise to the nuclei. The two limiting cases are called incomplete condensation and complete condensation regimes. In the first case, typical of growth on insulators above room temperature, the re-evaporation of adatoms is dominant over the nucleation *i.e.* the second term in Eq. (1) is dominant over the last ones (giving $n_1 = F\tau_a$ in steady state for adatom density). This gives rise to a sticking coefficient far lower than one as most of deposited atoms go back into vacuum. In the regime of complete condensation which is generally relevant for metal on metal growth or at low temperatures, re-evaporation of adatoms from the substrate onto the vapour is negligible. In the case of stable dimers, the maximum cluster density n_c is given by:

$$n_c = \eta (D_0/F)^{-1/3} \exp(E_d/3\,k_B T) \tag{3}$$

where η is a prefactor related to capture numbers and D_0 is the diffusion pre-factor. In this simple case, it is worth to notice that the slope of n_c versus T in an Arrhenius plot gives E_d. This latter prediction has been observed and used to

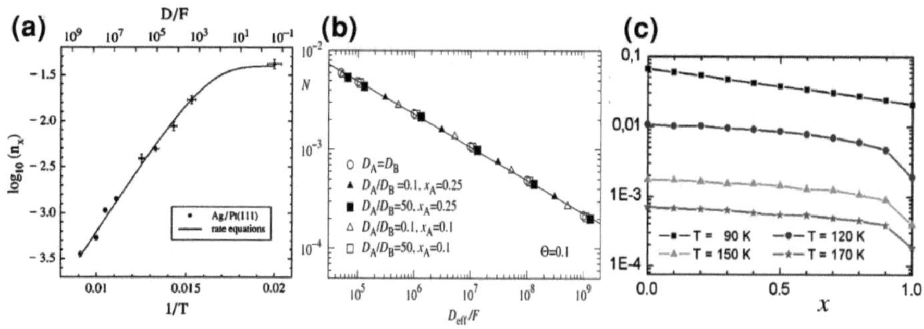

Fig. 11 **a** Critical Ag cluster density on Pt(111) as a function of temperature in an Arrhenius plot. The full line is the integration of rate equations that shows the expected linear trend for $D/F > 10^5$ (from Ref. [64]). **b** Simulated cluster densities of bimetallic alloys at a coverage of 0.1 ML as a function of an effective diffusion coefficient (from Ref. [65]). **c** Simulated cluster densities of a bimetallic alloy as a function of the composition for different temperatures of deposition

measure E_d in numerous examples of surface nucleation and growth, as shown in Fig. 11a for Ag/Pt(111). At higher temperature, the critical nucleus size generally increases, leading to a higher slope in the same Arrhenius representation.

3.2.2 Bimetallic particles

The nucleation and growth of bimetallic particles is more complex as many new parameters should be taken into account. Diffusion coefficients and re-evaporation times are specific of each element. Clusters and capture numbers are no more characterized by their size but also by their composition. Moreover, the critical cluster size can be different depending on the composition (consider for example the trivial case of stable AA and AB dimers and unstable BB dimers). The general trends are therefore very complex but typical behaviour can be discussed in some simple cases.

In the incomplete condensation regime for both species, the diffusion lengths are given by $(D_A\tau_{aA})^{1/2}$ and $(D_B\tau_{aB})^{1/2}$. The element with the smallest diffusion length (A) will have a lower nucleation rate and sticking coefficient to the surface. Therefore, nucleated clusters will be enriched with the other element (B) in the first stage. When the nucleation rate increases, the capture rate of A also increases leading to a change of composition with coverage, very similarly to what has been shown in Fig. 9. It is worth noting that this nucleation regime, generally corresponding to the deposition of metals on insulators above room temperature, has been studied for several bimetallic systems and is developed in more details in the Chap. 2.

In the complete condensation regime with stable dimers, the rate equations can be simplified and solved analytically [65]. An important mean field result is that the critical cluster density is given by the same formula than for monometallic nanoparticles (Eq. (3)) but with an effective diffusion coefficient such as $1/D_{eff} = x_A/D_A + x_B/D_B$. This can be understood in a simple time residence scheme where a 'mean' adatom stays a characteristic time on an atomic site before hopping which is the weighted average of the residence times of adatoms A and B. This finding has been checked by comparison with Kinetic Monte-Carlo simulations for different ratios of diffusion coefficients and different concentrations, as shown in Fig. 11b [65]. When looking at the cluster density as function of concentration, this gives a non-linear trend with the less diffusing element imposing a high density for a wide range of concentration. This behaviour is rather independent from the temperature as soon as the diffusion times for A and B are small as compared to the time of deposition. For lower temperatures, *i.e.* in the so-called post-nucleation regime, the behaviour is different, more linear (*cf.* Fig. 11c).

To our knowledge, such variations of the density with concentration and temperature have not been measured experimentally yet on model systems in the complete condensation regime. This would be of great interest as it would validate the mean field approach of rate equations for such complex systems.

Other important quantities for nanoalloys such as size and concentration distributions are generally not predicted by this mean field approach. It is therefore necessary to go beyond by using simulations, as discussed for example in Chap. 8.

4 Conclusion

As shown in this chapter, the realization of bimetallic nanoparticles can be done by using various techniques, either in solutions or in vacuum. The main challenges are generally the control of the composition and the narrow size distribution of nanoparticles. The out-of-equilibrium nucleation and growth process can lead to unexpected results in the case of bimetallic particles, such as size dependant composition and complex change of the particles density as a function of the composition. It is of particular importance to better understand and control these phenomena for a further development of nanoalloys based applications.

References

1. Zhang, Q., Xie, J., Yu, Y., Lee, J.Y.: Monodispersity control in the synthesis of monometallic and bimetallic quasi-spherical gold and silver nanoparticles. Nanoscale **2**, 1962 (2010)
2. Lamer, V.K., Barnes, M.D.: Monodispersed hydrophobic colloidal dispersions and light scattering properties. I. Preparation and light scattering properties of monodispersed colloidal sulfur. J. Colloid Sci. **1**, 71 (1946)
3. Beattie, J.K.: Monodisperse colloids of transition metal and lanthanide compounds. Pure Appl. Chem. **61**, 937 (1989)
4. Park, J., Joo, J., Kwon, S.G., Jang, Y., Hyeon, T.: Synthesis of monodisperse spherical nanocrystals. Angew. Chem. **46**, 4630 (2007)

5. Turkevich, J., Stevenson, P.C., Hilier, D.: A study of the nucleation and growth processes in the synthesis of colloidal gold. J. Discuss. Faraday Soc. **11**, 55 (1951)
6. Pileni, M.-P.: Reverse micelles as microreactors. J. Phys. Chem. **97**, 6961 (1993)
7. Petit, C., Pileni, M.P.: Synthesis of cadmium sulfide in situ in reverse micelles and in hydrocarbon gels. J. Phys. Chem. **92**, 2282 (1988)
8. Petit, C., Lixon, P., Pileni, M.P.: Synthesis of cadmium sulfide in situ in reverse micelles 2 influence of the interface on the growth of the particles. J. Phys. Chem. **94**, 1598 (1990)
9. Ingelsten, H.H., Bagwe, R., Palmqvist, A., Skoglundh, M., Svanberg, C., Holmberg, K., Shah, D.O.: Kinetics of the formation of nano-sized platinum particles in water-in-oil microemulsions. J. Colloid Interface Sci. **241**, 104 (2001)
10. Ingelsten, H., Beziat, J.-C., Bergkvist, K., Palmqvist, A., Skoglundh, M., Qiuhong, H., Falk, L.K.L., Holmberg, K.: Deposition of platinum nanoparticles, synthesised in water-in-oil microemulsions, on alumina supports. Langmuir **18**, 1811 (2002)
11. Tanori, J., Gulik-Krzywicki, T., Pileni, M.P.: Phase diagram of copper(II) bis(2-ethylhexyl) sulfosuccinate, $Cu(AOT)_2$−isooctane−water. Langmuir **13**, 632 (1997)
12. Lisiecki, I.: Size, shape, and structural control of metallic nanocrystals. J. Phys. Chem. **109**, 12231 (2005)
13. Petit, C., Taleb, A., Pileni, M.P.: Self-organization of magnetic nanosized cobalt particles. Adv. Mater. **10**, 259 (1998)
14. Petit, C., Lixon, P., Pileni, M.P.: In situ synthesis of silver nanocluster in AOT reverse micelles. J. Phys. Chem. **97**, 12974 (1993)
15. Salzeman, C., Zhai, W., Goubet, N., Pileni, M.P.: How to tune the Au internanocrystal distance in two- dimensional self-ordered superlattices. J. Phys. Chem. Lett. **1**, 149 (2010)
16. Petit, C., Rusponi, S., Brune, H.: Magnetic properties of cobalt and cobalt–platinum nano crystals investigated by magneto-optical Kerr effect. J. Appl. Phys. **95**, 4251 (2004)
17. Xiong, L., Manthiram, A.: Nanostructured Pt−M/C (M = Fe and Co) catalysts prepared by a microemulsion method for oxygen reduction in proton exchange membrane fuel cells. Electrochim. Acta **50**, 2323 (2005)
18. Shevchenko, E., Talapin, D., Rogach, A., Kornowski, A., Haase, M., Weller, H.: Colloidal synthesis and self-assembly of $CoPt_3$ nanocrystals. J. Am. Chem. Soc. **124**, 11480 (2002)
19. Yashima, M., Falk, L.K.L., Palmqvist, A.E.C., Holmberg, K.: Structure and catalytic properties of nanosized alumina supported platinum and palladium particles synthesized by reaction in microemulsion. J. Colloid Interface Sci. **268**, 348 (2003)
20. Duxin, N., Brun, N., Colliex, C., Pileni, M.P.: Synthesis and magnetic properties of elongated Fe-Cu alloys. Langmuir **14**, 1984 (1998)
21. Chen, D.H., Chen, C.J.: Formation and characterization of Au–Ag bimetallic nanoparticles in water-in-oil microemulsions. J. Mater. Chem. **12**, 1557 (2002)
22. Brust, M., Walker, M., Bethell, D., Schiffrin, D.J., Whyman, R.: Synthesis of thiol-derivatised gold nanoparticles in a two-phase liquid–liquid system. J. Chem. Soc. Chem. Comm. **7**, 801 (1994)
23. Brust, M., Bethell, D., Schiffrin, D.J., Kiely, C.J.: Novel gold-dithiol nano-networks with non-metallic electronic properties. Adv. Mater. **7**, 795 (1995)
24. Brust, M., Bethell, D., Kiely, C.J., Schiffrin, D.J.: Self-assembled gold nanoparticle thin films with non-metallic optical and electronic properties. Langmuir **14**, 5425 (1998)
25. Wilson, O.M., Scott, R.W.J., Garcia-Martinez, J.C., Crooks, R.M.: Synthesis, characterization, and structure-selective extraction of 1–3-nm diameter AuAg dendrimer-encapsulated bimetallic nanoparticles. J. Am. Chem. Soc. **127**, 1015 (2005)
26. Demortiere, A., Petit, C.: First synthesis by liquid–liquid phase transfer of magnetic Co_xPt_{100-x} nanoalloys. Langmuir **23**, 8575 (2007)
27. Demortiere, A., Losno, R., Petit, C., Quisefit, J.P.: Composition study of CoPt bimetallic nanocrystals of 2 nm. Anal. Bioanal. Chem. **397**, 1485 (2010)
28. Wikander, K., Petit, C., Holmberg, H., Pileni, M.P.: Size control and growth process of alkylamine-stabilized platinum nanocrystals: a comparison between the phase transfer and reverse micelles methods. Langmuir **22**, 4863 (2006)

29. Cheng, Y., Schiffrin, D.J.: Electrodeposition of metallic gold clusters at the water/1, 2-dichloroethane interface. J. Chem. Soc. Faraday Trans. **92**, 3865 (1996)
30. Demortiere, A., Launois, P., Goubet, N., Albouy, P.A., Petit, C.: Shape-controlled platinum nanocubes and their assembly into two-dimensional and three-dimensional superlattices. J. Phys. Chem. B **112**, 14583 (2008)
31. Templeton, A.C., Chen, S.W., Gross, S.M., Murray, R.W.: Water-soluble, isolable gold clusters protected by tiopronin and coenzyme a monolayers. Langmuir **15**, 66 (1999)
32. Teranishi, T., Kiyokawa, I., Miyake, M.: Synthesis of monodisperse gold nanoparticles using linear polymers as protective agents. Adv. Mater. **10**, 596 (1998)
33. Zheng, H., Fan, J., Stucky, G.D.: One-step one-phase synthesis of monodisperse noble-metallic nanoparticles and their colloidal crystals. J. Am. Chem. Soc. **128**, 6550 (2006)
34. Jana, N.R., Peng, X.G.: Single-phase and gram-scale routes toward nearly monodisperse au and other noble metal nanocrystals. J. Am. Chem. Soc. **125**, 14280 (2003)
35. Sun, S.: Recent advances in chemical synthesis, self-assembly, and applications of FePt nanoparticles. Adv. Mater. **18**, 393 (2006)
36. Ai, F., Yao, A., Huang, W., Wang, D., Zhang, X.: Synthesis of PVP-protected NiPd nanoalloys by modified polyol process and their magnetic properties. Physica E **42**, 1281 (2010)
37. Varanda, L.C., Jafelici, Jr, M.: Self-assembled FePt nanocrystals with large coercivity: reduction of the fcc-to-L10 ordering temperature. J. Am. Chem. Soc. **128**, 11062 (2006)
38. Smith, D.K., Korgel, B.A.: The influence of the CTAB surfactant on the colloidal seed medaited synthesis of gold nanorods. Langmuir. 24, 644, (2009)
39. Alloyeau, D., Prévot, G., Le Bouar, Y., Oikawa, T., Langlois, C., Loiseau, A., Ricolleau, C.: Ostwald ripening in nanoalloys: when thermodynamics drives a size-dependent particle composition. Phys. Rev. Lett. **105**, 255901 (2010)
40. Phillipot, K., Chaudret, B.: Organometallic approach to the synthesis and surface reactivity of noble metal nanoparticles. C.R. Chimie **6**, 1019 (2003)
41. Kwon, S.G., Hyeon, T.: Colloidal chemical synthesis and formation kinetics of uniformly sized nanocrystals of metals, oxides, and chalcogenides. Acc. Chem. Res. **41**, 1696 (2008)
42. Murray, C.B., Norris, D.J., Bawendi, M.G.: Synthesis and characterization of nearly monodisperse CdE (E = sulfur, selenium, tellurium) semiconductor nanocrystallites. J. Am. Chem. Soc. **115**, 8706 (1993)
43. Son, S.U., Yang, Y., Park, J., Na, H.B., Park, H.M., Yun, H.J., Lee, J., Hyeon, T.: Designed synthesis of atom-economical Pd/Ni bimetallic nanoparticle-based catalysts for Sonogashira coupling reactions. J. Am. Chem. Soc. **126**, 5026 (2004)
44. Chaudret, B.: Organometallic approach to nanoparticles synthesis and self-organization. C.R. Chimie **6**, 117 (2005)
45. Pan, C., Dassenoy, F., Casanove, M.-J., Philippot, K., Amiens, C., Lecante, P., Mosset, A., Chaudret, B.: A new synthetic method toward bimetallic ruthenium platinum nanoparticles; composition induced structural changes. J. Phys. Chem. B **103**, 10098 (1999)
46. Byrappa, K., Adshiri, T.: Hydrothermal technology for nanotechnology. Prog. Cryst. Growth Charact. Mater. **53**, 117 (2007)
47. Schukin, D.G., Radziuk, D., Möhwald, H.: Ultrasonic fabrication of metallic nanomaterials and nanoalloys. Ann. Rev. Mater. Res. **40**, 345 (2010)
48. Wu, J.H., Kim, Y.K.: Synthesis and microwave properties of highly permeable FeCo-based nano-alloys. Phys. Stat. Sol. (a) **204**, 4087 (2007)
49. Johnston, R.L., Ferrando, R. (eds.): Nanoalloys—From Theory to Applications, Faraday Discussions, vol. 138, RSC Publishing, Cambridge (2008)
50. Ferrando, R., Jellinek, J., Johnston, R.L.: Nanoalloys: from theory to applications of alloy clusters and nanoparticles. Chem. Rev. **108**, 845 (2008)
51. Penuelas, J., Andreazza-Vignolle, C., Andreazza, P., Ouerghi, A., Bouet, N.: Temperature effect on the ordering and morphology of CoPt nanoparticles. Surf. Sci. **602**, 545 (2008)
52. Alloyeau, D., Ricolleau, C., Mottet, C., Oikawa, T., Langlois, C., Le Bouar, Y., Braidy, N., Loiseau, A.: Size and shape effects on the order–disorder phase transition in CoPt nanoparticles. Nat. Mater. **8**, 940 (2009)

53. Dietz, T.G., Duncan, M.A., Powers, D.E., Smalley, R.E.: Laser Production of Supersonic Metal Cluster Beams. J. Chem. Phys. **74**, 6511 (1981)
54. Milani, P., deHeer, W.A.: Improved pulsed laser vaporization source for production of intense beams of neutral and ionized clusters. Rev. Sci. Instrum. **61**, 1835 (1990)
55. Tournus, F., Blanc, N., Tamion, A., Hillenkamp, M., Dupuis, V.: Dispersion of magnetic anisotropy in size-selected CoPt clusters. Phys. Rev. B **81**, 220405 (2010)
56. Gimenez, F., Chapon, C., Henry, C.: Nucleation and growth kinetics of Pd and CuPd particles on NaCl(100). New J. Chem. **22**, 1289 (1998)
57. Repain, V., Baudot, G., Ellmer, H., Rousset, S.: Two-dimensional long-range–ordered growth of uniform cobalt nanostructures on a Au(111) vicinal template. Europhys. Lett. **58**, 730 (2002)
58. Hamm, G., Becker, C., Henry, C.: Bimetallic Pd–Au nanocluster arrays grown on nanostructured alumina templates. Nanotechnology **17**, 1943 (2006)
59. N'Diaye, A.T., Gerber, T., Busse, T., Myslivecek, J., Coraux, J., Michely, T.: A versatile fabrication method for cluster superlattices. New J. Phys. **11**, 103045 (2009)
60. N'Diaye, A.T., Bleikamp, S., Feibelman, P.J., Michely, T.: Two-dimensional Ir cluster lattice on a graphene moire on Ir(111). Phys. Rev. Lett. **97**, 215501 (2006)
61. Repain, V., Rohart, S., Girard, Y., Tejeda, A., Rousset, S.: Building uniform and long-range ordered nanostructures on a surface by nucleation on a point defect array. J. Phys. Cond. Mat. **18**, S17 (2006)
62. Brune, H.: Microscopic view of epitaxial metal growth: nucleation and aggregation. Surf. Sci. Rep. **31**, 121 (1998)
63. Venables, J.: Introduction to Surface and Thin Film Processes. Cambridge University Press, Cambridge (2000)
64. Brune, H., Bales, G.S., Jacobsen, J., Boragno, C., Kern, K.: Measuring surface diffusion from nucleation island densities. Phys. Rev. B **60**, 5991 (1999)
65. Einax, M., Ziehm, S., Dieterich, W., Maass, P.: Scaling of island densities in submonolayer growth of binary alloys. Phys. Rev. Lett. **99**, 016106 (2007)

Bimetallic Nanoparticles, Grown Under UHV on Insulators, Studied by Scanning Probe Microscopy

Claude Henry and Clemens Barth

Abstract Nowadays scanning probe microscopies (atomic force microscopy and scanning tunnelling microscopy) are common techniques to characterize at the atomic level the structure of surfaces. In the last years, these techniques have been applied to study the nucleation and growth of metal clusters (mono or bimetallic). Basic elements of scanning probe microscopy will be presented. With the help of the atomistic nucleation theory and using some earlier results obtained by TEM we show that the growth rate and the composition evolution of bimetallic particles grown from two atomic vapours sequentially or simultaneously condensed on insulating substrates (bulk or ultrathin film) can be predicted. The published work on the growth of bimetallic particles studied by STM and AFM is presented in a comprehensive way giving simple rules to select the best method to obtain homogeneous assemblies of nanoparticles with given mean sizes and chemical compositions. Although the application of scanning probes microscopy to the growth of supported bimetallic particles is relatively young, recent development of AFM and STM techniques paves the way for a complete in situ characterization, including morphology and surface composition.

C. Henry (✉) · C. Barth
CINaM-CNRS, Campus de Luminy, Case 913,
13288 Marseille cedex 09, France
e-mail: henry@cinam.univ-mrs.fr

C. Barth
e-mail: barth@cinam.univ-mrs.fr

D. Alloyeau et al. (eds.), *Nanoalloys*, Engineering Materials,
DOI: 10.1007/978-1-4471-4014-6_2, © Springer-Verlag London 2012

1 Introduction

In this chapter we will introduce scanning probe microscopy, which can be used to characterize supported bimetallic nanoparticles in situ. For sake of consistency we make the choice to restrict this overview on bimetallic particles supported on bulk insulators (alkali halides, oxides) or on ultrathin oxide films on metallic substrates. The choice of an insulator substrate has two main advantages: electronic decoupling of the particles from the substrate and tendency to grow 3D clusters instead of 2D islands. The nanoparticles are grown by atomic deposition under UHV whereas STM and AFM are best suited techniques to study in situ nucleation and growth of single bimetallic particles. The aim of the chapter is to provide a practical and comprehensive overview on the preparation and on the structural and morphological characterization of the nanoalloys. In the first section a rapid introduction of the scanning probes is presented with a selection of articles, reviews and textbooks for deeper understanding of these techniques. We also focus on new methods useful for the characterization of nanoparticles down to atomic scales. In the second section, an introduction of nucleation theory is presented restricted to the practical case considered here: nucleation and growth of bimetallic nanoparticles by condensing at room temperature (RT) or moderate temperature two metallic vapors, sequentially or simultaneously on an insulating substrate. In these conditions the atomistic nucleation theory is valid and re-evaporation of adatoms takes place (incomplete condensation). Analytic expressions are presented that will guide the reader in choosing the best conditions for the preparation of homogeneous collections of nanoparticles. In the third section we will review the main results on the characterization of the nucleation, growth, structure, morphology and chemical composition of the bimetallic nanoparticles by using STM and AFM together with some additional complementary techniques. From the different bimetallic systems investigated we will try to rationalize the results in term of empirical rules and analytical results from nucleation theory.

2 Introduction to Scanning Probes Microscopies: STM and AFM

In this section the scanning tunneling microscopy (STM) and atomic force microscopy (AFM) are first briefly reviewed (Sects. 1 and 2). Section 3 concentrates on the application of both techniques for studying the shape of supported metal clusters whereas the following section (Sect. 4) discusses the application of Kelvin probe force microscopy (KPFM). The aim of these four sections is to introduce the reader into the domain of STM, AFM and KPFM and to supply the reader with reviewing literature, which can be used for a further, deeper reading.

2.1 Scanning Tunneling Microscopy

The introduction of scanning tunneling microscopy (STM) by Binning, Rohrer, Gerber and Weibel in 1982 [1] revolutionized experimental studies of conducting surfaces in many scientific fields, ranging from physics to chemistry and up to biology. This specific local imaging technique has been reviewed several times [2–4] and is also described from the theory point of view [5, 6]. The reader is referred to two well-known text books [7, 8], which discuss all aspects of the technique, whereas a brief overview is given in the following.

The principle of a scanning tunneling microscope is quite simple: If the surfaces of two metallic electrodes (tip and surface) are brought very close together, *e.g.* to a distance of about one nanometer and less, electrons can overpass the gap distance and *tunnel* from one conductor to the other one, which is thanks to the quantum mechanical tunneling probability [7]. The tunneling is done such that the electrons tunnel from the metal with the higher Fermi level to the other metal with the lower Fermi level. In order to create a difference of Fermi levels, a so-called *bias potential* is applied in STM between the two conductors (Fig. 1a). The innovative idea of Binnig and Rohrer was to shrink one of the surfaces to an atomic size such that the initial surface becomes a very sharp object called 'tip', which apex is formed by only a few atoms. With help of piezo-electric motors, the surface is scanned in X and Y directions and the flow of electrons (tunnel current) is kept constant by adjusting the tip-surface distance upon changes in the surface topography during scanning. The relative changes in the tip-surface distance are recorded in dependence on X and Y so that an image of the surface is created. Because the tip apex is formed by few atoms only and because the tunnel current depends exponentially on the tip-surface distance, it is almost the last atom of the tip that produces the contrast of STM images. In best cases, the true atomic resolution can be obtained (Fig. 1b). Note that apart from the latter *topography imaging mode*, a second mode exists: In the *constant height mode* the tip keeps a constant distance to the surface during scanning (see scanning mode explained in Ref. [7]) such that it does not follow the topography. Because of the varying topography and because the tip follows only the mean inclination of the sample, the tunnel current is varying during scanning and is recorded in dependence on X and Y such that the resulting *tunnel current image* contains all information of the surface. The constant height mode can be used only on flat surfaces but permits scanning surfaces with high speed even at video frequencies (see example in Ref. [9]).

STM has become the standard surface analysis technique in surface science, especially in surface physics and surface chemistry. Applications have shown true atomic resolution first on semiconductor surfaces [1] and later on a metal surfaces [12]. In the case of surfaces of metallic alloys the chemical identification of atoms is difficult because STM has in general no specific chemical sensitivity. However in some circumstances atomically resolved images can present a chemical sensitivity as it was first shown by Varga's group (see Fig. 2b) in 1993 on $Pd_{25}Ni_{75}(111)$ [13]. Later, the same group has succeeded in imaging chemical

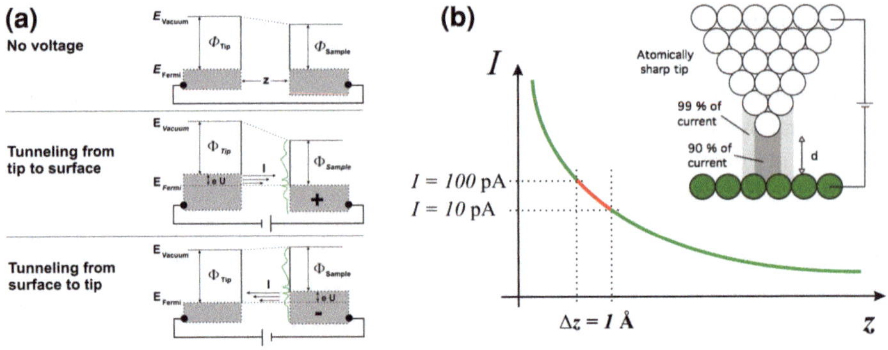

Fig. 1 The principle of STM. **a** The Fermi levels of two electrodes (tip and surface) are aligned if they are connected from behind. If the tip and surface are separated by a distance of $z \approx 1$ nm and if a potential U is applied between both electrodes, electrons tunnel from the electrode with the higher Fermi level (E_{Fermi}) to the other one with the lower Fermi level. If the tip is at positive potential in the case here, the electrons tunnel from the sample to the tip and vice versa. **b** The electron flow (tunnel current I) depends exponentially on the tip-surface distance z. A change of the distance by $\Delta z = 1$ Å changes the tunnel current by one order of magnitude. This is the reason why the tip can be precisely positioned with picometer precision above the surface. It also explains that it is almost the last atom of the tip that contributes to the image contrast. Because of the small dimension of the atom, atomic resolution can be obtained

atomic contrast in several surfaces of bulk metallic alloys like PtRh [11, 14], AgPd [15], CoPt [16], FeNi [17]. Other groups have imaged with a chemical contrast alloyed surfaces obtained by depositing a thin layer of a metal on the surface of another metal like in the case of CuPd [18, 19]. Ultrathin films of insulators supported on metal single crystals can also be imaged at atomic resolution by STM like NaCl/Cu [20], MgO/Ag [21], alumina/NiAl [22]. Chemical reactions on metal surfaces can be followed at atomic scale by STM [23], even at high-speed, which permits to produce movies of surface reactions [9]. Not only imaging can be done but also spectroscopy, which has become a large field in STM. Different spectroscopy modes exist like STS (scanning tunneling spectroscopy), constant current spectroscopy (CCS), constant separation spectroscopy (CSS) (see overview in Refs. [7, 8]). During a spectroscopy experiment the tip either scans the surface or is hold at one position above the surface. In STS, when the tip is hold at one point on the surface, the bias voltage V is swept and the tunnel current I recorded so that I(V) curves are obtained, which are mostly converted to dI/dV or $dI/dV/(I/V)$ curves. The latter two types of curves are a fingerprint for the local density of states (LDOS) at the Fermi level for the specific position on the surface. It helps to study the local electronic structure of the surface. More recently a new type of measurement which is like inelastic electron tunneling spectroscopy (IETS) but acquired with an STM is able to resolve vibration of an adsorbed isolated molecule on a surface [24]. In this case one records the second derivative of the current (d^2I/dV^2) as a function of the bias voltage. Vibrational/rotational modes of a single molecule can be excited for instance and electron induced modifications of

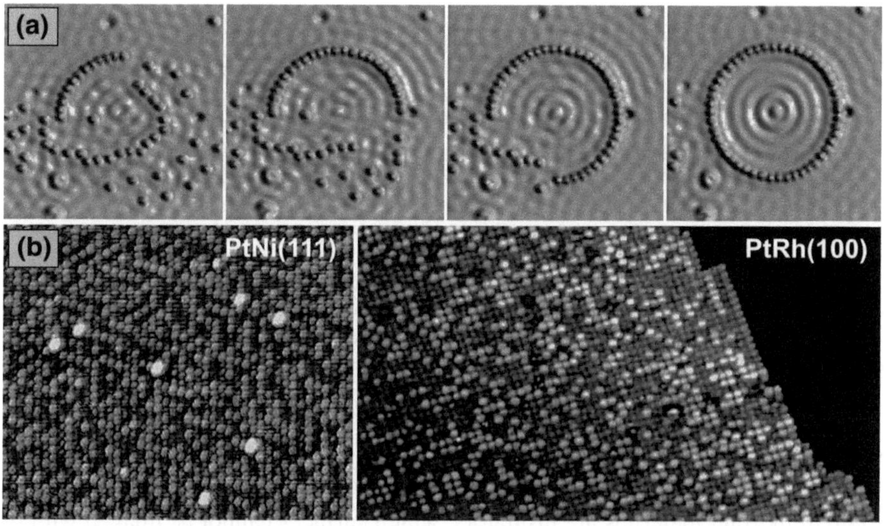

Fig. 2 **a** Series of STM images which were obtained in between several steps of atom manipulation. Thanks to the manipulation, 48 Fe ad-atoms could be placed into a circle on a crystalline Cu(111) surface at 4 K (from Ref. [10], reprinted with permission from AAAS and Copyright (1993) IBM). **b** Surfaces of metal alloys. In the NiPt case a *chemical* contrast can be seen, well-separating the two types of atoms. However, the contrast strongly depends on the tip composition (last tip atom). On PtRh(100) a chemical contrast can be always observed. The bright atoms are the Rh atoms whereas the dark ones are the Pt atoms (from Ref. [11], copyright (1999), with permission from Elsevier)

molecules (desorption, dissociation, hopping, chemical reactions) can be realized. In order to position the tip and its last atoms with utmost precision above single surface atoms or molecules, tunneling spectroscopy is mostly applied at liquid nitrogen or helium temperatures where the tip's last atom remains for a very long time of days above a single molecule for instance.

A breakthrough in STM was certainly the work of Eigler and Schweizer [25] who demonstrated for the first time that single Xe atoms adsorbed on a Ni surface can be manipulated such that they are moved on the surface by the STM tip. The most impressive example is the arrangement of 48 Fe ad-atoms in a circle on a Cu surface which represents a quantum corral where the confinement of electrons produced standing waves [10] (Fig. 2a). Manipulation of molecules can be used to functionalize the apex of the STM tip. It has been shown that the presence of a CO molecule can improve the topographic resolution in the STM image but it can also resolve vibrational modes in adsorbed molecules that are not visible with a bare tip [26]. Bimetallic AuPd atomic chains have been fabricated by atom manipulation on a metal substrate and STS has shown the evolution of the electronic structure by adding Pd atoms in a gold chain [27]. In combination with scanning tunneling spectroscopy (STS), the nano-manipulation has been used especially in surface chemistry for inducing chemical reactions by the tip like inducing an 'Ullmann' reaction [28] or controlling the hydrogenation of a single molecule [29].

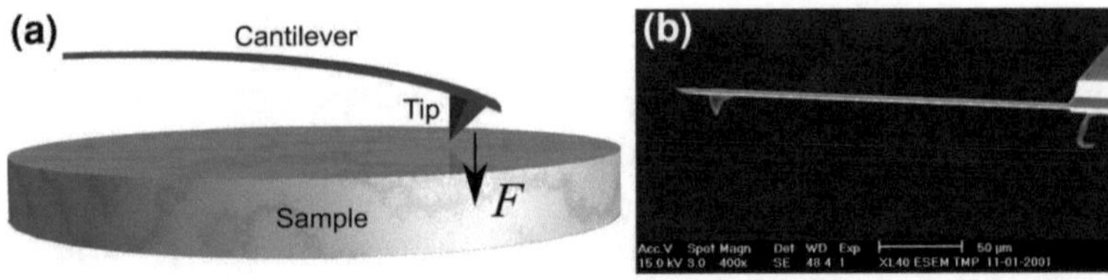

Fig. 3 The principle of AFM. **a** A pyramidal tip is attached to a beam, the *cantilever*. As soon as the tip is close to the surface ($z < 5$ nm) a force acts onto the apex of the tip attracting the tip and cantilever towards the surface. The result is that the cantilever bends, what can be measured by optical laser methods for instance. **b** A cantilever-tip system is nowadays produced from one single silicon or silicon nitride crystal by etching methods. Cantilevers with almost all possible characteristics (size, resonance frequency, coating etc.) are commercially available today

2.2 Atomic Force Microscopy (Scanning Force Microscopy)

Many important materials are insulators, which surfaces cannot be imaged with the STM. This concerns especially oxide materials like bulk Al_2O_3 or MgO, which surfaces are used to support metal clusters in catalysis for instance. This limitation has been soon realized after the introduction of the STM, and four years later, in 1986, Binnig, Quate and Gerber invented a second type of microscope, the atomic force microscope (AFM) [30]. The AFM allows nowadays obtaining almost the same type of resolution on almost any atomically flat and clean surface, even if the material is an insulator [31].

The principle of the AFM is somewhat similar to the one of the STM on the one hand but quite different on the other. Like in STM a sharp tip is used, which is, however, attached in perpendicular position at one end of a flexible beam called *cantilever* (Fig. 3). The other end of the cantilvever is fixed. If the tip is brought into a close distance to the surface ($z < 5$ nm), a force can be detected between tip and surface by measuring the bending of the cantilever. If the force is kept constant during scanning such that variations of the surface topography are compensated by adjusting the tip-surface distance by a regulation loop, an image is obtained representing the topography of the surface. In this *contact AFM mode* (c-AFM), the very end of the tip is always in hard contact with the surface due to the *jump-into-contact* mechanism (see Ref. [32] for further details). It has been shown that in this *static mode* the AFM cannot yield a high resolution at the atomic scale because the last tip atoms modify or even change the atoms of especially reactive surfaces [33, 34]. In 1991, the so-called *frequency modulated non-contact AFM mode* (FM nc-AFM or just nc-AFM) was introduced by Albrecht, Grütter, Horne and Rugar in order to overpass the limits of the c-AFM [35]. In FM nc-AFM, the cantilever and tip are excited to oscillation at their resonance frequency f_0. As soon as a force acts onto the tip, the resonance frequency changes (detunes) to a value f, and the tip-surface distance is adjusted upon changes which appear in the detuning $\Delta f = f - f_0$ during scanning the surface. The detuning Δf is therefore the same type of regulation signal as it is the force in c-AFM or the tunnel current in STM.

In nc-AFM, the tip's last atoms can be brought very close to the surface in a distance of some Angströms, at which the true atomic resolution is mostly obtained [36]. Although the atomic contrast formation in nc-AFM is different with respect to STM, one can roughly state that also in nc-AFM the last atoms of the tip apex produce the atomic contrast. Starting with the first-time atomic resolution in 1995 on Si(111)7 × 7 [37], it has been impressively shown in the last 15 years that the true atomic resolution can be obtained on any atomically flat and clean surface, independently if it is the surfaces of metals, alloys, semi-conductors, insulators or oxides (many applications are reviewed in Refs. [31, 38–41]). The nc-AFM catches successively up with its sister technique STM, which can be best seen by recent manipulation experiments in nc-AFM where single atoms and molecules could have been moved on the surface [36]. As in STM, spectroscopy exists also in nc-AFM, which is based on recording and interpreting the detuning and, with it, the force or potential in dependence on the tip-surface distance or even additionally on X and Y (3D force fields). In combination with theory, force spectroscopy is mostly used to identify single atoms on surfaces as it has been shown on Sn and Pb covered Si(111) surface [42, 43] (see Fig. 4a). In some cases an identification can be done also on real insulator surfaces by just imaging as impressively demonstrated for CaF_2(111) [44] or on the (001) surfaces of $CdCl_2$ or $MgCl_2$ doped NaCl [45] (see Fig. 4b). In all such experiments, the chemical contrast in images with atomic resolution strongly depends on the nature of the tip's last atoms.

2.3 Imaging the Shape of Metal Clusters with the STM and AFM

Most importantly for this chapter is the contribution of STM and AFM in the in situ imaging of the morphology of metal clusters supported on non-metallic substrates (see Ref. [46] for a review on the subject). Figure 5 presents Pd nanoparticles grown on MoS_2(0001), (ex situ) TEM and (in situ) STM images of the same sample are displayed in (a) and (b), respectively [47]. On the STM image the particles appears larger and the edges are rounded in comparison with the TEM image. The size distributions (Fig. 5c) from the two images indeed show an increase of the mean size in STM (14 against 12 nm by TEM). This enlargement in the STM images of 3D objects is typical and due to a deformation of the shape of the object by the scanning tip. This effect is schematically represented in Fig. 6. The imaged profile is enlarged by an amount which depends on the shape of the tip and on the slope of the facets: steeper is the facet smaller is the deformation. If the exact shape of the tip is known the original profile of the particle can be restored but with a loss of information at the bottom as seen on Fig. 6 (for example re-entrant angles cannot be imaged).

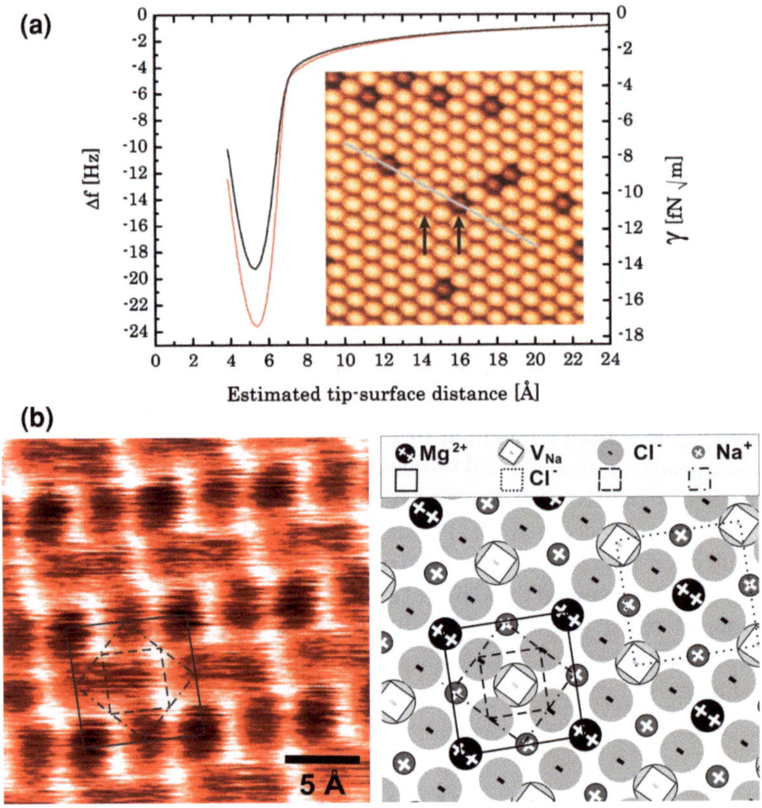

Fig. 4 **a** High resolution nc-AFM image of Sn on the Si(111) surface and set of force curves measured over Sn (*bright*) and Si (*dark*) atoms. Both type of atoms produce a clear chemical contrast in the image but also in the detuning (Δf) versus distance curves, which represent the tip-atom interaction (from Ref. [43], copyright (2006) by The American Physical Society). **b** Noncontact AFM images obtained on the (001) surface of a Cd^{2+} doped NaCl crystal. Thanks to the specific geometric structure of each sub-lattice formed by the Na^+, Cl^-, Cd^{2+} ions and the chemical contrast each ion is producing, all ionic species can be unambiguously identified (see left drawing) by just imaging (from Ref. [45], copyright (2008) by The American Physical Society)

The top facet is not distorted and if it is large enough and flat, atomic resolution can be obtained in STM as seen on Fig. 7b for a Pd nanoparticle on an ultrathin film of alumina on NiAl(110) [48] or for Pt nanoparticles on $TiO_2(110)$ [49]. If the particles are prepared at high temperature they can then get the equilibrium shape which is defined by the Wulff theorem for a free particle and the Wulff–Kaichew theorem from a supported one [46]. From a precise measurement of the particle height and of the top facet size, it is possible to deduce the adhesion energy of the particle [46, 48, 50].

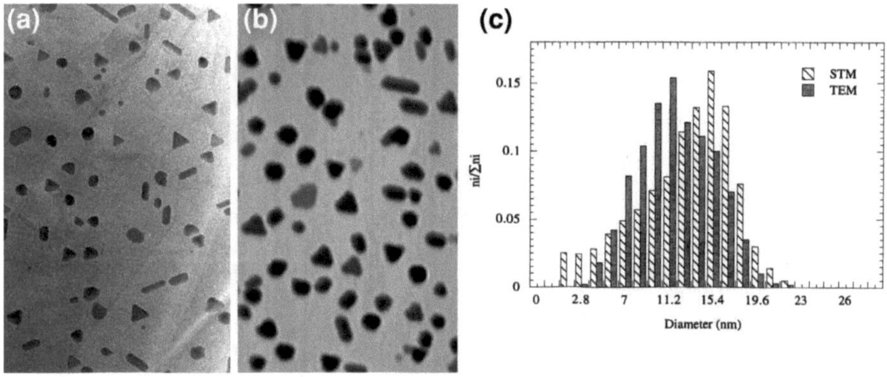

Fig. 5 TEM (**a**), STM (**b**) and corresponding size distributions (**c**) for Pd nanoparticles grown on $MoS_2(0001)$ under UHV. The STM images are recorded in situ after the deposition while TEM pictures are obtained ex situ after thinning of the sample by exfoliation. From Ref. [47], copyright (2000), with permission from Elsevier

Fig. 6 Profile view of the deformation of the image of a 3D particle due to the shape of the scanning tip. From Ref. [46], copyright (2005), with permission from Elsevier

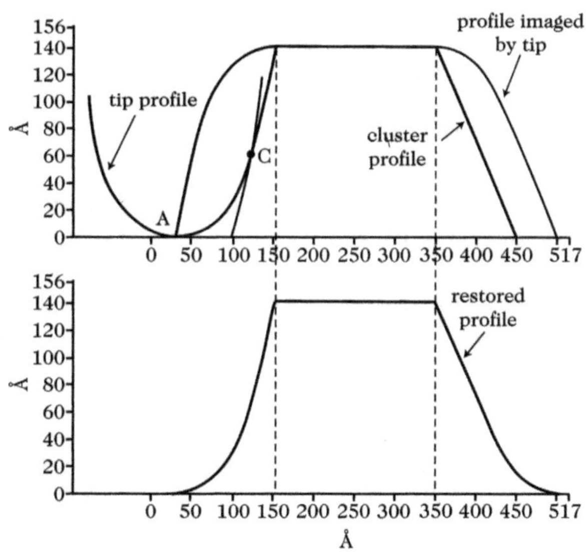

In the case of tiny clusters it has been possible to image at atomic resolution all the atoms on the surface of metal clusters [51]. Figure 7a shows an STM image of a Pd clusters on $MoS_2(0001)$ containing exactly 27 atoms (20 in the first layer and 7 in the second one) [51]. The top layer (sulphur atoms) of the substrate is also imaged at atomic resolution then the epitaxial orientation of the metal clusters can be revealed. The sulphur atoms at the periphery of the clusters appear higher probably through an electronic effect. Unfortunately such nice atomically resolved images of metal clusters are scarce because they need to have a very flat sample and an exceptional tip (the control of the tip in STM experiments is still a major issue).

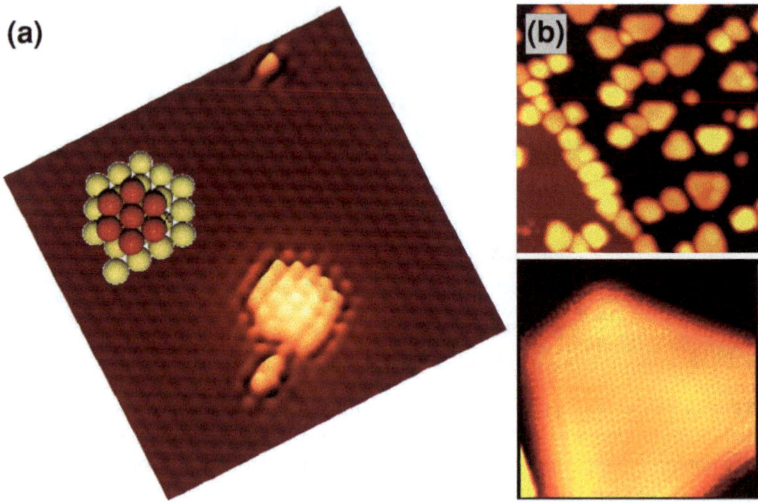

Fig. 7 STM images of Pd clusters showing atomic resolution. **a** Pd cluster on $MoS_2(0001)$ containing 27 atoms, a ball model (inset) shows the atomic arrangement of the atoms in the two layers (from Ref. [51], copyright (1997), with permission from Elsevier). **b** Palladium nanoparticles (5 nm) on a thin Al_2O_3 film supported on NiAl(110) showing atomic resolution on the top (111) facet (image at the bottom) (from Ref. [48], copyright (1999) by The American Physical Society)

In the case of imaging the shape of nanoparticles by AFM the problem of the deformation of the image by the tip shape is still present and even worth because commercial AFM tip are generally bigger than STM tips (see Ref. [46]). A solution to reduce this effect is to grow a nanotip at the apex of the AFM tip in a SEM or a FIB [46] or even to glue a carbon nanotube [52]. Contact AFM has been very often used to image nanoparticles supported on insulators but the large majority of these studies were performed in air. Some studies on the growth of metal nanoparticles have been performed in situ under UHV (*e.g.* Au/mica [53], Au/MgO [54]). In the case of large Au particles (around 10 nm) on mica (001), the atomic lattice of the top (111) facet has been imaged [53] but in contact mode the true atomic resolution cannot be obtained, that means that point defects cannot be imaged.

Almost as soon as the technique became available, nc-AFM has also been applied in studies of various adsorbed nanoclusters on different substrate surfaces as reviewed in Ref. [41]. One of the main aims in nc-AFM is to provide high resolution of both the cluster and the surface and to obtain atomic scale details of the clusters and adsorption sites. Efforts in this direction have generally focused on standard model systems, particularly Au nanoclusters on alkali halide surfaces [55, 57]. These studies often provided atomic resolution on the substrate. However, approaching closer to a nanocluster resolves it only as a fuzzy hemisphere (Fig. 8a) [55], rather than as its true geometric shape. However on

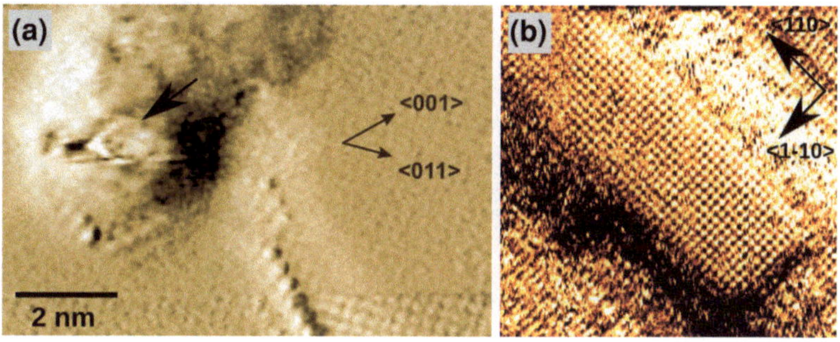

Fig. 8 Noncontact AFM images of supported gold particles. **a** Detuning image of a 2 nm gold cluster on KBr(001), the cluster appears like a fuzzy cloud while the top rectangular facet can bee seen. The KBr substrate is imaged at atomic resolution (from Ref. [55]). **b** Gold nanoribbon on InSb(001). The top facet is imaged at atomic resolution (from Ref. [56])

large gold ribbons supported on InSb(001) particles atomic resolution has been obtained [56] (Fig. 8b). The fuzzy shape of the 2 nm cluster in Fig. 8a is in fact the image of the apex of tip which appears at several places on large scale images [55]. This is another effect of the deformation of the image of small objects by the shape of the imaging tip [55, 58]. In the extreme case where the tip is much larger than the cluster, the cluster is rather a 'needle' and images in fact the tip apex (Fig. 8a). In such a case, the same type of tip-image can be observed at all clusters on the surface, which leads to very complex patterns if the cluster density is high (Fig. 9b).

Significant improvement in imaging the shape of nanoparticle by nc-AFM can be achieved by imaging the surface in the constant height mode as recently demonstrated [59]. In this specific mode the tip scans the surface in a constant height and only the very last nanometer of the tip is put into a close distance to the top facets of the clusters. Since the tip does not follow anymore the contours of the clusters the convolution is greatly reduced. An extensive experimental and theoretical study of Pd nanoclusters on MgO showed that the real shape of the clusters which is a square pyramid truncated on the top by a (001) facet [46] could be seen in constant height mode independently of the shape of the tip [60] (compare image (b) in Fig. 9 with image (c) of same clusters and with the TEM image (a)).

2.4 Kelvin Probe Force Microscopy in nc-AFM

A severe limitation for all scanning probe microscopy techniques is the lack of chemical sensitivity; one can rarely determine the chemical nature of surface atoms or supported molecules or clusters from imaging the surface topography alone (except in some peculiar circumstances where chemically resolved atomic

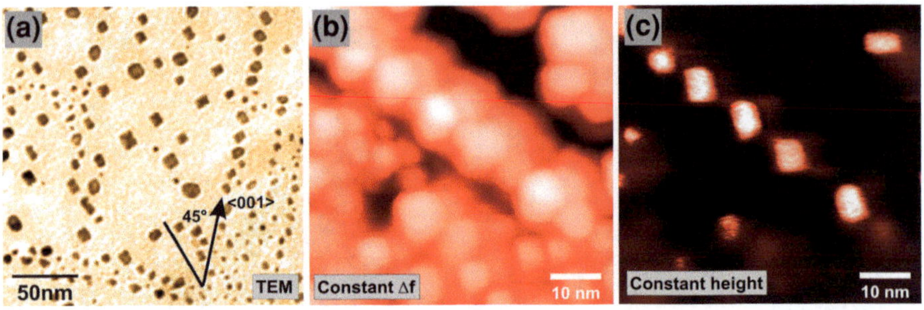

Fig. 9 Pd clusters (7 nm) on MgO(100) imaged in situ by noncontact AFM and ex situ by TEM. **a** TEM image obtained using the carbon replica transfer method. The clusters have the shape of a square pyramid and a mean size of 7 nm (from Ref. [46], copyright (2005), with permission from Elsevier). **b** Nc-AFM topography image obtained in the constant Δf mode. **c** Constant height image representing the detuning Δf of the same Pd clusters. The topography image shows a typical contrast, which was strongly influenced by the tip. However, the convolution could be reduced by imaging the clusters in the constant height mode (from Ref. [60], copyright (2008), American Institut of Physics)

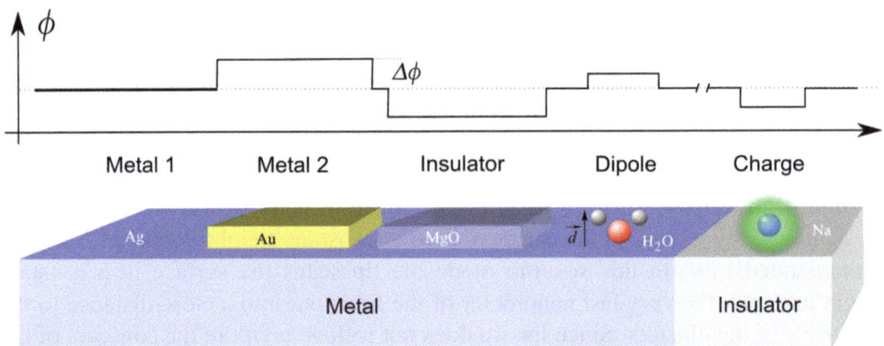

Fig. 10 Objects of different chemical compositions exhibit differences in the local work function on the surface (metal) or in the electrostatic surface potential (insulators), which can be used for the chemical identification in AFM. From Ref. [41]

contrast can be obtained *e.g.* Figs. 2b and 4a). However, in AFM the long-range electrostatic forces alone may assist chemical identification. As illustrated in Fig. 10, nano-objects of different chemical composition (metal or insulating films, ions, molecules, etc.) change the local electrostatic surface potential by changing the local work function (WF) of a metal surface [61–63]. Such objects also change the local electrostatic potential when supported on an insulating surface [64]. Measuring the local work function or electrostatic potential of a surface by AFM with a resolution in the mV range and at the nanometer scale can indeed be used for chemical identification of surface species. Apart from other electrostatic AFMs like Electrostatic Force Microscopy (see Refs. [65–67]), this can be especially achieved using Kelvin probe force microscopy (KPFM), which has become a well-established surface science tool in many scientific disciplines [41]. For a detailed description of KPFM the reader is referred to some papers [41, 68–70] whereas the principle of the KPFM is briefly summarized in the following.

In a KPFM measurement, a *dc* (U_{dc}) and an *ac* voltage (U_{ac}) with frequency f_{ac} are applied between tip and surface. Owing to the modulation of the electrostatic force by the *ac* voltage, the electrostatic contribution, oscillating at frequency f_{ac} and also at 2 f_{ac}, can be extracted (first and second harmonic) by a Lock-In amplifier, which is supplying the ac voltage at the same time. The first harmonic includes a term with $U_{dc} + U_{CPD}$, where the potential difference U_{CPD} is the contact potential difference for conducting tip-surface systems. By varying the dc voltage such that the first harmonic becomes zero ($U_{dc} = -U_{CPD}$), the total electrostatic tip-surface interaction is minimized at each point on the surface and a so-called *Kelvin image* of $U_{dc} = -U_{CPD}$ is obtained. The contact potential difference U_{CPD} is in fact proportional to the difference of WF between tip and surface ($U_{CPD} * e = \phi_{Surface} - \phi_{tip}$). If at two different places on the surface the contact potential difference U_{CPD} is obtained, the difference of the potentials, $U_{CPD,\,1} - U_{CPD,\,2}$, is proportional to the WF difference of the materials at the two places (($U_{CPD,\,1} - U_{CPD,\,2}) * e = \phi_{Surface,\,1} - \phi_{Surface,\,2}$). In other words, the contrast of Kelvin images ($U_{dc} = -U_{CPD}$) reflect variations of the surface WF. Since surface charges or dipoles can significantly change the condition $U_{dc} = -U_{CPD}$, Kelvin images represent also the distribution of surface charges and dipoles, which is especially the source of contrast in Kelvin images obtained on bulk insulator surfaces [71, 72].

A major goal for KPFM is to characterize supported metal clusters on surfaces. Metal nanoclusters on oxide surfaces for instance play a particularly important role in heterogeneous catalysis [73, 74] then it is particularly interesting to obtain information on the electronic structure of an individual clusters (local work function, charge state…). Quite often the support can change the electronic and therefore the catalytic properties of the clusters [74]. Since phenomena like charge transfer or polarization modifies the electronic properties (WF of large clusters), KPFM is a promising technique for studying all this with nanometer resolution.

Simple surface systems, demonstrating that KPFM works quite efficiently, are conducting surface systems (*e.g.* metal clusters on graphite or on low bandgap semiconductors) [75–77]. KPFM accurately represents the work function differences between large clusters and the substrate surface with a nanometer resolution and with an energy resolution in the mV range as shown for gold clusters on HOPG [75], Si [76] or InSb [77]. With respect to bulk insulators support, KPFM has been mainly used to study metal clusters on alkali halides surfaces like NaCl(001) [78] or oxide surfaces like TiO_2(110) [79] or MgO(001) [80].

In such experiments, the bulk insulator separates the clusters from a conducting support (metallic sample holder), so that no conducting channels exist between the clusters and the support. This aspect is important especially when charge is transferred between a cluster and the tip during contact. The charge stays for a very long time on the cluster and electrons can be transferred only to the neighboring clusters located at nanometer scale distances from the charged cluster [78]. When clusters are 'neutral' on the insulating surface, KPFM measures the WF difference between the insulator on the conducting sample holder and large clusters [79, 80]. However, in most cases, the clusters are influenced by the insulating support, which can be observed for gold clusters on alkali halide surfaces [78], palladium clusters on MgO(001) [80] or even for single platinum atoms on the $TiO_2(110)$ surface [81]. Important issues in KPFM are the lateral resolution, the accuracy of WF measurements and the resolution in voltage. Because KPFM is detecting only the electrostatic tip-surface interaction, the lateral resolution and accuracy of a measurement depend much more on the tip size and shape than it is the case in the standard topography imaging mode. The reason is that electrostatic forces are much more long-ranging (under specific conditions up to 50 nm and more) than forces like van der Waals and short-ranges forces, which are responsible for the topography contrast. If electrostatic forces are more long-ranging, they act on a larger tip volume and, as a consequence, the convolution with the tip is increased. These aspects have been studied in particular on ultra-thin insulating films supported on metal surfaces. These surfaces are well suited for such studies since the films lower the WF of the metal underneath by more than 0.5 eV and the surface is quite flat. As a rule of thumb it can be said that although a lateral resolution of some nanometer can be obtained on such flat surfaces the accuracy is mostly relatively low. Several 10 nm large structures are needed such that the WF is enough saturating. However, since it may happen that the tip may change due to, *e.g.*, tip-changes during scanning, the tip can reduce the size of its apex or so-called nanotips are formed at the tip apex. This greatly enhances the lateral resolution but also the accuracy of KPFM measurements [82]. The lateral resolution and accuracy increase as soon as nanometer large nano-objects like clusters are imaged. The reason is that when the tip is placed above a cluster, which has a height of some nanometers for instance, the tip does not 'feel' anymore contributions of the substrate surface. Lateral resolutions of a few nanometer can be obtained in best cases (Fig. 11).

3 Nucleation and Growth of Bimetallic Clusters on Insulators

3.1 Overview of Nucleation and Growth Theory

3.1.1 Nucleation Kinetics

Nucleation theory has been strongly developed in the 1960/1970 s in particular to understand the early stages of deposition of metallic thin films on insulators which were also used as model for epitaxial growth [83–85]. Classical nucleation theory was originally developed by Volmer [86]. From this theory a nucleus can spontaneously grow if it has a critical size i^*. Taking the case of a liquid droplet the radius of the critical nucleus is:

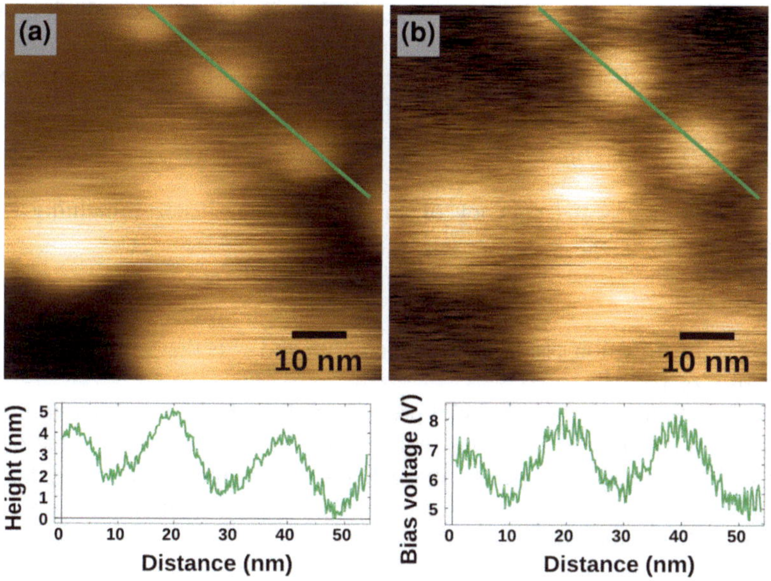

Fig. 11 Pd clusters grown on MgO(001): **a** topography, **b** Kelvin image. The line profile at bottom shows that the lateral resolution in the Kelvin mode is at least 5 nm. The mean value of the difference of contact potential between Pd cluster and MgO is 2.4 V which corresponds to the difference of work function of bulk materials. From Ref. [80], Copyright (2009) American Chemical Society

$$R^* = 2\gamma v/\Delta\mu \quad (1)$$

γ is the surface energy of the droplet, v is the atomic volume and $\Delta\mu$ is the variation of the chemical potential between gas and liquid phases which is expressed by:

$$\Delta\mu = kT\ln(P/P_\infty) \quad (2)$$

P/P_∞ is the supersaturation, P and P_∞ are the actual vapour pressure around the nucleus and of the infinite liquid phase, respectively. The nucleation rate is expressed by:

$$J = C\exp(-\Delta G^*/kT) \quad (3)$$

ΔG^* is the energy barrier (nucleation barrier) to form the critical nucleus and C is a constant.

However in the case of condensation of a metal vapour on an insulator substrate the supersaturation can be very high. Taking as an example the condensation of Pd on an MgO crystal at 700 K the supersaturation is around 10^{12} then from Eq. (1) the critical radius would be 0.17 nm. Thus in this case the critical nucleus is only one atom, which means that the dimer is already stable. In that case the classical nucleation theory is no longer available. The growth process occurs by accretion of adatoms (like a polymerization process). It is described by the so called 'atomistic nucleation theory' which has been developed by Zinsmeister [87] from basic ideas first expressed by Frenkel [88]. The rate equations given by Zinsmeister express the variation with time of the number of clusters of size i:

$$dn_i/dt = \omega_{i-1}n_{i-1} - \omega_i n_i \quad \text{for } i = 2, 3 \ldots \infty \tag{4}$$

ω_i is the attachment frequency of an adatom to a cluster containing i atoms which is expressed by:

$$\omega_i = \sigma_i D n_i \tag{5}$$

where D is the diffusion coefficient of an adatom and σ_i is the capture number for a cluster of size i.

To calculate the number of nuclei present on the substrate we have to integrate the system of differential equations which can reduced in only two equations [87] which can be solved numerically if we know the various capture numbers. For sake of simplicity Zinsmeister assumes that it is a constant (between 1 and 4). From this scheme the nucleation frequency is:

$$J = 2\omega_1 n_1 \tag{6}$$

Assuming that the growth is negligible the density of adatoms is equal to the stationary value:

$$n_1 = F\tau \tag{7}$$

where F is the flux of atoms impinging on the substrate and τ the life time of an adatom before desorption. Then combining Eqs. (5)–(7) the nucleation rate becomes:

$$J = 2\sigma_1 D F^2 \tau^2 \tag{8}$$

Then the nucleation rate is proportional to the square of the impinging flux for a homogeneous substrate without defects.

The diffusion coefficient is expressed by:

$$D = (\nu a^2/4)\exp(-E_d/kT) \tag{9}$$

The life time t is expressed by:

$$\tau = (1/\nu)\exp(E_a/kT) \tag{10}$$

where ν is the frequency factor, E_a and E_d are the adsorption energy and diffusion energies of an adatom. Combining Eqs. (8)–(10) the nucleation frequency becomes:

$$J = (\sigma a^2 F^2/2\nu)\exp[2E_a - E_d)/kT] \tag{11}$$

Generally insulating surfaces (like alkali halides, oxides) contain point defects which strongly bind adatoms. If an adatom is indefinitely trapped on a defect it is already a stable nucleus, then the critical size is $i^* = 0$. This case is called nucleation on defects. Robins and Rhodin have treated this case and compared with TEM measurements of the nucleation of Au on MgO(100) [89]. The nucleation rate is now expressed by:

$$J = \sigma D n_1 n_o \tag{12}$$

where n_o is the density of defects. Again assuming a negligible growth the nucleation frequency becomes:

$$J = (\sigma a^2 F/4) \exp[(E_a - E_d)/kT] \tag{13}$$

Now the nucleation rate varies linearly with the flux, this dependency allows the separation between nucleation on perfect surface and nucleation on point defects. This linear dependency has been experimentally observed for Au/MgO [89], Pd/MgO [90]. In that simple case the rate equations can be integrated (still assuming a negligible growth) and the density of clusters as a function of time reaches exponentially the density of defects [89]:

$$n(t) = n_o[1 - \exp(-tJ/n_o)] \tag{14}$$

This exponential behaviour has been observed for Au and Pd nanoparticles on MgO [89, 91] and also for Pd/NaCl and CuPd/NaCl [92]. However, contrary to this simple theory of nucleation on defects, the saturation density of clusters has been observed to decrease by increasing temperature for Au/MgO (in situ AFM study) [54], Pd/MgO studied in situ by AFM [93] and by He diffraction [91]. In fact, the assumption that defects are perfect sinks for adatoms is not true, due to the finite adsorption energy on a defect site, at high temperature adatoms can escape from the defects and the number of populated defect sites decreases with increasing temperature. This effect has been included in a more accurate treatment of point defect nucleation made by Venables and coworkers [93, 94] which has been compared with the AFM study of the nucleation of Pd/MgO [93]. From this model, the saturation density of clusters is equal to the density of defects only on a limited range of temperature, at high temperature the density of clusters decreases but also at low temperature the cluster density is larger due to the fact that nucleation on normal sites is no longer negligible (Fig. 13).

3.1.2 Growth Kinetics

By integration of the system of rate Eq. (4) one obtains the size distribution of the growing clusters at each time. Zinsmeister has solved this system of differential equation assuming a constant value for the attachment frequency (ω_i) [95]. However by this treatment several aspects of the growth of clusters are not taken into account: the direct impingement on the growing cluster (which is important at the late stages of growth) and the competition between clusters for the capture of diffusing adatoms (which is important at high density of clusters). Several authors have tried to treat more accurately the calculation of the attachment frequencies [96–101]. In the typical growth conditions we consider here (metal on insulator at temperature above RT) the diffusion of adatoms is limited by desorption and the diffusion length X_s of an adatom is:

$$X_s = (D\tau)^{1/2} = (a/2)\exp[(E_a - E_d)/2kT] \quad (15)$$

Then, one can consider that around each cluster exists a *collection zone* where every adsorbed will be captured by the growing cluster. In the case of an isolated cluster the width of the capture zone is close to X_s. Halpern has treated exactly the growth rate for an isolated cluster [96]. In order to take into account for the competition between growing clusters, Sigsbee and Stowell have used the lattice approximation [97, 98]. They consider that the growing clusters are sited on a regular square (or hexagonal) lattice, the growth conditions are the same for all clusters. In these conditions the density of adatoms around each cluster can be analytically solved [99]. Another way to treat this problem of the capture of adatoms: the uniform depletion approximation has been introduced by Lewis and Venables [99, 100]. In this treatment one considers a uniform density of clusters and the growth flux around each cluster is calculated assuming a uniform density of adatoms all over the substrate which is derived from the calculation of an isolated cluster by Halpern [96]. It has been shown that the uniform depletion approximation is a rather good approximation of the lattice model [101]. In a general case the growth rate of clusters can be calculated only numerically. However in some cases, the capture number (σ) can be expressed then an explicit growth law can be obtained, as shown by Kashchiev [102, 103]. In the case of an isolated cluster, *i.e.* $L/X_s \gg 1$, L is the mean half distance between two neighbouring clusters (incomplete condensation regime) and if the direct impingement is negligible (*i.e.* early stage of growth) the cluster radius varies with $t^{1/3}$ for 3D clusters [103] and $t^{1/2}$ for 2D clusters [102]. At late stage of growth (*i.e.* very large clusters) the clusters grow practically only by direct impingement then the radius vary linearly with time [103].

Another limiting case corresponds to the situation of a strong competition between clusters for capture of adatoms (*i.e.* $L/X_s \gg 1$: complete condensation regime); the cluster radius varies with $t^{1/3}$ or $t^{1/2}$ for 3D and 2D clusters, respectively. In a general case the growth rate of a cluster can be expressed by a power law of the deposition time:

$$R(t) = R_o t^p \quad (16)$$

The exponent for any experimental has been calculated in the case of the lattice approximation for 3D clusters [104]. Figure 12 displays the exponent p as a function of the reduced radius (R/X_s) and for various value of the reduced cluster lattice parameter (L/X_s).

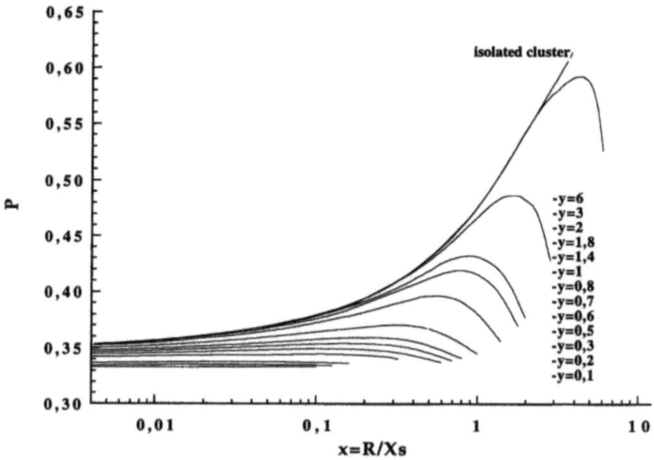

Fig. 12 Exponent of the growth power law as a function of the reduced radius (R/X_s) and for various reduced intercluster half-distance ($y = L/X_s$). Top curve represent the case of an isolated cluster. From Ref. [104], copyright (1998), with permission from Elsevier

From this figure we immediately recognize the limiting cases: exponent close to 1/3 for incomplete condensation (L/X_s small) and beginning of growth (R/X_s small) and for complete condensation (L/X_s large). However we can see on Fig. 12 that p is between 1/3 and 0.4 for a large range of values of L/X_s and R/X_s, in agreement with experimental measurements for Pd clusters on various insulating substrates [104].

3.1.3 Nucleation and Growth on a Regular Lattice of Point Defects

The case of nanostructured substrate exhibiting a regular array of defects is a quasi-ideal system. Indeed, nucleation on defects is very rapid and then due to regularly spaced nuclei the growth rate will be uniform and as a consequence the size dispersion will be very narrow. In the recent years several naturally nano-structured substrates have been discovered like ultrathin films of alumina on $Ni_3Al(111)$ [105–107], ultrathin film of titania on Pt(111) [108] and CoO film on Ag(001) [109]. These nanostructured oxide substrates have shown to be good templates to grow arrays of metal clusters [107–112]. The nucleation and growth of metal clusters on a lattice of defects have been studied by kinetic Monte Carlo simulation (KMC) in the case of 2D and 3D growth [113, 114]. The energetic parameters used for the simulation of 3D growth correspond to the case of Pd clusters on nanostructured alumina on $Ni_3Al(111)$ [105–107]. Figure 13 displays the variation of the saturation density as a function of the substrate temperature. The simulation shows a rapid nucleation on the defects until a complete occupation but only in the temperature range 240–300 K. At higher temperature some

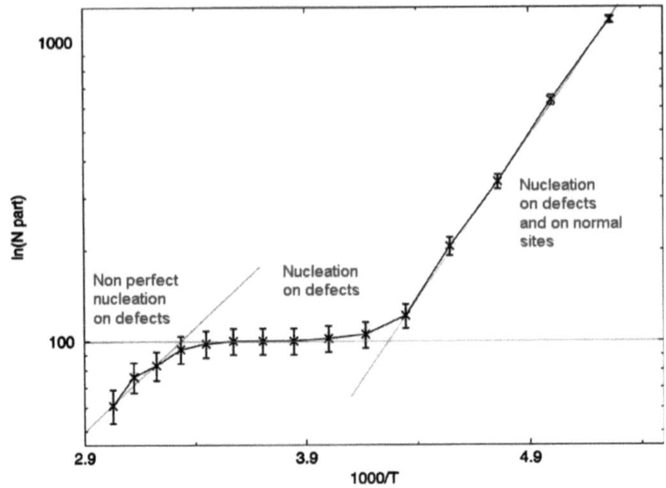

Fig. 13 Variation of the saturation density of clusters versus reciprocal temperature calculated by KMC simulation for Pd/alumina/$Ni_3Al(111)$. Adapted from Ref. [114], copyright (2010), with permission from Elsevier

defects remain empty because they are no longer perfect sinks for adatoms. At very low temperature nucleation occurs also on regular adsorption sites and the density of clusters is larger than the density of defects. After the nucleation stage clusters grow uniformly that results in a very sharp size distribution (size dispersion around 7% of the mean size) [114].

3.1.4 Nucleation and Growth of Bimetallic Clusters (AB)

Case of simultaneous deposition

The case of nucleation on a homogeneous substrate with a critical nucleus of one atom has been for the first time treated by Anton and Harsdorff [115]. One has to add the nucleation of AA, BB and AB dimers. However, now it is known that for insulator substrates we are interested here (metal oxides, alkali halides….) nucleation is controlled by defects.

Considering a nucleation on point defects the total nucleation rate is the sum of the nucleation of A and B which are given by Eqs. (12) and (13):

$$\begin{aligned} J &= J_A + J_B \\ &= (\sigma_A a^2 F_A/4)\exp[(E_{aA} - E_{dA})/kT] + (\sigma_B a^2 F_B/4)\exp[(E_{aB} - E_{dB})/kT] \end{aligned} \quad (17)$$

In a general case the energetic terms are different then nucleation of one species will dominate. The growth rate is calculated by summing the contribution of the two metals given by the growth kinetics of monometallic clusters (Sect. 3.1.2). Like in the case of a single metal, power laws are also expected [92, 116].

The time dependence of the chemical composition can be also obtained in integrating the growth equations but it can be also obtained more simply if the experimental growth law is known by adding the two contributions to the growth: direct impingement and capture of diffusion of adatoms. The first contribution is obvious and the second one can be calculated in the approximation of the *collection zone* [92].

Case of sequential depositions (A, then B)
Nucleation of A will take place. Still assuming nucleation on point defects, if at the end of the first deposition all defects sites are not occupied, during the second deposition there will be competition between nucleation of B and growth of bimetallic AB clusters. The relative rates of these two processes will depend on the relative values of the adsorption and diffusion energies which depend on the strength of the metal substrate interaction. If interaction of B with the substrate is weaker than those of A, growth will be favoured. It will also depend of the density of nuclei. A higher density will favour growth of bimetallic particles. The growth rate of bimetallic clusters, will be given by Eq. (16) with an initial condition which is the size of the pure A cluster at the end of the first deposition.

3.2 Nucleation and Growth of Bimetallic Clusters: Experiments

3.2.1 Simultaneous Deposition

The F_A and F_B fluxes of atoms A and B impinge simultaneously on the substrate. For the nucleation stage the two types of atoms compete for the occupation of nucleation sites: this competition depends on the flux ratio F_A/F_B and the relative values of the adsorption energy of the two types of atoms E_{aA} and E_{aB}. If adsorption energy of A is significantly larger than the one of B, the nucleation is dominated by the A species. Assuming nucleation on defects, rapidly all the defects are filled by A-rich clusters then bimetallic clusters will grow by capture of A and B adatoms (dominant at the early stages of growth) or by direct impingement (dominant at late stage of growth). Thus the composition of the clusters will start from pure A up to the nominal composition which $Z_{A\infty} = F_A/(F_A + F_B)$. Table 1 displays examples of growth studies of bimetallic clusters on insulators by simultaneous depositions.

The first systematic studies on the nucleation have been performed by the group of Anton [116–121]. For these early studies, scanning probe microscopy techniques were not yet invented and (ex situ) TEM was the main technique of characterization of the bimetallic nanoparticles. By TEM the size distribution, the structure and the morphology of the nanoparticles can be determined [46]. The composition can be determined with a TEM by X-Ray fluorescence [116, 122] and on the modern (S)TEM by HAADF and EELS [123] techniques at the level of a single particle. Anton's group has investigated several types of bimetallic

Table 1 Simultaneous deposition of the two metals (A,B). $\mathbf{\Delta H_{sublimation}}$ is the sublimation energy of the pure metal indicated in the order A/B. **Z = f(t)** means the chemical composition evolves during the deposition of the two metals, eventually after a long time the nominal composition can be reached. **A segreg.** means that A segregates at the particle surface

AB	FePt	AuPd	CuPd	AgPd	AuCu	AuAg
$\Delta H_{sublimation}$ (eV/atom)	5.85/4.29	3.94/3.78	3.94/3.50	3.94/2.91	3.78/3.50	3.78/291
Substrate	NaCl(100)	NaCl(100) CeO_2/Ru(0001) Fe_3O_4/Pt(111) MgO/Ag(100)	NaCl(100)	NaCl(100)	NaCl(100)	NaCl(100) KBr(100)
Techniques	AFM TEM/EDX	TEM/EDX IRAS,TPD	TEM/EDX	TEM/EDX	TEM/EDX	TEM/EDX
Results, references	FePt alloy [126]	AuPd alloy Z = f(t) Au segreg. [119–121, 128]	CuPd alloy Z = f(t) [92, 124, 125]	AgPd alloy Z = f(t) [119, 120]	AuCu alloy Z = f(t) [118]	AuAg alloy Z = f(t) [116, 117]

nanoparticles: AuPd [119–121], AgPd [119], AuCu [118], AuAg [116] on UHV-cleaved NaCl. The AuPd system has been studied in details. Figure 14 displays the evolution of the Au content in the nanoparticles as a function of the deposition time and for various ratios F_{Pd}/F_{Au}. It is clear that during the first stages of growth the nanoparticles contain mainly Pd [120]. This result is due to the much larger adsorption energy of Pd on NaCl compared to Au. By increasing the flux of gold atoms the initial concentration of Au in the particles increases but for all flux ratios the composition of the nanoparticles evolves with deposition time. The nominal composition $Z_{A\infty}$ is eventually reached for a ratio $F_{Au}/F_{Pd} = 16.3$ after about one hour of deposition that would correspond to very large particles ($D = 40$ nm [121]). The composition has been also determined from Monte Carlo simulation for the early stages of growth of the nanoparticles (the capture numbers are assumed to correspond to the isolated cluster limit, see Sect. 3.1.2) and a good agreement with experiment (Fig. 14) has been obtained with adsorption energies of 0.78 and 0.48 eV for Pd and Ag, respectively.

The growth of CuPd on UHV-cleaved NaCl has been also studied by TEM and EDX in the Henry's group [92, 124, 125]. Figure 15 displays the Cu concentration as a function of time for a flux ratio $F_{Cu}/F_{Pd} = 8$ [92]. Again, as the interaction of Pd with NaCl is much larger than those of Cu, nucleation is dominated by Pd and the particles become enriched in Cu during their growth. The nominal concentration (88.8% of Cu) is not reached after 900 s of deposition.

Figure 16 displays the nucleation kinetics and the growth rate of CuPd nanoparticles. The nucleation kinetics follows the nucleation on defects behaviour (see Eq. (13)): the density of clusters reached rapidly a saturation value. The nucleation kinetics is very close to the case of pure Pd [92], showing that for Pd and Cu simultaneous deposition the nucleation is controlled by Pd. The growth

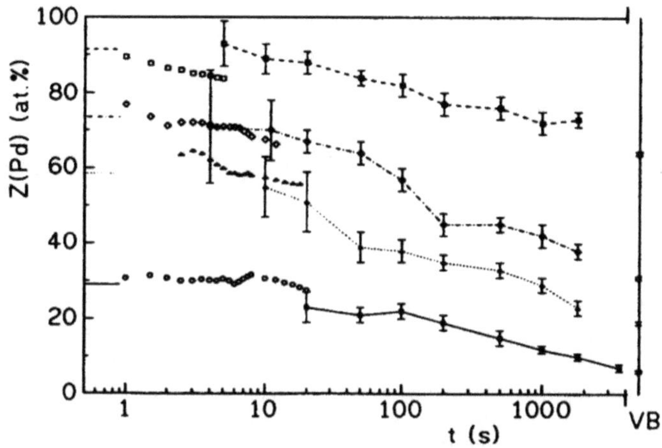

Fig. 14 Semi-logarithmic plot of the concentration of Pd in AuPd nanoparticles grown on NaCl(001) by simultaneous deposition of Au ($F_{Au} = 4.4 \times 10^{13}$ cm^{-2} s^{-1}) and Pd (from the top to the bottom curves $F_{Pd} = 7.9$, 2, 1 and 0.27×10^{13} cm^{-2} s^{-1}) as a function of deposition time. The solid symbols correspond to experimental measurement by EDX and the open symbols correspond to MC simulation. From Ref. [120], copyright (1990) by The American Physical Society

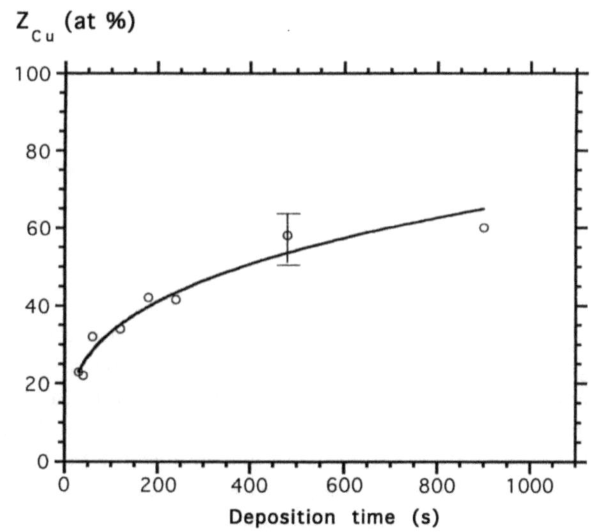

Fig. 15 Variation of the concentration of Cu in CuPd clusters grown on NaCl(001) by simultaneous deposition of Pd ($F_{Pd} = 1 \times 10^{13}$ cm^{-2} s^{-1}) and Cu ($F_{Cu} = 8 \times 10^{13}$ cm^{-2} s^{-1}) at 553 K as a function of deposition time, measured by EDX. From Ref. [92], reproduced by permission of the Royal Society of Chemistry

rate of CuPd clusters follows a power law with an exponent of 0.42 close to the case of pure Pd (0.40). These results prove two facts: (i) the growth mainly occurs by capture of atoms adsorbed on the NaCl substrate which is expected because the substrate coverage is low (smaller than 5%), (ii) the capture of Cu adatoms is weak.

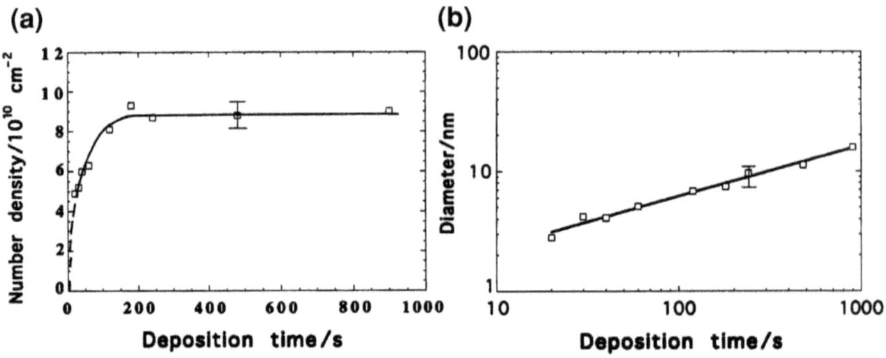

Fig. 16 Nucleation (**a**) and growth kinetics (**b**) of CuPd clusters on NaCl(001) by simultaneous deposition of Pd ($F_{Pd} = 1 \times 10^{13}$ cm^{-2} s^{-1}) and Cu ($F_{Cu} = 8 \times 10^{13}$ cm^{-2} s^{-1}) at 553 K as a function of deposition time. From Ref. [92], reproduced by permission of the Royal Society of Chemistry

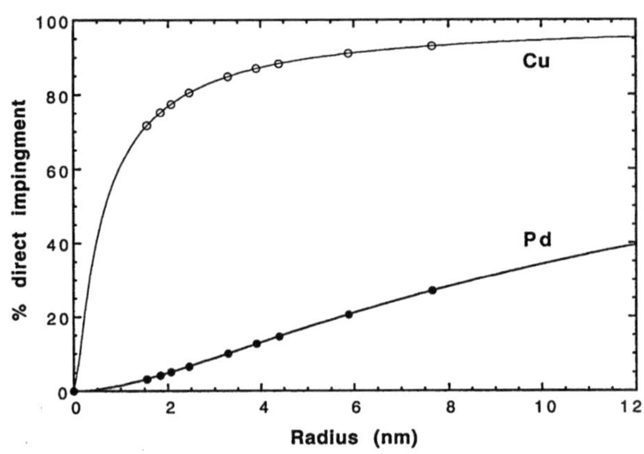

Fig. 17 Percentage of direct impingement for Cu and Pd during the growth of CuPd clusters on NaCl(001) by simultaneous deposition of Pd ($F_{Pd} = 1 \times 10^{13}$ cm^{-2} s^{-1}) and Cu ($F_{Cu} = 8 \times 10^{13}$ cm^{-2} s^{-1}) at 553 K as a function of deposition time. From Ref. [92], reproduced by permission of the Royal Society of Chemistry

In order to determine the energetic parameters (E_a and E_d) for this system the growth of the bimetallic particles has been simulated by using the lattice approximation (see Sect. 3.1.2). From the fit of the growth kinetics and of the time dependent composition with this growth model the ($E_a - E_d$) parameters of 0.38 and 0.07 eV have been determined for Pd and Cu, respectively [92]. Once these parameters have been determined it becomes possible to know the relative contribution of the two growth mechanisms (direct impingement and capture of adatoms) during the growth for the two types of atoms. Figure 17 shows the proportion of incorporation of Pd and Cu atoms by direct impingement during the growth of the CuPd clusters. We see clearly that Cu is mainly incorporated by direct impingement while Pd is mainly incorporated by capture of adatoms. This becomes clear if we consider that at the growth temperature (553 K) the mean diffusion length of Cu is very small (0.5 nm) while it is much larger for

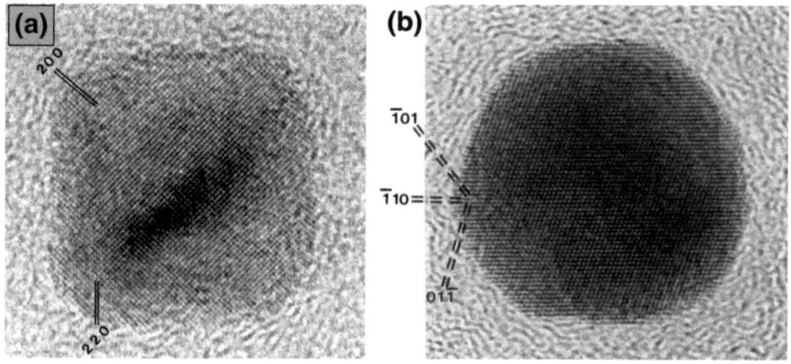

Fig. 18 HRTEM pictures of CuPd nanoparticles grown by simultaneous deposition on NaCl. **a** After growth at 553 K. **b** After annealing at 638 K. From Refs. [124, 125]

Pd (7.5 nm) in comparison with the size of the clusters (the relative contribution of the diffusion-capture process to the direct impingement is roughly equal to $2X_s/R$).

The structure of the nanoparticles can be determined by HRTEM. In the case of CuPd/NaCl(001) the particles grown at 553 K have a face centered cubic structure that corresponds to a solid solution *i.e.* no chemical order (see Fig. 18a). After annealing at 638 K the particles get the CsCl type structure corresponding to the ordered CuPd structure (see Fig. 18b) [125].

The growth of bimetallic nanoparticles by simultaneous deposition of the two metals has rarely been studied by AFM or STM [126–128]. In the case of FePt/NaCl [126] the nanoparticles were characterized by a combination of in situ STM, ex situ TEM, TED and EDX which show that the particles were bimetallic and epitaxied on the NaCl substrate, the average composition was determined.

In summary, for simultaneous deposition, the composition of the bimetallic particles evolves during the growth except at the late stage of growth when the direct impingement is the main growth mechanism where the concentration tends to the nominal one. The variation of the composition is due to the different incorporation of the two metals which is related to the width of the capture zone around the clusters which depends exponentially on the energetic parameter $E_a - E_d$ which varies with metal substrate interaction. In principle it could be possible to correct for this difference of capture rate by increasing the flux of the metal having the weaker interaction but in practice it is necessary to adjust continually the flux ratio that would be very difficult to manage.

3.2.2 Sequential Deposition

In the case of sequential deposition, A then B, the nucleation is controlled by A but during the second deposition metal B can nucleate new pure B clusters or only grow pre-existing A clusters. The competition between these two processes depends on the metal-substrate interactions but also on the distance between prenucleated clusters. If the prenucleated A clusters occupy all the defects sites which are separated by a distance much smaller than the diffusion length (X_s) of an adsorbed B atom the growth of AB clusters will dominate. If the mean diffusion length of B atoms is smaller than the distance between A clusters the growth of A clusters will be negligible. Finally if some defects are not occupied by A clusters nucleation of B clusters will occur. We will see all these possibilities by looking on the published works in the recent years (Table 2) which use mainly STM as a charaterization tool of the nanoparticles.

CoPd is the most studied systems [129–132]. The CoPd particles are grown on an ultrathin alumina thin film obtained by high temperature oxidation of NiAl(110) by depositing sequentially Co then Pd or Pd then Co. The particle nucleation and growth were investigated in situ by STM. By depositing first Co then Pd, the density of clusters stays unchanged after the second deposition and the mean size of the clusters increases, that means that no nucleation of pure Pd occurred and that Pd participates only in the growth of bimetallic clusters [129, 132]. In the reverse case, Pd deposition then Co deposition, after the second deposition the density of clusters increases by about 30% and the size of clusters increases [129, 132]. In this case new pure Co clusters are formed together with bimetallic CoPd clusters. For Fe deposited on Pd preformed clusters, pure Fe clusters are observed together with bimetallic ones, in the reverse order only bimetallic clusters are observed [133]. If now Au [111, 134] or Ag [135] is deposited on prenucleated Pd clusters one get only bimetallic particles (no new clusters are formed) on the contrary if Pd is deposited on Au prenucleated clusters one obtains pure Pd and bimetallic AuPd particles [136]. These different behaviours can be rationalized if one compare the sublimation energy of the different metals (see Table 2) which roughly scales with the interaction of the metal with the substrate. The sublimation energy of Pd is smaller than those of Co or Fe and larger than those of Ag or Au. During the first deposition the nucleation rate (*i.e.* the number of formed clusters) for Co or Fe will be much larger than for Pd and the nucleation of Pd will be larger than for Ag or Au. Assuming a predeposition of Pd clusters, in the second step the competition between nucleation of Co or Fe and growth of bimetallic clusters will be in favour of the nucleation of pure Co or Fe clusters. In the case of Au or Ag the nucleation of pure Au or Ag will be disfavoured by a low nucleation rate and fast diffusion of adatoms that will prefer to grow prenucleated Pd clusters.

However, in the case of nucleation on point defects and assuming that they act as perfect sinks for adatoms, nucleation of Au or Ag can occur except if all the defects were already occupied by Pd and then no nucleation could occur [111]. The same reasoning holds in the case of deposition of Pd on prenucleated Co or Fe clusters.

On Fig. 20 we see the effect of the cluster density after the first deposition (Ag) for the system Au/Ag on $TiO_2(110)$ which has been studied in situ by STM [137]. Figure 19 displays a series of STM images from the same area after deposition of Au on prenucleated Ag clusters. The observation of the same area is a clear advantage because it becomes possible to see the individual mechanisms:

Table 2 Sequential deposition of the two metals (A,B). **$\Delta H_{sublimation}$** is the sublimation energy of the pure metal indicated in the order A/B. For the deposition order **A/B** means that A metal is deposited on pre-nucleated B clusters, for the results AB means bimetallic particle, **A or B** means pure A or B particles. **A segreg.** means that A segregates at the particle surface

AB	PtRh	CoPd	FePd	AuPd	AgPd	AuAg
$\Delta H_{sublimation}$ (eV/atom)	5.85/5.75	4.39/3.94	4.29/3.94	3.94/3.78	3.94/2.51	3.78/2.91
Substrate	$TiO_2(110)$	Alumina/ NiAl(110)	Alumina/ NiAl(110)	Alumina/ $Ni_3Al(111)$, $TiO_2(110)$	Alumina/ NiAl(110)	$TiO_2(110)$, Alumina/ NiAl(110)
Techniques	STM, LEIS	STM, AES LEIS, IRAS TPD	STM, XPS IRAS,TPD	STM	STM, XPS IRAS, TPD	STM, photon emission
Deposition order, results, references	**Pt/Rh** PtRh+Pt, [139]	**Co/Pd** CoPd (alloy/core–shell) + Co **Pd/Co** PdCo(core–shell) [129–132]	**Fe/Pd** FePd+Pd **Pd/Fe** PdFe(core–shell) [133]	**Pd/Au** AuPd+Pd [136] **Au/Pd** AuPd (Au segreg.) [111, 134]	**Ag/Pd** AgPd (Ag segreg.) [135]	**Au/Ag** AuAg(core-shell) + Au [137] **Ag/Au** AuAg [154]

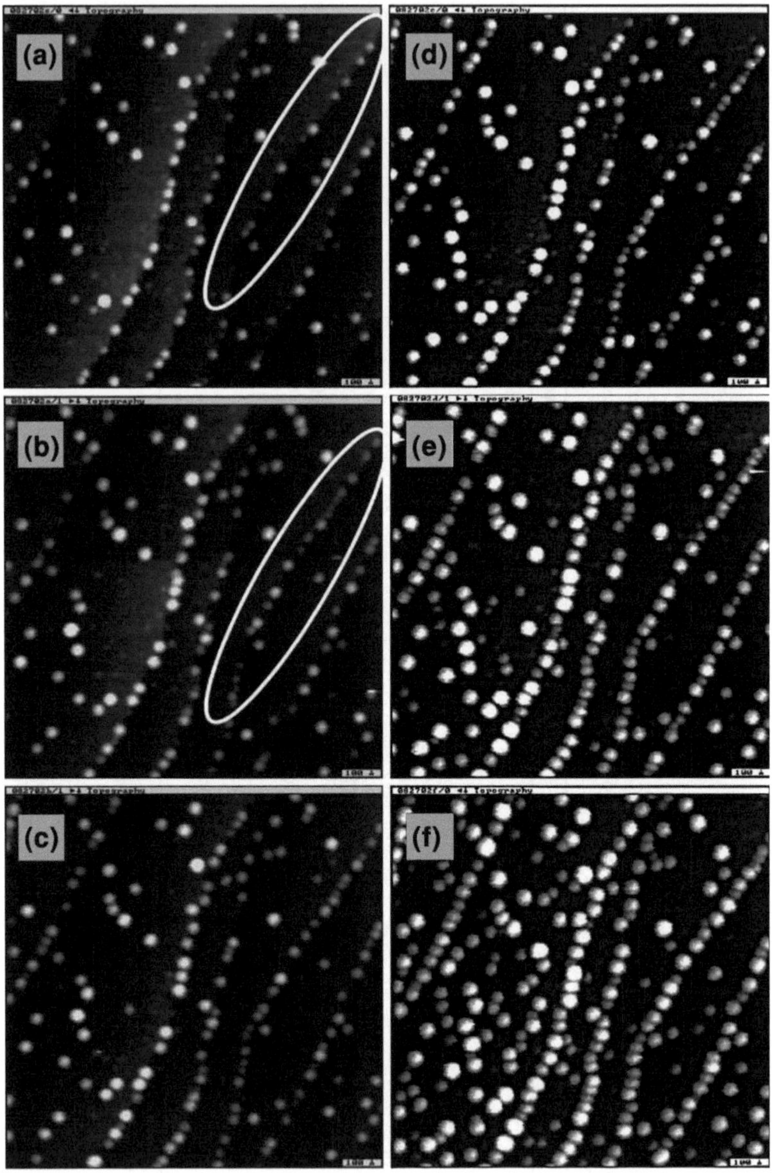

Fig. 19 STM images of a series of depositions of Au (**b** 0.17, **c** 0.34, **d** 0.51, **e** 0.85 and **f** 1.53 ML) on prenucleated Ag clusters (**a** 0.033 ML) on $TiO_2(110)$. From Ref. [137], copyright (2004), with permission from Elsevier

nucleation of a new cluster, cluster growth, cluster mobility, coalescence of two clusters. In this particular case it is also possible to distinguish between steps and terraces. By increasing the amount of deposited Au atoms it is clear that Ag

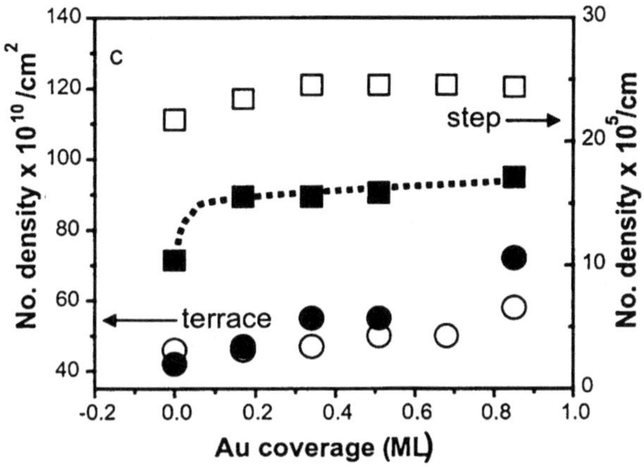

Fig. 20 Variation of the cluster density on steps (*square*) and terraces (*circle*) as a function of the amount of deposited Au on Ag clusters prenucleated on $TiO_2(110)$ from two different coverages: 0.08 (*open symbol*) and 0.033 ML (*solid symbol*). From Ref. [137], copyright (2004), with permission from Elsevier

clusters grow and that new pure Au clusters appear. This is not surprising because the adsorption energy is larger for Au than for Ag (if we still scale with the sublimation energy—see Table 1).

From this study the density of clusters has been measured as a function of the amount of deposited gold both on terraces and on steps. We see on Fig. 20 that the density of clusters is larger on steps than on terraces that is a well known effect which is responsible for the step decoration phenomenon already observed in the first studies of metal deposition on ionic crystals [138]. On steps, by increasing the Ag coverage the density of clusters strongly increases showing that after the thinner Ag deposit the saturation density was far to be reached. Then, during Au deposition the density of clusters notably increased. For the thick Ag deposit the density of clusters is close to saturation and the density of clusters is more or less constant after Au deposition. On terraces, after the deposition of Ag at the lowest coverage (0.033 ML) the saturation density of clusters is more or less reached then the density of clusters increases only slightly during Au deposition. The density of defects is certainly much larger on steps than on terraces.

In the case where the interaction energies of the two metals with the support are very close the use of STM is very important to see which mechanisms occurs during the second deposition. Taking the case of Pt deposited on preformed Rh clusters on $TiO_2(110)$ [139], the sublimation energies of Pt and Rh are 5.85 and 5.75 eV/atom, respectively. Figure 21 shows a series of STM pictures after increasing the amount of Pt deposited on prenucleated Rh clusters. The density of clusters is very large after the deposition of Rh, after Pt deposition the main phenomenon is an increase of the cluster size. However a close examination of the successive pictures from the same area shows that some nucleation of pure Pt occurs but also in the same time some bimetallic clusters disappear by coalescence. Increasing deposition temperature, favours cluster growth relatively to nucleation of pure Pt as expected from nucleation theory (see Sect. 1).

The competition between growth of AB clusters and pure B nucleation depends also on the distance between prenucleated A clusters. Indeed if two clusters are

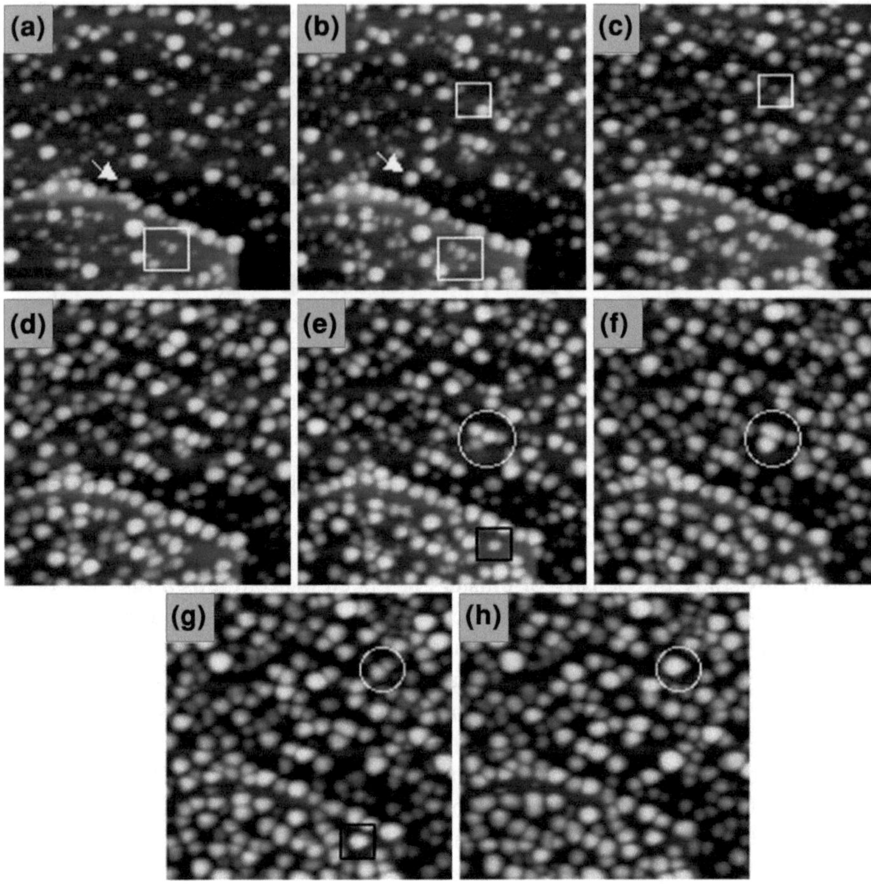

Fig. 21 STM images (50 nm × 50 nm) on the same area of increasing amount of Pt deposited (**b** 0.09, **c** 0.18, **d** 0.33, **e** 0.51, **f** 0.88, **g** 1.18 and **h** 1.6 ML) on prenucleated Rh clusters (**a** 0.3 ML) on TiO_2(110). From Ref. [139], copyright (2006), with permission from Elsevier

very close nucleation of a new cluster has less chance to occur than in the case where the clusters are far each other. Conversely, if two clusters are very close they strongly compete for capture of adatoms. Therefore the composition of the clusters will not be homogenous on the substrate in the typical situation of randomly distributed defects. A way to avoid these problems is to have a substrate presenting a regular array of point defects. This situation occurs in the case of nanostructured oxide ultrathin films [105–109]. In this case the nucleation will occur rapidly on all the defects and when saturation is reached, the clusters will grow homogeneously on the whole substrate. Such a case is presented on Fig. 22. By depositing Au on prenucleated Pd clusters (at 0.1 ML the saturation density is reached at RT), in situ STM observation shows that the density of clusters is constant and the size dispersion is weak [111, 140] (see Figs. 22 and 23). However

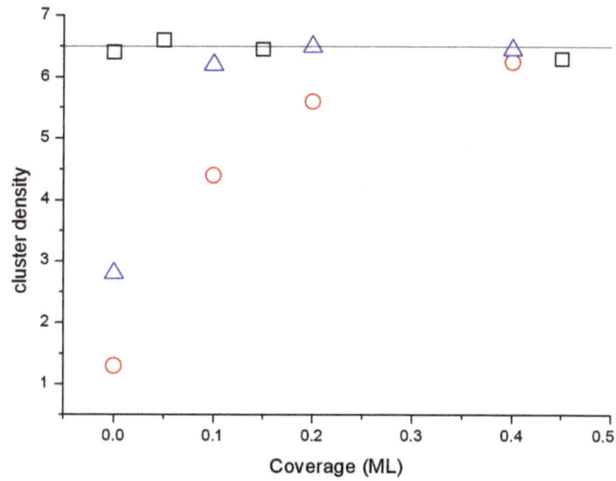

Fig. 22 Au/Pd on alumina on $Ni_3Al(111)$: density of clusters as a function of coverage. *Square* Au on pre-nucleated Pd, *circle* Pd on 0.1 ML pre-nucleated Au, triangle: Pd on 0.2 ML pre-nucleated Au. The straight line corresponds to the saturation of the defects (6.5×10^{12} cm^{-2}). From Refs. [136, 140], reproduced by permission of the Royal Society of Chemistry

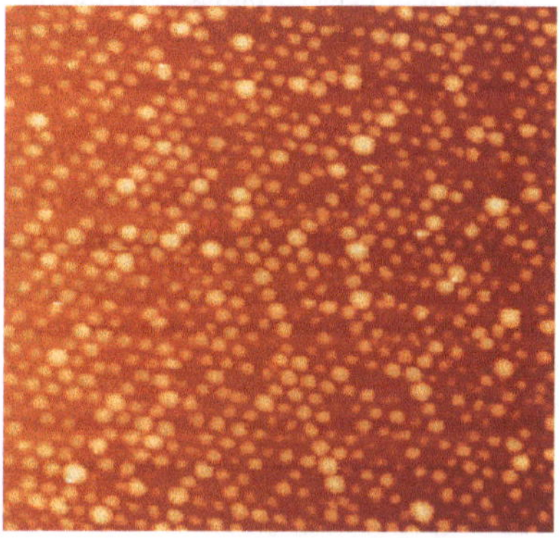

Fig. 23 STM image (200 nm × 200 nm) of AuPd clusters obtained by deposition of 0.04 ML of Au on pre-nucleated Pd clusters (0.02 ML) on alumina on $Ni_3Al(111)$. From Ref. [111]

in the reverse case when Pd is deposited on prenucleated Au clusters, we see on Fig. 22 that the saturation density is not reached because the point defects are not perfect sinks for Au adatoms at RT and the density of clusters increases by formation of pure Pd clusters until all defects are occupied [136]. However by increasing the amount of predeposited gold the proportion of pure Pd clusters decreases. Taking into account the fact that the point defects are not perfect sinks for Au adatoms, during the gold deposition on prenucleated Pd clusters below the saturation density, growth of bimetallic clusters will mainly occur because gold atoms can escape from the defects while they stick permanently to preformed Pd clusters [111].

Cluster Shape

STM and AFM are well suited techniques to study in situ the nucleation and growth of bimetallic clusters while TEM techniques are limited to ex situ observation. However scanning probe techniques (AFM, STM) are a severe limitation which is due to the deformation of the object by the shape of the scanning tip that

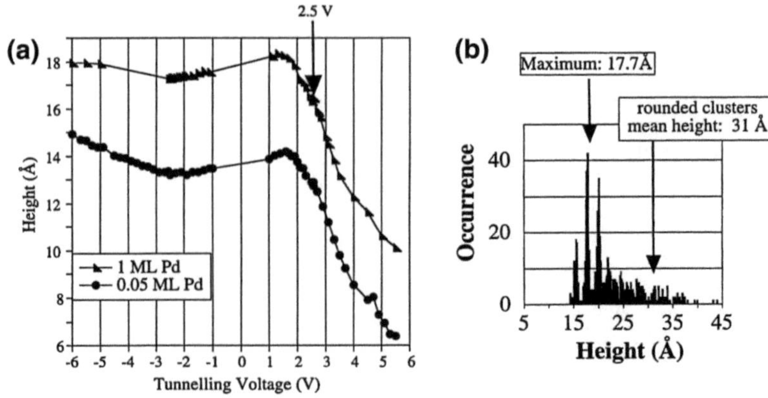

Fig. 24 STM imaging of Pd clusters grown on a ultrathin alumina film on NiAl(110). **a** Apparent height as a function of bias voltage for coverage of Pd of 1 and 0.5 ML. **b** Height distribution. The different peaks separated by roughly 0.22 nm correspond to integer numbers of layers. From Ref. [132], copyright (2007), with permission from Elsevier

leads to an increase of the apparent size and rounding of the shape. This effect and some alternatives are discussed in the first part of the paper (see Sect. 2.3). Often it is believed that height measurements are very accurate by STM. This is true for metallic samples after calibration of the microscope. However in the case of metal clusters supported on an insulting ultrathin film the situation is not so clear. Figure 24a displays the measured height as a function of bias voltage for two deposits of Pd on alumina on NiAl(110) [132]. At negative or low positive value of the bias voltage the cluster height is nearly constant but at voltage larger than 2 V the height strongly decreases. In fact at positive voltage higher than 2.2 V the tip probe unoccupied states of the oxide film then the measured height is no longer representative of the true height of the Pd clusters. However below the band gap of the oxide the tunnelling current comes only from Pd states from the clusters. In order to calibrate accurately the height scale, one can measure the height histogram from a large number of clusters (see Fig. 24b). After these calibration a systematic correction of 0.3 nm is applied for height measurements at a bias voltage of 2.5 V [132]. A similar dependence of the apparent cluster height as a function of the bias voltage has been observed for Pd clusters on alumina on $Ni_3Al(111)$ [141].

Figure 25 displays an STM image of PdCo clusters on an ultrathin alumina film on NiAl(110) which are obtained by depositing 1 ML of Co on prenucleated Pd clusters (1.25 ML). Two types of clusters are visible: large facetted ones corresponding to bimetallic clusters and small round ones that correspond to pure Co clusters [132]. The large bimetallic clusters have a top truncated tetrahedron shape. The top facet is flat and atomic resolution have been obtained (see Fig. 25b) it corresponds to a (111) plane [132]. The atomically resolved image presents bright atoms and dark ones. They are identified from previous studies of Co atoms

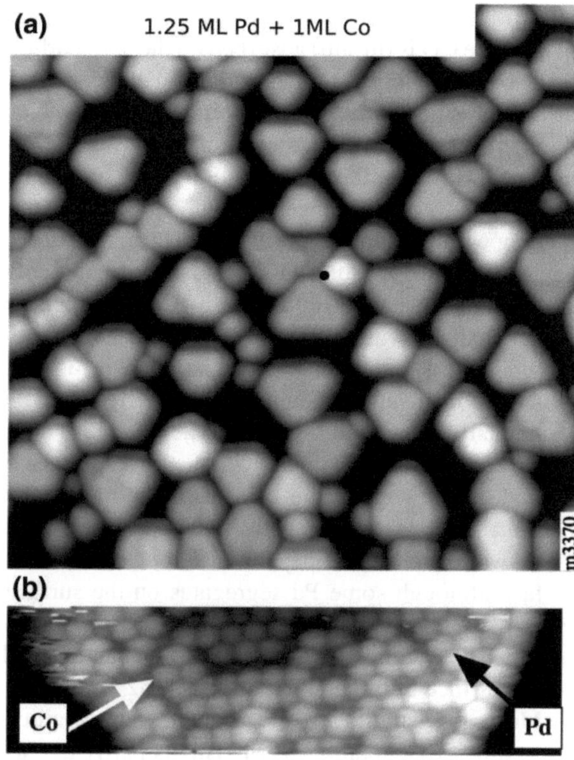

Fig. 25 STM images of 1 ML of Co deposited on prenucleated Pd clusters (1.25 ML) on a ultrathin alumina film on NiAl(110). (**a**) 100 nm × 100 nm area (**b**) atomically resolved image of a top (111) facet of a PdCo cluster showing chemical sensitivity. From Ref [132], copyright (2007), with permission from Elsevier

deposited on Pd(111) to Pd and Co atoms, respectively [132]. This image shows that by depositing Co on Pd clusters at RT alloy is formed instead of core shell structure. Thus segregation of Pd toward cluster surface is possible at RT while in the bulk segregation occurs above 300°C [142]. Atomic resolution with chemical sensitivity on top facets of bimetallic particles is the ideal method to study surface composition at the level of one particle. However, it is very difficult task and from the best of our knowledge it has been reported only once for PdCo nanoparticles [132]. Other techniques like LEIS, MEIS can provide such information on a collection of bimetallic particles. LEIS is the acronym for low energy ion scattering uses ion energies between 1 and 5 keV and probes only the top surface of the nanoparticles (see Ref. [143] for details on the technique). By LEIS it has been shown in the case of AuPd bimetallic particles that Au has a tendency to segregate at the surface of the particle [144] as expected from surface energy consideration and observed on single crystal alloys. MEIS (medium energy ion scattering) uses more energetic ions (50–200 keV) and thus provides information about the surface and the bulk composition of the nanoparticles [145]. It has been confirmed by MEIS that in AuPd nanoparticles Au segregates on the surface (in agreement with AuPd particles prepared by simultaneous deposition [128]) and that subsurface layer is enriched in Pd [146]. Surface composition can also be indirectly studied by IRAS (infrared reflection absorption spectroscopy) and TPD (thermal programmed desorption). In this case CO is (in situ) adsorbed on the nanoparticles and the stretching frequency or the binding energy of CO adsorbed on the A and B metal atoms are probed by IRAS or TPD, respectively. If the stretching frequencies or the binding energies are different for the two types of atoms in principle, the surface composition can be obtained. However the interpretation of the spectra is not so straightforward. Firstly, CO adsorption can induce surface segregation (the metal which has the largest binding with CO tends to be at the surface). CO adsorption induced segregation has been observed for CuPd [147] and AuPd nanoparticles [148]. Secondly, the CO stretching frequency is influenced by CO coverage and environment of a given atom which changes with surface composition. For these reasons these two techniques provide mainly qualitative information like for example presence of a core shell structure or an alloy. Combination of IRAS and TPD has been used to study sequentially grown PdCo [130, 131], PdFe [133] and AgPd [135] nanoparticles. For the three systems if the metal with the lower surface energy [149] is deposited in the second step it stays at the surface. If the metal with the higher surface energy is deposited in the second step, the first deposited metal segregates on the surface as for Fe/Pd while it is not exactly the case for Co/Pd where part of Co stays at the particle surface although some Pd segregates on the surface and eventually Co cover the PdCo particles at very high Co coverage (2 nm Co/0.1 nm Pd) [131]. In Co/Pd, it was observed by chemically resolved STM images and by LEIS that Pd tends to segregate on the surface of bimetallic particles [132].

Much less studies on the growth of bimetallic nanoparticles have been performed by AFM than by STM. This is partly due to the fact that STM is an older technique than AFM but also by the fact that for in situ surface studies AFM is generally restricted to bulk insulator for which STM is inapplicable because STM is easier to operate and atomic resolution is much more difficult to reach by AFM than by STM. However in the recent years non contact AFM has made major advances [41] and in some cases nc-AFM has a better resolution than STM, for instance in imaging atoms of an adsorbed molecules [150]. Moreover, chemical identification of atoms can be performed at RT with systematic investigation of Δf versus distance curves [43] (see Sect. 2.2 and Fig. 4a). Concerning the bimetallic nanoparticles on insulators few studies using AFM have been published in the recent years. The first one concerned the growth of FePt nanoparticles on NaCl surfaces [126]. Figure 26a displays an AFM picture of the bimetallic nanoparticles obtained by simultaneous deposition of the two metals. The particles appear more or less round and some are square, their apparent mean size is around 8 nm and their mean height is 5.5 nm (Fig. 26b). By TEM the particle size is between 6 to 7 nm and their outline is square or rectangle (Fig. 26c). The shape observed by AFM is less precise than by TEM because of the effect of the convolution with the tip shape, as we have seen previously, but AFM brings an accurate measure of the particle height that is not possible with TEM except in transverse view or by tomography [151].

The second study reports on ex situ AFM observations of AuPd nanoparticles grown on silica/Si(100) through an ice buffer layer [127]. Another study concerns

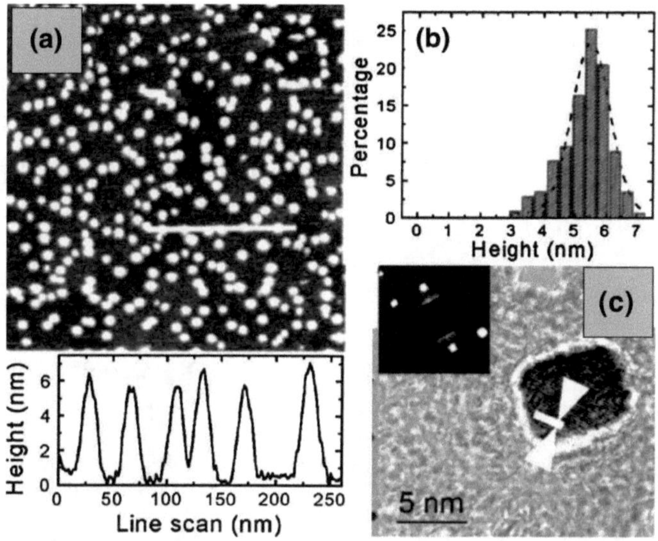

Fig. 26 FePt bimetallic particles on NaCl(001) **a** AFM picture with a line profile at the *bottom* **b** height histogram **c** TEM picture of a single particle. From Ref. [126], copyright (2005), American Institute of Physics

bimetallic AuIn particles on a semiconductor surface [152]. Au is deposited at RT on InSb(001) and square Au islands are formed as observed in situ by nc-AFM. After annealing at 600 K the gold nanoparticles coalesce and larger square particles are obtained (see Fig. 27, top). In the KPFM mode (see Sect. 2.4) the particles show a contact potential difference of 10–20 mV (Fig. 27b, top). After annealing to 650 K the particle shape is the same but the difference of contact potential increases to 50–100 mV (see Fig. 27b, bottom). This large change in the contact potential difference is due to the formation of AuIn alloy nanoparticles. This interesting observation opens the way for a measurement of the composition of individual bimetallic nanoparticles. Indeed the contact potential difference measures (in the absence of permanent charges) the difference of work function between the substrate and the bimetallic particle (see Sect. 2.4). The work function of a bimetallic particle is expected to vary with its composition.

In summary some general rules can be drawn in the case of sequential deposition. In the case of deposition of B on pre-deposited A clusters:

- choose E_{ad}(A) > E_{ad} (B)
- A coverage sufficiently large to saturate all defects
- large F_A to have fast nucleation
- choose a low deposition temperature of A to have fast nucleation but not too low to avoid homogenous nucleation
- choose a high deposition temperature of B to favour growth of prenucleated clusters.

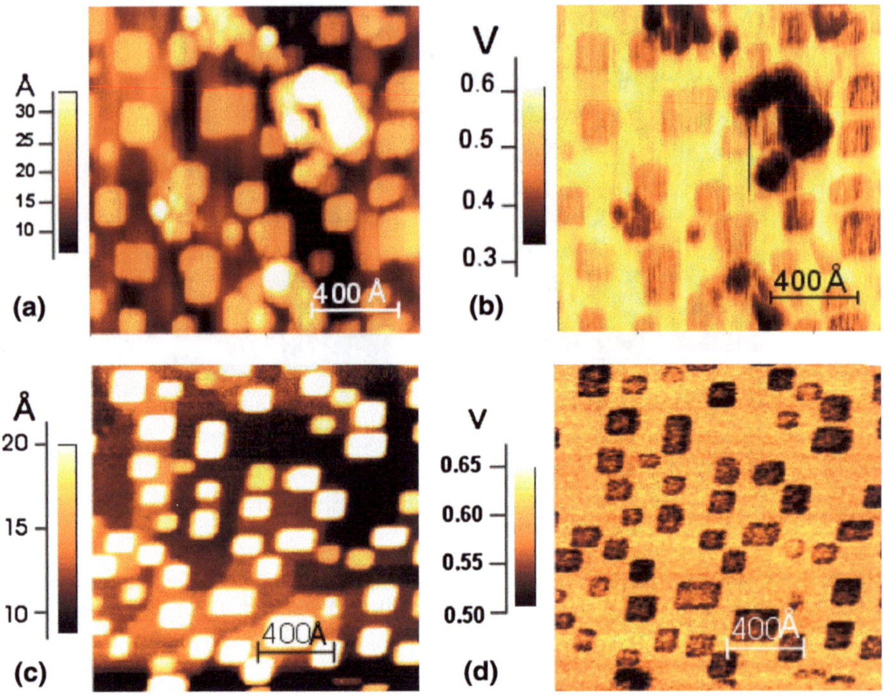

Fig. 27 Non contact AFM images of Au particles grown at RT on InSb(001): top after annealing at 600 K, bottom after annealing at 650 K; topography (**a**) and Kelvin images (**b**). From Ref. [152], copyright (2004), with permission from Elsevier

The growth rate can be calculated by knowing $(E_a - E_d)_B$, the radius of the prenucleated clusters R_A and their density n_A by using the growth model (see Sect. 3.1.2 and Ref. [104]).

Conclusion and Perspectives

STM and AFM are best suited techniques to study in situ the nucleation and growth of supported bimetallic clusters (STM is limited to bulk conducting substrates and ultrathin insulating films supported on metals whereas AFM can be used on any substrate surface). The morphology of the bimetallic nanoparticles can be studied by these two techniques but with some limitations. The size of the particles is enlarged and the shape is rounded on the edges by the interaction with the tip (tip-object convolution). This effect is often increased in AFM because the commercial tips are generally bigger than the STM ones. In non-contact AFM the actual shape can be revealed (at least the top facet) by using the constant height mode. The shape characterization by scanning probe microscopies can be completed by using integral techniques like GISAXS (grazing incidence small angle X-ray scattering) which can also be operated in situ during growth [153] or ex situ by TEM. The substrate surface can be easily imaged with atomic resolution by AFM and STM but it is more difficult on the nanoparticles except for large facetted particles. The resolution of the facet structure could certainly be improved by working with functionalized tips. The determination by STM and AFM of the surface chemical composition of the bimetallic particles is still a big issue, but recent works on surface alloys have shown that atomic resolution with chemical sensitivity can be achieved with both techniques. Again a better understanding and control of the scanning tip is necessary to achieve routinely chemical sensitivity at atomic scale. Otherwise KPFM could become a technique to analyse surface composition on large facets. Nevertheless complementary integral surface science techniques are still necessary to fully characterize supported alloy nanoparticles, like LEIS (surface composition), MEIS (surface and bulk composition). Other techniques can provide some information on the surface composition of the nanoparticles after adsorption of molecules (typically CO) like IRAS or TPD but care has to be taken in order to avoid artefacts like surface segregation induced by the adsorbate. At the level of a single particle ex situ TEM techniques (EDX, EELS, HAADF) can provide quantitative information on the bulk composition of the particles.

From theory side, atomistic nucleation theory is well adapted to treat accurately the nucleation and growth of bimetallic clusters by deposition of atoms on an insulator substrate. It can provide analytic kinetic laws for nucleation and growth and the evolution of the chemical composition can be predicted. From the two modes of deposition: simultaneous and sequential the latter one is preferred because the composition can be more easily controlled. However for sequential composition it is important to choose correctly the order of deposition of the two metals and the deposition parameters (flux, substrate temperature). Again atomistic nucleation and growth theory can guide these choices. Two crucial parameters of this theory are the adsorption energy and the diffusion energy of adatoms. They are related to the interaction of the metal with the substrate. For a same substrate, the trends for different metals can be given by the cohesion energy of the metal. Besides nucleation-growth theory atomistic simulations can provide precious information about chemical order and segregation in bimetallic nanoparticles that can be different that for their bulk counterpart (see Chap. 8).

References

1. Binnig, G., Rohrer, H., Gerber, Ch., Weibel, E.: Surface studies by scanning tunneling microscopy. Phys. Rev. Lett. **49**, 57 (1982)
2. Hansma, P.K., Tersoff, J.: Scanning tunneling microscopy. J. Appl. Phys. **61**, R1 (1986)
3. Feenstra, R.M.: Scanning tunneling spectroscopy. Surf. Sci. **299–300**, 965 (1994)
4. Komeda, T.: Chemical identification and manipulation of molecules by vibrational excitation via inelastic tunneling process with scanning tunneling microscopy. Prog. Surf. Sci. **78**, 41 (2005)
5. Tsukada, M., Kobayashi, K., Isshiki, N., Kageshima, H.: First-principles theory of scanning tunneling microscopy. Surf. Sci. Rep. **13**, 265 (1991)

6. Hofer, W.A., Foster, A.S., Shluger, A.L.: Theories of scanning probe microscopes at the atomic scale. Rev. Mod. Phys. **75**, 1287 (2003)
7. Wiesendanger, R.: Scanning Probe Microscopy and Spectroscopy. Cambridge University Press, Cambridge (1994)
8. Chen, C.J.: Introduction to Scanning Tunneling microscopy. Oxford University Press, New York (2008)
9. Africh, C., Kohler, L., Esch, F., Corso, M., Dri, C., Bucko, T., Kresse, G., Comelli, G.: Effects of lattice expansion on the reactivity of a one-dimensional oxide. J. Amer. Chem. Soc. **131**, 3253 (2009)
10. Crommie, M.F., Lutz, C.P., Eigler, D.M.: Confinement of electrons to quantum corrals on a metal surface. Science **262**, 218 (1993)
11. Varga, P., Schmid, M.: Chemical discrimination on atomic level by STM. Appl. Surf. Sci. **141**, 287 (1999)
12. Hallmark, V.M., Chiang, S., Rabolt, J.B., Swalen, J.D., Wilson, R.J.: Observation of atomic corrugation on Au (111) by scanning tunneling microscopy. Phys. Rev. Lett. **59**, 2879 (1987)
13. Schmid, M., Stadler, H., Varga, P.: Direct observation of surface chemical order by scanning tunnelling microscopy. Phys. Rev. Lett. **70**, 1441 (1993)
14. Hebenstreit, E.L.D., Hebenstreit, W., Schmid, M., Varga, P.: $Pt_{25}Rh_{75}$(111), (110) and (100) studied by scanning tunnelling microscopy with chemical contrast. Surf. Sci. **441**, 441 (1999)
15. Wouda, P.T., Schmid, M., Nieuwenhuys, B.E., Varga, P.: STM study of the (111) and (100) surface of AgPd. Surf. Sci. **417**, 292 (1998)
16. Gauthier, Y., Baudoing-Savois, R., Bugnard, J.M., Hebenstreit, W., Schmid, M., Varga, P.: Segregation and chemical ordering in the surface layers of $Pd_{25}Co_{75}$(111): a LEED/STM study. Surf. Sci. **466**, 155 (2000)
17. Ondracek, M., Maca, F., Kudrnovsky, J., Redinger, J., Biedermann, A., Fritscher, C., Schmid, M., Varga, P.: Chemical ordering and composition fluctuations on the (001) surface of the $Fe_{64}Ni_{36}$ invar alloy. Phys. Rev. B **74**, 235437 (2006)
18. Murray, P.W., Stesgaard, I., Laegsgaard, E., Besenbacher, F.: Growth and structure of Pd alloys on Cu(100). Surf. Sci. **365**, 591 (1996)
19. Hoster, H.E., Filonenko, E., Richter, B., Behm, R.J.: Formation and short-range order of two-dimensional Cu_xPd_{1-x} monolayer surface alloys on Ru(0001). Phys. Rev. B **73**, 165413 (2006)
20. Hebenstreit, W., Schmid, M., Redinger, J., Podloucky, R., Varga, P.: Bulk terminated NaCl(111) on aluminum: a polar surface of a ionic crystal? Phys. Rev. Lett. **85**, 5376 (2000)
21. Schintke, S., Schneider, W.-D.: Insulators at the ultrathin limit: electronic structure studied by scanning tunnelling microscopy and scanning tunnelling spectroscopy. J. Phys. Condens. Matter **16**, R49 (2004)
22. Kresse, G., Schmid, M., Napetshnig, A., Shishkin, M., Köhler, L., Varga, P.: Structure of ultrathin aluminum oxide film on NiAl(110). Science **308**, 1440 (2005)
23. Wintterlin, J., Völkening, S., Janssens, T.U.V., Zambelli, T., Ertl, G.: Atomic and macroscopic reaction rates of a surface-catalyzed reaction. Science **278**, 1931 (1997)
24. Stipe, B.C., Razaei, M.A., Ho, W.: Single-molecule vibrational spectroscopy and microscopy. Science **280**, 1732 (1998)
25. Eigler, D.M., Schweizer, E.K.: Positioning single atoms with a scanning tunnelling microscope. Nature **344**, 524 (1990)
26. Hahn, J.R., Ho, W.: Single-molecule imaging and vibrational spectroscopy with a chemically modified tip of scanning tunnelling microscope. Phys. Rev. Lett. **87**, 196102 (2001)
27. Nilius, N., Wallis, T.M., Ho, W.: Building alloys from single atoms: Au-Pd chains on NiAl(110). J. Phys. Chem. B **108**, 14616 (2004)

28. Hla, S.W., Bartels, L., Meyer, G., Rieder, K.H.: Inducing all steps of a chemical reaction with the scanning tunneling microscope tip: toward single molecule engineering. Phys. Rev. Lett. **85**, 2777 (2000)
29. Katano, S., Kim, Y., Hori, M., Trenary, M., Kawai, M.: Reversible control of hydrogenation of a single molecule. Science **316**, 1883 (2007)
30. Binnig, G., Quate, C.F., Gerber, Ch.: Atomic force microscope. Phys. Rev. Lett. **56**, 930 (1986)
31. Morita, S., Wiesendanger, R., Meyer, E.: Noncontact Atomic Force Microscopy. Springer, Berlin (2002)
32. Meyer, E., Hug, H.-J., Bennewitz, R.: Scanning Probe Microscopy: The Lab on a Tip. Springer, Berlin (2004)
33. Shluger, A.L., Livshits, A.I., Foster, A.S., Catlow, C.R.A.: Models of image contrast in scanning force microscopy on insulators. J. Phys. Condens. Matter **11**, R295 (1999)
34. Howald, L., Haefke, H., Lüthi, R., Meyer, E., Gerth, G., Rudin, H., Güntherodt, H.-J.: Ultrahigh vacuum scanning force microscopy: atomic-scale resolution at monoatomic cleavage steps. Phys. Rev. B **49**, 5651 (1994)
35. Albrecht, T.R., Grütter, P., Horne, D., Rugar, D.: Frequency modulation detection using high-Q-cantilevers for enhanced force microscope sensitivity. J. Appl. Phys. **69**, 668 (1991)
36. Oyabu, N., Custance, Ó., Yi, I., Sugawara, Y., Morita, S.: Mechanical vertical manipulation of selected single atoms by soft nanoindentation using near contact atomic force microscopy. Phys. Rev. Lett. **90**, 176102 (2003)
37. Giessibl, F.J.: Atomic resolution of the silicon (111)−(7×7) surface by atomic force microscopy. Science **267**, 68 (1995)
38. Giessibl, F.J.: Advances in atomic force microscopy. Rev. Mod. Phys. **75**, 949 (2003)
39. Bonnell, D.A., Garra, J.: Scanning probe microscopy of oxide surfaces: atomic structure and properties. Rep. Prog. Phys. **71**, 044501 (2008)
40. Lauritsen, J.V., Reichling, M.: Atomic resolution non-contact atomic force microscopy of clean metal oxide surfaces. J. Phys. Condens. Matter **22**, 263001 (2010)
41. Barth, C., Foster, A.S., Henry, C., Shluger, A.L.: Recent trends in surface characterization and chemistry with high-resolution scanning force methods. Adv. Mater. **23**, 477 (2011)
42. Sugimoto, Y., Pou, P., Abe, M., Jelinek, P., Pérez, R., Morita, S., Custance, Ó.: Chemical identification of individual surface atoms by atomic force microscopy. Nature **446**, 64 (2007)
43. Sugimoto, Y., Pou, P., Custance, Ó., Jelinek, P., Morita, S., Pérez, R., Abe, M.: Real topography, atomic relaxations, and short-range chemical interactions in atomic force microscopy: the case of the α-Sn/Si(11)-($\sqrt{3}$x$\sqrt{3}$)R30° surface. Phys. Rev. B **73**, 205329 (2006)
44. Barth, C., Foster, A.S., Reichling, M., Shluger, A.L.: Contrast formation in atomic resolution scanning force microscopy on CaF_2(111): experiment and theory. J. Phys. Condens. Matter **13**, 2061 (2001)
45. Barth, C., Henry, C.: Imaging Suzuki precipitates on NaCl:Mg^{2+}(001) by scanning force microscopy. Phys. Rev. Lett. **100**, 096101 (2008)
46. Henry, C.: Morphology of supported nanoparticles. Prog. Surf. Sci. **80**, 92 (2005)
47. Perrot, E., Humbert, A., Piednoir, A., Chapon, C., Henry, C.: STM and TEM studies of a model catalyst: Pd/MoS_2(0001). Surf. Sci. **445**, 407 (2000)
48. Hansen, K.H., Worren, T., Stempel, S., Lægsgaard, E., Bäumer, M., Freund, H.-J., Besenbacher, F., Stensgaard, I.: Palladium nanocrystals on Al2O3: structure and adhesion energy. Phys. Rev. Lett. **83**, 4120 (1999)
49. Dulub, O., Hebenstreit, W., Diebold, U.: Imaging cluster surfaces with atomic resolution: the strong metal-support interaction state on Pt supported on TiO_2(110). Phys. Rev. Lett. **84**, 3646 (2000)
50. Silly, F., Castell, M.R.: Selecting the shape of supported metal nanocrystals: Pd huts, hexagons or pyramids on $SrTiO_3$(001). Phys. Rev. Lett. **94**, 046103 (2005)
51. Piednoir, A., Perrot, E., Granjeaud, S., Humbert, A., Chapon, C., Henry, C.: Atomic resolution on small 3D metal clusters by STM. Surf. Sci. **391**, 19 (1997)

52. Dai, H., Hafner, J.H., Rinzler, A.G., Colbert, D.T., Smalley, R.: Nanotubes as nanoprobes in scanning probe microscopy. Nature **384**, 147 (1996)
53. Ferrero, S., Piednoir, A., Henry, C.: Atomic scale imaging by UHV-AFM of nanosized gold particles on mica. Nano Lett. **1**, 227 (2001)
54. Højrup-Hansen, K., Ferrero, S., Henry, C.: Nucleation and Growth kinetics of gold nanoparticles on MgO(100) studied by UHV-AFM. Appl. Surf. Sci. **226**, 167 (2004)
55. Barth, C., Henry, C.: High resolution imaging of gold clusters on KBr(100) surfaces investigated by dynamic scanning force microscopy. Nanotechnology **15**, 1264 (2004)
56. Szymonski, M., Goryl, M., Krok, F., Kolodziej, J.J., de Mongeot, F.B.: Metal nanostructures assembled at semiconductor surfaces studied with high resolution scanning probes. Nanotechnology **18**, 044016 (2007)
57. Pakarinen, O.H., Barth, C., Foster, A.S., Henry, C.: High resolution scanning force microscopy of gold nanoclusters on the KBr(001) surface. Phys. Rev. B **73**, 235428 (2006)
58. Fain, Jr, S.C., Polwarth, C.A., Tait, S.L., Campbell, C.T., French, R.H.: Simulated measurement of small metal clusters by frequency-modulation non-contact atomic force microscopy. Nanotechnology **17**, S121 (2006)
59. Barth, C., Pakarinen, O.H., Foster, A.S., Henry, C.: Imaging nanoclusters in the constant height mode of the dynamic SFM. Nanotechnology **17**, S128 (2006)
60. Pakarinen, O.H., Barth, C., Foster, A.S., Henry, C.: Imaging the real shape of nanoclusters in scanning force microscopy. J. Appl. Phys. **103**, 054313 (2008)
61. Hill, I.G., Rajagopal, A., Kahn, A., Hu, Y.: Molecular level alignment at organic semiconductor-metal interfaces. Appl. Phys. Lett. **73**, 662 (1998)
62. Giordano, L., Cinquini, F., Pacchioni, G.: Tuning the surface metal work function by deposition of ultrathin oxide films: density functional calculations. Phys. Rev. B **73**, 045414 (2005)
63. Zhang, C., Yoon, B., Landman, U.: Predicted oxidation of CO catalyzed by Au nanoclusters on a thin defect-free MgO film supported on a Mo(100) surface. J. Amer. Chem. Soc. **129**, 2228 (2007)
64. Gross, L., Mohn, F., Liljeroth, P., Repp, J., Giessibl, F.J., Meyer, G.: Measuring the charge state of an adatom with noncontact atomic force microscopy. Science **324**, 1428 (2009)
65. Girard, P.: Electrostatic force microscopy: principles and some applications to semiconductors. Nanotechnology **12**, 485 (2001)
66. Berger, R., Butt, H.J., Retschke, M.B., Weber, S.A.L.: Electrical modes in scanning probe microscopy. Macromol. Rapid Commun. **30**, 1167 (2009)
67. Palermo, V., Palma, M., Samori, P.: Electronic characterization of organic thin films by kelvin probe force microscopy. Adv. Mater. **18**, 145 (2006)
68. Melitz, W., Shen, J., Kummel, A.C., Lee, S.: Kelvin probe force microscopy and its application. Surf. Sci. Rep. **66**, 1 (2011)
69. Weaver, J.M.R., Abraham, D.W.: High resolution atomic force microscopy potentiometry. J. Vac. Sci. Technol. B **9**, 1559 (1991)
70. Nonnenmacher, M., O'Boyle, M.P., Wickramasinghe, H.K.: Kelvin probe force microscopy. Appl. Phys. Lett. **58**, 2921 (1991)
71. Barth, C., Henry, C.: Kelvin probe force microscopy on surfaces of UHV-cleaved ionic crystals. Nanotechnology **17**, S155 (2006)
72. Barth, C., Henry, C.: Surface double layer on (001) surfaces of alkali halide crystals: a scanning force microscopy study. Phys. Rev. Lett. **98**, 136804 (2007)
73. Henry, C.: Surface studies of supported model catalysts. Surf. Sci. Rep. **31**, 235 (1998)
74. Heiz, U., Landman, U. (eds.): Nanocatalysis. Springer, Berlin (2007)
75. Glatzel, T., Sadewasser, S., Lux-Steiner, M.C.: Amplitude or frequency modulation-detection in Kelvin probe force microscopy. Appl. Surf. Sci. **210**, 84 (2003)
76. Kitamura, S., Iwatsuki, M.: High-resolution imaging of contact potential difference with ultra high vacuum noncontact atomic force microscope. Appl. Phys. Lett. **72**, 3154 (1998)

77. Krok, F., Sajewicz, K., Konior, K., Goryl, M., Piatkowski, P., Szymonski, M.: Lateral resolution and potential sensitivity in the Kelvin probe force microscopy: towards understanding of the sub-nanometer resolution. Phys. Rev. B **77**, 235427 (2008)
78. Barth, C., Henry, C.: Gold nanoclusters on alkali halide surfaces: charging and tunnelling. Appl. Phys. Lett. **89**, 252119 (2006)
79. Sasahara, A., Pang, C.L., Onishi, H.: Local work function of Pt clusters vacuum-deposited on a TiO_2 surface. J. Phys. Chem. B **110**, 17584 (2006)
80. Barth, C., Henry, C.: Kelvin force microcopy on MgO(001) surfaces and supported nanoclusters. J. Phys. Chem. C **113**, 247 (2009)
81. Sasahara, A., Pang, C.L., Onishi, H.: Probe microscope observation of platinum atoms deposited on the TiO_2(110)-(1x1) surface. J. Phys. Chem. B **110**, 13453 (2006)
82. Bieletzki, M., Hynninen, T., Soini, T.M., Pivetta, M., Henry, C., Foster, A.S., Esch, F., Barth, C., Heiz, U.: Topography and work function measurements of thin MgO(001) films on Ag(001) by nc-AFM and KPFM. Phys. Chem. Chem. Phys. **12**, 3203 (2010)
83. Niedermayer, R., Mayer, R.: Basic problems in thin film physics. Vandenhoeck & Ruprecht, Göttingen (1966)
84. Mathews, J.W.: Epitaxial Growth Part B. Academic press, New York (1975)
85. Kern, R., Lelay, G., Métois, J.J.: Basic mechanisms in the early stages of epitaxy. In: Kaldis, E. (ed.) Current Topics in Materials Science, vol. 3, p. 131. North Holland, Amsterdam (1979)
86. Volmer, M.: Kinetic der Phasenbildung. T. Steinkopff Verlag, Leipzig (1939)
87. Zinsmeister, G.: A contribution to Frenkel's theory of condensation. Vacuum **16**, 529 (1966)
88. Frenkel, J.: Theorie der adsorption und verwandter erscheinungen. Z. Physik **26**, 117 (1924)
89. Robins, J.L., Rhodin, T.N.: Nucleation of metal crystals on ionic surfaces. Surf. Sci. **2**, 346 (1964)
90. Henry, C.: Growth, structure and morphology of supported metal clusters studied by surface science techniques. Cryst. Res. Technol. **33,** 1119 (1998)
91. Meunier, M., Henry, C.: Nucleation and growth of metallic clusters on MgO(100) by helium diffraction. Surf. Sci. **307**, 514 (1994)
92. Gimenez, F., Chapon, C., Henry, C.: Nucleation and growth kinetics of Pd and CuPd particles on NaCl(100). New J. Chem. **22**, 1289 (1998)
93. Haas, G., et al.: Nucleation and growth of supported clusters at defect sites: Pd/MgO(001). Phys. Rev. B **61**, 11105 (2000)
94. Venables, J.A., Harding, J.H.: Nucleation and growth of supported metal clusters at defect sites on oxide and halide (001) surfaces. J. Cryst. Growth **211**, 27 (2000)
95. Zinsmeister, G.: Theory of thin film condensation part C: aggregates size distribution in island films. Thin Solid Films **4**, 363 (1969)
96. Halpern, V.: Cluster growth and saturation island densities in thin film growth. J. Appl. Phys. **40**, 4627 (1969)
97. Sigsbee, R.A.: Adatom capture and growth rates of nuclei. J. Appl. Phys. **42**, 3904 (1971)
98. Stowell, M.J.: Capture numbers in thin film nucleation theories. Phil. Mag. **26**, 349 (1972)
99. Lewis, B.: Migration and capture processes in heterogeneous nucleation and growth I. Theory Surf. Sci. **21**, 273 (1970)
100. Venables, J.A.: Rate equation approaches to thin film nucleation kinetics. Phil. Mag. **27**, 697 (1973)
101. Lewis, B., Rees, G.: Adatom migration, capture and decay among competing nuclei on a susbtrate. Phil. Mag. **29**, 1253 (1974)
102. Kashchiev, D.: Mean thickness at which vapour-deposited thin films reach continuity. Thin Solid Films **55**, 399 (1978)
103. Kashchiev, D.: Growth of crystallites in deposition from vapours. Phys. Status Solidi (a) **64**, 715 (1981)
104. Henry, C., Meunier M.: Power laws in the growth kinetics of metal clusters on oxide surfaces. Vacuum **50**, 157 (1998)

105. Rosenhahn, A., Schneider, J., Becker, C., Wandelt, K.: Oxidation of $Ni_3Al(111)$ at 600, 800 and 1050 K investigated by scanning tunnelling microscopy. J. Vac. Sci. Technol. A **18**, 7 (2000)
106. Hamm, G., Barth, C., Becker, C., Wandelt, K., Henry, C.: Power laws in the growth kinetics of metal clusters on oxide surfaces. Phys. Rev. Lett. **97**, 126106 (2006)
107. Schmid, M., Kresse, G., Buchsbaum, A., Napetschning, E., Gritschneder, S., Reichling, M., Varga, P.: Nanotemplate with holes: Ultrathin alumina on $Ni_3Al(111)$. Phys. Rev. Lett. **99**, 196104 (2007)
108. Sedona, F., Granozzi, G., Barcaro, G., Fortunelli, A.: Defect evolution in oxide nanophases: the case of a zigzag- like TiO_x phase on Pt(111). Phys. Rev. B **77**, 115417 (2008)
109. Torelli, P., Soares, E.A., Renaud, G., Gragnaniello, L., Valeri, S., Guo, X.X., Luches, P.: Self-organized growth of Ni nanoparticles on a cobalt-oxide thin film induced by a buried misfit dislocation network. Phys. Rev. B **77**, 081409 (2008)
110. Becker, C., Rosenhahn, A., Wiltner, A., von Bergmann, K., Schneider, J., Pervan, P., Milun, M., Kralj, M., Wandelt, K.: Al_2O_3-films on $Ni_3Al(111)$: a template for nanostructured cluster growth. New J. Phys. **4**, 75 (2002)
111. Hamm, G., Becker, C., Henry, C.: Pd-Au nanocluster arrays grown on nanostructured alumina templates. Nanotechnology **17**, 1943 (2006)
112. Barcaro, G., Fortunelli, A., Granozzi, G., Sedona, F.: Cooperative phase transformation in self-assembled metal-on-oxide arrays. J. Phys. Chem. C **113**, 1143 (2009)
113. Rohart, S., Baudot, G., Repain, V., Girard, Y., Rousset, S., Bulou, H., Goyhenex, C., Proville, L.: Atomistic mechanisms for the ordered growth of Co nanodots on Au(788): a comparison between VT-STM experiments and multiscaled calculations. Surf. Sci. **559**, 47 (2004)
114. Sitja, G., Unac, R.O., Henry, C.: Kinetic Monte Carlo simulation of the growth of metal clusters on a regular array of defects on the surface of an insulator. Surf. Sci. **604**, 404 (2010)
115. Anton, R., Harsdorff, M.: Extension of the kinetic nucleation model to binary alloys. Thin Solid Films **44**, 341 (1977)
116. Anton, R., Harsdorff, M., Martens, T.H.: Nucleation and growth of binary alloys on substrates. Thin Solid Films **57**, 233 (1979)
117. Anton, R., Dröske, R.: Nucleation of Au-Ag binary alloys from the vapour on KBr single crystals. Thin Solid Films **124**, 155 (1985)
118. Kortekamp, T., Anton, R., Harsdorff, M.: Nucleation and growth of Au-Cu binary alloys from the vapour phase on NaCl single crystals. Thin Solid Films **145**, 123 (1986)
119. Anton, R., Schmidt, A., Schünemann, V.: Heterogeneous nucleation of binary alloy particles. Vacuum **41**, 1099 (1990)
120. Schmidt, A., Schünemann, V., Anton, R.: Monte Carlo simulation of the nucleation of binary-alloys particles of Au, Ag and Pd on NaCl substrates. Phys. Rev. B **41**, 11875 (1990)
121. Schmidt, A., Spode, M., Heinrich, J., Anton, R.: The nucleation and growth of Pd-Au alloy particles on NaCl(100). Thin Solid Films **196**, 253 (1991)
122. Alloyeau, D., Prévot, G., Le Bouar, Y., Oikawa, T., Langlois, C., Loiseau, A., Ricolleau, C.: Ostwald ripening in nanoalloys: when thermodynamics drives a size-dependent particle composition. Phys. Rev. Lett. **105**, 255901 (2010)
123. Herzing, A.A., Watanabe, M., Edwards, J.K., Conte, M., Tang, Z.R., Hutchnings, G.J., Kiely, C.J.: Energy dispersive X-ray spectroscopy of bimetallic nanoparticles in an aberration corrected scanning transmission electron microscope. Faraday Discuss. **138**, 337 (2008)
124. Gimenez, F., Chapon, C., Giorgio, S., Henry, C.: Bimetallic Pd-Cu clusters particles supported on NaCl. In: Proceedings ICEM 13, Paris, p. 351 (1994)
125. Gimenez, F.: Nucléation, croissance, composition et structure d'agrégats bimétalliques suopportés. Ph.D. thesis, Marseille (1997)
126. Gai, Z., Howe, J.Y., Guo, J., Blom, D.A., Plummer, E.W., Shen, J.: Self-assembled FePt nanodots arrays with mono-dispersion and -orientation. Appl. Phys. Lett. **86**, 023107 (2005)

127. Gross, E., Popov, I., Ascher, M.: Chemical reactivity of Pd-Au bimetallic nanoclusters grown via amorphous solid water as buffer layer. J. Phys. Chem. C **113**, 18341 (2009)
128. Abbott, H.I., Aumer, A., Lei, Y., Asokan, C., Meyer, R.J., Sterrer, M., Shaikhutdinov, S., Freund, H.-J.: CO adsorption on monometallic and bimetallic Au-Pd nanoparticles supported on oxide thin films. J. Phys. Chem. C **114**, 17099 (2010)
129. Heemeier, M., Carlsson, A.F., Naschitzki, M., Schmal, M., Bäumer, M., Freund, H.-J.: Preparation and characterization of a model bimetallic catalyst: Co-Pd nanoparticles supported on Al_2O_3. Angew. Chem. Int. Ed. **41**, 4073 (2002)
130. Carlsson, A.F., Naschitzki, M., Bäumer, M., Freund, H.-J.: The structure and reactivity of Al_2O_3-supported cobalt-palladium particles: a CO-TPD, STM and XPS study. J. Phys. Chem. B **107**, 778 (2003)
131. Carlsson, A.F., Bäumer, M., Risse, T., Freund, H.-J.: Surface structure of Co-Pd bimetallic particles supported on Al_2O_3 thin films studied using infrared reflection absorption spectroscopy of CO. J. Chem. Phys. **119**, 10885 (2003)
132. Napetschnig, E., Schmid, M., Varga, P.: Pd, Co and Co-Pd clusters on the ordered alumina film on NiAl(110). Contact angle, surface structure and composition. Surf. Sci. **601**, 3233 (2007)
133. Felicissimo, M.P., Martyanov, O.N., Risse, T., Freund, H.-J.: Characterization of Pd-Fe bimetallic model catalysts. Surf. Sci. **601**, 2105 (2007)
134. Han, P., Goodman, D.W.: Controlling the size and distribution of Pd-Au nanoparticles on TiO_2(110). J. Phys. Chem. C **112**, 6390 (2008)
135. Khan, N.A., Uhl, A., Shaikhutdinov, S., Freund, H.-J.: Alumina supported model Pd-Ag catalysts: a combined STM, XPS, TPD and IRAS study. Surf. Sci. **600**, 1849 (2006)
136. Marsault, M., Hamm, G., Wörz, A., Sitja, G., Barth, C., Henry, C.: Preparation of regular arrays of bimetallic clusters with independent control of size and chemical composition. Faraday Discuss **138**, 407 (2008)
137. Santra, A.K., Yang, F., Goodman, D.W.: The growth of Ag-Au bimetallic nanoparticles on TiO_2(110). Surf. Sci. **548**, 324 (2004)
138. Bassett, G.A.: A new technique for decoration of cleavage and slip steps on ionic crystal surfaces. Phil. Mag. **3**, 1042 (1958)
139. Park, J.B., Ratliff, J.S., Ma, S., Chen, D.A.: In situ scanning tunneling microscopy studies of bimettalic cluster growth: Pt-Rh on TiO2(110). Surf. Sci. **600**, 2913 (2006)
140. Marsault M.: Réseaux réguliers d'agrégats mono et bimétalliques sur des substrats d'alumine nanostructurée. Ph.D. thesis, Marseille (2010)
141. Degen, S., Becker, C., Wandelt, K.: Thin alumina films on Ni_3Al(111): a template for nanostructured Pd clusters growth. Faraday Discuss. **125**, 343 (2003)
142. Krawczyk, M., Zommer, L., Lesiak, B., Jablonski, A.: Surface composition of the CoPd alloys studied by electron spectroscopies. Surf. Interface Anal. **25**, 356 (1997)
143. Diebold, U., Pan, J.M., Madey, T.E.: Growth mode of of ultrathin copper overlays on TiO_2(1101). Phys. Rev. B **47**, 3868 (1993)
144. Luo, K., Wei, T., Yi, C.W., Axnanda, S., Goodman, D.W.: Preparation and characterization of silica supported Au-Pd model catalysts. J. Phys. Chem. B **109**, 23517 (2005)
145. Gustafson, J., Haire, A.R., Baddeley, C.: Depth-profiling the composition of bimetallic nanoparticles using medium energy ion scattering. Surf. Sci. **605**, 220 (2011)
146. Haire, A.R., Gustafson, J., Trant, A.G., Jones, T.E., Noakes, T.C.Q., Bailey, P., Baddeley, C.J.: Influence of preparation conditions on the depth-dependent composition of AuPd nanoparticles grown on planar oxide surfaces. Surf. Sci. **605**, 214 (2011)
147. Bradeley, J.S., Hill, E.W., Chaudret, B., Duteil, A.: Surface chemistry on colloidal metals. Reversible adsorbate-induced surface composition changes in colloidal palladium-copper alloys. Langmuir **11**, 693 (1995)
148. Gao, F., Wang, Y., Goodman, D.W.: Reaction kinetics and polarization-modulation infrared reflection absorption spectroscopy (PM-IRAS) investigation of CO oxidation over supported Pd-Au alloy catalysts. J. Phys. Chem. C **114**, 4036 (2010)

149. Vitos, L., Ruban, A.V., Skriver, H.L., Kollar, J.: The surface energy of metals. Surf. Sci. **411**, 186 (1998)
150. Gross, L., Mohn, F., Moll, N., Liljeroth, P., Meyer, G.: The chemical structure of a molecule resolved by atomic force microscopy. Science **325**, 1110 (2009)
151. Langlois, C., Alloyeau, D., Le Bouar, Y., Loiseau, A., Oikawa, T., Mottet, C., Ricolleau, C.: Growth and structural properties of CuAg and CoPt bimetallic nanoparticles. Faraday Discuss. **138**, 375 (2008)
152. Goryl, M., Krok, F., Kolodziej, J.J., Piatkowski, P., Such, B., Szymonski, M.: Surface structure of Au/InSb(001) system investigated with scanning force microscopy. Vacuum **74**, 223 (2004)
153. Renaud, G., Lazzari, R., Revenant, C., Barbier, A., Noblet, M., Ullrich, O., Leroy, F., Borensztein, Y., Jupille, J., Henry, C., Deville, J.P., Scheurer, F., Mane–Mane, J., Fruchart, O.: In situ GISAXS towards a real time modelling of growing nanoparticles. Science **300**, 1416 (2003)
154. Benten, W., Nilius, N., Ernst, N., Freund, H.-J.: Photon emission spectroscopy of single oxide-supported Ag-Au alloy clusters. Phys. Rev. B **72**, 045403 (2005)

Probing Nanoalloy Structure and Morphology by X-Ray Scattering Methods

Pascal Andreazza

Abstract Because the nanoalloy properties are directly connected to their structural and chemical arrangements, the experimental investigations of these features by dedicated techniques is of primary importance, especially in order to understand and to control the nanoalloy formation or evolution mechanisms. In this chapter, we focus on the X-ray scattering techniques which can be performed ex- or in situ as well as for supported or embedded nanoalloys. While X-ray absorption methods are well known and ab initio calculations allow the simulation and the analysis of experimental data from many years, the X-ray scattering techniques applied to nanoalloys recently know a huge development. In particular, the recent interest for these techniques comes from the emergency of spatially or chemically-selective methods using resonant effects between scattering and reflectivity or scattering and absorption (like grazing incidence or anomalous scattering, respectively), as well as the development of theoretical approaches allowing scattering pattern simulations and consequently the fine interpretation of experimental data. Furthermore, in situ investigations which combine several techniques give new opportunities to follow structural transitions at different scales.

P. Andreazza (✉)
Centre de Recherche sur la Matière Divisée, Université d'Orléans, CNRS,
1bis rue de la Férollerie, 45071 Orleans Cedex 2, France
e-mail: pascal.andreazza@univ-orleans.fr

D. Alloyeau et al. (eds.), *Nanoalloys*, Engineering Materials,
DOI: 10.1007/978-1-4471-4014-6_3,

1 Introduction

1.1 Which Features of Nanoalloys are Relevant to Know?

Excepted the properties linked to the composition, much of the interesting features of nanoalloys are also those of pure nanoparticles. Among those ones, the structure (Fig. 1), i.e. the atom arrangement inside the particle, is the main feature, which controls, beside size effects, the most of physical or chemical properties, as presented in this book. For most of transition metals, the atomic structure is close-packing based with almost isotropic morphology e.g. truncated octahedra TOh (Fig. 1c) from face-centered cubic (fcc) bulk structure. However, at ultra-small size, non-crystalline structures [1], such as icosahedral Ih (Fig. 1b) or decahedral Dh, which optimized the atom packing, but expensed the internal strain, are also possible. Their stability depends on a critical balance between surface and volume (strain) contributions to the total energy that decreases as the size increases [2, 3]. The icosahedral or polyicosahedral morphology are expected to be the most favored at small size, then decahedra in an intermediate size range and fcc truncated octahedra at larger sizes.

However, this classical sequence of stability can be modified by the methods and conditions of fabrication. In addition, kinetic processes involved during their elaboration can yield metastable nanoparticles such as large icosahedra or decahedra [4, 5], or amorphous packing (Fig. 1a). Finally, surface contamination and interface effect with external support or matrix have been identified as factors inducing a change of the particle structure and morphology [1, 6–9].

Besides structure considerations, the new physical or chemical properties of nanoparticles, which arise from surface and confinement effects, intimately depend on their shapes and their spatial organization. For instance, the variation of shape, i.e. the nature of surface planes, edge or corner atom number, induce changes in the reactivity of catalysts. Moreover, the extension or reduction of one dimension, i.e. the shape anisotropy, modulate the magnetic anisotropy of single domain nanoparticles. In the case of supported nanoparticles, the aspect ratio, which corresponds to the ratio of the size in the substrate plane with respect to the out-of-plane size (height), is the result of the competition between interface effects and growth kinetic considerations. Finally the interesting properties are rarely due to single particles, but to an assembly of particles. In this context, the 2D or 3D spatial organization is a feature of primary importance to explain the mutual interaction effects in a collection of particles. In the concentrated system case, the correlation between particles can take different forms from the fully disordered to lattice organization (anisotropic inter-particle spacing to supra-crystalline structure).

With respect to monometallic particles, the addition of one or several metals in nanoparticles induces a larger diversity in their structure and morphology [3]. Combined to the composition (A_xB_{1-x}) effect, the finite quantity of matter could lead to a more various family of structures than those observed in bulk, such as disordered/ordered, segregated, core–shell structure, chemically-induced strained

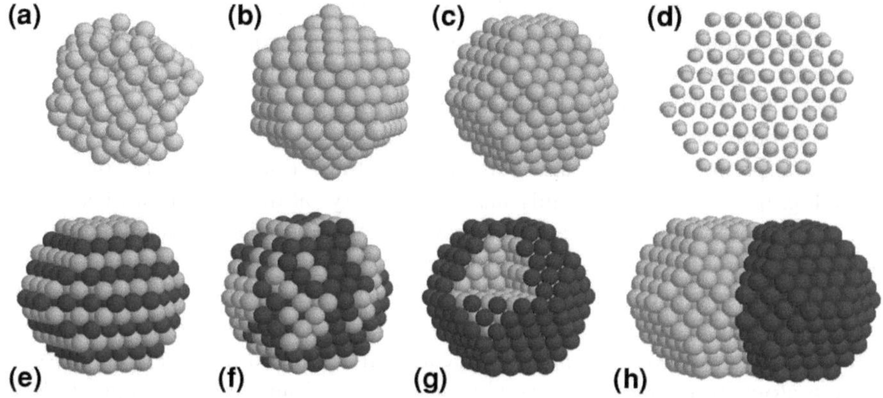

Fig. 1 Snapshots of possible nanoparticle structures. **a** Amorphous. **b** Icosahedral. **c** Truncated octahedral atom arrangements. **d** Static disorder with respect to perfect position (the apparent atom size is reduced for a better view). Nanoalloyed structures: **e** Chemically ordered. **f** Disordered (solid solution). Segregated structures: **g** Core-shell. **h** Demixing

particle (Fig. 1e–g). The capability to detect if an alloyed material is chemically ordered or disordered (randomly mixed) at small size, is not a trivial objective. Furthermore, the challenge is also to define if the partial order can take place within the particle or in the particle collection and then to differentiate these two configurations. Moreover, both thermodynamical and experimental factors can induce segregation between metals, leading either to core–shell particles by surface segregation of one metal or to dual particles A–B by demixing of metals (Fig. 1g, h, respectively). Consequently, it is of primary importance to investigate these effects in order to understand and to control the nanoalloys formation or evolution mechanisms since their physical and chemical properties (magnetic, optical or catalytic…), are directly connected to this A and B atom arrangement.

These few examples are illustrative of the need to characterize and control the structure and morphology of nano-objects in correlation with the fabrication strategies. In particular, a better understanding of the formation and stability mechanisms of nanoparticles is a prerequisite for the optimization of their application on a large extent, but passes through a fine characterization. This point is especially relevant in nanoalloys. Their properties are highly dependent on the size, shape, composition, structural and chemical arrangements, strain, interface and spatial organization. Any change of these parameters over the assembly of particles gives rise to a broadening of the distribution of properties.

1.2 Trends in the Nanoparticle/Nanoalloy Characterizations

The literature reports the huge development in the last decade of the characterization tools dedicated to nanoalloys [10–13]. The morphological and structural properties of nanomaterials are usually addressed by a wealth of techniques such as X-ray Diffraction (XRD), Transmission Electron Microscopy (TEM), High resolution Scanning Electron Microscopy (SEM), Scanning Tunneling Microscopy (STM) or Atomic Force Microscopy (AFM). Electron microscopy and X-ray-based techniques are probably the most commonly used to analyze the structure, the chemistry and the morphology of materials. With nanostructures, traditional X-ray or electron crystallography techniques mainly based on the crystal diffraction, fail because of their substantially limited length of structural coherence, i.e. the spatial extent of longer-range atomic order. Different strategies [14] exist using various high-resolution electron microscopy techniques (see Chap. 4) and various X-ray analysis tools based on X-ray absorption or scattering techniques. For example, the wide angle X-ray scattering (WAXS) technique appears better adapted to investigate inhomogeneous atom arrangement observed in nano-objects, than X-ray diffraction (XRD). The principal difficulty with the application of these methods to solve the nanoscale atomic arrangement is that, in general, any one technique does not give sufficient information to deduce a unique structural solution. A coherent strategy is required to combine the input from multiple experimental methods and theoretical models in a self-consistent global optimization scheme.

Among those, wide angle and small angle X-ray scattering (SAXS) are invaluable tools to study the bimetallic particles structure (long range-order, crystalline/non-crystalline/chemical order, segregation/alloying) and morphology (particle size and shape, inter-particle distance distributions) respectively. When performed at grazing incidence, these scattering techniques at wide angles and at small angles (GIWAXS and GISAXS, respectively) become powerful and sensitive tools for the study of thin surface layers composed of nanoparticles [15–17]. Complementarily, X-ray absorption spectroscopy (XAS) and anomalous X-ray scattering are two element-selective techniques that are of great value to discriminate the position of each species in nanoparticles, by tuning the X-ray wavelength across element absorption edges. Extended X-ray absorption fine structure (EXAFS) is well known for its ability to probe the chemical and short-range order (local structure), i.e. type, number and spacing between neighbors [18–20]. Chemical sensitivity can be also obtained in X-ray scattering experiments, through the anomalous effect. Thank to the strong variation of the atomic scattering factor of elements (the anomalous effect) close to its absorption edges [13, 21], probing of the chemical segregation in the non-alloyed particle morphology becomes possible.

In this chapter, we focus on these X-ray scattering methods which can be performed ex- or in situ as well as for supported or embedded nanoalloys, thank to the emergence of dedicated synchrotron radiation experiment set-up. While the X-ray absorption methods are well known and ab initio calculations allow the simulation and the analysis of experimental data from many years, the X-ray scattering methods applied to nanoalloys recently know a huge development due to the possibility to combine several characterization methods. Furthermore, their interest comes from the use of spatially or chemically-selective methods, as well as the development of theoretical approaches allowing scattering patterns simulations and consequently the fine interpretation of experimental data.

Complementarily, due to numerous works in bimetallic nanocatalyst for several decades, X-ray absorption spectroscopy methods applied to nanoalloys are well known and available in several reviews articles and books [19, 22–25].

1.3 Advantages and Drawbacks of X-Ray Techniques

The most widely used techniques to study the structure and morphology of nanostructured material are electron microscopies (TEM or SEM) and near field microscopies (STM or AFM), which can provide some of the required information in the studies of such materials. In addition, HAADF-HRSTEM (high angle annular dark field—high resolution scanning TEM), EELS (Electron Energy Loss Spectroscopy) and EFTEM (energy filtered TEM) techniques are particularly well adapted for multi-element nanostructures, such as nanoalloys (see Chap. 4), and know successful developments since the arrival of new generation of electron microscopes, this last decade.

However, these observations or analyses often suffer from slowness, sampling limits, and possible artifacts, such as those due to the inevitable convolution with the tip in the cases of STM and AFM, or the risk of sample modifications induced by the necessary sample preparation for buried nanostructures or induced by the electron beam (TEM-based techniques). In addition, they are either difficult to use in situ, e.g. in ultra high vacuum (UHV) or in liquid media during the growth of nano-objects. Recently, a relevant review [12] has been published detailing the advantages and drawbacks of X-ray scattering techniques for nanostructures in this context. X-rays are non-destructive, adapted to any kind of materials without suffering from charging effects; they provide depth sensitivity and a statistical averaging over the whole sample area; they can be used at any pressure or temperature, and in any kind of sample environment, during formation or evolution of nanoalloys and sometimes in real time. The main drawbacks of X-rays are the nearly unavoidable use of synchrotron radiation to get a reasonable counting time on nanostructures and data analysis that relies on modeling of reciprocal space measurements. However, in some cases, X-ray measurements on nanoalloys are able in laboratory set-up [26] with limited, but accurate performances. Because the determination of the atomic structure at the nanoscale is a complex problem, methods that can probe long- or short-range features as well as element-selective or average data, provide highly complementary information about the structure and morphology of nanoalloys and are much powerful when used together.

1.4 From the X-Ray/Matter Interaction to the Analysis Techniques

Photon-matter interaction in the X-ray range occurs via two fundamental ways: absorption and scattering processes, which are mainly photon-electron interaction mechanisms. In the first case, the photon energy is lost within the target atoms and transferred to an electron, which is ejected from the atom. In the second case, the wave is scattered by the atom electrons in a deviated direction, without change of wavelength in the case of elastic scattering. Depending on whether one is interested in the energy of electrons in the matter, i.e. in the atom nature, or in the spatial distribution of electrons, i.e. the atom position, the X-ray interaction process, which must be studied, will be the absorption or the scattering, respectively. Therefore, the analysis of these two processes provides a complete signature of the electron density of matter, from the atomic scale to the material scale, via two complementary techniques: the X-ray absorption spectroscopy and the X-ray scattering.

The difference $\mathbf{q}$ between the scattered wave vector $\mathbf{k}$ to the incident wave vector $\mathbf{k_i}$, is the key parameter of the scattering process. This vector $\mathbf{q}$ which corresponds to the momentum elastic transfer between the two waves without any interaction with other waves (kinematic theory), is called "scattering vector": $q = 4\pi \sin(\theta)/\lambda$ where λ is the wavelength and 2θ is the scattering angle [27, 28]. During an experiment, the measured quantity is the differential cross-section of the scattered wave defined into a unit solid angle in the given direction $\mathbf{q}$. This cross-section, also called scattered intensity, measures the effect of the charge (electrons) distribution in an irradiated matter volume on the incident wave through two factors:

$$\left.\frac{d\sigma(\mathbf{q})}{d\Omega}\right) = \left.\frac{d\sigma}{d\Omega}\right)_e |A(\mathbf{q})|^2 \tag{1}$$

The first factor $(d\sigma/d\Omega)_e$ is the Thomson electron scattering cross-section, or the scattering power of one electron and the second factor comes from the scattered wave amplitude:

$$A(\mathbf{q}) = \int \rho(\mathbf{r}) e^{-i\mathbf{q}.\mathbf{r}} d^3\mathbf{r} \tag{2}$$

where the integral is summed over all possible values of the position vector $\mathbf{r}$ within the volume of matter and $\rho(\mathbf{r})$ is the corresponding density of electrons. $A(\mathbf{q})$ is the Fourier transform of the electron density distribution of the irradiated volume where $\mathbf{q}$ represents the point (the zone) in reciprocal space where the scattering power is measured. The intensity can be expressed from the square modulus introducing a double sum [28, 29]:

$$|A(\mathbf{q})|^2 = \left| \int \rho(\mathbf{r}_1) e^{+i\mathbf{q}.\mathbf{r}_1} d^3\mathbf{r}_1 \int \rho(\mathbf{r}_2) e^{-i\mathbf{q}.\mathbf{r}_2} d^3\mathbf{r}_2 \right| \tag{3}$$

This expression evidences the importance of the correlation between different regions of the irradiated volume in the scattering phenomena. A suitable variable change allows the introduction of the correlation function $\gamma(\mathbf{r})$

$$|A(\mathbf{q})|^2 = \iint \rho(\mathbf{r}_1)\rho(\mathbf{r}+\mathbf{r}_1) d^3\mathbf{r}_1 e^{-i\mathbf{q}.\mathbf{r}} d^3\mathbf{r} = \int \gamma(\mathbf{r}) e^{-i\mathbf{q}.\mathbf{r}} d^3\mathbf{r} \tag{4}$$

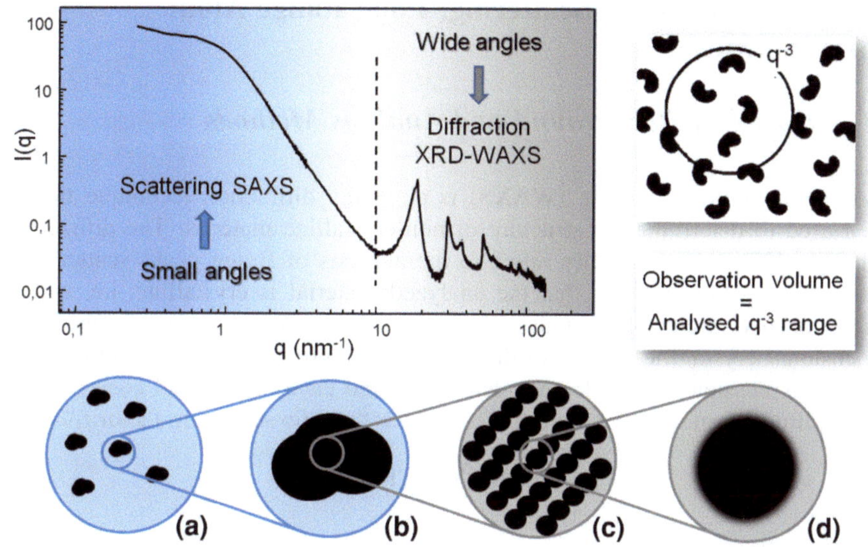

Fig. 2 Scattering and diffraction ranges versus momentum transfer q range in correspondence to the observation volume at decreased scale, from **a** the particle assembly to **d** the atom scale. Courtesy of N. Cohaut, results from Ref. [30]

The function $\gamma(\mathbf{r})$ defines the probability to find at a given position **r** in the volume, a surrounding region with the same electron density. Each electron of the volume acts in coherence only with a surrounding region. In the case of a nanoparticle, we can consider two regions at two different scales, the atom volume and the particle volume. And consequently, $\gamma(\mathbf{r})$ describes directly the geometry and the nature of these regions. The second term $exp(-i\mathbf{q}.\mathbf{r})$ is a phase term which take into account the spatial organization of these different regions, such as atoms or particles.

In the case of nanostructured materials, like an assembly of nanoparticles, the heterogeneity of electron density occurs at several scales (Fig. 2a–d). The characteristic lengths are: at large scale, the inter-particle distance and the particle size, and at small scale, the inter-atomic distance. These different scales in the real space (**r** ranges) correspond to several scales in the reciprocal space, or several ranges of **q** values (Fig. 2). Therefore, we can consider that the scattering vector acts as the magnification of a microscope. At very small values of q ($q < 1\ nm^{-1}$), i.e. small scattering angles, the scattering intensity corresponds to density fluctuations at large scales and thus depends on correlations between particles. While at large values of q, i.e. wide scattering angles, it depends on correlation between atoms. Since the larger the scattering angle the smaller the probed length scale, wide angle X-ray scattering (WAXS) is used to determine nanoparticle structure at the atomic length scale while small-angle X-ray scattering (SAXS) is used to explore structure at the assembly length scale.

2 Wide-Angle X-Ray Scattering: Long Range Atom Arrangement

2.1 Theoretical Background and Analysis Methods

Wide angle X-ray scattering (WAXS) is an X-ray diffraction technique that is often used to determine the structure of non-crystalline material. The diffraction technique [27, 31] specifically refers to the analysis of Bragg peaks scattered at wide angles, which implies that the analyzed material is crystalline, i.e. with a translation periodicity of the unit cell (containing a group of atoms) in the three dimensional crystalline lattice. In this case, it is more convenient to introduce a discrete expression of the electron density around each atom j and consequently the amplitude scattered by an atom j, called the "atomic scattering factor f_j" [32]:

$$\rho_{tot}(\mathbf{r}) = \sum_j \rho_j(\mathbf{r} - \mathbf{r}_j) \tag{5}$$

$$f_j(\mathbf{q}) = \int \rho_j(\mathbf{r} - \mathbf{r}_j)\mathrm{e}^{-i\mathbf{q}.(\mathbf{r}-\mathbf{r}_j)} d^3\mathbf{r} \tag{6}$$

The periodicity of the unit cell (nature and position of atoms in the crystalline lattice [**a**, **b**, **c**] defined by the vector $\mathbf{R}_n = n_1\mathbf{a} + n_2\mathbf{b} + n_3\mathbf{c}$) leads to the classical expression of the scattered amplitude as a summation on the atoms j of the unit cell and on the unit cells n of lattice:

$$A(\mathbf{q}) = \sum_j^{all.atoms} f_j(\mathbf{q})\mathrm{e}^{-i\mathbf{q}.\mathbf{r}_j} = \sum_{n \in crystal} \sum_{j \in cell} f_j\, \mathrm{e}^{-i\mathbf{q}.\left(\mathbf{r}'_j + \mathbf{R}_n\right)} = F(\mathbf{q}) \sum_{n \in crystal} \mathrm{e}^{-i\mathbf{q}.\mathbf{R}_n} \tag{7}$$

where $\mathbf{r}_\mathrm{j}$ can be decomposed in a sum of the atom position vector $\mathbf{r}'_\mathrm{j}$ in the unit cell and the unit cell position vector $\mathbf{R}_\mathrm{n}$ in the crystal lattice.

The first term $F(\mathbf{q})$ is the structure factor, i.e. the scattered amplitude of the unit cell of volume v. This term, depending on the position and the kind of atoms in the unit cell, reveals the symmetry of the unit cell. The second term which reveals the symmetry of the crystalline lattice, is nonzero in the Bragg/Lauë diffraction condition: $\mathbf{q}.\mathbf{R}_\mathbf{n} = 2\pi$ is verified, i.e. when $\mathbf{q}$ is a reciprocal lattice vector $\mathbf{G}_\mathrm{hkl} = h\mathbf{a}^* + k\mathbf{b}^* + l\mathbf{c}^*$ or $2\mathrm{d}_\mathrm{hkl} \sin(\theta_{hkl}) = \lambda$ where d is the distance between lattice plane (hkl) and θ_{hkl} is the corresponding Bragg angle (see Ref. [33]). However, in the case of nanoparticles, the use of this condition is frequently unadapted because of the small size and the particular structure of the diffracting objects.

Indeed, when the object has a finite size, but is crystalline, the intensity can be written:

$$I(\mathbf{q}) = A^*(\mathbf{q})A(\mathbf{q}) = \frac{|F(\mathbf{q})|^2}{v^2}\left[|\Im(\mathbf{q})|^2 * \sum_{\mathbf{G}} \delta(\mathbf{q} - \mathbf{G}_{hkl})\right] \tag{8}$$

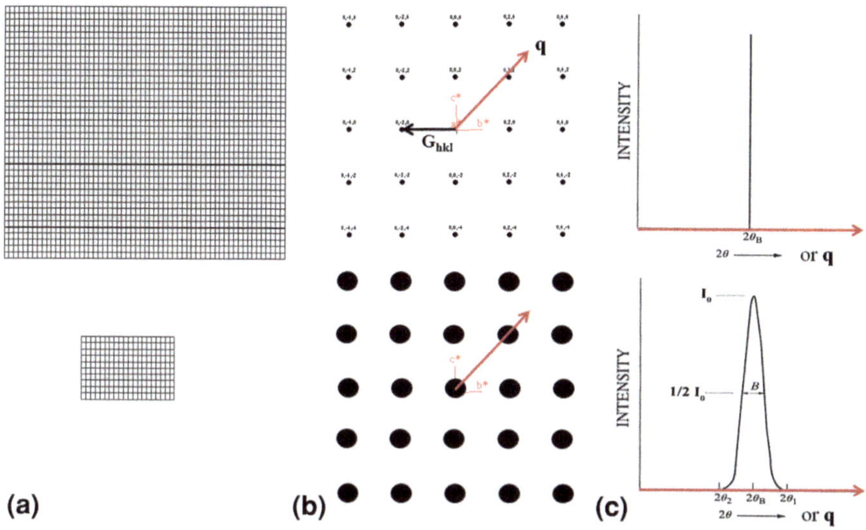

Fig. 3 Example of the size effect. **a** Shape and extent of diffracting crystals. **b** Corresponding reciprocal space map (Bragg spots). **c** q scan in the *arrow* direction to evidence the broadening of Bragg peaks

where $\Im(\mathbf{q})$ is the Fourier transform of the object shape function $\sigma(\mathbf{r})$ limiting the lattice extent. As smaller the particles broader the diffracting Bragg spots defined by the $\mathbf{q} = \mathbf{G}_{\mathrm{hkl}}$ reciprocal space condition (Fig. 3). This behavior is very accentuated for nanosized particles and is observed in surface diffraction when the analyzed thickness of matter is very small [34, 35].

If the scattering object is not crystalline (no underlying translational symmetry), as observed in nanoparticles with a five-fold symmetry (multi-twinned particles or with strong distortions of the atom packing), the conventional analysis of the diffraction patterns cannot be made from the Bragg peaks positions or shapes. With nano-objects, traditional X-ray diffraction, based on crystallography fails because of their substantially limited length of structural coherence and is replaced by the so-called "wide angle X-ray scattering" [36–38]. The intensity expression must take into account the correlation of all atom pairs ij.

$$I(\mathbf{q}) = \sum_i \sum_j f_i^*(\mathbf{q}) f_j(\mathbf{q}) \, \mathrm{e}^{-i\mathbf{q}\cdot(\mathbf{r}_j - \mathbf{r}_i)} \tag{9}$$

When it is assumed that particles are randomly oriented with respect to the incident beam, such as in preformed particles or particles prepared by atom deposition on amorphous substrate, the diffraction pattern is radially symmetric (Fig. 4a). The profile of the scattered intensity along a radial section of the pattern can by written as:

$$\langle I(q) \rangle = D(q) \sum_i \sum_j f_i(q) f_j(q) \, \sin\,(q r_{ij}) / q r_{ij} \tag{10}$$

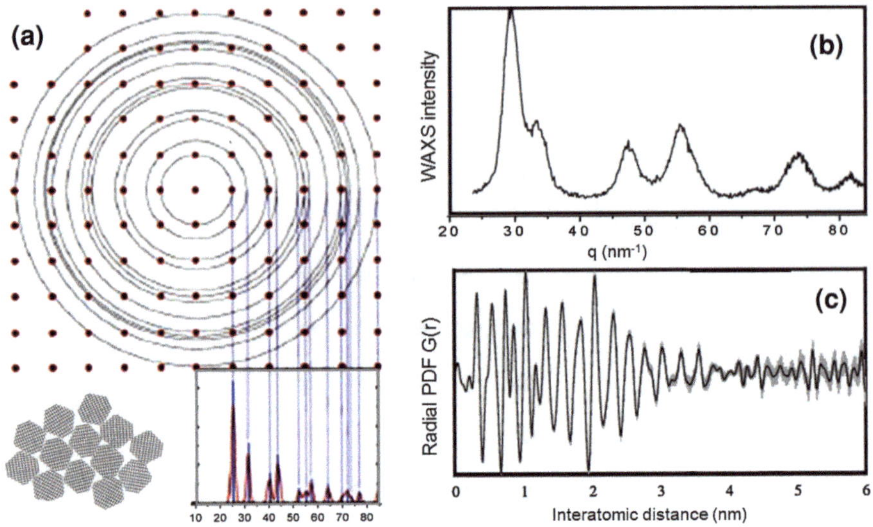

Fig. 4 **a** Effect of the uniform and random orientation of particles on the diffraction pattern which induce the symmetric halo formation (powder effect). **b** WAXS pattern of Au crystalline particle sample in isotropic orientation. **c** The corresponding pair distribution function PDF. From Ref. [39], IUCr's copyright permission

This expression, called "the Debye equation" of the kinematic approximation [27] can include a Debye–Waller factor *D(q)* which expresses an mean degree of dynamic (thermal vibrations) or static (defects) disorder.

$$D(q) = \exp(-q^2\sigma_{ij}^2) \quad \text{and} \quad r_{ij} = \left|\mathbf{r}_j - \mathbf{r}_i\right| \tag{11}$$

where σ_{ij} is the mean standard deviation of interatomic distances r_{ij}.

The resulting intensity is the signature of the radial distribution of atom pairs, through the so-called "pair distribution function" *G(r)* describing the number of atoms per unit of volume located at a distance between r and $r + dr$ from an origin atom. The main consequence in the scattering pattern is the formation of maxima of intensity for the most abundant distances (nearest neighbor, second…) independently of their orientation in the particle.

Practically, in a direct method, the WAXS pattern can be calculated from the atom positions obtained by a geometric construction [36, 40, 41] or by atomistic simulations [42–44] and subsequently compared with experimental patterns for structure identification. In this case, the instrumental and polarization factors, and the incoherent intensity I_{incoh} which expresses the intensity from inelastic scattering (Compton scattering) or from non–particle elastic scattering (matrix, substrate, atmosphere…) must be considered to formalize the experimental conditions [28, 35, 45].

Another method, the indirect method, is to extract the atomic pair distribution function (PDF) from the experimental data and to compare it with the *G(r)* calculated function of nanoparticle models. In this method [46, 47], the data analysis is achieved in the real-space with a direct demonstration of the atom position, but with a more complex procedure. In addition, data acquisition with high accuracy is required up to large values of the momentum transfer q. The PDF is obtained from the Fourier transform of the experimentally observable structure functions, $S(q)$, which is a normalized expression of the total scattered intensity as follows:

$$S(q) - 1 = \left\lfloor 1/(N\langle f\rangle^2) \right\rfloor \sum_{i\neq j} \sum_j f_i(q) f_j(q) \sin(qr_{ij})/qr_{ij} \tag{12}$$

where $<f>$ is the average scattering factor, N being the total number of atoms.

Hence, the corresponding pair distribution function is obtained as follows:

$$G(r) = 2/\pi \int_0^\infty q[S(q) - 1] \sin(qr) dq \tag{13}$$

It peaks at characteristic distances separating pairs of atoms and thus reflects the atomic-scale structure (Fig. 4c). In reality, the measurement is achieved in a finite q window (angular scan), due to limitations of experimental set-up and X-ray wave length (from q_{min} to finite q_{max} values). This effect induces spurious oscillations which are negligible if the q_{max} value is very high [39]. As in the direct method, polarization, self-absorption and background corrections must be also applied. A fast analysis of the experimental PDF provided an approximate measurement of the metal–metal bond length and of the order extent inside the particles [48, 49]. However, best values of the parameters defining the nanoparticle structure were estimated from the agreement reached between experimental and computed PDF [26, 50].

2.2 Size and Structural Effects

At small size (up to several nanometer), the diffraction pattern of a particle gives the appearance of an amorphous material. However, it is not the case; although the particles lose the crystalline order, they take highly symmetric structures. Furthermore, despite dominant size effects, which likely hide information about atom arrangement, structural effects can have also a significant impact on the diffraction pattern in nanometer-sized particles. The challenge is to discriminate the size and structure effects, which are frequently correlated, in the stability and transition mechanisms, as developed in the Chaps. 6 and 8. In this section and the following, three types of structure (fcc truncated octahedra TOh, icosahedra Ih and decahedra Dh) will be used as examples while reviewing the various effects.

Single crystaline nanoparticles, or nanocrystals exhibit features in a diffraction pattern (Fig. 5a, b) that are size-dependent in term of feature shape and

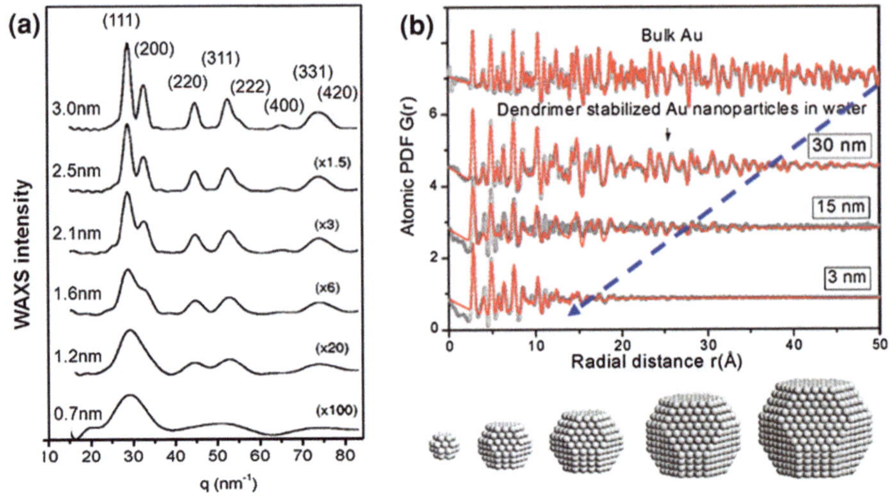

Fig. 5 Size effect **a** in the WAXS pattern of crystalline fcc Pt nanoparticle for about 0.7, 1.2, 1.6, 2.1, 2.5, 3 nm diameters, **b** in the experimental atomic PDFs (*symbols*) for Au nanoparticles with an average size of 3, 15, 30 nm, and for bulk. PDFs calculated from particle model are shown as *solid lines*. From Ref. [51], copyright (2005) by The American Physical Society

size-independent in term of feature position. The origin of these maxima of intensity is equivalent to diffraction conditions, as the Bragg law, and consequently, the position of these peaks is lattice parameters-dependent. In this case, the structure identification is easy (lattice, shape, interatomic distance). However, the major remark in the Fig. 5a observation is the considerable broadening and the consequent overlap in the maxima (peaks) of the scattering profiles, which increase the difficulty in peak identification, and thus the structure identification. In this context, the extracted PDFs for nanoparticles can give better results to solve the structure though the coherent length (radial extent of the PDF) is reduced by the small particle size [39, 50, 51].

Besides this effect, at small size, the interatomic relaxation due to confinement or surface effects may induce a uniform strain or more frequently a strain distribution, which consequently yield slight shifts in the position peaks or anomalous peak heights and widths [42, 43, 52]. In addition, the interface effects, like that due to ligands or matrix may induce a significant surface disorder, particularly in objects with a high surface-area-to-volume ratio. Figure 5b shows that this effect further reduces the coherence length, as the size effect. Figure 6a displays the relaxation effect in a WAXS pattern between rigid particles and relaxed particles from Monte Carlo (MC) simulations using a semi-empirical many-body potential [43, 53], showing that the main observed effects in the scattering profiles are the shift, broadening and intensity decay of the peak intensity.

Non-crystalline nanoparticles (Ih and Dh) scaterring profiles exhibit features that are strongly size-related [54, 55]. Indeed, the distribution of atom–atom pair

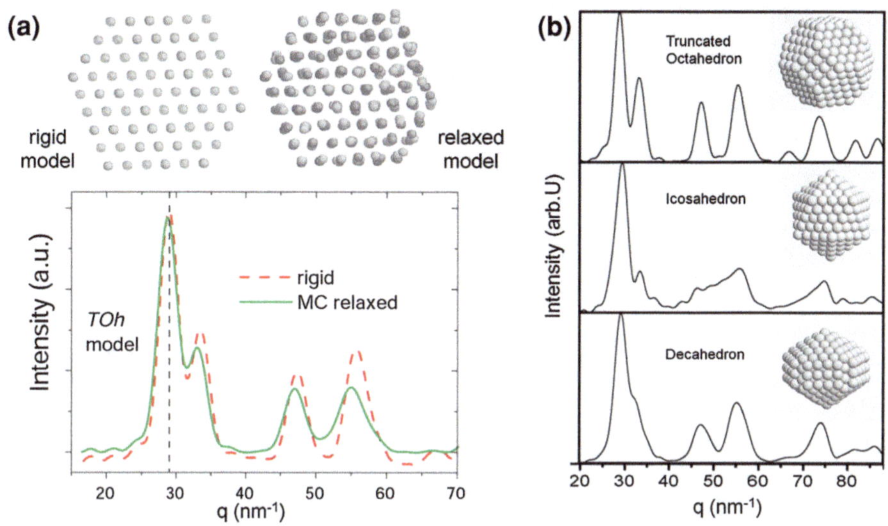

Fig. 6 **a** Effect of the relaxation of atom position (CoPt MC relaxed structure versus rigid crystalline structure of 2 nm): Particle snapshots and corresponding patterns (from Ref. [43]). **b** Scattering patterns for fcc TOh, Ih and Dh structures for particles, approximately 2.5 nm in diameter

distances (pair correlation) in Dh or Ih nanoparticles changes with the size, unlike the crystalline structure (see Chap. 8). These structures exhibit two or three different interatomic distances (with distance variation up to 5% in the Ih case), due to the elastic distortion of tetrahedra, which composed these multi-twinned particles [1]. With respect to the crystalline structure, the distortions involve less change in the Dh than in the Ih structure pattern, except in the region of main intensity maximum (close to 30 nm^{-1} in Fig. 6b). These wide angle X-ray scattering pattern provide a accurate signature of the structure of the particles at small size, allowing the identification of crystalline or non-crystalline structures.

Moreover, it is important to consider the consequences of the size distribution in sample when interpreting wide-angle scaterring data. In this case, the support of complementary technique to provide morphological data is appreciable, like TEM (Fig. 7a) or SAXS techniques. This size polydispersity effect is not restricted to a simple distribution of peak width but to a distribution of peaks position (for example as the shoulder indicated by an arrow in the Fig. 7c) especially in the icosahedral case [11, 17].

The total intensity from an assembly of nanoparticles is a weighted summation of the intensities from several sizes n and structure types i [36, 41]:

$$I(q) = \sum_{i \sim types} x_i \sum_{n \sim sizes} w_{i,n} I_{i,n}(q) \tag{14}$$

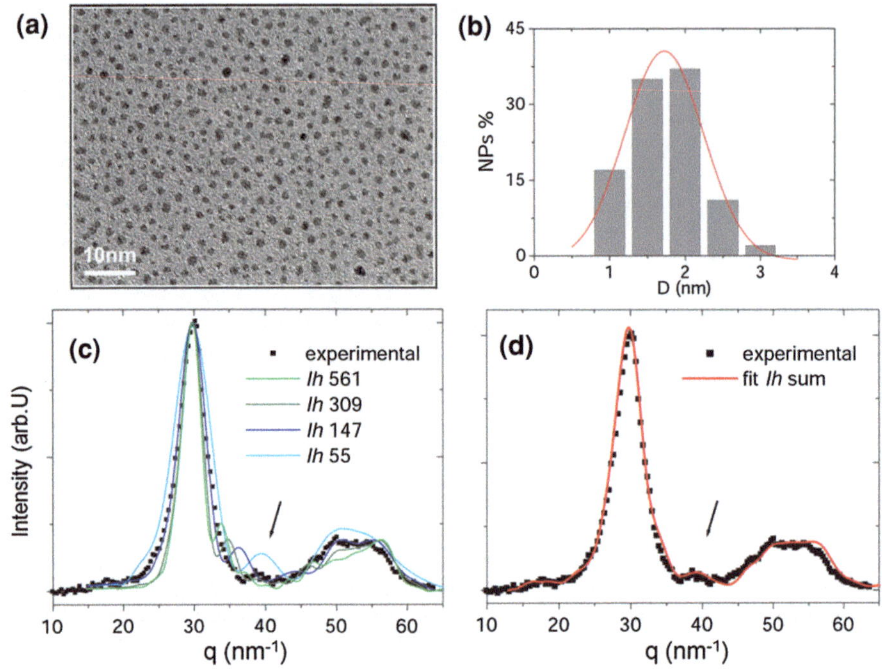

Fig. 7 **a** TEM micrograph. **b** Size distribution histogram of CoPt alloyed supported particles. **c** Corresponding experimental WAXS pattern with several calculated pattern from Ih structure of different size (number of atoms). **d** Best fit with an Ih linear combination in agreement with the size histogram. The arrow indicates the size-dependent shoulder of the Ih pattern

x_i is the number fraction of each structure type i and $w_{i,n}$ the size distribution function. In the case of single type of particles ($i = 1$), Ih for the example of Fig 7, the excellent fit of this intensity results from a calculated size distribution in good agreement with the histogram obtained by TEM. However, the more the number of coexisting structures increases, the more difficult the results become to interpret [39, 55, 56], especially in the bimetallic particles. With the support of size information coming from other techniques, the efforts can be concentrated in the structure identification.

2.3 Mixing and Segregation in Nanoalloys

The case of multi-metallic particles is more complex, because the addition of one or several metals induces a larger diversity of structure [3]. As in bulk materials, the composition variation (A_xB_{1-x}) leads to an atom intermixing in form of solid solution structures (continuous randomly chemically disordered at all compositions as for Ni–Pt, Rh–Pt, Ag–Au, Co–Rh...) or of ordered structures as Cu–Pd,

Cu–Au, Co–Pt, Fe–Pt... (with a continuous fcc solid solution at high temperature). However, for a given composition, the finite matter quantity or the surface-induced-strain as observed in small particles could lead to surface segregation or core–shell arrangements which modify the uniformity of the structure from nanoparticle center to surface. When the intermixing is not favorable as in Co–Ag, Cu–Ag, Pd–Au, Ni–Ag... systems, the nanoparticle could exhibit subcluster (dual) segregated (A and B type clusters) or core–shell arrangements.

The X-ray diffraction method is a characterization technique extremely dependent of the symmetry in the atom arrangement. Consequently, a change in the chemical order, i.e. in the layout of atoms in the particle (the alternation of atoms) is easier to detect than local disorder or surface segregation.

In nanoalloys, a first indicator of the mixing effect is the interatomic distance. Analysis of the experimental PDF provides good estimates of the mean metal–metal bond length and the short range structure as showed for example in the works of Lecante and coworkers [48]. Experimental results in Co_xRh_{1-x} clusters ($\sim$2.5 nm) are summarized in Fig. 8a, b. While the Rh-rich particles display a typical face-centered cubic structure, the Co-rich particles show a structural disorder. The second peak (indicated by a dash line in Fig. 8a), gradually decreases with increasing cobalt content, which indicates the weakening of the octahedral symmetry of the atomic sites in the structure. In Fig. 8b, the evolution of the measured first interatomic distance d_0, is reported with the composition of the particles. By comparison with the monotonous, although nonlinear, evolution of this distance in the corresponding bulk alloys (dashed line), in the particles d_0 strongly increases between pure Co and $Co_{0.5}Rh_{0.5}$ and then remains almost constant up to pure Rh. Thank to the PDF analysis, the structural behavior of Co_xRh_{1-x} particles with the composition could be explained by relaxation effect and/or surface segregation of Co.

The Cu–Au type alloys, like Co–Pt or Fe–Pt are model systems to evidence the chemical order and disorder effect by X-ray diffraction. These alloys (A_xB_{1-x}) based on the fcc A1 structure can be obtained over a wide range of compositions as bulk [58]. For either A- or B-rich alloys, A_3B and AB_3, a chemically ordered structure (Ll_2 cubic structure) which is described as alternate pure (rich atom type) and mixte AB planes in the 3 directions [001], [010] and [100], can be observed. For the compositions near 50–50, a chemical ordering appears leading to a tetragonal Ll_0 ordered phase by deformation of the cubic lattice. This latter structure can be described as alternate pure A and pure B (001) planes, giving rise to a crystalline anisotropy along the (001) axis.

For the $L1_0$ structure, the ordering of atoms lowers the overall symmetry of the structure with respect to A1 structure and the anisotropy induces a structure tetragonalization ($c < a$) up to 8% for CuAu [59]. These effects can be observed on diffraction patterns with more or less strong consequences. Figure 8c shows the ordering effect with the presence of superlattice peaks due to alternation of two kinds of planes in Fe–Pt particles [57].When the particles are very small (less than 3 nm) or very disordered (atom position disorder), it is difficult to separate overlapped contributions of the splitting effect (Fig. 8c, B, C and D patterns) due

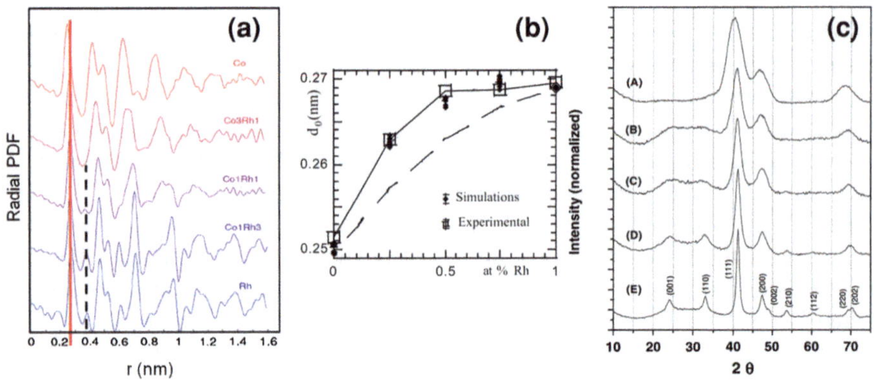

Fig. 8 **a** Experimental PDF patterns of Co_xRh_{1-x} clusters of increasing Co concentration. **b** Corresponding evolution of deduced nearest neighbor distance d_0 in the bulk (*dashed line*) and in the particles (*square*) (from Ref. [48], copyright (2004) by The American Physical Society). **c** Experimental diffraction patterns ($\lambda = 1.54056$ Å) of 4 nm FePt equiatomic particle assemblies of A1 (A) and $L1_0$ structures annealed at increasing temperature: (B) 450°C, (C) 500°C, (D) 550°C, and (E) larger particles (from Ref. [57]. Reprinted with permission from AAAS)

to the tetragonalization. The evidence of the ordered state is due only to the stronger superlattice peaks at smaller q value, i.e. the (001) and (110) peaks [57].

To evaluate the chemical order in a quantitative way, a calculation of the long-range (LRO) chemical order parameter S is necessary either from atom sites as explained in the Chap. 6 or from experimental diffraction data. Commonly in crystalline materials of large size (bulk or thin film), the LRO parameter is obtain from the ratio of superlattice-to-fundamental (001)/(002) peak intensities [60, 61], taking into account atomic scattering factors and correction factors (set-up geometry, Debye–Waller). However, the overlapping between large peaks in small particles (Fig. 8c) may induce a large uncertainty on the S value [57, 62]. A more convenient method is to fit experimental data with simulated patterns of relaxed ordered clusters (obtained by numerical simulation) with different S parameters. Figure 9a displays two phenomena in the wide-angle scattering profiles of different calculated clusters of same size (1,289 atoms, 3.2 nm) corresponding to increasing LRO. The effects of the transition are clearly distinguishable in the patterns of the ordered clusters: the increasing of the (001) and (110) peaks corresponding to the formation of alternation of planes more and more rich in Pt and Co atoms and the splitting of (200)/(002) and (220)/(202) peaks due to the tetragonalization. Figure 9b shows the comparison between the experimental pattern and simulated patterns of a partial ordered cluster (LRO = 0.44) and a mix of ordered-disordered clusters [63]. In both cases, the main requirement of the fitting was the intensities scaling adjustment of the superlattice peaks (110) and (001) and the main (111) peak with the experimental pattern. An excellent matching is obtained with a bimodal distribution of fully chemically ordered NPs

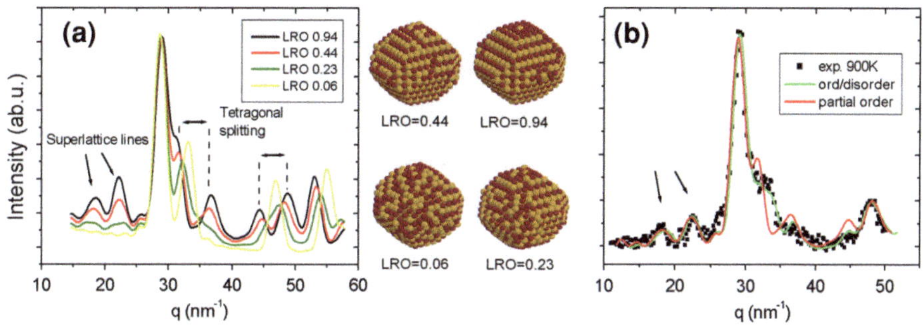

Fig. 9 **a** Calculated patterns of four different MC relaxed truncated octahedron CoPt clusters of same size (1,289 atoms) corresponding to LRO of 0.06, 0.23, 0.44, 0.94 with their associated snapshots. **b** Comparison between experimental (*square marks*) and calculated (*solid line*) diffraction patterns (partial ordered cluster LRO $= 0.44$ and a mix of ordered-disordered clusters, in *green* and *red* color, respectively). From Ref. [63], copyright (2010) by The American Physical Society

and fully disordered ones rather than with a distribution of partially ordered particles. These results show the difficulty to transfer the X-ray diffraction analysis procedures from bulk to nano-sized materials. Once again, the support of simulations of nanoparticle models provided an unambiguous identification of the nanoalloy structure. However, this can be achieved through a morphological description (size, shape) coming from other complementary techniques, as discussed in Sect. 4.3.

However, the capability to detect if a bimetallic particle is alloyed or segregated, is a more difficult problem. In fact, the challenge is also to discriminate, if the particles are in a core–shell or in a dual arrangement. To simplify the problem, we can consider a binary system without gradient of structure or composition. Thus, several cases can be considered: firstly, the shell structure is different from that of the core; secondly, the atom arrangement of the shell is structurally coherent with the core one.

In the former case, the shell and core contributions in the diffraction patterns are separated as in the case of oxide shell on a metallic core [54, 64] or as in the case of immiscible metallic system. Figure 10a shows the evolution of Ag diffraction contributions after annealing of 3.5 nm (FePt)Ag particles [65]. The separated shoulder at low temperature indicates the formation of Ag–Ag pairs by surface segregation. However, at higher temperature, the Ag contribution shape is too narrow to fit with a shell of one or 2 monolayers of Ag around the FePt core, and the coalescence of Ag atoms in particles is very probable. This example illustrates that it is necessary to take into account all features of the X-ray pattern and not only the peak position.

In the latter case, the interpretation of scattering pattern is ambiguously without the use of simulated patterns. The possible epitaxy of the segregating metal in the core surface must be in this case taken into account. Systems with high mismatches between interatomic distance of the two metals are easier to identify. Figure 10b shows a study on 4 nm PtRu nanoparticles [66] obtained by successive

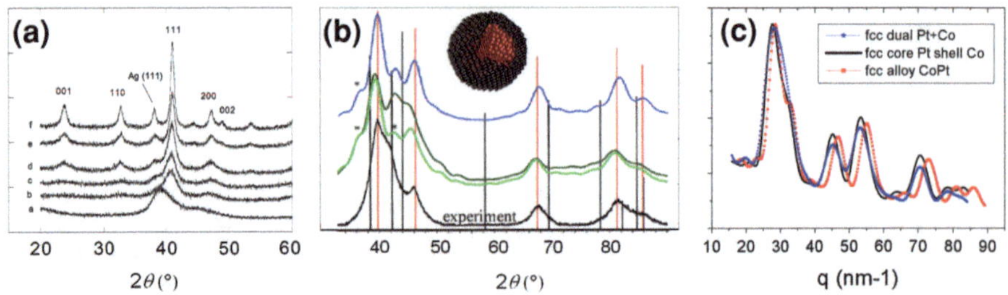

Fig. 10 **a** WAXS data of self-assembled $[FePt]_{88}Ag_{12}$ nanoparticles: as-prepared (*curve a*) and after heat treatment at 300°C (*curve b*), 350°C (*curve c*), 400°C (*curve d*), 450°C (*curve e*), and 500°C (*curve f*) (from Ref. [65], copyright (2002) American Chemical Society). **b** WAXS pattern showing Debye simulations from a 4.0 nm Ru@Pt model cluster with a 1ML Pt shell (*top*, *blue*), a Ru@Pt model cluster with a 2ML Pt shell (*middle*, *dark green*), a 2ML Pt shell alone (*middle*, *light green*), and the experimental diffraction data (*bottom*, *black*). The *red* and *black* stick patterns are the diffraction positions for bulk Pt and Ru metals, respectively (from Ref. [66], copyright (2009) American Chemical Society). **c** Simulated WAXS pattern of 2 nm fcc CoPt nanoparticles in segregated dual and core/shell configuration, and in alloyed mixing. Courtesy of H. Khelfane

chemical synthesis of a Ru core covered with a shell of Pt atoms. A simple analysis of the scattering pattern shows an enlargement and a shift of peaks with respect to the alloy particle pattern. In this case, the epitaxy of a Pt monolayer on the Ru core leads to a double effect, one due to strains induced by the shell deposition and the other one due to the alloying at the core–shell interface.

In the Fig. 10c, the WAXS patterns of different particle configurations (alloyed, core–shell and separated particles, as represented in Fig. 1f–h) in the case of very small size (2 nm) are presented. These results show that it is difficult to distinguish the alloy to the coherent core–shell arrangement at this size in term of profile shape. However, the intensity maxima position reveals the alloying CoPt, while the Pt core -Co shell particles keep the core inter-atomic distance. Although, the dual-particle arrangement should be easier to identify, because the resulting intensity is the addition of Co and Pt particle contributions. The peak widths become broader than that of the initial core or alloy particles because the inter-atomic distances of Co–Co and Pt–Pt are different. However, this effect is only observed in the main peak, the overlapped (111) and (200) peak group.

The works of Kaszkur on Pd–Au nanocrystalline systems supported on silica exposed to different gas environments show subtle changes detected in the diffraction patterns [67]. These changes were interpreted in terms of an inversion of concentration profile in Pd–Au particles from Pd segregation induced by oxygen chemisorption to Au segregation for the clean surface.

In summary, the wide angle scattering can be suitable to evidence segregation in nanoparticles with respect to alloyed arrangement. However, this technique requires the support of simulated pattern especially when the particle size is ultra small (<2–3 nm). In this case, chemical selective techniques using absorption effects at the edge of atoms composing the particles, as anomalous X-ray scattering (see Sect. 4) or X-ray absorption spectroscopy seem well adapted.

3 Small-Angle X-Ray Scattering: Particle Dispersion and Morphology

3.1 Theoretical Background and Fast Analysis Method

Because the wide angle X-ray scattering technique doesn't provide directly the size and shape of nanoparticles, a complementary technique is necessary. Small-Angle X-ray Scattering (SAXS) is capable to give information on the morphological features of particles of nanometer to submicrometer size [68] as well as their spatial correlation (in volume or on a surface). In this technique [29], the scattering intensity is collected close to the incidence direction at small angle (small q values), corresponding to large scale inhomogeneities in the electron density, like nanoparticles in a media with a more or less high density.

Two main cases can be considered (from Eqs. (1) and (4) for the calculation and interpretation of the intensity coming from an assembly of particles of average electron density ρ_p dispersed in a media ρ_m (two phase system: particles p in a dispersed media m): the diluted and the concentrated cases.

For an assembly of N diluted identical particles, i.e. completely uncorrelated particles, the intensity is the sum of individual particle intensities:

$$I(\mathbf{q}) = NV_{part}^2 \langle \Delta\rho \rangle^2 P(\mathbf{q}) \tag{15}$$

$$P(\mathbf{q}) = \frac{1}{V_{part}^2 \langle \Delta\rho \rangle^2} \int_V \gamma_{part}(\mathbf{r})\, \mathrm{e}^{-i\mathbf{q}.\mathbf{r}}\, dV \tag{16}$$

$P(\mathbf{q})$ is the form factor expression only dependent on the geometric shape of particles with V_{part} the volume of one single particle and $\Delta\rho = |\rho_p - \rho_m|$, the density contrast. $\gamma_{part}(\mathbf{r})$ is a correlation function defined in Eq. (4). $P(\mathbf{q})$ contains the size and shape information, and in addition, their possible distribution. Experimentally, form factors can only be accurately measured in this dilute regime where particles can be considered as independent scatterers without any interactions. In the spherical shape case, the determination of the particle size is qualitatively obtained with the Guinier approximation analysis [28, 69] through the intensity slope at low q value (Fig. 11a) or more quantitatively from the location of minima (Fig. 11b). The greater the contrast intensity (maximum to minimum ratio), the greater the determination of size will be easy.

When the particles concentration increases, the wave scattered by different particles can interfere in a way that depends on their spatial organization, as the atoms in the case of wide angle scattering. The intensity becomes:

$$I(\mathbf{q}) = NV_{part}^2 \langle \Delta\rho \rangle^2 P(\mathbf{q}) S_{\mathrm{int}}(\mathbf{q}) \tag{17}$$

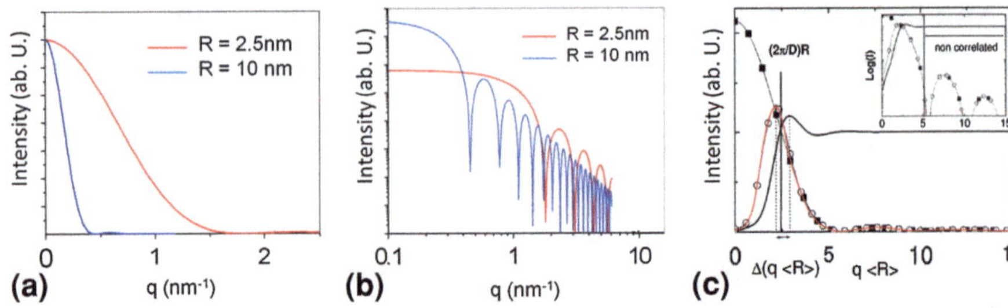

Fig. 11 **a** SAXS intensity profiles for two particle sizes in linear–linear scale. **b** In a log–log scale. **c** Total SAXS intensity profile (*Open circle*) of an assembly of correlated particles, with corresponding form factor (*filled squares*) and interference function (*continuous line*) with a distance between nanoparticles of D = 2.4 R. The inset shows the same evolution with the y-axis on a logarithmic scale (from Ref. [12], copyright (2009), with permission from Elsevier)

with

$$S_{\text{int}}(\mathbf{q}) = \frac{1}{N}\sum_{i}\sum_{j} e^{-i\mathbf{q}.(\mathbf{r}_j - \mathbf{r}_i)} \tag{18}$$

where $S_{int}(\mathbf{q})$ is the interference function or structure factor (correlation between i–j pairs of particles) irrespective to the particles shape and size, as a first approximation. For a randomly oriented system, it is the Fourier transform of the pair correlation function $g(r)$ of the particles position, which defines the organization of particles in the sample.

This intensity is the product of the form factor by the interference function, which are strongly correlated at small q values as illustrated in Fig. 11c. Consequently, the classical Guinier approximation cannot be applied to extract the size parameters. Nevertheless, in disordered isotropic systems, the interference function tends to one as q increases, and the scattering intensity is then entirely determined by the form factor. At high q value, the location of minima (insert in Fig. 10c) gives the size of the particles. In contrast, at small q value, the determination of the interference function is size-dependent. Indeed, while the average inter-particle distance D is considered as inversely proportional to the position of the intensity maximum ($D = 2\pi/q_m$), this determination is not exact mainly due to the influence of the form factor $P(\mathbf{q})$. The larger the particle size, the greater the uncertainty.

As a consequence, in order to get an accurate determination of the mean inter-particle distance and their size and shape, it is extremely important to fit the experimental scattered intensity with an adequate model for the interference function and for the shape.

3.2 Form Factors and Size Distribution

The scattering amplitude of a particle is the Fourier transform (FT) of the electron density distribution, i.e. the product of the average density and the FT of particle shape σ(r) when the particle is homogeneous.

$$A_{part}(\mathbf{q}) = \int \rho_{tot}(\mathbf{r})e^{-i\mathbf{q}.\mathbf{r}}d^3\mathbf{r} = \langle\Delta\rho\rangle \int_{\sigma(r)} e^{-i\mathbf{q}.\mathbf{r}}d^3\mathbf{r} = \langle\Delta\rho\rangle\Im_{part}(\mathbf{q}) \qquad (19)$$

with

$$\left|\Im_{part}(\mathbf{q})\right|^2 = P(\mathbf{q}) \times V_{part}^2 \qquad (20)$$

In general, the most commonly observed particles have a spherical, ellipsoidal, cubic or rod-like shape [28, 70–72] with main parameters, such as the diameter and the aspect ratio. If the particles are anisotropic and preferentially oriented (by epitaxial growth on a substrate, by self-organization, by magnetic effect, etc.…), the use of 2D detector reveals frequently the anisotropy of the shape, as shown in Fig. 12a, b.

In the case of anisotropic shape, each particle can be described in a Cartesian frame. For rod shape particle (cylinder shape of R radius and H height), with its origin at the centre of the bottom of the rod, its x and y axes in the basal side of the rod, and its z axis pointing upwards (cylinder axis), the form factor (Fig. 12a, b) shows anisotropic pattern in the q_{xy} (basal plane) versus q_z.(rod axis).

$$P_{rod}(q) = 4\frac{(J_1(q_{xy}R))^2}{(q_{xy}R)^2}\frac{\sin^2(q_zH/2)}{(q_zH/2)^2} \qquad (21)$$

with $J_1(x)$ is the Bessel function of first order and $q^2 = q_{xy}^2 + q_z^2$.

The mathematical expression for the spherical form factor becomes as follows.

$$P_{sphere}(q) = 9\frac{(\sin(qR) - qr\cos(qR))^2}{(qR)^6} \qquad (22)$$

This morphology provides an isotropic scattering pattern with fringes associated to zero of the *P(q)* function (Fig. 11c). It becomes clear that not only the size is available but others quantities like the particle shape, the particle density and the size and orientation distribution are hidden in the scattering profile.

For example, if the rod shape particles are disoriented, it is necessary to formalize the orientation of objects, i.e. to take into account the angle between the cylinder axis and the scattering vector **q**. and to calculate the integral over this angle, randomly or not [28]. Consequently, the intensity contrast between minima and maxima decreases, as the aspect ratio and the orientation disorder increase.

Furthermore, the assemblies of particles are more or less polydispersed in size and usually, the observed lateral size distribution (TEM observation, Fig. 7a, b) is well described by a lognormal or Gaussian distribution (Fig. 12e). Figure 12f

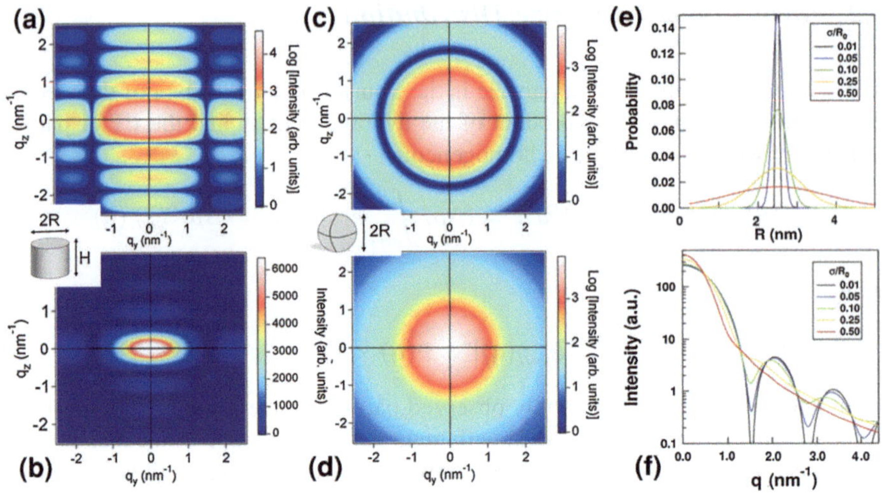

Fig. 12 2D SAXS pattern of a rod shape (cylinder: R = 2.5 nm and H = 10 nm) in **a** logarithmic and **b** linear intensity scale. 2D pattern of a spherical shape for: **c** a single size particle and **d** a distribution of size σ/R = 0.5. Courtesy of D. Babonneau. **e** Gaussian size distribution (R_0 = 2.5 nm) with various broadening σ/R). **f** Corresponding SAXS intensity (from Ref. [72], IUCr's copyright permission)

summarizes the main effects of the broadening of the particle size distribution [69, 72]. The intensity expression becomes:

$$I(\mathbf{q}) = \langle\Delta\rho\rangle^2 \int N(R)V^2(R)P(\mathbf{q}, R)dR \tag{23}$$

where N(R) is the size distribution function.

Sharp fringes of interference are observed for monodisperse particles (Fig. 12c). The obvious effect of the size distribution (see Fig. 12d) is to smooth the scattering curve (contrast of intensity). Hence, in order to discriminate between different shapes and accurately determine the nanoparticle size and size distribution, the intensity must be (i) measured far from the origin of the reciprocal space, over several orders of magnitude and (ii) fitted with simulated pattern to extract significant parameters. This requires experimental data with a signal background as low as possible because the form factor decreases rapidly with increasing q values.

Fitting procedures [73, 74] could be applied to deduce the size distribution from the lineshape of the scattering. The probability distributions for the morphological parameters (particle radius R, aspect ratio H/R, and orientation) can be formalized by analytical functions and adjusted with experimental patterns.

Furthermore, the particular case of nanoalloys must be considered with more or less success in the small angle scattering analysis. For example, if the particles are homogeneously alloyed, the form factor remains unchanged for the same size and shape. Only the electron density contrast $< \Delta\rho >$ changes with the particle composition A_xB_{1-x}, which changes the scale of intensity without modification of the scattering profile shape.

If the assembly of particles is a combination of two kinds of distribution: pure particles of A atoms and pure particles of B atoms, differences in the scattering pattern can be detected only if the distributions are separated in size or in shape. In this case, the analysis is facilitated if the electron densities are enough different.

If the particle forms a core–shell arrangement, the amplitude can be expressed as [29, 74]:

$$A_{core-shell}(\mathbf{q}) = \langle\Delta\rho_{core}\rangle\Im_{core}(\mathbf{q}) + \langle\Delta\rho_{shell}\rangle\left[\Im_{part}(\mathbf{q}) - \Im_{core}(\mathbf{q})\right] \tag{24}$$

with the average electron density contrasts $<\Delta\rho_{core}>$ or $<\Delta\rho_{shell}>$ in core and shell parts of the particle, respectively, (defined with respect to those of dispersion media ρ_m, like vacuum, solvent, matrix…). The combinations of the two terms with different periods and different prefactors in the total amplitude makes it possible to obtain relatively exotic scattering form factors, which deviates a lot from that of a simple homogeneous particles. This specific behavior should be allowing the possibility to identify the core–shell arrangement. The oscillations (minima of intensity) form non-regular q position with respect to homogeneous particles (Fig. 13a). However, it is important to remark that the scattering profiles are efficiently sensitive to the ratio of core to shell radii, only when the contrast ratio is strong $\Delta\rho_{core}/\Delta\rho_{shell} > 2$ [75–77].

For size-polydispersed particles, the results of an experimental pattern analysis have the same complexity and validity that those for homogeneous ones, except in the case of a distribution of ratio of core to shell radii. If the volume fraction of the shell becomes larger than that of the core, the scattering signal will be (depending on the ratio of contrasts) dominated by the contribution of the shell. In this complex situation, the resulting model can be ambiguous without complementary information from others characterization techniques. Figure 13b shows the case of Pt@Co particles (3.2 nm in average size) obtained at low temperature by sequential deposition of Co on Pt core. The experimental pattern fits very well with $R = 0.8$ nm Pt core/$\Delta R = 0.8$ nm Co shell model with respect to alloyed particles of the same size, which demonstrates the kinetic trapping of the core–shell structure. Although the volume of Co shell is high ($V_{co}/V_{Pt} \sim 8$), its scattering contribution is the same order of magnitude than the Pt core contribution due to the weak scattering of Co atoms with respect to Pt atoms.

3.3 Assemblies of Particles and Interference Function

When the particles are concentrated, i.e. the distance between particles is of the same order of magnitude than their size (up to several times their size), two cases could be considered: the ordered systems (characterized by a long range order,

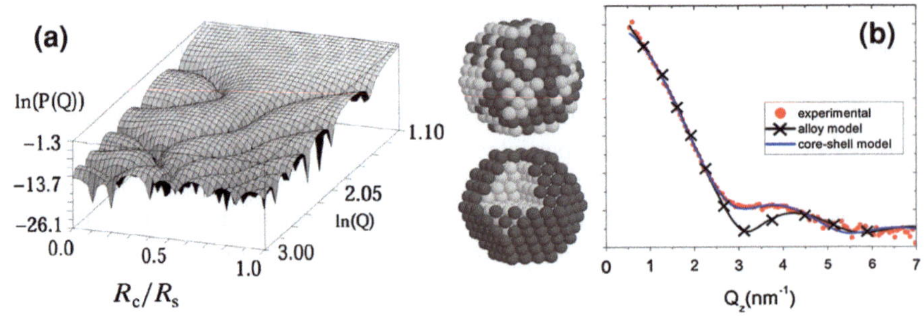

Fig. 13 **a** Dependence of the form factor for monodisperse core–shell particles on the scattering vector, q, and the ratio of core to shell radii, R_c/R_s, at the contrast ratio $\Delta\rho_{core}/\Delta\rho_{shell} = 5$ (from Ref. [77], IUCr's copyright permission). **b** Comparison of experimental core–shell Pt–Co scattering with core–shell and alloyed models of form factor (Courtesy of H. Khelfane)

as in crystals of particles or supracrystals) or the disordered systems characterized by the loss of long range order (characterized by a short range correlation).

In the former case, as in conventional crystallography, scattering or more precise diffraction by an ordered lattice of particles gives rise to Bragg peaks at the nodes of the reciprocal lattice. Figure 14a–c show the hexagonal 2D and face centered-cubic 3D supra-structure of cobalt nanoparticles of 7 nm in size [78]. The small angle diffraction reveals the well long range ordered organization of dry single supra-crystal of particles. X-ray profile can be extracted from the 2D pattern showing the first and second diffraction order (111 and 222). Peaks position and decay of peak intensities with q values are indicative of the suprastructure and the fluctuation in the interparticle spacing [79]. In the case of study in growth media, in solution [80], the small angle X-ray diffraction of gold nanoparticles provides powder-like intensity profiles that reveal the local self-assembly of three-dimensional superlattices in isotropic oriented domains (Fig. 14d). The peaks can be readily indexed to Bragg positions of a face cubic centered (fcc) crystals of particles. Differences with respect to fcc theoretical profiles come from the shape factor effect (oscillations) in the intermediate q region and a slight polydispersity.

In the latter case, paradoxically, it is more difficult to evaluate the disorder degree in the particle position than the order. When the particles do not present a long-range order, the only relevant statistical quantity in the interference function is the pair correlation function. The main feature for this type of function is the first-neighbour peak (position and shape). However, even when the particle arrangement is random with a monodisperse size, the main peak q_m position is not directly linked to the nearest-neighbor distance D between particles ($D = 2\pi/q_m$). Indeed, in addition to a broadening, the main peak is shifted as the particle position disorder increases [29, 69]. Moreover, when the arrangement is more complex, like for distance–size correlated arrangements, some arbitrary pair correlation functions, like the hard-core Debye model or hard core Percus–Yevick model [74], must be used to interpret the interference function effect and fit with the

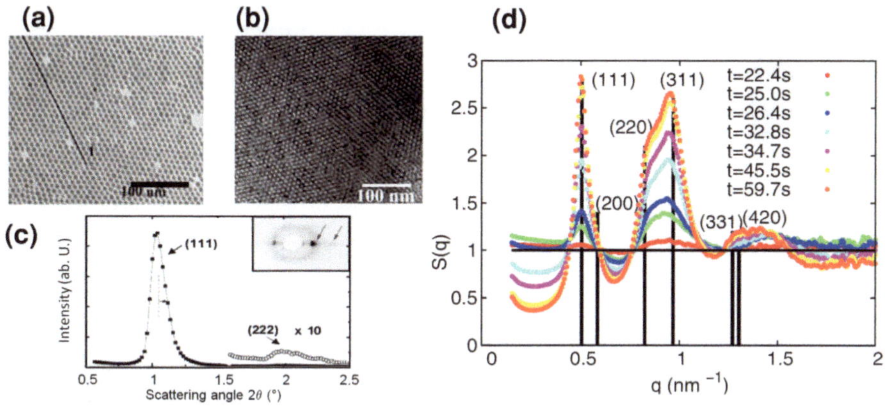

Fig. 14 TEM micrographs of Co particles organized **a** in a compact 2D hexagonal lattice and **b** in a 3D fcc lattice, **c** with a typical small angle diffraction pattern in inset and the intensity profile in the (111) direction (from Ref. [78], Copyright (2004) American chemical Society). **d** Apparent structure factors S(q) obtained during the self-organization of gold nanoparticles in solution (from Ref. [80], Copyright (2008) by The American Physical Society)

experimental pattern. The choice of particular function is linked to the physical mechanisms which govern the interparticle distance, as interaction pair potential in the case of charged particles or colloidal media, as kinetic parameter in the case of atom condensation, etc. [12, 29, 81, 82].

An intermediate state between the regular lattice (2D or 3D) and the fully disordered structures is the model of a paracrystal [83, 84], the long-range order is destroyed gradually in a probabilistic way. In the isotropic case, the interference function is the average over all direction, within a one-dimensional paracrystal theory, with the mean interparticle distance and the corresponding standard deviation (like in Gaussian distribution).

However, in some cases, none of these models could satisfactory reproduce the exact profile of the SAXS data. The pair correlation function g(r) or chord distribution must be deduced from TEM views by image processing, when it is possible [85–87] and applied for experimental scattering intensity interpretation.

3.4 *Grazing Incidence for Supported Nanoparticles*

For nanoparticles dispersed in the whole volume of a solvent, a colloid or a solid matrix, conventional transmission X-ray scattering geometry is used. Unfortunately, for a layer of particles on a substrate the signal is typically 10^{-6} times weaker (depending on the substrate type and thickness) than the scattering from the substrate, resulting in a low signal to noise ratio (Fig. 15a).

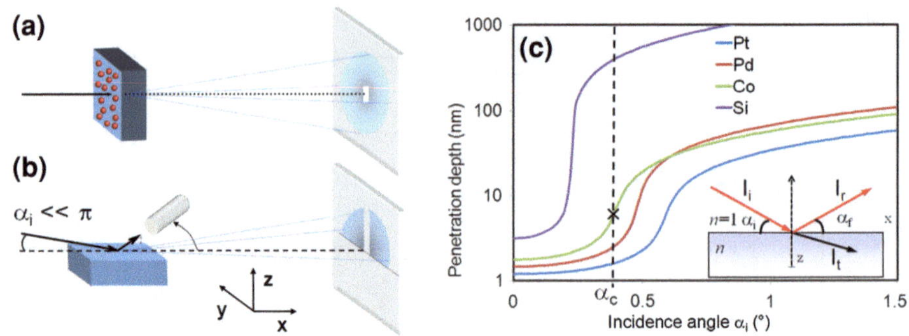

Fig. 15 Schematic drawing of the X-ray scattering geometry **a** in transmission, **b** in grazing incidence for supported or included nanoparticles in a substrate. **c** Penetration depth versus incidence angle obtained at 8 keV for Pt, Pd, Co and Si material

This intensity ratio can be improved if grazing incidence techniques are applied (Fig. 15b). By selecting an incidence angle α_i on the sample surface close and even below the angle of total external reflection α_c of X-rays (few tenths of degrees), the wave-field penetration depth is considerably decreased down to few nanometers thus enhancing the surface or subsurface signal compared to the volume one [88]. Considering the refractive index of the sample in the X-ray range, the incident beam (I_i) reflects in the specular direction (I_r), and produces also a transmitted beam (I_t) in the refractive direction in the material. The penetration depth of the X-rays can be tuned versus the incidence angle (Fig. 15c), thus probing only the sample volume in which the nanoparticles occur, from the surface (supported particles) to the bulk (embedded particles). Thus, any discontinuity in the local electronic density (surface roughness, islands, inclusions, etc.) scatters either the transmitted or the reflected beam.

Within the last years, grazing incidence small-angle X-ray scattering (GISAXS) emerged to be a versatile and frequently used analysis technique in the field of nano-structured thin films and surfaces [12, 89–92]. GISAXS is used for the characterization of correlations and shape of objects at surfaces or at buried interfaces. Experimentally, the GISAXS intensity is recorded perpendicular to the x axis, in an adapted framework (Fig. 16a), as a function of the in-plane q_y and out-of-plane q_z components respectively perpendicular and parallel to the substrate [92]. For supported nanoparticle samples, the 2D pattern reveals the anisotropy of in-plane and out-of-plane organization and shape. In the example of Fig. 16, with a size distributed NPs assembly as observed in TEM (Fig. 7a), the well adapted form factor corresponds to a truncated quasi-spherical shape (inset in Fig. 16a). Anisotropic form and structure factors are simultaneously fitted with calculated scattering profiles from at least two experimental cross sections, in the q_z and q_y directions (Fig. 16b, c).

However at these grazing angles, the reflected intensity is strong enough to modify the small angle scattering intensity with respect to a conventional

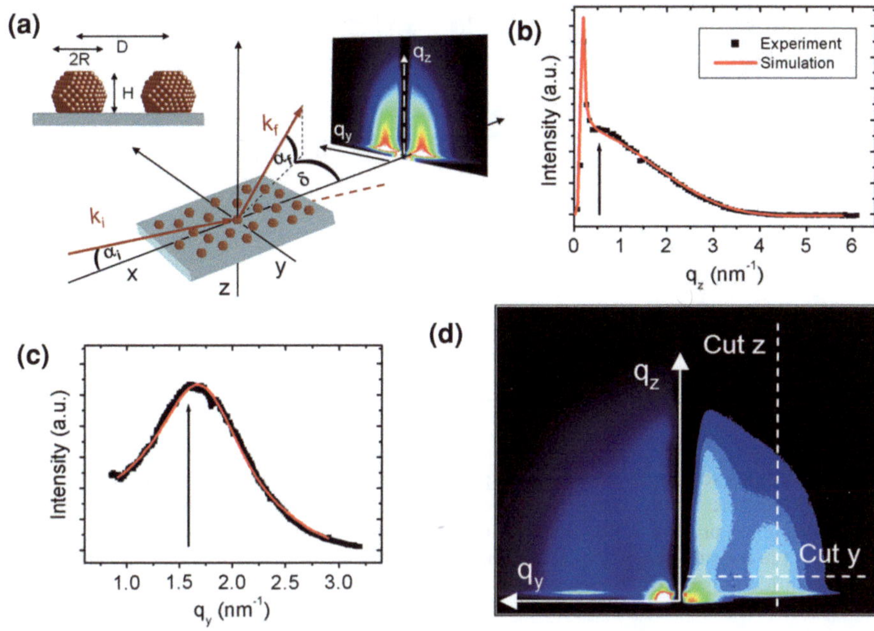

Fig. 16 **a** Scattering geometry: at a fixed grazing incidence α_i, the scattered intensity is recorded on a two-dimensional detector as a function of the out-of-plane angle α_f with respect to the substrate surface and of the in-plane angle δ. The components of the wavevector transfer $\mathbf{q} = \mathbf{k_i} - \mathbf{k_f}$, defined by the incident k_i and the scattered k_f wave vectors are $q_x = k_i\,(\cos\alpha_f \cos\delta - \cos\alpha_i)$, $q_y = k_i\,(\cos\alpha_f \sin\delta)$ and $q_z = k_i\,(\sin\alpha_f + \sin\alpha_i)$ in the laboratory frame. The two cuts of intensity, i.e. cross sections, in the **b** q_z and **c** q_y directions, are extracted from a 2D GISAXS pattern, as in the sample of CoPt deposited nanoparticles (from Ref. [63], copyright (2010) by The American Physical Society). The two cuts (*marks*) selected in the lobe intensity region (*dash lines*) in **d** are simultaneously fitted with a dedicated code (*red line*); the arrow shows the position of the other perpendicular cut

measurement (e.g. SAXS in transmission) far from the total reflection conditions. Because the surface acts as a mirror, multiple scattering effects (with reflected and scattered beams) come into play [12]. Hence, the kinematic approximation (Born approximation BA theory) used in SAXS analysis becomes inadequate. These effects affect the shape factor as described by the distorted wave Born approximation (DWBA) as shown in Fig. 17. An enhancement of the intensity appears at $\alpha_f = \alpha_c$ (Figs. 16d or 17b), called "the Yoneda peak", linked to the reflected beam. Except the scale difference in the parallel direction (in-plane cut in Fig. 17c) the changes appear mainly in the perpendicular direction. The resulting scattering minima become less pronounced and their q positions are shifted as the incidence value increases. These latter effect fails from incidences $\alpha_i > 2\ \alpha_c$. Consequently, except the case of monodisperse particles embedded in a thin film, the analysis of GISAXS patterns require formalism to model the correlation and shape effects at grazing incidence. To date, two dedicated programs are available to simulate and

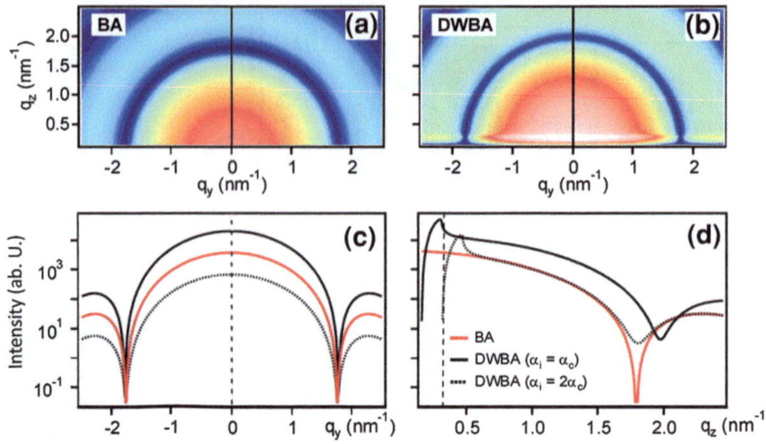

Fig. 17 GISAXS scattering patterns of a spherical particle on a silicon substrate (R = 2.5 nm) **a** in the kinematic theory (BA), **b** in the multi-scattering theory (DWBA) at $\alpha_i = \alpha_c$; and corresponding cut profiles. **c** I(q_y) in the parallel direction, **d** I(q_z) in the perpendicular direction with respect to the substrate plane. The dashed vertical lines correspond to q_y and q_z values of the cut position (courtesy of D. Babonneau)

fit GISAXS patterns: IsGISAXS [72] restricted to nano-objects deposited on a substrate or arranged in a single particle layer (two dimensions) and FitGISAXS [70] more adapted to scattered objects arranged in two or three dimensions in a stratified medium.

For example, Maurizio et al. [93] present a structural investigation on Co–Ni alloy nanoclusters obtained near-surface of silica slides by sequential ion implantation. The study based on small angle X-ray scattering is performed to obtain the cluster size distribution and volume fraction. A systematic comparison with size distribution obtained from transmission electron microscopy points out the potential of the technique for investigating these composite glasses [93]. A study of the $(C/FePt)_{20}$ granular multilayers (20 periods) prepared by ion-beam sputtering reveals the possibility of GISAXS technique to investigate the in-plane and in-depth structure of alloyed nanoparticles after annealing. GISAXS pattern of the as-deposited 3 nm-size particles (Fig. 18a) shows an intense nonspecular diffuse scattering at low q_y giving rise to three transverse Bragg peaks. Their presence at well-defined q_z-positions is typical of a highly periodic system with a period of 5.4 nm corresponding to the nominal bilayer thickness (C+FePt). Conversely, the absence of transverse Bragg peaks in the 2D GISAXS pattern of the multilayer annealed at 800°C (Fig. 18b) is characteristic of a poor vertical organization of the particles, in agreement with TEM observations.

Long range order stability in the 2D arrangement of supported particles of CoPt or $CoPt_3$ obtained by colloidal chemistry [94, 95] were investigated in order to study numerous effects: the temperature, the ligands length, the surface chemistry or structuration. In the example of Fig. 18c, d, the influence of the functional

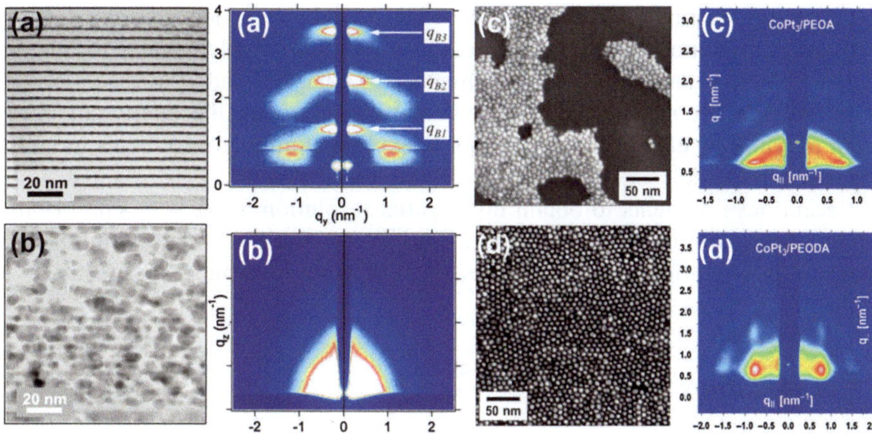

Fig. 18 TEM cross-sections and corresponding experimental GISAXS patterns of sputtering (C/FePt) multilayers: **a** as-deposited. **b** Post-annealed at 800°C (from Ref. [70]). **c** and **d** SEM in-plane images and corresponding GISAXS pattern of chemical synthesis $CoPt_3$ nanoparticles on two different polymeric surfaces (from Ref. [95], copyright (2006), with permission from Elsevier)

polymeric surface was revealed through the degree of defects in the hexagonal lattice for two type of polymeric sub-layer (poly-ethylene–glycol and -oxide). As in the former case, the vanishing of parallel oscillation in the 2D GISAXS pattern and the broadening of the main correlation peak (Fig. 18c) is characteristic of a poor 2D (in-plane) organization of the particles, in agreement with TEM observations.

Specific application of this technique in situ and in quasi real time conditions, for example during particles growth or chemical reactions are presented in the next part of this chapter (Sects. 4.2 and 4.3).

4 Trends in the Synchrotron Radiation Experiments for Nanoalloys

The particular properties of synchrotron radiation such as the wide spectral range and the exceptional X-ray flux have led to the emergence of new techniques and new types of experiments. Third-generation synchrotron facilities offer unprecedented opportunities for ultra-small, ultra-fast and in situ measurements with low to high-energy range, high-brilliance, X-ray beams. The capabilities provided by synchrotron radiation have had an enormous impact on resolving forefront scientific issues in the area of phenomena occurring at nanoscale in nanoparticles.

The high brilliance of these modern synchrotron radiation sources facilitates in situ studies, which provide direct structure–function relationships with both spatial and time resolution; this is especially effective when applied combined X-ray methods quasi-simultaneously or in combination with complementary techniques such as magnetic and optical measurements (see Chaps. 9 and 10 respectively) or mass and vibrational spectroscopies. Furthermore, tuning the energy of X-ray beam provides an element-selective tool to explore atom rearrangements in multi-element nanoparticles.

Recent developments to obtain high spatial resolution using very small beams (up to 10 nm) are in progress [96, 97]. Used also with a high time resolution, this opens up possibilities for a new range of single particle in situ experiments.

4.1 Anomalous Scattering

Anomalous small angle X-ray scattering (ASAXS) is shown to be an ideal technique to investigate the segregation effect of multi-metallic materials at the atomic scale. Indeed, in the nanocomposite materials, i.e. particles of A atoms buried in a matrix M (M could be metallic or non-metallic), when the electron density of particles is close to those of the matrix, the electron contrast becomes too small to detect a signal of small angle scattering from the particles. Nevertheless, a contrast variation can be obtain exploiting the variation in the atomic scattering factor of elements (A and M) near their X-ray absorption edges, and this is useful for the characterization of multi-metallic systems providing element specificity [98, 99].

In this technique [100], the signal from nanoparticles on the surface of a substrate [101, 102], buried in a layer [103] or within a matrix [13] can be separated from that of their dispersion media. Furthermore, ASAXS is also well adapted to reveal core–shell systems [104]. A high electron density contrast between the core and the shell is not required, but a high variation of the atomic scattering factor is necessary near the absorption edge of elements.

In the small angle conditions (q the scattering vector is close to zero), we can approximate the scattering factor as the atomic number of the element. However, we cannot do this approximation when the X-ray energies are adjusted near the absorption edge of the element; in this case, a complex anomalous dispersion correction is needed and the scattering factor is expressed by Creagh et al. [105]:

$$f(q,E) = f_0(q) + f'(E) + if''(E) \tag{25}$$

where f_0 is the Fourier transform of the electron density of the atom equal to Z at small angle, f' and f'', respectively, the real and imaginary part of the dispersion correction become independent to q parameter. The imaginary part of the dispersion correction is related to the absorption of the X-rays. From this equation it can be seen that the dispersion correction factors are energy dependent and in the vicinity of the absorption edge these factors show strong variations (Fig. 19a).

The scattering intensity depends on the correlation between atoms through the Debye formula which considers the scattering factors of atoms of the particle, and

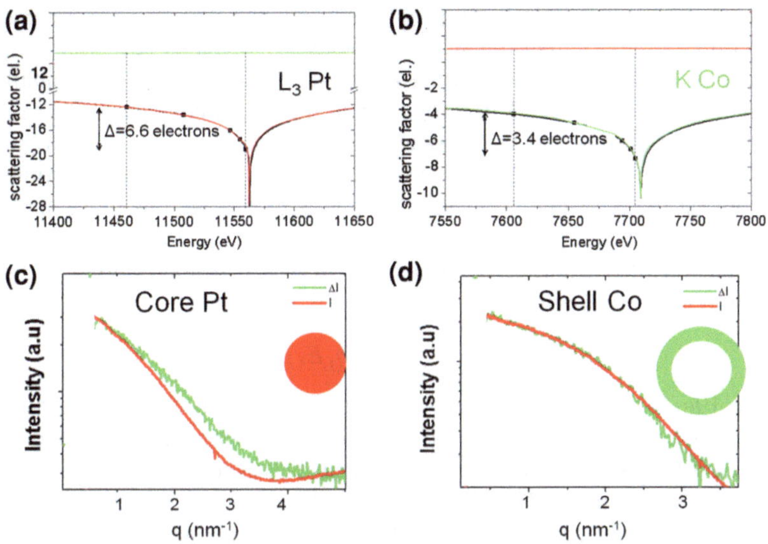

Fig. 19 Atomic scattering factor variation for Co (*green line*) and Pt (*red line*) near **a** the L_{III}- Pt and **b** the K-Co edges. Normalized differential intensity and total intensity of core/shell particle (~2.5 nm), near **c** the edge of Pt core and **d** the edge of Co shell. No effect is observed at the shell edge. From Ref. [100]

the distance between them. When the system is composed of two different types of elements, e.g. type A and type B, it can be separated in three parts, the so called partial structure factors (PSFs). Here this factors will be called $S_{AA}(\mathbf{q})$, $S_{BB}(\mathbf{q})$ and $S_{AB}(\mathbf{q})$ and they will describe, respectively, the partial structure of element A, element B and partial structure factor of the mixed elements A and B. These factors contain the shape information of homogeneous domains, such as segregation resulting domains (e.g. pure A and pure B). Then the X-ray scattering intensity for this system can be written by Lyon et al. [99]:

$$I(\mathbf{q},E) = F_A^2(E)S_{AA}(\mathbf{q}) + 2F_A(E)F_B(E)S_{AB}(\mathbf{q}) + F_B^2(E)S_{BB}(\mathbf{q}) \tag{26}$$

$$F_i(E) = |f_i(E) - f_m| = |f_i| \tag{27}$$

where f_m is the scattering factor of the matrix.

The calculation or extraction of the PSFs is far from trivial and is done using computational analysis from at least three sets of measurements corresponding to three different values of the scattering factor (at three energies). To minimize the statistic errors, much energy must be used to solve PSF with a good accuracy. Practically, for a three phase system, for example, two segregated metals A and B (core–shell, dual particle, two kind of particles) and the matrix, SAXS data must be obtained at different X-ray energies slightly below an absorption edge of each metal A and B, where the metal scattering strength changes a lot, typically, the K or L_{III} edges for transition metals (Fig. 19a). If the bimetallic particles are alloyed, the different PSF exhibit the same profile, while if the particles are unmixed, the PSF are different and reveal the shapes of A and B domains [21, 99].

In the case of Pt core- Co shell particles presented in Fig. 19, the small angle scattering is measured at the K-Co edge and at the L_{III}-Pt edge (SWING beam line at Synchrotron Soleil—St Aubin, France). First of all, it is interesting to consider the intensity difference between two energies varying close to the absorption edge of the core (at the edge and 100 eV below the edge), as well as those at the shell edge. As at the Pt edge, the intensity coming from the Co atom doesn't change with the energy, the resulting difference reveal in a first approximation, only the contribution of the Pt region of the particle, i.e. the Pt core:

$$\Delta I_{Pt}(q) = (I(q, E_2) - I(q, E_1)) \propto S_{PtPt}(q) \tag{28}$$

Figure 19c, d show the difference between the intensity from the particles and ΔI at both edges [100]. The displayed effect is significant at the core edge, and not at the shell edge, which have been confirm in the reverse case Co core- Pt shell particles (not shown here). This differential method is only qualitative because S_{AB} PSF is neglected, and quantitative results can be obtained only with PSF analysis method, which show that the particles of 2.5 nm in size exhibit a shell of 0.35 nm. To our knowledge, below this thickness of shell, anomalous scattering doesn't allow an accurate characterization of a core–shell arrangement between two metals with very different scattering factor as Co and Pt.

ASAXS was used to characterize electrochemical Cu dissolution and dealloying processes of a carbon-supported $Pt_{25}Cu_{75}$ electrocatalyst precursor in acidic electrolytes [106]. By performing ASAXS at both the Pt and Cu absorption edges, detailed information were obtained on the changes in the size distribution function of the Pt atoms and Cu atoms with the temperature. The results allow suggesting a surface dealloying through the formation of a enriched Pt shell surrounding a Pt–Cu core.

Recently, anomalous grazing incidence SAXS have been performed to separate the scattering contributions of two types of metallic nanoparticles (Au and Cu particles of 2–3 nm in carbon layer). This study [13] shows that the quality of the analysis depends on a precise monitoring of the incident beam, a camera correction of 2D pattern (dark counts, flat field and camera distortion) and a subtraction of non-anomalous scattering. Furthermore, when the incidence angle is above the critical angle, the particular multiple scattering effect at grazing incidence are fairly weak compared with the anomalous contrast variations, which simplify the scattering analysis and open new opportunities to study supported nanoalloys [100].

4.2 In Situ and Real Time Experiments

Elucidation of the reasons which govern the formation or the structural or morphological changes in nanoalloys, is a primary importance to validate as well fundamental theories than practical applications. During the formation of

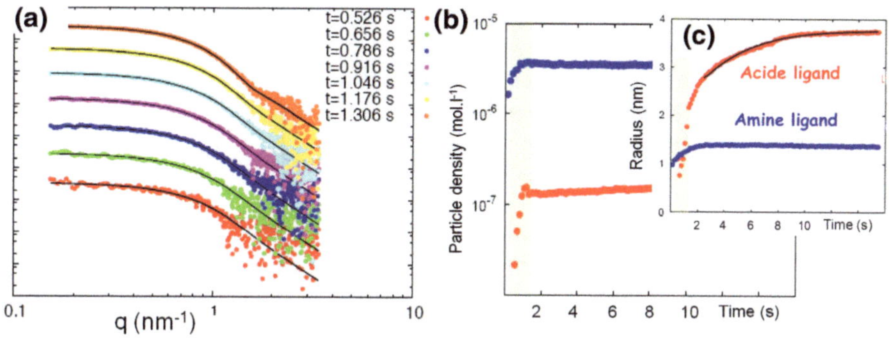

Fig. 20 **a** In situ SAXS patterns for the first instants of the gold nucleation reaction for the case of an acid ligand. **b** Deduced particle density. **c** Average radius as a function of time for the two different ligands. From Ref. [117], copyright (2007) American Chemical Society

nanoparticles, the time and length scales of nucleation and growth processes and their inherent transient nature hinds the possibility of real time investigation. Furthermore, few methods appropriate for the in situ investigation of these phenomena have been established thus far. Among those, the X-ray scattering techniques can be applied in various types of environment from ultra-high vacuum to liquid media, even during chemical reactions, in situ and in quasi real-time when kinetic phenomena are involved like during growth [17, 107, 108], annealing [63, 109] or a catalytic reaction [110]. The combination of fast 2D detector and high brilliance of synchrotron radiation enables the millisecond-interval observation of structure, shape and organization process of nanoparticles.

Abecassis et al. report in situ and real time studies of the nucleation and growth of gold nanoparticles by Au salt reduction in toluene [80, 117]. The use of a fast-mixing stopped-flow device enables the assessment of the whole particle formation process with a 200 ms time resolution. The number of particles, their size distribution, and the yield of the reaction is determined in real time through the quantitative analysis of the SAXS data on an absolute scale. The Au nanoparticles form in a few seconds and the experimental setup enables the monitoring of their formation from the very beginning of the reaction. The role of two types of ligands (acid and amine) is revealed in term of nucleation rate and the subsequent growth rate and final size of particles (Fig. 20). In addition, the formation of three-dimensional superlattices of these gold nanoparticles has been followed directly in situ also by means of small angle X-ray scattering (Fig. 14e). These assemblies spontaneously form in a dilute solution providing the particles are large enough to generate a van der Waals driven attraction sufficient to counterbalance the thermal energy. The superlattices appear very soon after the formation of the individual particles (few seconds) and their growth kinetics is slower than predicted by a mechanism of simple diffusion of the nanoparticles towards the superlattices. In the two cases, time resolved experiments were performed at the ID2 beam line at the European synchrotron radiation facility (ESRF) by the same SAXS technique.

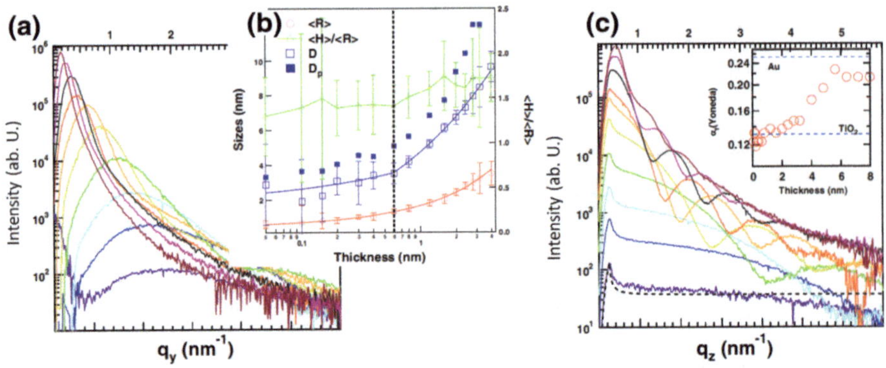

Fig. 21 GISAXS profiles obtained from 2D patterns in **a** the parallel q_y and **c** the perpendicular q_z direction, respectively. The equivalent thicknesses are, from bottom to top, 0.05, 0.15, 0.3, 0.6, 1.2, 2.0, 2.8, 4.0, 5.6, and 8.0 nm. **b** evolution of the GISAXS extracted morphological parameters as function of the deposited thickness: central radius R, central height over radius H/R, and average spacing between islands D and the value D_p as deduced from the location of the correlation peak $D_p = 2\pi/q_y$. The *vertical dashed line* marks the limit between the two growth mode. From Ref. [113], copyright (2007) by The American Physical Society

Numerous recent works using in situ time resolved X-ray scattering emerge in wet-chemical synthesis or self-assembling of gold and alloys particles [111, 112]. In particular, recent works provide flow-based, in situ SAXS analysis to improve time resolution and increase signal-to-noise ratios and to reduce the potential for X-ray beam damage or radiation-induced reactivity.

In another domain, vapor deposition of metals leads to the formation of 3D nanoparticles on structured or amorphous non-metallic substrate. The growth process depend of various parameters, like the surface free energies (substrate, metals), the interaction potentials between atom pairs (homo or heteroatomic), their respective electronic and crystalline structures, the lattice parameters and kinetic factors which govern the shape, the size and the structure of the metallic nanoalloys. One of the challenges is to separate the extrinsic or kinetic mechanisms to the thermodynamic intrinsic behaviors of particles at the nanometer scale, with a suitable analysis method. However, due to the high surface to volume ratio and the high reactivity of numerous transition metals, such as Co, Fe, Ni, Rh, etc., in situ analysis under ultrahigh vacuum (UHV) is absolutely required to prevent any influence of the environment during the formation. Surface contamination and contact effect with external support or matrix have been identified as factors inducing a change of the particle structure and morphology [6–8].

Figure 21 reveals an interesting study on the Au/TiO_2 system using GISAXS technique [113]. The evolution of the morphological parameters with deposited thickness can be used to examine the type of nucleation, the growth modes, the characteristics of the coalescence, as well as the equilibrium shape, the interfacial energy and the adhesion energy. In this case, upon increasing coverage, (i) main peak of interference function shifts at small q values in the parallel direction (Fig. 21a), (ii) interference fringes appear in the perpendicular direction up to show three bounces (Fig. 21c) and (iii) the Yoneda's peak becomes round and shifts from the critical angle of TiO_2 to a value close to the critical angle of gold (inset of Fig. 21c). The evolution of radius, height and spacing between supported particles extracted from these GISAXS profiles reveal that the growth occurs in two stages: a growth at constant particle density limited by surface diffusion, followed by a coalescence mechanism at constant equilibrium shape (Fig. 21b). The coalescence involves a change in shape from flat 3D to 3D of the particles. They demonstrate the sensitivity of the GISAXS technique to particle layer morphology, especially in real time conditions, and also the richness of information extracted from morphological parameters.

Many studies by small- and/or wide-angle scattering have been done for monometallic Pd, Co, Au, Ag nanoparticles [87, 108, 113] but only few papers present bimetallic NPs X-ray scattering investigations in real-time and in situ conditions. Nucleation and growth of ultrasmall supported CoPt nanoparticles obtained by thermal evaporation were studied in UVH and in real time by grazing incidence X-ray diffraction [17]. The amorphous carbon-coated SiO_2 substrate was chosen to limit the particle–substrate interactions and to allow comparisons with ex-situ TEM results. The deposition rates ($0.2–0.6 \times 10^{15}$ atoms/cm^2/h) were selected as slow as possible in order to limit the kinetic effects on the bimetallic particles during the growth. Figure 22 shows the evolution of the wide angle scattering pattern during the growth in the 1–4 nm size range at 500°C substrate temperature. The diffraction patterns are fitted using different relaxed Co_{50} Pt_{50} and Co_{25} Pt_{75} cluster structures obtained by Monte Carlo (MC) simulations (disordered truncated octahedra, decahedra and icosahedra) within a semi-empirical tight-binding potential. The different patterns [43] show that the cluster structure evolution is size-dependent yielding an icosahedral structure at the earlier stage of formation to fcc structure transition from 2 nm in size. Indeed, besides the size effect, in the same growth conditions, by varying the composition from the equi-concentration to a Pt-rich phase, the nanoparticle transition from icosahedral to fcc occurs at smaller size. These in situ and real time results demonstrate the capability to finely interpret the wide angle scattering results with the support of atomistic simulations.

4.3 Combined Small and Wide Angle Scattering Experiments

The main difficulty to describe structure, morphology and spatial organization of nanoalloys is that, in general, any one technique does not give sufficient information to deduce all these characteristics. A coherent strategy is required to combine the input from multiple experimental methods as well as theoretical models. Commonly, one of the X-ray scattering methods (wide or small angles) are combined with non-X-ray techniques, like TEM-based techniques [10, 26, 64, 66, 114, 115], AFM-STM techniques [34], UV–visible spectroscopy [107, 116], etc., or with absorption X-ray techniques, like EXAFS-XANES [8, 48] or X-ray

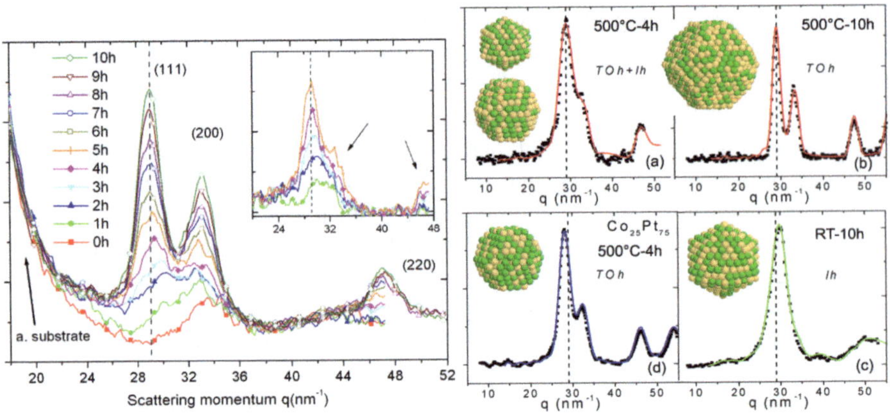

Fig. 22 Evolution of GIWAXS patterns as a function of deposition time. The inset displays the first scans up to 1.8 ML CoPt. Selected experimental (*square*) and simulated (*line*) diffraction patterns of nanoparticles assemblies during the growth: at 500°C **a** R = 1 nm. **b** R = 1.6 nm. **c** Growth at room temperature with R = 1.2 nm. **d** Growth of $Co_{25}Pt_{75}$ at 500°C, R = 1 nm. From Ref. [17], Copyright (2008) by The American Physical Society

magnetic circular dichroism (XMCD) in the case of magnetic particles (see Chap. 9). In particular, structure data, obtained by wide angle X-ray scattering or absorption techniques are augmented with morphological information from other sources such as electron or near-field microscopies. The combination of several techniques leads to a complete description of nanoalloy assemblies provided that the objects and their organization remain the same in the different investigations. Indeed, in most cases, these techniques are used *ex situ* in different set-up, the best strategy is to investigate the same sample. Many works [117–120] search to compare and to validate same information (morphological or structural) through several techniques (TEM and SAXS, AFM and GISAXS or HRTEM and WAXS-XRD, etc.). However, good agreement between the results from different methods is not always encountered or presented in the literature, especially in nanoscale objects with multi-element composition.

For in situ or real time investigation, probing simultaneously with complementary techniques is requested but is possible only in a limited number of technique combinations, like in small and wide angle scattering techniques [17, 102, 121, 122]. This combination can be easily performed in the same experimental set-up with similar conditions, in transmission geometry as well as in grazing incidence one. Furthermore, in situ investigations which combine several techniques give new opportunities to follow structure transitions at different scales.

For example, structural and order/disorder phase transitions in bimetallic FePt or CoPt [63] nanoparticles induced by annealing have been investigated combining in situ and in real time GISAXS and GIWAXS. Figure 15a shows a schematic view of the experiment geometry. GISAXS measurements provide morphological features of nanoparticles as a function of size, shape and correlation distance

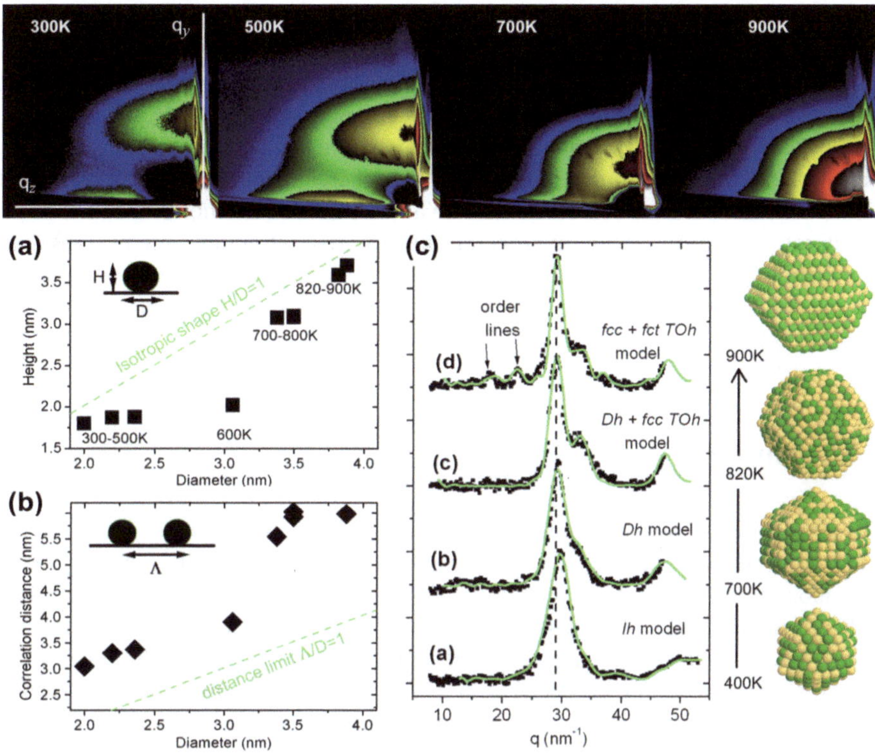

Fig. 23 GISAXS patterns of annealing CoPt sample at selected temperatures (in the *top*) and correlation between the diameter D and **a** the height H or **b** the average interparticle distance during the temperature evolution (extracted from GISAXS experimental cross sections). **c** Selected experimental and calculated GIWAXS spectra using Monte Carlo relaxed CoPt clusters at increasing annealing temperature. The insets show the snapshots of different simulated clusters (Co and Pt atoms are represented as light and dark spheres respectively). A *vertical dashed line* corresponding to the fcc (111) line is given for comparison. From Ref. [63], copyright (2010) by The American Physical Society

between particles, while GIWAXS allows the determination of the atomic structure. In the former study, the sample were fabricated ex situ forming protected (C/FePt) particles and the scattering experiments were performed post-deposited, while in the latter case, the study were carried in the bare supported CoPt particles from the initial growth stage to the annealing stage using the same UHV setup allowing molecular beam deposition at French ESRF BM32 beamline

The results (Fig. 23) reveal several structural transitions (from Ih to TOh models) during annealing at increasing temperature, yielding chemically disordered clusters at low temperature, followed by an $L1_0$ ordering at higher temperature and larger size [63]. Futhermore, a coalescence process of as-grown icosahedral particles (2 nm) induced by annealing was detected at low temperature (<300°C), higher temperatures (>400°C) are necessary to the formation of decahedral structure by atomic rearrangement, followed by a transition to the fcc morphology and finally, a chemical ordering around 900 K at fixed particle size. The particle mobility and coarsening is revealed by GISAXS measurements (Fig. 23a, b) through the interparticle distance and size evolution, which explain the coalescence-induced structural transition from Ih to Dh and, then, in the transition towards the equilibrium atomic fcc structure, has been clearly identified. The combined information from GISAXS and GIWAXS shows that no direct structural transition from non-crystalline to $L1_0$ structure seems possible. The combination of these techniques provide a straightforward tool to evaluate the particle and atom mobility in growth or annealing processes as well as chemical reactions [110].

5 Conclusion

The experimental investigations of morphology (size, shape and spatial organization) and structure (order, disorder, crystalline or non-crystalline arrangement) of nanoalloys by dedicated techniques is of primary importance, especially in order to understand and to control the nanoalloy formation or evolution mechanisms. In this chapter, we have focussed on the X-ray scattering techniques which can be performed ex- or in situ as well as for supported or embedded nanoalloys. While X-ray absorption methods are well known and ab initio calculations allow the simulation and the analysis of experimental data from many years, the X-ray scattering techniques applied to nanoalloys recently know a huge development. In particular, the recent interest for these techniques comes from the use of spatially or chemically-selective methods, like grazing incidence geometry or anomalous effect, as well as the development of theoretical approaches allowing scattering patterns simulations and consequently the fine interpretation of experimental data. Furthermore, in situ investigations which combine several techniques give new opportunities to follow structure transitions at different scales.

Acknowledgments We would like to thank C. Andreazza-Vignolle, J. Penuelas, N. Bouet, H. Khelfane, C. Mottet, H. Tolentino, M. De Santis, O. Lyon, A. Ramos, R. Felici, for their help during measurement and/or analysis of wide and small angle scattering and Y. Garreau, D. Babonneau, D. Thiaudière, R. Lazarri, G. Renaud, O. Spalla, P. Lecante, S. Billinge for discussion or contribution about scattering theory or measurement examples.

References

1. Ino, S.: Epitaxial growth of metals on rocksalt faces cleaved in vaccum. 2. Orientation ans structure of gold particles formed in ultra-high vacuum. J. Phys. Soc. Jpn. **21**(2), 346 (1969)
2. Baletto, F., Ferrando, R.: Structural properties of nanoclusters: energetic, thermodynamic, and kinetic effects. Rev. Mod. Phys. **77**(1), 371–423 (2005)

3. Ferrando, R., Jellinek, J., Johnston, R.L.: Nanoalloys: from theory to applications of alloy clusters and nanoparticles. Chem. Rev. **108**(3), 845–910 (2008)
4. Rodriguez-Lopez, J.L., Montejano-Carrizales, J.M., Pal, U., Sanchez-Ramirez, J.F., Troiani, H.E., Garcia, D., Miki-Yoshida, M., Jose-Yacaman, M.: Surface reconstruction and decahedral structure of bimetallic nanoparticles. Phys. Rev. Lett. **92**(19), 196102 (2004)
5. Wang, R.M., Dmitrieva, O., Farle, M., Dumpich, G., Ye, H.Q., Poppa, H., Kilaas, R., Kisielowski, C.: Layer resolved structural relaxation at the surface of magnetic FePt icosahedral nanoparticles. Phys. Rev. Lett. **100**(1), 017205 (2008)
6. Anders, S., Toney, M.F., Thomson, T., Thiele, J.U., Terris, B.D., Sun, S.H., Murray, C.B.: X-ray studies of magnetic nanoparticle assemblies. J. Appl. Phys. **93**(10), 7343–7345 (2003)
7. de la Presa, P., Rueda, T., Hernando, A., Ramallo-Lopez, J.M., Giovanetti, L.J., Requejo, F.G.: Spontaneous oxidation of disordered fcc FePt nanoparticles. J. Appl. Phys. **103**(10), 103909 (2008)
8. Favre, L., Dupuis, V., Bernstein, E., Melinon, P., Perez, A., Stanescu, S., Epicier, T., Simon, J.P., Babonneau, D., Tonnerre, J.M., Hodeau, J.L.: Structural and magnetic properties of CoPt mixed clusters. Phys. Rev. B **74**(1), 014439 (2006)
9. Ouerghi, A., Penuelas, J., Andreazza-Vignolle, C., Andreazza, P., Bouet, N., Estrade-Szwarckopf, H.: Chemical and structural aspects of CoPt silicide nanostructures grown on Si(100). J. Appl. Phys. **100**(12), 124310 (2006)
10. Miyazaki, T., Kitakami, O., Okamoto, S., Shimada, Y., Akase, Z., Murakami, Y., Shindo, D., Takahashi, Y.K., Hono, K.: Size effect on the ordering of L1(0) FePt nanoparticles. Phys. Rev. B **72**(14), 144419 (2005)
11. Penuelas, J., Andreazza-Vignolle, C., Andreazza, P., Ouerghi, A., Bouet, N.: Temperature effect on the ordering and morphology of CoPt nanoparticles. Surf. Sci. **602**(2), 545–551 (2008)
12. Renaud, G., Lazzari, R., Leroy, F.: Probing surface and interface morphology with grazing incidence small angle X-ray scattering. Surf. Sci. Rep. **64**(8), 255–380 (2009)
13. Simon, J.-P., Babonneau, D., Drouet, M., Lyon, O.: Anomalous grazing incidence small-angle X-ray scattering of carbon-encapsulated Au and Cu clusters. J. Appl. Crystallogr. **42**, 312–322 (2009)
14. Billinge, S.J.L., Levin, I.: The problem with determining atomic structure at the nanoscale. Science **316**(5824), 561–565 (2007)
15. Lazzari, R., Leroy, F., Renaud, G.: Grazing-incidence small-angle x-ray scattering from dense packing of islands on surfaces: development of distorted wave born approximation and correlation between particle sizes and spacing. Phys. Rev. B **76**(12), 125411 (2007)
16. Naudon, A., Babonneau, D., Thiaudiere, D., Lequien, S.: Grazing-incidence small-angle X-ray scattering applied to the characterization of aggregates in surface regions. Phys. B **283**(1–3), 69–74 (2000)
17. Penuelas, J., Andreazza, P., Andreazza-Vignolle, C., Tolentino, H.C.N., De Santis, M., Mottet, C.: Controlling structure and morphology of CoPt nanoparticles through dynamical or static coalescence effects. Phys. Rev. Lett. **100**(11), 115502 (2008)
18. Alayoglu, S., Zavalij, P., Eichhorn, B., Wang, Q., Frenkel, A.I., Chupas, P.: Structural and architectural evaluation of bimetallic nanoparticles: a case study of Pt-Ru core-shell and alloy nanoparticles. ACS Nano **3**(10), 3127–3137 (2009)
19. Frenkel, A.: Solving the 3D structure of metal nanoparticles. Zeitschrift Fur Kristallographie **222**(11), 605–611 (2007)
20. Krayzman, V., Levin, I., Woicik, J.C., Yoder, D., Fischer, D.A.: Effects of local atomic order on the pre-edge structure in the TiK X-ray absorption spectra of perovskite CaTi1-xZrxO3. Phys. Rev. B **74**(22), 224104 (2006)
21. Lyon, O., Guillon, I., Servant, C.: Determination of displacements from nanoprecipitates in a Cu-Ni-Fe single crystal using anomalous small-angle and large-angle X-ray scattering. J. Appl. Crystallogr. **34**, 484–492 (2001)
22. Bazin, D., Kovacs, I., Lynch, J., Guczi, L.: Ru-Co/NaY bimetallic catalysts: In situ EXAFS study at CoK- and RuK-absorption edges. Appl Catal a-Gen **242**(1), 179–186 (2003)

23. Hwang, B.J., Sarma, L.S., Chen, J.M., Chen, C.H., Shih, S.C., Wang, G.R., Liu, D.G., Lee, J.F., Tang, M.T.: Structural models and atomic distribution of bimetallic nanoparticles as investigated by X-absorption spectroscopy. J. Am. Chem. Soc. **127**(31), 11140–11145 (2005)
24. Iwasawa, Y.: Xas for Catalysts and Surfaces. World Scientific, Singapore (1996)
25. Nashner, M.S., Frenkel, A.I., Somerville, D., Hills, C.W., Shapley, J.R., Nuzzo, R.G.: Core shell inversion during nucleation and growth of bimetallic Pt/Ru nanoparticles. J. Am. Chem. Soc. **120**(32), 8093–8101 (1998)
26. Dassenoy, F., Casanove, M.J., Lecante, P., Pan, C., Philippot, K., Amiens, C., Chaudret, B.: Size and composition effects in polymer-protected ultrafine bimetallic PtxRu1-x ($0 < x < 1$) particles. Phys. Rev. B **63**(23), 235407 (2001)
27. Guinier, A.: X-Ray Diffraction in Crystals, Imperfect Crystals, and Amorphous Bodies. Freeman, San Francisco (1963)
28. Guinier, A., Fournet, G.: Small-angle Scattering of X-ray. Wiley, New York (1955)
29. Glatter, O., Kratky, O.: Small Angle X-ray Scattering. Academic Press, London (1982)
30. Pikunic, J., Clinard, C., Cohaut, N., Gubbins, K.E., Guet, J.M., Pellenq, R.J.M., Rannou, I., Rouzaud, J.N.: Structural modeling of porous carbons: constrained reverse Monte Carlo method. Langmuir **19**(20), 8565–8582 (2003)
31. Jenkins, R., Snyder, R.L.: Introduction to X-ray Powder Diffraction. Wiley Interscience, New York (1996)
32. Cowley, J.M.: Diffraction Physics. North-Holland Physics Publishing, Amsterdam (1981)
33. Shmueli, U.: International Tables for Crystallography. Wiley, New York (2006)
34. Leroy, F., Renaud, G., Letoublon, A., Rohart, S., Girard, Y., Repain, V., Rousset, S., Coati, A., Garreau, Y.: Kink ordering and organized growth of Co clusters on a stepped Au(111) surface: a combined grazing-incidence X-ray scattering and STM study. Phys Rev B **77**(4), 045430 (2008)
35. Robach, O., Garreau, Y., Aid, K., Veron-Jolliot, M.B.: Corrections for surface X-ray diffraction measurements using the Z-axis geometry: finite size effects in direct and reciprocal space. J. Appl. Crystallogr. **33**, 1006–1018 (2000)
36. Reinhard, D., Hall, B.D., Ugarte, D., Monot, R.: Size-independent fcc-to-icosahedral structural transition in unsupported silver clusters: an electron diffraction study of clusters produced by inert-gas aggregation. Phys Rev B **55**(12), 7868–7881 (1997)
37. Rodriguez, A., Amiens, C., Chaudret, B., Casanove, M.J., Lecante, P., Bradley, J.S.: Synthesis and isolation of cuboctahedral and icosahedral platinum nanoparticles. Ligand-depend. struct. Chem. Mater. **8**(8), 1978–1986 (1996)
38. Vogel, W., Rosner, B., Tesche, B.: Structural investigations of Au_{55} organometallic complexes by X-ray-powder diffraction and transmission electron microscopy. J. Phys. Chem. **97**(45), 11611–11616 (1993)
39. Hall, B.D., Zanchet, D., Ugarte, D.: Estimating nanoparticle size from diffraction measurements. J. Appl. Crystallogr. **33**, 1335–1341 (2000)
40. Andreazza, P., Andreazza-Vignolle, C., Rozenbaum, J.P., Thomann, A.L., Brault, P.: Nucleation and initial growth of platinum islands by plasma sputter deposition. Surf. Coat. Technol. **151**, 122–127 (2002)
41. Koga, K., Takeo, H., Ikeda, T., Ohshima, K.: In situ grazing-incidence X-ray-diffraction and electron-microscopic studies of small gold clusters. Phys. Rev. B **57**(7), 4053–4062 (1998)
42. Kaszkur, Z.: Nanopowder diffraction analysis beyond the Bragg law applied to palladium. J. Appl. Crystallogr. **33**, 87–94 (2000)
43. Penuelas, J., Andreazza, P., Andreazza-Vignolle, C., Mottet, C., De Santis, M., Tolentino, H.C.N.: Real-time icosahedral to fcc structure transition during CoPt nanoparticles formation. Eur. Phys. J-Spec. Top. **167**, 19–25 (2009)
44. Pundt, A., Dornheim, M., Guerdane, M., Teichler, H., Ehrenberg, H., Reetz, M.T., Jisrawi, N.M.: Evidence for a cubic-to-icosahedral transition of quasi-free Pdâ€ H-clusters controlled by the hydrogen content. Eur. Phys. J. D **19**(3), 333–337 (2002)
45. Furukawa, K.: The radial distribution curves of liquids by diffraction methods. Rep. Prog. Phys. **25**, 395 (1962)

46. Dinnebier, R.E., billinge, S.J.L.: Powder Diffraction: Theory and Practice. Royal Society of Chemistry. Royal Society of Chemistry, London (2008)
47. Farrow, C.L., Billinge, S.J.L.: Relationship between the atomic pair distribution function and small-angle scattering: implications for modeling of nanoparticles. Acta Crystallogr. Sect. A **65**, 232–239 (2009)
48. Fromen, M.C., Lecante, P., Casanove, M.J., Guillemaud, P.B., Zitoun, D., Amiens, C., Chaudret, B., Respaud, M., Benfield, R.E.: Structural study of bimetallic CoxRh1-x nanoparticles: size and composition effects. Phys.Rev. B **69**(23), 235416 (2004)
49. Ruan, C.-Y., Murooka, Y., Raman, R.K., Murdick, R.A.: Dynamics of size-selected gold nanoparticles studied by ultrafast electron nanocrystallography. Nano Lett. **7**(5), 1290–1296 (2007)
50. Gilbert, B.: Finite size effects on the real-space pair distribution function of nanoparticles. J. Appl. Crystallogr. **41**, 554–562 (2008)
51. Petkov, V., Peng, Y., Williams, G., Huang, B.H., Tomalia, D., Ren, Y.: Structure of gold nanoparticles suspended in water studied by X-ray diffraction and computer simulations. Phys. Rev. B **72**(19), 195402 (2005)
52. Longo, A., Martorana, A.: Distorted f.c.c. arrangement of gold nanoclusters: a model of spherical particles with microstrains and stacking faults. J. Appl. Crystallogr. **41**, 446–455 (2008)
53. Rossi, G., Ferrando, R., Mottet, C.: Structure and chemical ordering in CoPt nanoalloys. Faraday Discuss. **138**, 193–210 (2008)
54. Kim, K.-E., Lee, M.-K., Sung, Y.-M., Kim, T.G.: Enhanced L1(0) chemical ordering and FePt/Fe3O4 core/shell structure formation in Zn-doped FePt nanoparticles. Appl. Phys. Lett. **90**(17), 173117 (2007)
55. Cervellino, A., Giannini, C., Guagliardi, A.: Determination of nanoparticle structure type, size and strain distribution from X-ray data for monatomic f.c.c.-derived non-crystallographic nanoclusters. J. Appl. Crystallogr. **36**, 1148–1158 (2003)
56. Hall, B.D.: Debye function analysis of structure in diffraction from nanometer-sized particles. J. Appl. Phys. **87**(4), 1666–1675 (2000)
57. Sun, S.H., Murray, C.B., Weller, D., Folks, L., Moser, A.: Monodisperse FePt nanoparticles and ferromagnetic FePt nanocrystal superlattices. Science **287**(5460), 1989–1992 (2000)
58. Leroux, C., Loiseau, A., Broddin, D., Vantendeloo, G.: Electron microscopy study of coherent two phase mixtures $L1_{0L10} + L1_2$ on CoPt alloys. Philos. Mag. B-Phys. Condens. Matter. Stat. Mech. Electron. Opt. Magn. Prop. **64**(1), 57–82 (1991)
59. Kuznetsov, V., Tsai, K., Kaminskii, P., Turkebaev, T.: Calculation of the thermodynamic properties of noble-metal alloys in the model electron-density functional method. Russ. Phys. J. **40**(5), 411–419 (1997)
60. Grange, W., Ulhaq-Bouillet, C., Maret, M., Thibault, J.: Chemical long-range ordering in a CoPt alloy film grown by molecular beam epitaxy. Acta Mater. **49**(8), 1439–1444 (2001)
61. Whang, S.H., Feng, Q., Gao, Y.Q.: Ordering, deformation and microstructure in L1(0) type FePt. Acta Mater. **46**(18), 6485–6495 (1998)
62. Cb, Rong, Li, D., Nandwana, V., Poudyal, N., Ding, Y., Wang, Z.L., Zeng, H., Liu, J.P.: Size-dependent chemical and magnetic ordering in L1(0)-FePt nanoparticles. Adv. Mater. **18**(22), 2984 (2006)
63. Andreazza, P., Mottet, C., Andreazza-Vignolle, C., Penuelas, J., Tolentino, H.C.N., De Santis, M., Felici, R., Bouet, N.: Probing nanoscale structural and order/disorder phase transitions of supported Co-Pt clusters under annealing. Phys. Rev. B **82**(15), 155453 (2010)
64. Zeng, H., Li, J., Wang, Z.L., Liu, J.P., Sun, S.H.: Bimagnetic core/shell FePt/Fe3O4 nanoparticles. Nano Lett. **4**(1), 187–190 (2004)
65. Kang, S., Harrell, J.W., Nikles, D.E.: Reduction of the fcc to L1(0) ordering temperature for self-assembled FePt nanoparticles containing Ag. Nano Lett. **2**(10), 1033–1036 (2002)
66. Alayoglu, S., Nilekar, A.U., Mavrikakis, M., Eichhorn, B.: Ru-Pt core-shell nanoparticles for preferential oxidation of carbon monoxide in hydrogen. Nat. Mater. **7**(4), 333–338 (2008)

67. Kaszkur, Z.: Direct observation of chemisorption induced changes in concentration profile in Pd-Au alloy nanosystems via in situ X-ray powder diffraction. Phys. Chem. Chem. Phys. **6**(1), 193–199 (2004)
68. Simon, J.-P.: Contribution of synchrotron radiation to small-angle X-ray scattering studies in hard condensed matter. J. Appl. Crystallogr. **40**, S1–S9 (2007)
69. Zemb, T., Lindner, P.: In neutrons, In neutrons, X-ray and Light: Scattering Methods Applied to Soft Condensed Matter. Elsevier, New York (2002)
70. Babonneau, D.: FitGISAXS: software package for modelling and analysis of GISAXS data using IGOR Pro. J. Appl. Crystallogr. **43**, 929–936 (2010)
71. Feigin, L.A., Svergun, D.I.: Structure Analysis by Small-angle X-ray and Neutron Scattering. Plenum Press, New York (1987)
72. Lazzari, R.: IsGISAXS: a program for grazing-incidence small-angle X-ray scattering analysis of supported islands. J. Appl. Crystallogr. **35**, 406–421 (2002)
73. Martelli, S., Di Nunzio, P.E.: Particle size distribution of nanospheres by Monte Carlo fitting of small angle X-ray scattering curves. Part. Part. Syst. Charact. **19**(4), 247–255 (2002)
74. Pedersen, J.S.: Analysis of small-angle scattering data from colloids and polymer solutions: modeling and least-squares fitting. Adv. Colloid Interface Sci. **70**, 171–210 (1997)
75. Mykhaylyk, O.O., Ryan, A.J., Tzokova, N., Williams, N.: The application of distance distribution functions to structural analysis of core-shell particles. J. Appl. Crystallogr. **40**, S506–S511 (2007)
76. Ottewill, R.H., Hanley, H.J.M., Rennie, A.R., Straty, G.C.: Small-angle neutron—scattering studies on binary-mixture of charged particles. Rep. Prog. Phys. **11**(10), 3757–3765 (1995)
77. Wagner, J.: Small-angle scattering from spherical core-shell particles: an analytical scattering function for particles with Schulz-Flory size distribution. J. Appl. Crystallogr. **37**, 750–756 (2004)
78. Lisiecki, I., Albouy, P.A., Pileni, M.P.: Face-centered cubic "upracrystals" of cobalt nanocrystals. Adv. Mater. **15**(9), 712–716 (2003)
79. Robach, O., Quiros, C., Valvidares, S.M., Walker, C.J., Ferrer, S.: StructureStructure and Pt magnetism of FePt nanoparticles investigated with X-ray diffraction. J. Magn. Magn. Mater. **264**(2–3), 202–208 (2003)
80. Abecassis, B., Testard, F., Spalla, O.: Gold nanoparticle superlattice crystallization probed in situ. Phys. Rev. Lett. **100**(11), 115504 (2008)
81. Blasing, J., Strassburger, G., Eberbeck, D.: Determination of particles—size distribution and correlation of particles in ferrofluids under the influence of magnetic fields. Phys Status Solidi a-Appl Res **146**(2), 595–602 (1994)
82. Fukasawa, T., Sato, T.: Versatile application of indirect Fourier transformation to structure factor analysis: from X ray diffraction of molecular liquids to small angle scattering of protein solutions. Phys. Chem. Chem. Phys.**13**(8), 3187–3196 (2011)
83. Hosemann, R., Bagchi, S.N.: Direct Analysis of Diffraction by Matter. North Holland, Amsterdam (1962)
84. Leroy, F., Lazzari, R., Renaud, G.: Effects of near-neighbor correlations on the diffuse scattering from a one-dimensional paracrystal. Acta Crystallogr. Sect. A **60**, 565–581 (2004)
85. Levitz, P.: Off-lattice reconstruction of porous media: critical evaluation, geometrical confinement and molecular transport. Adv. Colloid Interface Sci. **76**, 71–106 (1998)
86. Levitz, P., Tchoubar, D.: Disordered porous solids—from chord distribution to small-angle scattering. J. Phys. I **2**(6), 771–790 (1992)
87. Revenant, C., Renaud, G., Lazzari, R., Jupille, J.: GrowthGrowth of Ag on MgO(001) studied in situ by grazing incidence small angle X-ray scattering. Nucl. Instrum. Methods Phys. Res. Sect. B-Beam Interact. Mater. Atoms **246**(1), 112–117 (2006)
88. Daillant, J., Gibaud, A.: X-ray and Neutron Reflectivity: Principle and Applications. Lect Notes in Physics, Springer, Paris (1999)

89. Babonneau, D., Petroff, F., Maurice, J.L., Fettar, F., Vaures, A., Naudon, A.: Evidence for a self-organized growth in granular Co/Al(2)O(3) multilayers. Appl. Phys. Lett. **76**(20), 2892–2894 (2000)
90. Levine, J.R., Cohen, J.B., Chung, Y.W., Georgopoulos, P.: Grazing-incidence small-angle X-ray scattering: new tool for studying thin film growth. J. Appl. Crystallogr. **22**(6), 528–532 (1989)
91. Naudon, A., Thiaudiere, D.: Grazing-incidence small-angle scattering. Morphology of deposited clusters and nanostructure of thin films. J. Appl. Crystallogr. **30**(2), 822–827 (1997)
92. Revenant, C., Leroy, F., Lazzari, R., Renaud, G., Henry, C.R.: Quantitative analysis of grazing incidence small-angle X-ray scattering: Pd/MgO(001) growth. Phys Rev B **69**(3), 035411 (2004)
93. Maurizio, C., Longo, A., Martorana, A., Cattaruzza, E., D''capito, F., Gonella, F., de Julian, C., Mattei, G., Mazzoldi, P., Padovani, S., Boesecke, P.: Grazing-incidence small-angle X-ray scattering and X-ray diffraction from magnetic clusters obtained by Co plus Ni sequential ion implantation in silica. J. Appl. Crystallogr. **36**, 732–735 (2003)
94. Flege, J.I., Schmidt, T., Aleksandrovic, V., Alexe, G., Clausen, T., Gehl, B., Kornowski, A., Bernstorff, S., Weller, H., Falta, J.: Grazing-incidence small-angle X-ray scattering investigation of spin-coated COPt3 nanoparticle films. Nucl. Instrum. Methods Phys. Res. Sect. B-Beam Interact. Mater. Atoms **246**(1), 25–29 (2006)
95. Aleksandrovic, V., Greshnykh, D., Randjelovic, I., Fromsdorf, A., Kornowski, A., Roth, S.V., Klinke, C., Weller, H.: Preparation and electrical properties of cobalt-platinum nanoparticle monolayers deposited by the Langmuir-Blodgett technique. ACS Nano **2**(6), 1123–1130 (2008)
96. Riekel, C., Burghammer, M., Davies, R.: Progress in micro- and nano-diffraction at the ESRF ID13 beamline. Synchrotron radiation in polymer science, vol. 14, IOP Conference Series-materials Science and Engineering (2010)
97. Schroer, C.G., Lengeler, B.: Focusing hard X rays to nanometer dimensions by adiabatically focusing lenses. Phys. Rev. Lett. **94**(5), 054802 (2005)
98. Goerigk, G., Haubold, H.G., Lyon, O., Simon, J.P.: Anomalous small-angle X-ray scattering in materials science. J. Appl. Crystallogr. **36**(1), 425–429 (2003)
99. Lyon, O., Simon, J.P.: Anomalous samll-angle-X-ray scattering determination of the partial structure factors and kinetic-study of unmixed CiNiFe alloys. Phys. Rev. B **35**(10), 5164–5174 (1987)
100. Andreazza, P., Khelfane, H., Lyon, O., Andreazza-Vignolle, C., Ramos, A., Samah, M.: Trends in anomalous small-angle X-ray scattering in grazing incidence for supported nanoalloyed and core-shell metallic nanoparticles. Eur. Phys. J.S.T. **208**, 237–250 (2012)
101. Lee, B., Seifert, S., Riley, S.J., Tikhonov, G., Tomczyk, N.A., Vajda, S., Winans, R.E.: Anomalous grazing incidence small-angle X-ray scattering studies of platinum nanoparticles formed by cluster deposition. J. Chem. Phys. **123**(7), 074701 (2005)
102. Leroy, F., Renaud, G., Letoublon, A., Lazzari, R.: Growth of Co on Au(111) studied by multiwavelength anomalous grazing-incidence small-angle X-ray scattering: from ordered nanostructures to percolated thin films and nanopillars. Phys. Rev. B **77**(23), 235429 (2008)
103. Huang, T.W., Yu, K.L., Liao, Y.F., Lee, C.H.: Anomalous grazing-incidence small-angle X-ray scattering investigation on the surface morphology of an FePt magnetic nanoparticle monolayer on functional modulated substrates. J. Appl. Crystallogr. **40**, S480–S484 (2007)
104. Haug, J., Kruth, H., Dubiel, M., Hofmeister, H., Haas, S., Tatchev, D., Hoell, A.: ASAXS study on the formation of core-shell Ag/Au nanoparticles in glass. Nanotechnology **20**(50), 505705 (2009)
105. Creagh, D., Hubbell, J.H., Creagh, D., McAuley, W.J.: International Tables for X-ray Crystallography, vol. C. Kluwer Academic, Dordrecht (1992)
106. Yu, C., Koh, S., Leisch, J.E., Toney, M.F., Strasser, P.: Size and composition distribution dynamics of alloy nanoparticle electrocatalysts probed by anomalous small angle X-ray scattering (ASAXS). Faraday Discuss. **140**, 283–296 (2008)

107. Abecassis, B., Testard, F., Spalla, O., Barboux, P.: Probing in situ the nucleation and growth of gold nanoparticles by small-angle x-ray scattering. Nano Lett. **7**(6), 1723–1727 (2007)
108. Renaud, G., Lazzari, R., Revenant, C., Barbier, A., Noblet, M., Ulrich, O., Leroy, F., Jupille, J., Borensztein, Y., Henry, C.R., Deville, J.P., Scheurer, F., Mane-Mane, J., Fruchart, O.: Real-time monitoring of growing nanoparticles. Science **300**(5624), 1416–1419 (2003)
109. Nguyen, H.L., Howard, L.E.M., Stinton, G.W., Giblin, S.R., Tanner, B.K., Terry, I., Hughes, A.K., Ross, I.M., Serres, A., Evans, J.S.O.: Synthesis of size-controlled fcc and fct FePt nanoparticles. Chem. Mater. **18**(26), 6414–6424 (2006)
110. Saint-Lager, M.C., Bailly, A., Mantilla, M., Garaudee, S., Lazzari, R., Dolle, P., Robach, O., Jupille, J., Laoufi, I., Taunier, P.: Looking by grazing incidence small angle x-ray scattering at gold nanoparticles supported on rutile TiO_2(110) during CO oxidation. Gold Bull. **41**(2), 159–166 (2008)
111. Henkel, A., Schubert, O., Plech, A., Sonnichsen, C.: Growth kinetic of a rod-shaped metal nanocrystal. J. Phys. Chem. C **113**(24), 10390–10394 (2009)
112. Jiang, Z., Lin, X.M., Sprung, M., Narayanan, S., Wang, J.: Capturing the crystalline phase of two-dimensional nanocrystal superlattices in action. Nano Lett. **10**(3), 799–803 (2010)
113. Lazzari, R., Renaud, G., Jupille, J., Leroy, F.: Self-similarity during growth of the Au/TiO2(110) model catalyst as seen by the scattering of X-rays at grazing-angle incidence. Phys. Rev. B **76**(12), 125412 (2007)
114. Klemmer, T.J., Liu, C., Shukla, N., Wu, X.W., Weller, D., Tanase, M., Laughlin, D.E., Soffa, W.A.: Combined reactions associated with L1(0) ordering. J. Magn. Magn. Mater. **266**(1–2), 79–87 (2003)
115. Sra, A.K., Schaak, R.E.: Synthesis of atomically ordered AuCu and $AuCU_3$ nanocrystals from bimetallic nanoparticle precursors. J. Am. Chem. Soc. **126**(21), 6667–6672 (2004)
116. Lantiat, D., Babonneau, D., Camelio, S., Pailloux, F., Denanot, M.F.: Evidence for capping-layer effects on the morphology and plasmon excitation of Ag nanoparticles. J. Appl. Phys. **102**(11), 113518 (2007)
117. Borchert, H., Shevehenko, E.V., Robert, A., Mekis, I., Kornowski, A., Grubel, G., Weller, H.: Determination of nanocrystal sizes: a comparison of TEM, SAXS, and XRD studies of highly monodisperse COPt3 particles. Langmuir **21**(5), 1931–1936 (2005)
118. Gontard, L.C., Dunin-Borkowski, R.E., Ozkaya, D., Hyde, T., Midgley, P.A., Ash, P.: Crystal size and shape analysis of Pt nanoparticles in two and three dimensions. In: EMAG-NANO 2005: imaging, analysis and fabrication on the nanoscale, Journal of Physics Conference Series, vol. 26, pp. 367–370 (2006)
119. Jensen, H., Pedersen, J.H., Jorgensen, J.E., Pedersen, J.S., Joensen, K.D., Iversen, S.B., Sogaard, E.G.: Determination of size distributions in nanosized powders by TEM, XRD, and SAXS. J. Exp. Nanosci. **1**(3), 355–373 (2006)
120. Thomann, A.L., Rozenbaum, J.P., Brault, P., Andreazza-Vignolle, C., Andreazza, P.: Pd nanoclusters grown by plasma sputtering deposition on amorphous substrates. Appl. Surf. Sci. **158**(1–2), 172–183 (2000)
121. Olander, J., Lazzari, R., Jupille, J., Mangili, B., Goniakowski, J., Renaud, G.: Size- and temperature-dependent epitaxy for a strong film-substrate mismatch: the case of Pt/MgO(001). Phys. Rev. B **76**(7), 075409 (2007)
122. Torelli, P., Soares, E.A., Renaud, G., Gragnaniello, L., Valeri, S., Guo, X.X., Luches, P.: Self-organized growth of Ni nanoparticles on a cobalt-oxide thin film induced by a buried misfit dislocation network. Phys. Rev. B **77**(8), 081409 (2008)

Transmission Electron Microscopy: A Multifunctional Tool for the Atomic-scale Characterization of Nanoalloys

Damien Alloyeau

Abstract Nanoalloys are attracting increasing attention because of the immense technological potential that arises from combination of size effects with composition effects. However, the design of nanosystems with new and tunable properties requires understanding the complex phenomena that influence the size, shape, composition, and atomic structure of multimetallic alloy cluster. In that regard, Transmission Electron Microscope (TEM) is one of the most complete characterization tools for studying nanoalloys. Here we will detail how TEM gives access to morphological, structural and chemical information on individually analyzed nanoparticles. We will describe the principle, advantages and limits of the different presented techniques. To finish, we will shed light on the outstanding performances of the recently developed aberration corrected microscopes that provide unprecedented opportunities to analyze dynamical processes at high resolution and atomic scale-chemistry in nanoalloys.

1 Introduction

Transmission electron microscopy is a technique designed for the observation and the physical and chemical characterization of matter developed in 1932 by Knoll and Ruska [1]. The latter received the Nobel Prize in Physics in 1986 for this invention. Since then, the race for technical developments has been going on relentlessly, enriching the functionalities of microscopes with many complementary

D. Alloyeau (✉)
Laboratoire Matériaux et Phénomènes Quantiques (UMR CNRS 7162),
Bâtiment Condorcet, Case courrier 7021, Université Paris Diderot—Paris 7,
75205, Paris cedex 13, France
e-mail: damien.alloyeau@univ-paris-diderot.fr

D. Alloyeau et al. (eds.), *Nanoalloys*, Engineering Materials,
DOI: 10.1007/978-1-4471-4014-6_4, © Springer-Verlag London 2012

characterization devices. Indeed, the exploitation of the many phenomena occurring simultaneously during the interaction between an electron beam and matter makes it possible to analyze bi- or three-dimensional morphology, crystal structure and the composition of a sample using the same instrument. Due to this polyvalence, the electron microscope is a tool perfectly suited to the characterization of nanoalloys, where, as already mentioned, the size, shape, composition and crystal structures are closely related to their physical and chemical properties.

In a general manner, transmission electron microscopy has played a crucial role in material science research. Consequently, the principles, performances and applications of the transmission electron microscope (TEM) have been deeply described for more than 40 years, in particular in the following famous references [2–9]. In this chapter, we will examine how the potential of the TEM can be used to obtain the quantitative information necessary to understand the properties of bimetallic nanoparticles. In this broad overview, we will put forward the idea that the analysis of individual nano-objects is a crucial aspect in studying nanoalloys, where particles from the same sample often exhibit different properties. Finally, we will describe the amazing performances of the new electron microscopes which, thanks to the development of aberration correctors for electromagnetic lenses and extremely coherent electron sources, open up new prospects for the characterization and understanding of nanoalloys.

2 Basic Principles

2.1 Electron–Matter Interactions

Electron-matter interactions are explained by the wave/corpuscle duality of electrons, proposed by de Broglie in 1924. In a microscope, electrons are accelerated to high energy (about several hundred keV) that prompts interactions with electron shells and the nuclei of irradiated matter (according to the corpuscular aspect). With an accelerating voltage of 200 kV, the speed of the electrons is 0.695 times the speed of light and their associated wavelength is 0.00251 nm. In this sub-nanometric wavelength range, diffraction phenomena occur on crystal lattice, which allow the observation of interference patterns (according to the undulatory aspect). It is important to note that electron-matter interaction is about 10^4 to 10^5 times higher than with X-rays or neutrons. This means that electrons are strongly absorbed by matter. This absorption depends on the speed of the electrons imposed by the acceleration voltage, and on the thickness and nature of the material they go through. This low capacity of penetration of primary electrons is actually one of the limiting factors of electron microscopy, requiring the preparation of sufficiently thin samples ($\leq$ 100 nm) for a signal to be transmitted [10]. However this is not a problem in practice for the study of nanoalloys, excepted in the case of clusters embedded in a bulk matrix. It is convenient to classify electron—matter interactions in two categories.

2.1.1 Elastic Interactions

During an elastic process, the electron has the same kinetic energy—hence the same wavelength—before and after the interaction. The incident electron wave, perpendicular to the wave plane, can be defined by its wave vector $\mathbf{k_0}$ ($|\mathbf{k_0}| = 1/\lambda$). It is influenced by the potential of the matter, that is, by its charge density (nucleus and electrons). The term used in this situation as with X-rays, is the scattering of the incident wave. If the material under study is a crystal (i.e. with a periodic atomic distribution), the elastically scattered electron waves interfere, with high intensity in certain directions: this is known as Bragg diffraction, whose principle has been described in the Chap. 3. Among the elastic electrons that have moved through the sample, a distinction is made between the electron beam that has been transmitted parallel to the incident beam (transmitted or unscattered beam), and scattered electron beams forming an angle 2θ relative to the incident beam—θ being the scattering angle.

2.1.2 Inelastic Interactions

Electrons can interact with matter by yielding a part of their energy, causing emission of electrons or electromagnetic radiations according to whether the interaction with the atoms of the material under study involves atomic nuclei, core electrons or valence electrons. We will not consider an exhaustive list of the various types of electron-matter interactions; but only those involved in the techniques described further on.

Energy Losses by Primary Electrons

The energy lost by a primary electron interacting with core electrons or valence electrons is a characteristic of the nature of the material. It can be measured on a microscope with a spectrometer dispersing electrons with a different angle according to their kinetic energy, using a so-called 'magnetic prism'. We will see that the spectroscopic examination of energy losses allows in particular a quantitative and spatially resolved analysis of the chemical composition and the chemical bonds of the material. The probability for a primary electron to undergo an inelastic interaction with the sample depends on its initial energy and on the composition and thickness of the sample.

Electromagnetic Radiation

Primary electrons can interact with core electrons by ejecting one of them. The hole thus formed is immediately filled by an electron from an outer shell. The difference in energy of the electron moving to another shell is transformed into an X-photon with an energy which is a characteristic of the excited atom. As in an X-ray tube, the X-ray radiations emitted by a sample consist of a continuous spectrum due to Bremsstrahlung processes, with additional sharp peaks at certain energies. As we will show in the section on EDX quantitative analysis (Sect. 4.1), the characterization of the emitted X-radiation, in terms of energy and intensity, can be used for the chemical analysis of materials.

Damage from Irradiation

Strong electron-matter interaction may induce various problems, in particular when observing nano-objects. The first is overheating of the sample by the electron beam. The variation in temperature depends on the energy of the primary electrons and the thermal conductivity of the material. Phase transformations can thus appear prematurely under the action of an overly condensed beam. Moreover, the focusing of the beam may induce carbon contamination, caused by the hydrocarbons that are present in the chamber of a TEM. The electron beam reacts with stray hydrocarbons in the beam's path to create hydrocarbon ions which then condense and form carbon-rich polymerized film on the area being irradiated, making the sample less transparent. It is not uncommon either to observe the apparition of defects in the crystals if the energy of the electrons is higher than the threshold of atom displacement [11]. Lastly, when observing nanoparticles for several minutes, it is common to see cluster moving which deeply influence the structural study of a single nanoparticle.

2.2 *Microscopes*

Transmission electron microscope relies on the undulatory nature of electrons, whose small wavelength allows for very high resolution. The three technical requirements to build such an apparatus are: an electron sources, suitable lenses, and plane detectors sensitive to electrons. The main components of a TEM are shown in Fig. 1. Transmission electron microscopes differ by their type of sources, their optics and their auxiliary equipments, but they share the same principle of operation.

2.2.1 Electron Sources

The electron beam is produced by an electron gun whose main features are its energy dispersion and its brightness (density of current per unit of solid angle and by unit of energy). Before the 1970s, these guns were only based on the thermionic emission of a filament or a single-crystal tip, heated by Joule effect. The main problems of this kind of gun are the increase in Coulomb interactions between electrons, called the Boersch effect (in which some electrons speed up, and others slow down), that aggravates as emission increases. The energy dispersion of this

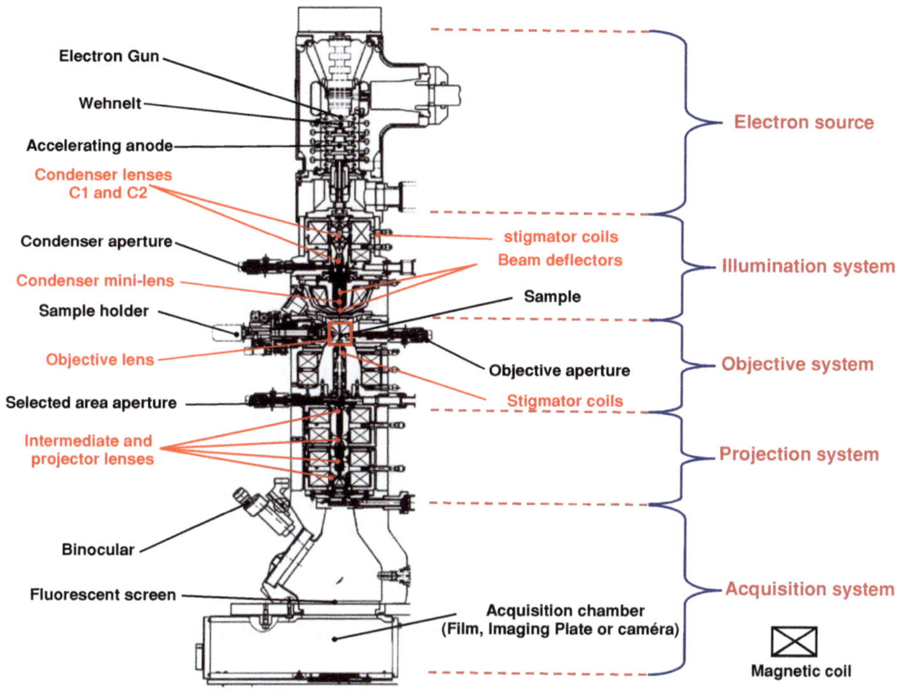

Fig. 1 Schematic representation of a column in a transmission electron microscope (JEOL 2100F)

type of electron gun ($\sim$1 eV) degrades imaging because of chromatic aberrations of the lenses, and bars energy loss spectroscopy studies on nano-objects. In the 1990s, field emission guns (FEGs), where the source is a small needle-shaped tungsten crystal, became a standard in TEM. Electrons are emitted via an electric field applied between the cathode and an extraction anode located right under the tip. This is the principle of field emission, called tunnel effect. The intensity of the electric field depends on the electron work function for the material making up the tip. This is why in Schottki-FEGs the emitting tip is covered with zirconia (ZrO_2), which has a lower work function than tungsten, allowing for higher electron emission. In guns using field emission, an electrostatic lens located under the extraction anode performs a cross-over (virtual source) whose size (50 nm) is several hundred times less than in a thermionic sources. The geometry of this virtual source makes it possible to increase the brightness to afford better visibility of the signals emitted by the sample. The energy dispersion of this kind of gun is 0.7 eV. Moreover, in FEG, the probe diameter can be as small as 0.2 nm, which is out of reach of traditional guns; this is quite useful to study objects a few nanometers in size. Lastly, the developments of cold FEGs or monochromators reduced the energy dispersion down to 0.2 eV, while maintaining the extraordinary spatial resolution of the microscope [12–15]. We will see why such a level of performance revolutionizes the atomic scale imaging and chemical analysis of nanoalloys. Whatever the type of gun used, the electrons are then accelerated by a series of cascaded anodes to reach their final energy ranging from 80 to 1,000 kV depending on the microscope.

2.2.2 The Optical System of the Column

The electromagnetic lens, designed by Hans Buch in 1923, is, along with the theoretical discovery made by de Broglie, at the root of the development of the TEM. These lenses, made up of a copper coil, create a magnetic field between two pole pieces. This field, modulated through the intensity of the current going through the coils, directly affects the trajectory of the electrons by changing the focusing of the beam on the optical axis of the microscope. Thanks to these lenses with variable focal distances, the laws of geometrical optics also apply in electron optics, hence the obvious analogies between electron and light microscopes. Just like a glass lens, electromagnetic lenses have an image plane where the image of the object is formed and a back focal plane where all the beams parallel with the optical axis before the lens converge at the focal point. In the case of an electromagnetic lens, the focal distance (f) can be related to the magnetic field (B) by the following equation:

$$\frac{1}{f} = \frac{e^2}{4mv^2} \int_{\mathrm{L1}}^{\mathrm{L2}} B^2(z)\mathrm{d}z \tag{1}$$

where v is the speed of the primary electrons, m their mass, e their charge and z the direction defined by the optical axis. The integration terminals L1 and L2 are defined by the z-position of the pole piece. The magnetic field is proportional to the current going through the coil and consequently, as evidenced in Eq. (1), the stronger the current of a lens, the smaller the focal distance.

The optical system of a microscope consists of three groups of lenses

(i) The first group is the *illumination system* which forms the image of the source on the sample. It includes two condenser lenses C1 and C2, a condenser mini-lens CM and the upper objective lens (Pre-Field Objective lens or PFO). The currents applied on these lenses control the size of the probe (spot size) and the angle of convergence (2α) of the beam on the sample. These two parameters are also influenced by the condenser aperture on C2 which reduces the angular opening of the electron source, thus increasing the coherence of the beam. The two main operation modes available on an electron microscope differ by the convergence and size of the beam which interacts with the sample. Scanning transmission electron microscopy (STEM mode) consists in scanning a nanometric and extremely convergent electron probe on the sample (Fig. 2a). Later in this chapter, we will show that the fine control of the position of this nano-probe is the main advantage of the STEM mode, which offers unique imaging and nano-analysis capabilities for analyzing the structure and local chemistry of nanoalloys. In the

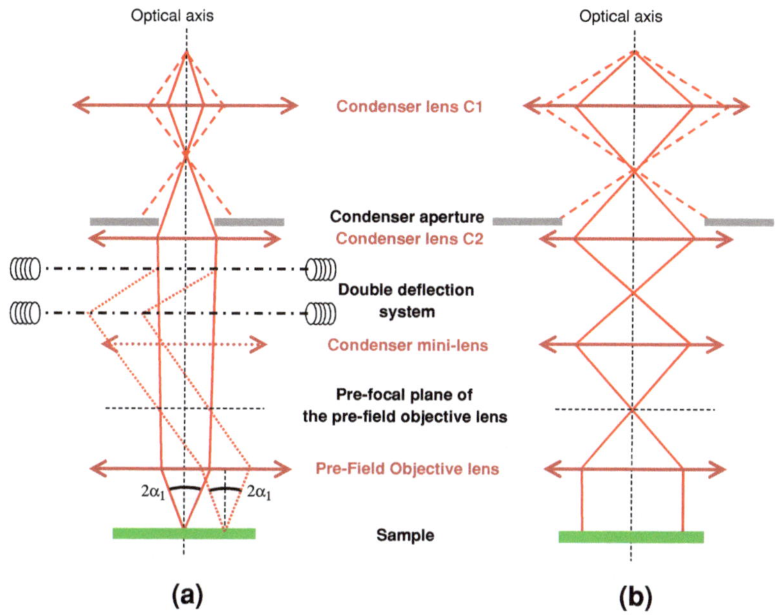

Fig. 2 Functional diagram showing the role of the illumination system. **a** In STEM illumination condition, the condenser mini-lens is off, and the focal lengths of C1 and C2 are fixed to form a beam almost parallel with a propagation direction along the optical axis. The quasi-parallelism of the beam after C2 has two consequences. Firstly, it avoids the distortion of the beam during the scanning process performed by a double deflection system. It preserves the size and the convergence angle of the probe whatever the position of the beam in the plane of the sample. Secondly, this almost parallel beam is focused on the sample by the PFO inducing a large convergent angle characteristic of the STEM mode. **b** In TEM mode a large parallel beam is obtained by using the CM to focus the image of the electron source in the object plane of the upper objective lens

following of this first section, we will describe the conventional TEM mode which has a similar optical configuration to an optical microscope. Therefore, in TEM mode the purpose of the illumination system is to illuminate the sample with a large and quasi-parallel beam ($\alpha \leq 0$–4 mrad) (Fig. 2b).

(ii) The *objective system*, which appears in a frame on Fig. 1, is the key element of the TEM. The wave plane on the output surface of the material is transferred by the lower objective lens which provides the first magnified image of the object. The resolution capacity of a microscope depends on the aberrations of this lens, especially its spherical aberration. We will revisit this problem of resolution in the section covering high-resolution imaging (Sect. 3.1).

A basic principle of the optics of converging lenses states that under Fraunhofer conditions (i.e. with an infinitely remote source), the Fourier transform of the object is formed in the back focal plane of the lens. This is why the diffraction pattern of the sample is formed in the back focal plane of the objective lens, while

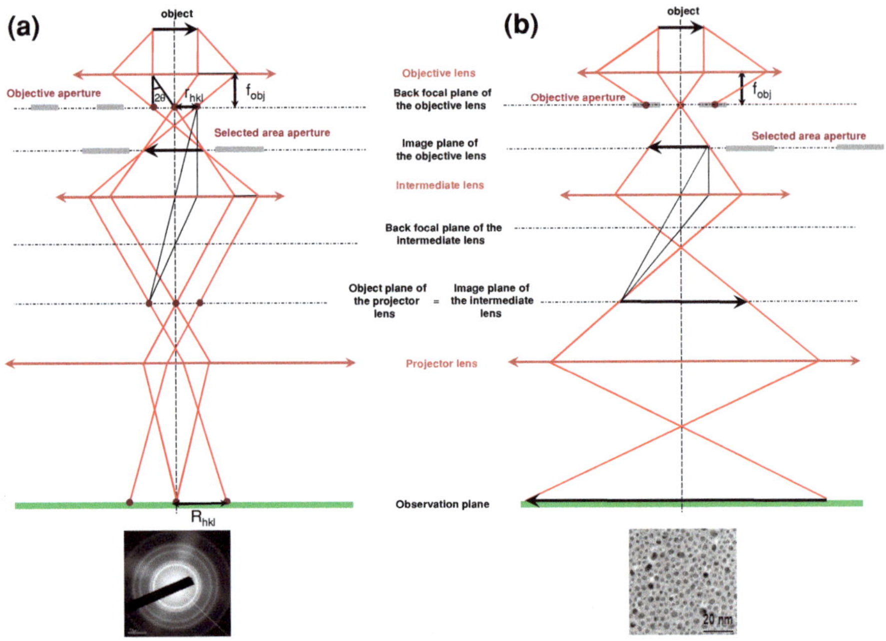

Fig. 3 Simplified illustrations of the principle of operation of the objective, intermediates and projection lenses. **a** Object observed in diffraction mode. **b** Object observed in image mode

the magnified image of the object is formed by reverse Fourier transform on the image plane of the lens (Fig. 3).

(iii) The *intermediate and projection lenses* make up the last group of lenses of the microscope. By modifying the currents of the intermediate lenses, one can conjugate the object plane of the projection lens either with the back focal plane of the objective lens (diffraction mode, Fig. 3a), or with the image plane of the objective lens (image mode, Fig 3b). The image or the diffraction pattern of the object are magnified by the projection lens and projected in the observation plane. In image mode, the magnification can reach a factor of 1,500,000.

Additionally, several deflecting coils at various levels of the column of the microscope make it possible to modify the shape of the beam and to move it (beam tilt and beam shift) relative to the optical axis. They have important roles, including gun alignment and correction of the astigmatism of the condenser, objective and intermediates lenses (in which case they are called 'stigmators').

2.2.3 Detection Systems

The TEM has long been considered to be an excellent observation tool but hardly able to provide any quantitative information. The development of increasingly powerful detection systems for high-energy electrons played a major part in the emergence of quantitative electron microscopy. Three types of detection systems sensitive to electrons are in use today. They differ by their image elements abusively called pixels. These pixels are characterized by their size (resolution), their sensitivity threshold (minimal number of electrons detected), their dynamic range of intensity (in which the response of a pixel is proportional to the number of electrons irradiating its surface), and saturation threshold.

First of all, traditional film planes commonly called negatives, based on the principle of photographic emulsion, have long been the only support in use. They now tend to give way to digital detectors. These film planes have a very good resolution thanks to the small size of the layered silver crystals coding image data (approx. 20 μm per pixel). The storage capacities of a pixel, approximately 500 electrons before the saturation threshold, are largely sufficient for a qualitative interpretation of the images. However their dynamic range of intensity, from approximately 100–200 electrons, limits quantitative contrast analyses. Moreover, the transfer of the images to digital format requires scanning the photographs, which adds a stage in the data processing chain.

Significant efforts have been devoted over the last years to equip the most recent microscopes with more quantitative means of detection. Detection systems called imaging plates (IPs) are among the latest developments in this field. These flexible supports are similar to a traditional plane film and are placed in the same way in the plane films handling system under the column of the microscope (Fig. 1). The dynamic range of intensity going from 1 to 2^{20} electrons (for a pixel size of 15 through 50 μm) of IPs make it possible to produce diffraction images with long exposure times, thus increasing the statistics of electron counts without saturating the image. This technology is thus perfectly appropriate to quantitative studies by electron diffraction; the counterpart of the high image quality is that the time required to "develop" IPs restricts the quantity of analyzable data.

CCD (Charge Coupled Devices) type cameras are one of the most important technical evolutions in the optimization of signal detection. They rely on networks of photodiodes manufactured in the form of a 1024×1024, 2048×2048, or even 4096×4096 matrix. The dynamic intensity range of CCD cameras extends from 1 to 2^{16} electrons, pixel size varies from 15 to 25 μm, depending on models. Yet CCD cameras have a slight drawback: due to the small size of their detection surface, they have a more restricted imaging field than IPs or negatives for a given magnification. However, the increase in the size of photodiode matrices tends to mitigate this problem. Using CCDs for data acquisition, the data-processing interface of current microscopes is so effective that the computer can take charge of microscope alignment. This technology is undoubtedly the best compromise between ease of acquisition and image quality. It is also closely associated to the developments of energy loss spectroscopy on TEMs.

2.3 Image Formation: Diffraction and Conventional Imaging

2.3.1 Diffraction Pattern

We showed in the preceding chapter that the amplitude scattered by a crystal depends on the structure and shape of the crystal volume. Of course the same goes for electron diffraction. Thus, if the crystal has a finite number of unit cells parallel to the z direction for example, as it is the case for electron microscopy samples, the volumes of diffraction are extended along the reciprocal axis z*. With a nanoparticle, diffraction volumes thus extend in all directions and one observes increasingly diffuse spots as the size of nanoparticles decreases.

As with X-rays, the geometry of electron diffraction is explained by constructing the Ewald's sphere. The aim of this graphic construct is to determine which lattice planes will result in a diffracted signal for a given wavelength of incident radiation. Indeed, the diffraction spots visible on a diffraction pattern correspond to the nodes of the reciprocal lattice intercepted by the Ewald's sphere (Fig. 4a). In the case of X-rays (with a wavelength of approximately 0.1 nm), the Ewald's sphere generally goes through a small number of nodes of the reciprocal lattice. In the case of electrons, because of their small wavelength, the radius of the sphere is very large compared to dimension of the reciprocal lattice. It is thus possible to override locally the sphere by its tangent plane. This is why the electron microscope allows the direct observation of a cross-section of the reciprocal space in which many hkl reflections are simultaneously excited. This phenomenon is accentuated by the extension of the diffraction volumes, so that a reflection remains visible even if it does not strictly fulfill Bragg's condition (Fig. 4a).

A diffraction diagram of an assembly of nanoparticles depends largely on the orientation of the diffracting objects. If the nanoparticles are fabricated by vapor phase deposition or by electro-deposition on a crystalline substrate (epitaxial growth) [16], one will observe the diffraction of a monocrystal, i.e. an arrangement of spots with certain distances and angles which immediately suggest a logical relation between this geometry and the symmetry of the crystal (Fig. 4b). Conversely, for randomly oriented nanoparticles (obtained by synthesis on an amorphous substrate or chemical preparation), one can see concentric rings around the transmitted beam (at the origin of the reciprocal network) (Fig. 4c). The reciprocal space is made up by the superposition of the reciprocal lattice of each particle (powder diagram). By measuring the distances between a given hkl reflection and the transmitted beam (R_{hkl}) on a diffraction pattern, one can use the following equation to precisely determine the inter-reticular distance of the corresponding hkl lattice plane (d_{hkl}), provided the camera length of the microscope (L) has been appropriately calibrated.

$$d_{hkl} = \frac{L\lambda}{R_{hkl}} \tag{2}$$

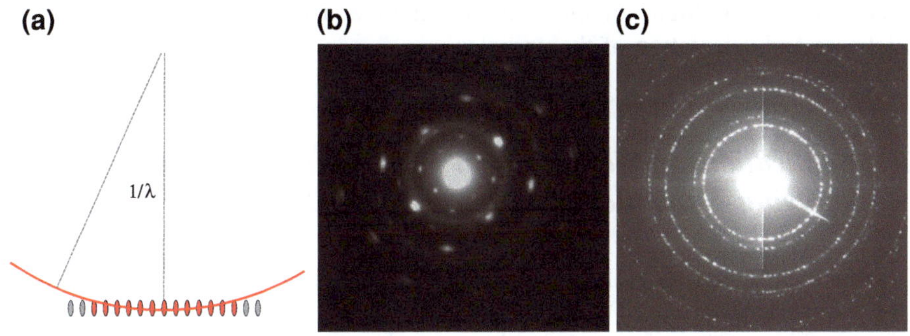

Fig. 4 **a** Representation of the Ewald's sphere intercepting a plane of the reciprocal lattice. **b** Diffractogram of an assembly of epitaxied nanoparticles. **c** Diffractogram of an assembly of randomly oriented nanoparticles

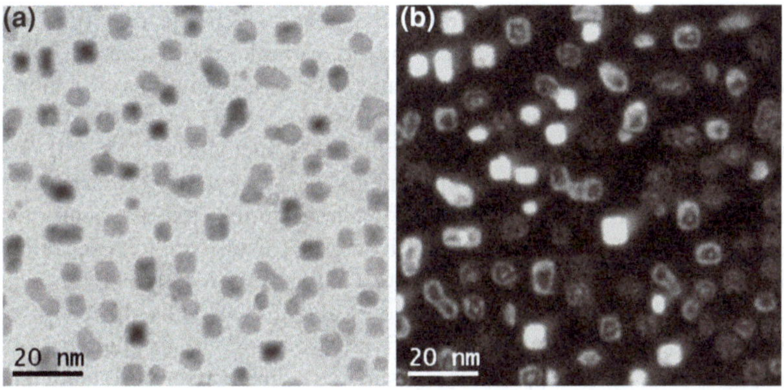

Fig. 5 **a** *Bright field* image of epitaxied nanoparticles. **b** *Dark field* image of epitaxied nanoparticles. Image acquired on the same area

2.3.2 Conventional Imaging

There are two modes of conventional imaging on a TEM in which a scattered beam is selected by an aperture (objective aperture) in the back focal plane of the objective lens (diffraction plane). A bright field image is obtained by selecting only the unscattered beam with the objective aperture (Fig. 5a). The weakly scattering parts of the sample produce a bright image. This is the case for areas rich in light elements (amorphous carbon film), and for nanoparticles situated far from any Bragg positions. Conversely, a dark field image is obtained by selecting a scattered beam with the objective aperture (Fig. 5b). In these conditions, the nanocrystals in a Bragg orientation for the selected scattered beam, produce a bright image while the rest is dark. The contrasts obtained in these two modes of imaging are directly related to the amplitude of the selected beams and thus depend on the nature, thickness and orientation of the particles.

2.4 Applications of Conventional Electron Microscopy: Studying Nanoalloys Under Different Conditions

Conventional imaging and diffraction are basic tools of the electron microscope which are commonly used to obtain the size distribution and the crystal structure of nano-objects. These fast and simple to use characterization techniques are ideal to observe and optimize the results of nanoparticle synthesis [17–24].

2.4.1 In Situ Heating Experiment

Less classically, these two techniques can be used for the real time study of nanoalloys thermodynamic properties. Indeed, using a heating sample holder, one can directly observe the morphological and structural evolutions of nanoparticles according to the temperature. Information on the kinetics of the mechanisms observed is crucial for the synthesis and use of nanoalloys [25, 26]. Figure 6 illustrates an in situ annealing experiments performed on CoPt nanoparticles obtained by pulsed laser ablation [25]. Surprisingly, the nanoparticles after synthesis are in a Face Centered Cubic (FCC) disordered phase, called 'non equilibrium' phase (Fig. 6a), because it does not comply with the phase diagram of CoPt in bulk state. By heating the particles above 600°C the diffusion of metal species within particles is sufficiently activated to prompt their chemical ordering, which is identified by the appearance of superstructure reflections characteristic of $L1_0$ ordered structure on the diffraction pattern (Fig. 6b, c).

The coalescence mechanisms through which the morphology of nanoparticles evolves towards increasingly large aggregates can be observed from a temperature of 700°C (Fig. 6c). The FCC disordered structure is found when the particles are heated above the phase transition temperature (Fig. 6d). This experiment makes it possible to determine temperature ranges in which kinetic phenomena like the ordering and the coalescence of CoPt nanoparticles occur. These results, which are extremely important for the synthesis of nanoparticles, show that the FCC disordered structure of the nanoparticles elaborated below 600°C is related to the kinetics of formation of the nanoparticles, which is faster than the ordering kinetics in this range of temperature. In other words, below 600°C, it takes more time to order the particles than to form them. This means that a wise choice of the temperature of the substrate and/or the deposition rates of the metal species makes it possible to control the size of the particles (< 10 nm) and to stabilize the $L1_0$ structure during the synthesis process. The chemically ordered structure is precisely the one required in this kind of magnetic nanoalloys, because of its strong magnetic anisotropy [27].

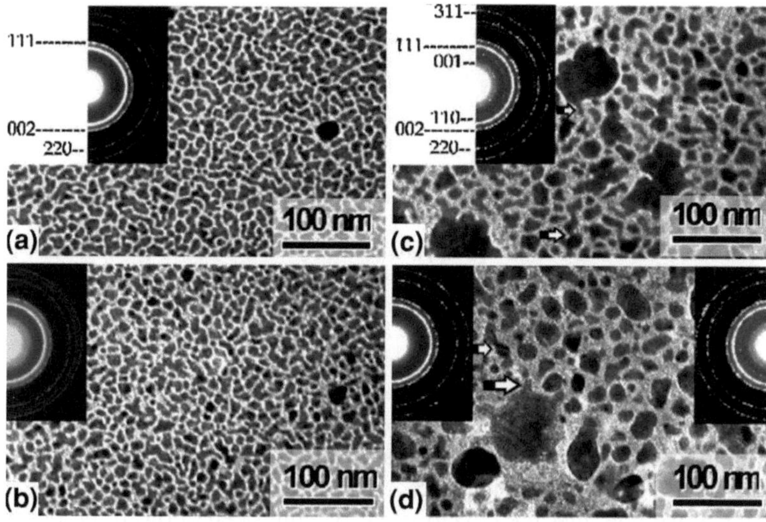

Fig. 6 Morphological (*bright field* images) and structural (diffraction in the insert) evolutions of $Co_{45}Pt_{55}$ nanoparticles according to the temperature. **a** 25°, NPs with FCC structure; **b** At 600°C, appearance of superstructure reflections characteristic of $L1_0$ chemical order; **c** At 740°C, $L1_0$ ordered nanoparticles coalescing (the *arrows* show the bridges of inter-diffusion of atoms between NPs.); **d** At 750°C, phase transitions. *Left insert*: rise in temperature, after the phase transition from order to disorder. *Right insert*: decrease in temperature, after the phase transition from disorder to order. From Ref. [25]

2.4.2 Environmental TEM

This in situ heating experiment illustrates the capabilities of TEM for probing dynamical process. The development of environmental sample holders has also opened up the unique possibility to monitor nanostructures during exposure to reactive gas environment and elevated temperature. Therefore Environmental Transmission Electron Microscopy (ETEM) gives insights into the structural evolution of catalytic nanoparticles under reaction conditions [28, 29]. Figure 7 shows the reversible shape transformations of large Pt particles observed by ETEM during oxidation reduction cycles [30]. The particle shape under 3 mbar of O_2 (Fig. 7a) is close to cube limited by (100) faces but truncated at the corners by (111) faces. The reduction of the particle in H_2 induces the development of (111) faces and the decrease of the (100) faces, resulting in a cuboctahedron shape (Fig. 7b). Finally, with the adsorption of O_2 the Pt particles recover their initial shape (Fig. 7c). This example highlights the necessity to develop in situ probes, because the drastic morphological changes observed under reaction conditions, cannot be deduced from *postmortem* TEM observation. So far, ETEM analysis has mainly been performed on monometallic catalysts [31, 32], but in addition to morphological changes, catalytic reactions on nanoalloys may induce segregation effects, which could also be observed directly in the microscope. More generally,

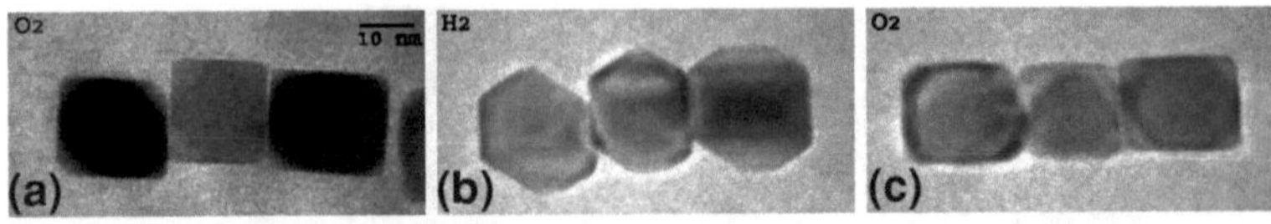

Fig. 7 **a** Set of three square Pt particles mainly limited by six (001) faces and truncated by (111) facets at the corners, observed in O_2 at 3 mbar. **b** The same set of three Pt particles in H_2 at 3 mbar. The particles are mainly limited by (111) faces, truncated by (001) facets, and observed in the [110] direction. **c** The same three Pt particles in O_2 at 3 mbar. The particles have got their original morphology mainly limited by (001) faces and truncated by (111) faces at the eight corners. From Ref. [30], copyright (2010) American Chemical Society

the ability to image—with atomic resolution [31, 32] (Sect. 3.1)—the size, shape and detailed surface structure of heterogeneous catalysts in their functioning state is of primary importance for the understanding of the many dynamic processes occurring at the gas-particle interface.

To finish, we note that the development of liquid-cell allowed the observation of clusters in solution, which can be a breakthrough for the understanding of the liquid-catalyst reactions [33] or the colloidal growth mechanisms [34, 35].

3 Correlating the Size and Structure of Nanoparticles

The specific properties of nanoalloys are often associated with the well-known "size effects", i.e. the influence of the size of the bimetallic nanoparticles on a property of the material (optical, magnetic, catalytic…). These variations of properties can often be explained by the close relationship between the size and the crystalline structure of a nanoparticle. Taking into account the inevitable polydispersity of size in a sample of nanoparticles, the use of diffraction techniques in analyzing the structure of an assembly of nanoparticles is not recommended to study in detail the size/structure relation. Indeed, to determine in which size range the structural properties of a nanoparticle differ from those of bulk material, it is essential to use techniques that make it possible to determine the structure and the size of individual nano-objects. Size effects are to be expected in metals for nanoparticles under 10 nm. The techniques of electron microscopy that we will describe in this section are adapted to highlight these phenomena, which have mostly been demonstrated theoretically so far.

3.1 High-Resolution Imaging

The idea of atom was born near the shores of the Aegean Sea, almost 2,500 years ago. The Greek philosopher Leucippus and his disciple Democritus (460–370 BCE) were the first to suggest that all matter was made up of minute particles, invisible to the naked eye. Through this conception of matter, which was only philosophical so far, Democritus called such a particle *atomos*, which means “that cannot be cut”. Although nuclear physics has now showed that atoms can in fact be broken down, the word remained. The wave/corpuscle duality, the wavelength of the electron (under 0.1 nm) and the development of electron optics allow today the visualization of these atoms.

The principle of high-resolution transmission electron microscopy (HRTEM) is to produce an image of interferences between the scattered and unscattered beams. The way to achieve this is to select several beams by means of an aperture placed in the back focal plane of the objective lens. Unlike conventional imaging, which produces an image whose contrast varies with the amplitude of the single selected beam, the contrast of high-resolution images is related to the relative phase shifts of the various beams. This is why it is also called phase contrast imaging.

The phase shifting of the beams depends on the interaction of electrons with the sample but also on the optics of the microscope, especially the passage through the objective lens. Taking into account the origin of contrast, the distribution of the minima and maxima of intensity observed on HRTEM images does not represent in general the position of the atoms, but rather the periodic modulation of the phase of the wave front at the exit of the sample. A simulation of the interference pattern is necessary to interpret contrasts quantitatively. However, HRTEM images, after a simple Fourier transform operation, make it possible to derive information on crystal structure and defects (grain boundaries, dislocations, twin crystals…). As shown in Fig. 8, HRTEM contrasts can be influenced and even reversed, by modifying the optical parameters of the microscope (here, its focus). However the power spectra of these two images show identical reflections, because the periodicity of the maxima (or minima) of intensity is directly related to the periodicity of the lattice planes. These digital diffractograms make it possible, just as with diffraction, to determine the structure of a nanoparticle.

High-resolution imaging is obviously a technique of choice for the study of the structural properties of nanoparticles, since size and structure are observable simultaneously. As X-ray scattering methods (Chap. 3), HRTEM in combination with contrast simulations have been intensively exploited to study the structural stability of small non-crystalline nanostructures according to their size and synthesis parameters [19, 36–42]. For example, the structural transformation from icosahedra (Fig. 8c) to decahedra (Fig. 8d) in gas-phase prepared FePt multiply twinned particles was evidenced at increasing temperatures and particle sizes [43]. When the particles grow due to inter-particle collisions and coalescence during the gas-phase sintering, the icosahedral structures become energetically disfavored with respect to decahedra. It has been observed that the formation of $L1_0$ ordered FePt particles is mediated by this first structural transition, since chemically ordered regions grow inside the tetrahedral sub-units of the decahedra.

Reinforced by digital processing of the images, high-resolution imaging is also a very interesting approach to study the atomic structure of nanoparticles of non-miscible alloys. Digital contrast imaging is obtained by calculating the inverse

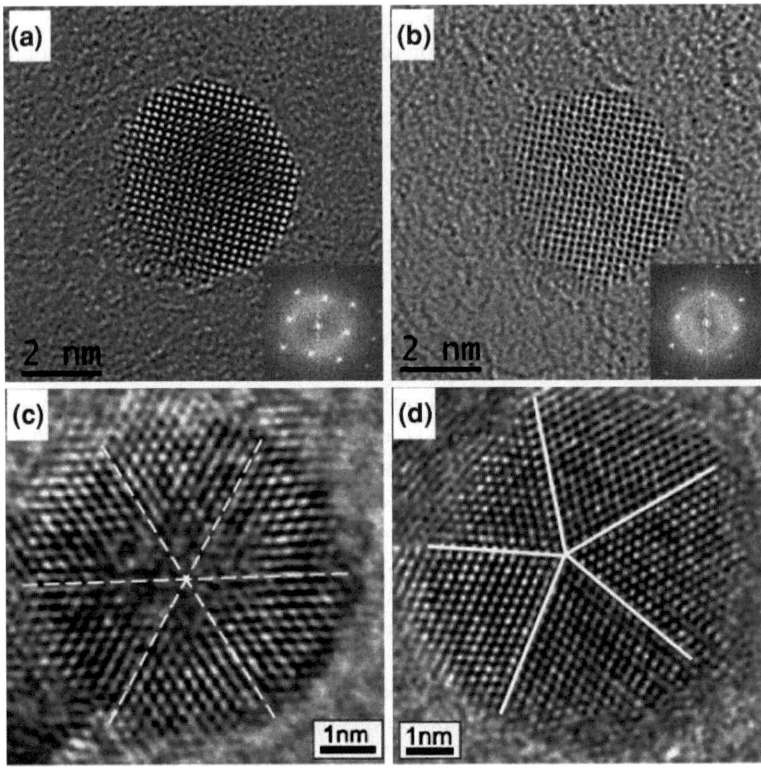

Fig. 8 HRTEM Images. **a** and **b** Single crystal FCC gold nanoparticle acquired under various defocusing conditions. **c** FePt icosahedral nanoparticle. The lattice contrast exhibits distinct patterns of threefold symmetries (*dashed lines*), which are typical for icosahedral particles lying with one of their threefold symmetry axes parallel to the electron beam. **d** FePt decahedral nanoparticle along its fivefold symmetry. The twin boundaries between adjacent tetrahedra within the decahedra are indicated by *white lines* (from Ref. [43], copyright (2004), with permission from Elsevier)

Fourier transform of the power spectrum of an HRTEM image, on which a well defined spatial frequency is selected with a low-pass filter (Fig. 9c). This technique is called digital dark field imaging since the low pass filter plays the same role as the objective aperture in conventional dark field imaging. By selecting a suitable reflection for the crystal lattice of one of the two elements, one obtains a chemical mapping exhibiting the distribution of the two metal species in the nanoparticles (Fig. 9d, e). One displays in fact the part of the nanocrystal where distances between interference fringes (i.e. atomic planes) correspond to the selected reflection on the power spectrum. We note that such a chemical imaging required a clear visibility of the atomic structure on the whole studied nanocrystal. Indeed, the chemical mapping obtained by digital dark field imaging is very easily affected by crystalline defects. In fact, any loss of structural information induced by a possible problem of orientation of the structure will cause a loss of chemical

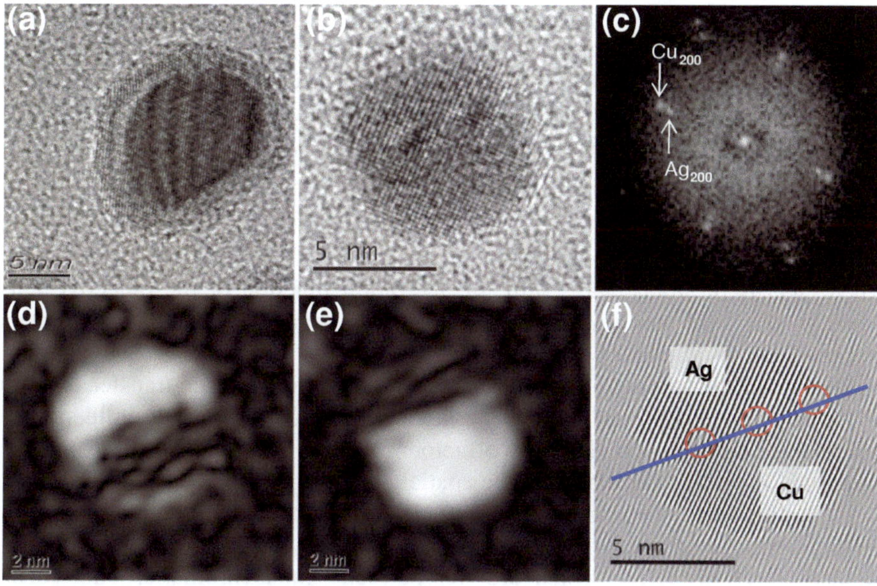

Fig. 9 HRTEM image of Cu–Ag nanoparticles. **a** Core-shell configuration (from Ref. [44], copyright (2008) by Springer Science + Business). **b** Janus configuration (from Ref. [19], reproduced by permission of the Royal Society of Chemistry). **c** Power spectrum of image *b*. Digital processing of the particle in Janus configuration is performed by computing an inverse Fourier transform of image *c*, with selection of **d** the 200 reflection of silver, **e** the 200 reflection of copper and **f** both 200 reflections of copper and silver. This kind of digital processing is used to visualize the dislocation network at the Cu–Ag interface (*encircled in red*)

information, which can be misleading. In addition, the lattice parameters of the two elements must be different enough to select the corresponding reflections individually with the low pass filter (Fig. 9c). In this example on Cu–Ag nanoparticles [19], high-resolution imaging were used to show that according to their size and the systhesis conditions, Cu–Ag nanoparticles had a core–shell configuration (Fig. 9a), or a 'Janus' (two-faced) configuration (Fig. 9b), in which demixtion occurs on both sides of the cluster. By selecting the reflections of the two species in the same direction, one clearly observes the structure of the interface between the two crystal structures. As illustrated on the Fig. 9f, the relaxation of the crystal structure of silver on that of copper induced the presence of dislocations at the interfaces between the two metals (indicated by red circles). The distribution of these dislocations was unexpectedly the same as with a Cu–Ag interface in bulk material.

Paradoxically, high-resolution imaging on a microscope not corrected for spherical aberration is limited by its resolution, which prevents the analysis of a large number of nanoparticles. The performance of an electron microscope is expressed via the contrast transfer function (CTF), introduced by Otto Scherzer in 1949 [45]. The periodic modulation of the phase of the wave front is transferred as an intensity modulation in the image which is affected by the CTF: $T(\nu)$. The latter depends on the conditions of observation controlled by the operator (defocusing Δf : infinitesimal variation of the focal length of the objective), the aberrations of the objective lens (spherical aberration C_s and chromatic aberration C_c) and of the spatial frequency ν. It is expressed in the following way:

$$T(\nu) = \exp(-i\chi(\nu)) = \exp\left(-i\pi\lambda\left(C_s\lambda^2\frac{\nu^4}{2} + \Delta f\nu^2\right)\right) \tag{3}$$

where $\chi(\nu)$ is the total phase shift, or aberration function of the objective lens, which modifies the amplitude and the phase of the beams between the object plane and the back focal plane of the lens.

Outside of the geometric aberrations taken into account in the transfer function of the lens, the spatial and temporal coherence of the beam highly influence the information transfer, too. These two factors result in envelope functions with an amplitude inversely correlated to the spatial frequency:

(i) Partial time coherence is related to electronic instability, like fluctuations of the high-voltage source (ΔV) or lens current (ΔI_{obj}) and the energy dispersion of incident electrons (ΔE), which are not perfectly monochromatic. This partial coherence translates into a spread of defocus of the image δ, as follows:

$$\delta = C_c\left(4\left(\frac{\Delta I_{obj}}{I_{obj}}\right)^2 + \left(\frac{\Delta E}{V}\right)^2 + \left(\frac{\Delta V}{V}\right)^2\right)^{\frac{1}{2}} \tag{4}$$

where C_c is the chromatic aberration coefficient.

The envelope function corresponding to partial temporal coherence is expressed by:

$$E_c(\nu) = \exp\left(-\frac{1}{2}(\pi\lambda\delta)^2\nu^4\right) \tag{5}$$

The expression of this envelope function highlights the importance of the beam coherency and electrical stabilities of the microscope. Similarly, mechanical instabilities (specimen drift and vibration) are critical factors for image quality. They can be taken into account through additional envelope functions. By optimizing all these parameters, the damping due to these envelope functions is minimized and one pushes the information limit to higher spatial frequency.

(ii) Partial spatial coherence is related to the non-parallelism of the incident beam. It depends on the divergence of the beam emitted by the gun and is characterized by the half angle of convergence on the sample (α). The corresponding envelope function is expressed by:

$$E_s(\nu) = \exp\left(-\frac{(\pi\alpha)^2}{\ln(2)}\left(C_s\lambda^2\nu^3 + \Delta f\nu\right)^2\right) \tag{6}$$

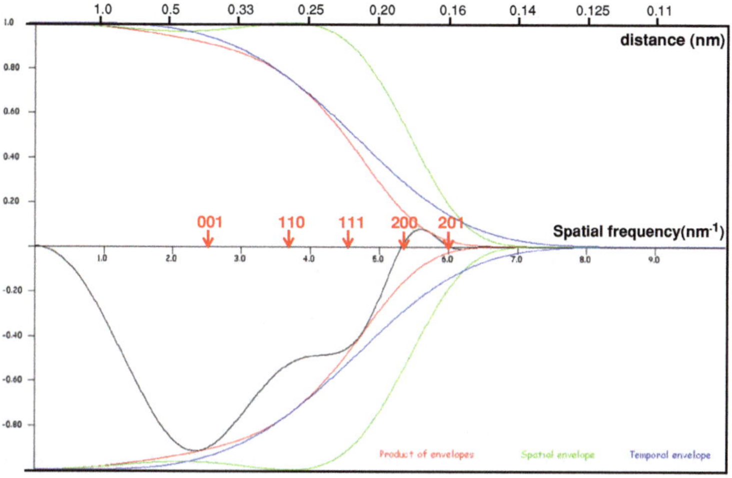

Fig. 10 Total CTF (*black curve*), corresponds to the product of the CTF $T(v)$ by the partial spatial coherence (E_s, *green curve*) and the partial temporal coherence (E_c, *blue curve*) envelopes. The *red curve* corresponds to the total envelope function. Microscope parameters: acceleration voltage of 200 kV, $C_s = 0.5$ mm, $\Delta f = -42$ nm (Scherzer defocus), $\delta = 10$ nm. The inter-reticular distances from the CoPt system appear in *red*

As seen on Fig. 10, the total, or partially coherent, transfer function of the microscope is the product of the coherent transfer function $T(v)$ by the envelope functions just presented (E_c and E_s).

Defocus is obviously a crucial parameter to fully exploit the capabilities of an electron microscope in HRTEM mode. At Scherzer defocus the CTF presents a wide band where low spatial frequencies are transferred into image intensity with a similar phase (Fig. 10). This particular defocus condition depends on both spherical aberration and electron wavelength in the following way:

$$\Delta f_{Schezer} = -1,2 \times (C_S\lambda)^{1/2} \tag{7}$$

The first cut-off frequency v_0 gives the point-to-point resolution of the microscope d_O (0.19 nm in Fig. 10), which determines the instrument ability to resolve a family of atomic planes with an inter-reticular distance d_{hkl}. At Scherzer defocus the point to point resolution of the microscope is optimum and defined as:

$$d_0 = \frac{1}{v_0} = \frac{1}{\sqrt{2}} C_S^{1/4} \lambda^{3/4} \tag{8}$$

Comparing the bandwidth of the CTF with the values of d_{hkl} in metals, shows that only the planes with the lowest indices are viewable with an aberration uncorrected microscope. Thus the conditions of orientation for which one can observe the structure of a metal nanoparticle is limited to a few zone axes. It is thus difficult to determine by HRTEM the structure of a large number of nanoparticles when working on randomly oriented clusters. This limitation is a hindrance for the study of size effects, which requires statistical results on the structure of the nanoparticles according to their size.

3.2 Nanobeam Diffraction

The nanobeam diffraction mode (NBD) consists in forming a nano-probe parallel to optical axis (Fig. 12b). This mode is a standard feature of most microscopes. It makes it possible to obtain the diffraction of a single particle (Fig. 11b). The size of the particle can be observed before focusing the beam (Fig. 11a). This method is thus well suited to the investigation of size effects, because it makes it possible to analyze the atomic structure of objects while having access to information on the morphology of the analyzed particle. The NBD mode is not limited by the resolution of the microscope, making it possible to analyze more nanoparticles that high-resolution imaging. Quantitative analysis of electron diffraction pattern consists in comparing diffraction spot intensities with simulations calculated in the framework of the dynamical theory [46–48]. Such investigations give access to quantitative information on the atomic structure of materials [3]. The determination of the chemical order parameter (i.e. degree of order) in magnetic alloy nanoparticles perfectly illustrates the importance and complexity of quantitative structural study in nanostructures. If several TEM methods have been proposed to determine the intrinsic atom order in bimetallic nanoparticles [49–51], the quantitative analysis of nanobeam diffraction patterns is so far, the most relevant and precise strategy [26, 52–54]. Indeed the intensity ratio between superstructure and fundamental reflections can be, under precise conditions, very sensitive to the chemical order. However, the first challenge of quantitative structural analysis is to determine all the parameters which influence simultaneously diffraction contrasts: electron optic parameters, both crystal thickness and orientation and of course, its atomic structure. Therefore, measuring the order parameter in a single nanocrystal requires determining very precisely the thickness, composition and orientation of the analyzed particle. It has been clearly demonstrated that along high index zone axis orientations, such as [−114] or [−116] (Fig. 11b), the order parameter measurements are less sensitive to thickness variations and consequently, more precise, because these specific crystal orientations minimize dynamical effects [52, 54]. The use a high tension voltage ($\geq$300 kV) also optimizes order parameter measurements [53]. As it remains very challenging to orientate small nanocrystal along a given zone axis, it is highly recommended to perform quantitative analysis on epitaxied nanoparticles. Although very challenging, NBD quantitative analysis have shown the influence of the particle size on the order parameter in magnetic alloys nanoparticles (Fig. 11c) [52–54].

As NBD technique consists in a manual acquisition of diffraction patterns, it presents two important drawbacks. Firstly, it is a time consuming technique which is a limiting factor to study size effect in nanostructures. Secondly, the correlation between the image and the diffraction of an analyzed particle is not always simple

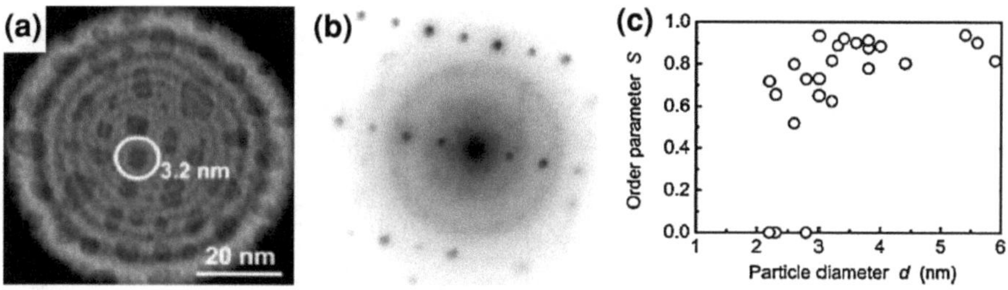

Fig. 11 Nanobeam diffraction of a 3.2 nm FePt nanoparticle. **a** Image of the analyzed nanoparticle (*encircled*). **b** NBD pattern along the [−116] zone axis orientation (*inverted contrast*). **c** Relationship between the order parameter S and particle diameter d. From Ref. [52], copyright (2005) by The American Physical Society

to obtain. Indeed, it is difficult to know precisely on which particle the beam is focused during the acquisition of the diffraction, due to sample drift. This is even more difficult with smaller particles with uniform morphology.

3.3 STEM/NBD Technique

A recent development of electron optics solves the problem of the lack of data in structural studies of a single nano-object via electron microscopy. This technique, called STEM/NBD, is motivated by the study of randomly oriented nanoalloys on a substrate. It consists in using the microscope in STEM mode with a parallel nano-probe, similar to the one use in NBD mode. In conventional STEM mode, the sample is scanned with an extremely convergent electron probe (Fig. 12a).

STEM detectors, generally located above the observation screen, collect and amplify the signal generated in any point of the zone scanned by the beam. A computerized acquisition system can reconstruct the image point by point. According to its position, a detector can acquire the transmitted beam (bright field detector), the scattered beams (dark field detector), or the beams scattered at high angles (High Angle Annular Dark Field). The main advantage of the STEM mode is the accurate control of the probe position on the sample. The interface between the acquisition system of the images and the microscope makes it possible to control the electron probe and to position it on a nanoparticle (spot mode) selected directly on the screen of the computer. It is even possible, on the same image, to select several nanoparticles and then probe them individually. The position of the probe on a particle is maintained in real time by a system correcting the drift of the sample. The only problem to carry out a structural investigation in STEM mode is that the great convergence of the probe (several mrad) is incompatible with the acquisition of a diffraction pattern because of the increase in the diameter of the bragg reflections (as the superposition of reflections prevents structure determination). To solve this problem while preserving the advantages of the STEM mode, the microscope must be set up in NBD mode first, then toggled in STEM

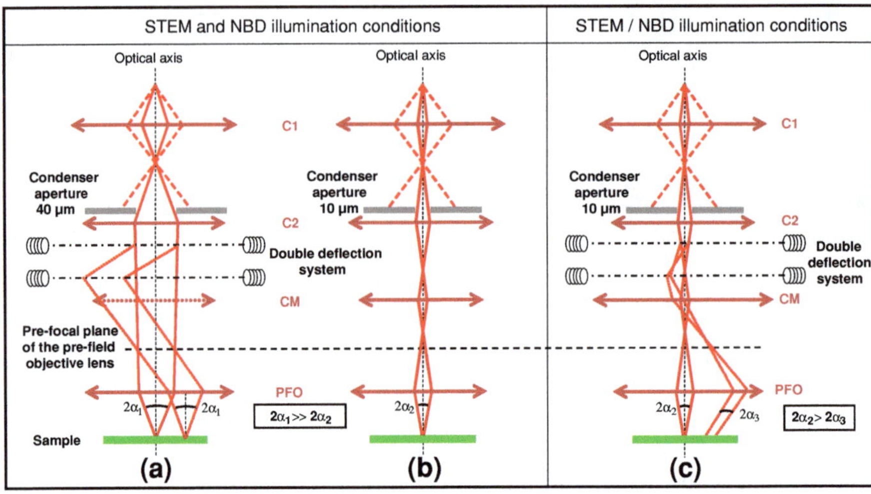

Fig. 12 Ray diagrams of the illumination system (C1 and C2, condenser lenses; CM, condenser mini-lens; PFO, pre-field objective lens). **a** Conventional STEM conditions, **b** conventional NBD conditions, **c** STEM/NBD conditions (*large dashed lines*: stopped beams by the condenser aperture; *dotted lines*: deflected beams by the scanning system). From Ref. [55], copyright (2008), with permission from Elsevier

mode but maintaining the lens currents of the illumination system of the NBD mode. The STEM/NBD method (Fig. 12c) [55], makes it possible to acquire with a parallel nano-probe the diffraction of all the nanoparticles selected on the image (Fig. 13). Automatically, one can simultaneously determine the structure and the size of several hundreds of nano-objects in one day of work. This technique opens new ways for localized structural analysis on a nanometric scale [56]. In particular, one can demonstrate and explain experimentally the existence of a size effect on the structural properties of nanoalloys. As illustrated on Fig. 13a, contrary to larger CoPt particles, sub-3 nm particles are not ordered after high temperature annealing [55]. This size effect, which matches Monte Carlo simulations (Fig. 13d) [57], is explained by a reduction in the phase transition temperature of the small particles, going from 825°C in bulk CoPt to 500–600°C for nanoparticles between 2 and 3 nm. Obtaining ordered nanoparticles with a size lower than 3 nm requires lower-temperature annealing to stay below this new phase transition temperature, and for a longer time to take into account the relation between temperature and ordering kinetics [57, 58].

4 Probing Nanoalloy Chemistry

The concept of composition gives nanoalloys an additional dimension compared to mono-metallic nanoparticles. This dimension fans out the potential of application of nanoalloys, but correlatively it complexifies their study and the exploitation of

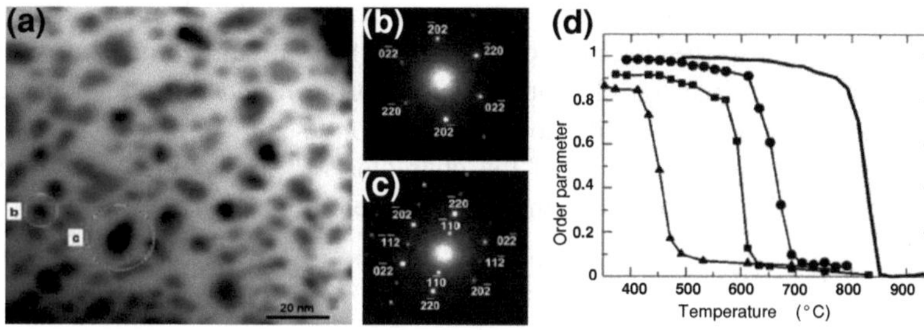

Fig. 13 **a** STEM/NBD image of CoPt NPs annealed for 1 h at 700°C. **b** Nanodiffraction of particle b (size 3 nm) *circled* in image **a**. The particle is oriented in the [111] direction, and the absence of superstructure reflections confirms its FCC structure. **c** Nanodiffraction of the particle c (size 7 nm) *circled* on image **a**. The particle is also oriented in the [111] direction and its structure is $L1_0$-ordered (from Ref. [55], copyright (2008), with permission from Elsevier). **d** Long-range order parameter computed by Monte Carlo simulation as a function of the temperature: bulk CoPt (*plain lines*), spherical particles with a size of 3 nm (*circles*), 2.5 nm (*squares*) and 2 nm (*triangles*). The phase transition temperature is given by the position of the inflection point of the curves (results from [57])

their properties. Consequently, the quantitative analysis and the visualization of the distribution of chemical species become crucial information. Electron microscopy proposes several techniques for these purposes.

4.1 Energy Dispersive X-Ray Spectroscopy

We have already indicated that electrons of the incident beam can eject a core electron from an atom of the sample, because they have much more energy than the binding energy connecting an electron on a core shell (n = K, L or M) to the atomic nucleus. This ionization process will cause the emission of a X photon with characteristic energy when the atom returns to its fundamental state: an electron on a n' shell, higher than the ionized level (n'> n) moves back to orbital n. It gains an energy $E_{XR} = E_{n'} - E_n$ which is reemitted in the form of an X photon characteristic of the target atom. The principle of EDX analysis consists in using a detector made of a semiconductor diode, placed above the sample, to count and analyze the energy of the X-rays emitted by the irradiated area of the sample. The result appears as an energy spectrum of emitted X-rays ranging from 0 to 20 keV on which the characteristic spectral lines of the elements are easily identifiable.

Absolute quantification of the concentration of an element in a sample in the absence of a reference sample introduces a lot of imprecision, which can reach 15% in some cases. It is preferable to resort to relative quantification, by comparing the intensities of X-ray lines characteristics of two (or more) elements present in sample (A and B). This quantification method is thus perfectly adapted to the investigation of nanoalloys. For thin samples, Cliff and Lorimer [59] have developed a technique to determine the concentration of the chemical elements of a sample according to the intensity obtained by choosing an element of reference, with the assumption that the phenomena of absorption and fluorescence are negligible in thin samples. The Cliff-Lorimer equation uses a single sensitivity factor k_{AB} correlating the intensities (I) of the characteristic X-ray lines of elements A and B with the relative composition of the two elements. It reads:

$$\frac{C_A}{C_B} = k_{AB}\frac{I_A}{I_B} \tag{9}$$

where C_i is the mass percentage of the element.

Although it is possible to assay the elements of a given sample based on the theoretical values of the Cliff-Lorimer factors, it is strongly advised to determine experimentally these factors k_{AB} on one's own equipment. This requires a homogeneous reference sample containing precisely known concentrations of elements of interest [60].

After this calibration stage, EDX analysis makes it possible to measure the relative composition of an assembly of bimetallic nanoparticles with a precision of about $\pm$ 1%. In nanoalloys study, determining the global composition value of particle samples is obviously indispensable. However it is also essential to check if this composition is homogeneous, using measurements on single particles (EDX nano-analysis). Just as for the structural analysis of single nanoparticles, it is practical to exploit the performance of the STEM mode in order to perfectly control the position of the nanometric probe. Note that in this case it is not necessary to scan the clusters with a parallel probe—conventional STEM mode can be used. With a microscope equipped with a FEG, the precision of measurements on a bimetallic nanoparticle with a size lower than 10 nm remains lower than 5%. Yet, because of the irradiation damage caused by the focused probe, it is difficult to analyze particles smaller than 1.5 nm.

EDX nano-analysis studies performed on bimetallic nanoparticles provided evidence that the relative composition of nanoalloys can be strongly modified during nanoparticle coarsening [61]. As demonstrated on Fig. 14, if the composition of as-grown CoPt nanoparticles is found to be $Co_{50}Pt_{50}$ ($\pm$ 5%), the composition of the particles after high temperature annealing presents a clear size dependant behavior. Indeed, the largest particles formed by the growth mechanisms during annealing, present a large excess of cobalt.

This phenomenon observed in several bimetallic nanosystems [62, 63] is due to Ostwald ripening which becomes more complex when considering the coarsening of metallic alloy clusters. This thermoactivated process has been broadly studied because it plays a determinant role in the evolution of cluster size during both chemical and physical synthesis of nanoparticles [64]. It causes large particles to grow, drawing material from the smaller particles, which shrink. The size-dependent composition of bimetallic nanoparticles observed after annealing, originates from the fact that the evaporation rate of atoms from particles is about a few orders of magnitude higher for one of the two elements. Consequently, in all

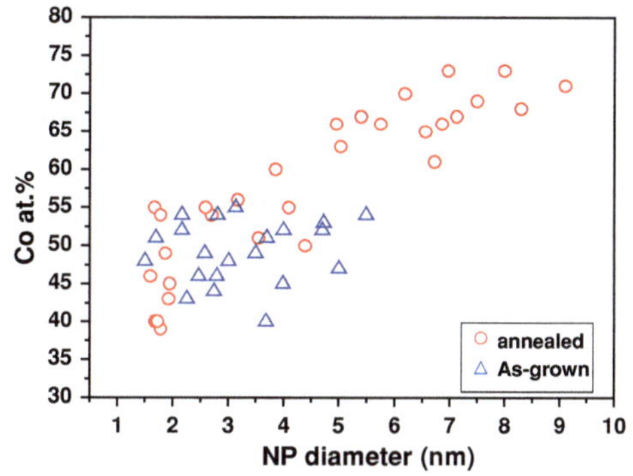

Fig. 14 Single particle composition measured by EDX nanoanalysis, as a function of their size. *Blue triangle*: as grown NPs; *red circle*: NPs after 1 h at 750°C

alloy clusters containing species with different mobilities, it is difficult to maintain the initial composition of the particles during annealing. These works illustrate the complexity of controlling together size and composition in nanoalloys, which is nevertheless crucial for understanding and exploiting their physical and chemical properties.

4.2 Electron Spectroscopy and HAADF-STEM Imaging

During inelastic interaction between a primary electron and matter, the energy loss of the electron is a characteristic of the atom with which it interacts. Thus, localized and quantitative studies of the composition can be based on measurements of the electron energy loss spectrum. The most common tool to measure the energy loss of the electrons is a post-column filter located where the signal is collected. The electrons are then distributed spatially according to their speed **v** by a dispersive magnetic field **B** which is orthogonal to their direction of propagation. The force **F** exerted on these achromatic electrons is given by the relation:

$$\mathbf{F} = -e\mathbf{v} \wedge \mathbf{B} \tag{10}$$

where e is the electron charge.

The electron follows a circular trajectory with a radius R in the plane perpendicular to **B** given by:

$$R = \gamma \frac{m}{eB} v \tag{11}$$

with

$$\gamma = \frac{1}{\sqrt{1 - \frac{v^2}{c^2}}} \tag{12}$$

The trajectory of an electron is thus directly connected to its kinetic energy, which makes it possible to know its energy loss during its interaction with the sample. Thus one obtains the spectrum of energy losses in the so-called dispersion plane.

Considering the various processes involved in electron-matter interactions, an energy loss spectrum features three zones of interest. The first is located around an intense peak centered on 0 eV (the 'zero loss peak') due to elastic electrons. The second zone of interest is the low loss region (from 0 to 50 eV), which covers the inter-band transitions (from the valence band to the conduction band) and the plasmon resonances. Lastly, in the core loss region (> 50 eV) one finds the absorption thresholds corresponding to the individual excitations of the core electrons (K, L, M shells). The energy of these absorption thresholds is tabulated for various materials.

There exist two methods to derive chemical information from the energy loss of the electrons.

4.2.1 Energy Filtered Transmission Electron Microscopy

Energy Filtered Transmission Electron Microscopy (EFTEM) consists in using a variable-size slit, measured in eV, in order to form an image with only the electrons which have a given energy. The three windows method is the most commonly used to obtain the chemical mapping of an element in a sample. It consists in acquiring an image with a slit centered on the absorption threshold of an element, as well as two pre-threshold images, to extract the continuous background under the absorption threshold. EFTEM is an outstanding technique for the chemical characterization of immiscible alloy nanoparticles (Fig. 15c). Figures 15a, b show a comparison of the chemical mappings of silver on a CuAg nanoparticle, obtained by digital dark field produced from a high-resolution image (Sect. 3.1) and by EFTEM, respectively [44].

Spatial resolution of EFTEM ($\sim$1 nm) is less precise than the atomic resolution of high-resolution imaging, but it is much less misleading since EFTEM imaging is completely independent from the structure orientation. On this example it is then possible to display unambiguously the presence of a thin layer of silver around the nanoparticle. Moreover, EFTEM can apply to core–shell nanostructure even if the lattice parameters of the two elements are too close to be separated by Bragg filtered imaging, like the Au–Pd alloy, which is very popular in catalysis.

4.2.2 Energy Loss Spectroscopy

A particularly powerful technique that combines the imaging and analytical capabilities of the microscope is known as electron energy loss spectroscopy in Spectrum imaging mode (EELS SI). The first step in the EELS SI process is to collect a STEM image of the area of interest. This image is then divided into an

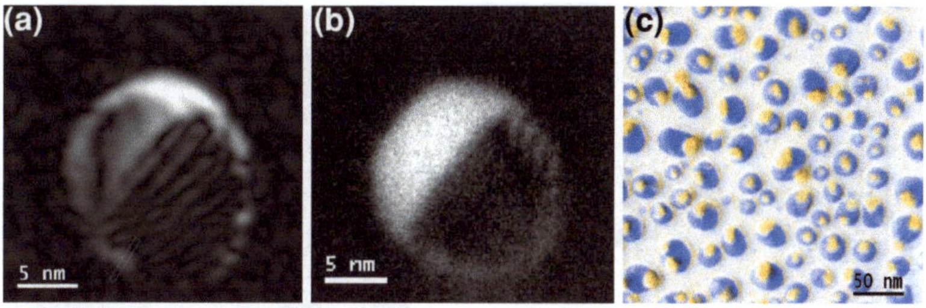

Fig. 15 Chemical analysis of CuAg nanoparticles. **a** Digital dark field obtained by selecting the (200) reflection of the structure of Ag on the FFT of the high-resolution image. **b** EFTEM image obtained on the same nanoparticle via the three windows method on the threshold of silver. **c** Chemical mapping of copper (*yellow*) and silver (*blue*). From Ref. [44], copyright (2008) by Springer Science + Business

array of pixels with a user defined resolution. The STEM probe is then scanned over the same area of the sample that was used to form the image, but this time it is stopped for a pre-determined time interval, or dwell time, at each pixel. During this time interval an entire energy loss spectrum is acquired. The multidimensional data set thus obtained is called a "datacube". Each point of the image contains spectral information from which physical and chemical information can be extracted. Using post-acquisition processing techniques, it is possible to obtain a chemical mapping from the sample, similar to that obtained by EFTEM [65]. However, both acquisition and treatment of the immense amount of data contained in a data cube are time consuming. Therefore, although EELS SI has a better spatial resolution [66], it is more convenient to use EFTEM to obtain a complete chemical map of a nanoparticle assembly (Fig. 15c), whereas STEM–EELS is most commonly used to detect the presence of an element in a defined point of the sample (point analyses modes).

One of the main advantages of EELS is the possibility to simultaneously measure energy loss spectra and acquire an image on a STEM High Angle Annular Dark Field (HAADF) detector. The electrons collected on this ring detector have a large scattering angle because they interact with atom nuclei and consequently the intensity of HAADF images is proportional to $\rho.t.Z^{\alpha}$, where α is a constant ranging between 1.5 and 2; ρ, t and Z, are the density, thickness and atomic number of the material respectively. HAADF-STEM, known as Z-contrast imaging, is thus a chemically sensitive imaging which is highly interesting for the study of nanoalloys. Figure 16 illustrates how Z-contrast imaging and EELS can be associated to study the configuration of core–shell nanostructures [67]. Differences of contrast on the HAADF-STEM image show the structural configuration of the nanoparticle (here a core–shell structure), while the energy loss spectra make it possible to identify the nature of the core (here iron). It is important to note that HAADF-STEM contrast is also influenced by the channeling effect which corresponds to the tendency for the electron beam to stay close to the atomic columns when it crosses the sample [4]. Thus, channeling effect maximizes the intensity of nanocrystals in zone axis-orientation and prohibits a direct interpretation of the

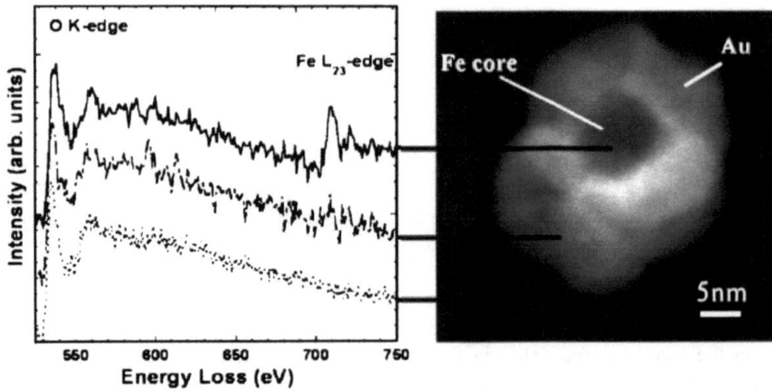

Fig. 16 Z-contrast image of an Au-coated Fe nanoparticle obtained by scanning transmission electron microscopy, the corresponding oxygen K-edge and the Fe L_{23}-edge spectra acquired from the center (*solid*) and surface (*dashed*) of the Fe/Au nanoparticle, and the silica film support (*dotted*). The nanoparticle core is composed predominantly of a Fe metallic phase. From Ref. [67], copyright (2005) American Chemical Society

images in HAADF-STEM images in terms of chemical localization by Z-contrast. To restore an image contrast linked to the chemical composition within the nanoparticles large collection angles must be used in order to make images free of dynamical interference effects [68].

It must also be noted that electron energy loss also occurs through the excitation of plasmons. Such low energy losses are used to study the optical properties of nanoobjects at the nanometer scale. For example, both EFTEM and EELS enable direct imaging of surface plasmon resonance on single nanoparticles [69–72].

5 Determining the Three-Dimensional Morphology of Clusters

As all the types of imaging which use penetrating radiations, traditional electron microscopy can provide a translucent display of the sample. The microscopist has the advantage of observing the structure inside the object; however, the structural details present at various degrees of depth are superimposed in a two-dimensional projection (2D image). Such images, in which the depth dimension is lost, are difficult to interpret and can even be counter-intuitive when the observer is interested in the surface or the thickness of the analyzed material. This lack of information on the shape of the nano-objects is an actual problem for the study and exploitation of their properties, because of the widely acknowledged shape effects on the magnetic [73], structural [57], optical [74], and even catalytic [75] properties of a material.

Among the techniques developed on the TEM to obtain three-dimensional information on a nanometric scale, tomography is without any doubt the simplest and most effective [76], The principle of this technique is to reconstruct a three-dimensional image of an object from a series of projected images obtained while varying the angle between the object and the incident beam. Electron tomography thus proceeds in two stages: first acquiring images at various tilts of a precise zone of the sample, then reconstructing a 3D image via a back-projection algorithm very similar to the one used in magnetic resonance imaging.

5.1 Acquisition

The quality and resolution of a tomographic reconstruction (tomogram) are entirely dependent on three parameters: the angular range of visualization of the object, the signal/noise ratio, and the alignment of the images. Thus, the technical developments specific to tomography mainly related to the design of sample holder with large rotation ability between the two pole pieces of the objective lens (± 80°). There are two ways to collect the images: fixed tilt increments (with an angular step of one or half a degree) or graduated tilt increments where the tilt increment is proportional to the cosine of the tilt angle. This method is often referred to as cosine rule or Saxton rule tilting [77], The disadvantage of fixed increment tilting is that it underweight the high tilt data, which give the depth information. The Saxton scheme is then used to gives correct balance to the high tilt and the low tilt data. For the study of nanoalloys, the signal/noise ratio of images is often excellent, but it is recommended to acquire the series of images manually, as automatic acquisition frequently fails during the process because of the lack of a specific reference mark on an assembly of identical nanometric particles. In any case, it is essential to apply digital image alignment algorithms after acquisition.

Three imaging modes are used for tomographic acquisition in TEM: conventional TEM, HAADF–STEM and EFTEM imaging. A 3D object can be represented by a varying density of matter distributed in space. This density of matter translates numerically into grey level intensity. The intensity at a given point of a two-dimensional projection expresses the density of a matter column, while the intensity in a point of a 3D representation is directly correlated to the density of matter at this point in the object. Consequently, there is a fundamental problem with using conventional TEM imaging for tomographic acquisition. Indeed, bright field and dark field TEM imaging are both influenced by diffraction phenomena which enhance the contrast of nanoparticles in Bragg orientation. However, if one does not select a beam in the diffraction plane with the objective aperture, the undesirable effects of diffraction contrasts in tomography are weak, since for a given cluster these contrast enhancements are only observed on a few images of the tilt series. Therefore, TEM tomography is commonly used, because it remains very efficient for obtaining detailed and accurate 3D structures of nano-objects. Z-contrast imaging is obviously the most logical mode for tomographic acquisition, since HAADF-STEM images are free of diffraction contrast. In addition, as EFTEM imaging [78], the chemical sensitivity of this technique can be used to recover three-dimensional chemical information.

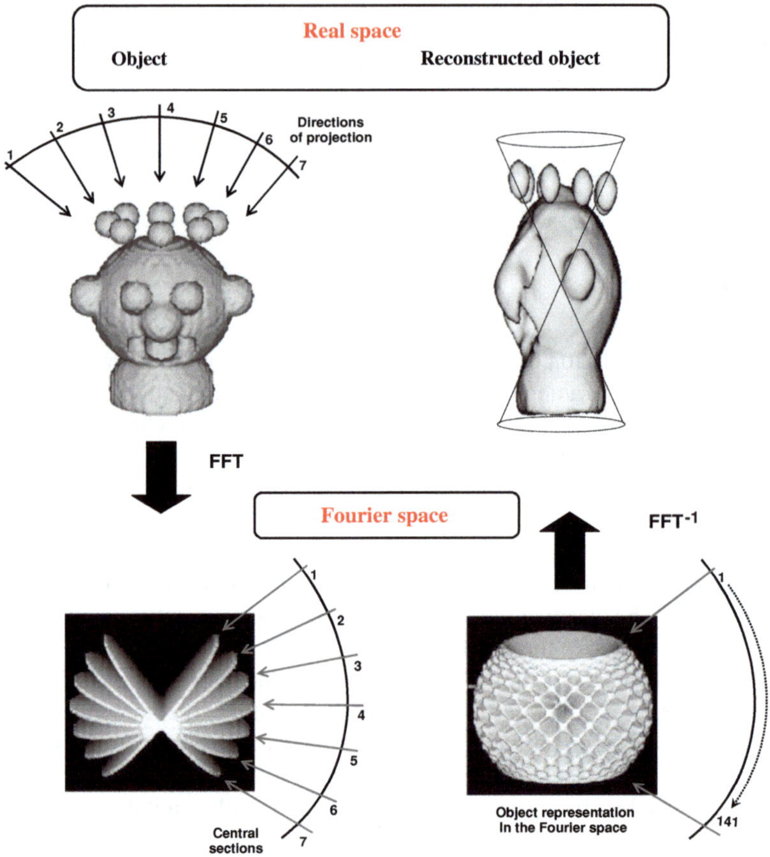

Fig. 17 Basic principle of electron tomography. The two-dimensional projections of a 3D object correspond to a central section in its 3D Fourier transform. When their number increases, this tends to form a 3D representation of the object in the Fourier space whose reverse transform provides the 3D reconstruction of the object. From Ref. [80]

5.2 Reconstruction

The algorithm used to reconstruct a 3D image from a series of 2D images rearranges in a volume the information on the density of matter included in each image. This operation is carried out through back-projection (Radon transformation) [79] of each image obtained as if the electron beam were propagated in the opposite direction. This weighted back projection method is performed in the reciprocal space, in which any 2D projection of an object corresponds to a central section in the 3D Fourier transform of the object. As can be seen on Fig. 17, each central section is perpendicular to the direction of projection.

By repeating this operation for a large number of tilt values of the object, one forms a representation of the 3D object in the Fourier space. Iterative reconstruction methods (SART, SIRT…) consist in rearranging the resulting 3D volume to recompute a better tomogram [81]. These techniques are very effective but also very expensive in CPU time; yet they are usable in practice today thanks to the development of graphic processing units (GPU) able to reconstruct a tomogram made up of 160 1024 × 1024 images in a few minutes.

Figure 17 evidences the main technical limitation of electron tomography. The angular range of the projected images of the object is never complete: there is always some part of the volume of the object which is not represented in the Fourier space. The so-called missing cone causes a lengthening of the reconstructed volume along the axis of the cone. The lengthening observed in the direction perpendicular to the substrate depends on the angular range ($\pm\alpha$) usually expressed in radians. The corresponding lengthening factor (e) is given by the following equation [82, 83]:

$$e = \sqrt{\frac{\alpha + \sin\alpha\cos\alpha}{\alpha - \sin\alpha\cos\alpha}} \tag{13}$$

Thus, for a tomography of nanoparticles with an angular range of $\pm 80°$ the lengthening factor is $e = 1.13$. Several teams are currently working on the development of reconstruction algorithms that would compensate by iteration for the effects of the missing cone [81].

5.3 Applications

Electron tomography proves to be a very powerful technique to reveal the shape of nanometric objects, even smaller than 5 nm. The results of these experiments are observed through a video file where the selected assembly of nanoparticles rotates, in order to analyze the 3D shape of the aggregates from various points of view. Moreover, thanks to the video processing utilities of reconstruction software and to the uniformity of the scales in the three (x, y, z) directions of the tomogram, it is possible to measure the thickness of the nanoparticles (Fig. 18). The estimated accuracy is ±0.5 nm [84], not taking into account the inaccuracies of the tomogram itself, which are very difficult to assess since they can have several causes (missing cone, incorrect image alignment, diffraction contrast…). Resolution as computed by the Fourier Shell correlation method [85, 86] is often used as a metrics of the quality of a tomogram, to validate the measurements. On nanoparticles made up of heavy elements like nanoalloys, it is possible to reach resolutions under 1 nm [84]. To date, only electron tomography allows the observation of nano-object 3D morphology with such accuracy.

As described in the chapter on the optical properties of nanoalloys (Chap. 10), knowledge of the 3D morphology as well as the precise thickness of the shell is needed in order to understand the surface plasmon resonances (SPRs) of core shell nanostructures. Indeed a very small change in the geometry of such heterogeneous

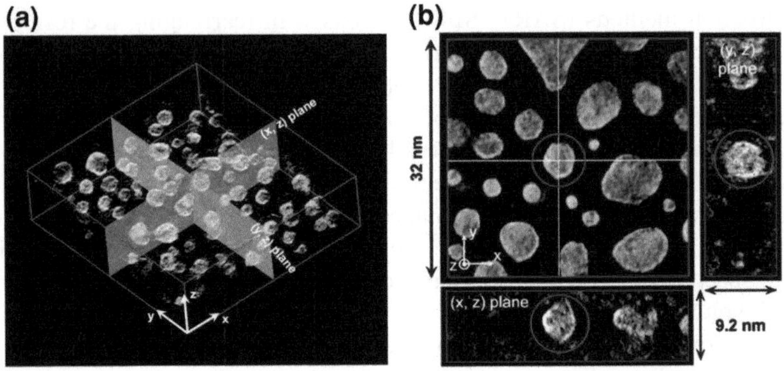

Fig. 18 Measuring the thickness of a nanoparticle by electron tomography acquired in bright field TEM mode. **a** Selection of the position of planes (x, z) and (y, z) on the tomogram. **b** Image of the particles according to the planes (x, y), (y, z) and (x, z). On these images, it is possible to measure the thickness and size in the plane of the substrate of the particle *circled in red*. From Ref [84]

clusters leads to dramatic changes in their SPRs. Figure 19 sheds light into the outstanding performances of HAADF STEM tomography for accessing three-dimensional chemical information in core–shell nanoparticles [87]. This full 3D analysis allows precise measurements of the core–shell dimension and composition and enables measured opto-electronic properties to be related to the particle morphology.

6 Aberration Corrected Microscopy: A New Generation of Instruments

The development of the aberration correctors and coherent electron sources has boosted the impact of electron microscopy on the characterization of matter at the atomic scale. This last section describes the unprecedented performances of imaging and analysis of these new microscopes, so that the non-specialist reader can become aware of the new prospects for the characterization of nanoalloys. Spherical aberration occurs on all optical systems containing spherical lenses. As glass lenses, electromagnetic lenses focus electron beams more tightly if they enter it far from the optic axis than if they enter closer to the axis. It therefore does not produce a perfect focal point and the image of a point is thus a blurry spot. This phenomenon, although identified a long time ago, has limited the resolution of optical devices like telescopes or electron microscopes until the end of the 1990s. Today the development and commercial availability of spherical aberration correctors on electron microscopes are revolutionizing their potential. It is important to note that there are correctors of aberration for the STEM mode as well as for the TEM mode. For the TEM mode, they correct the aberration of the objective lens,

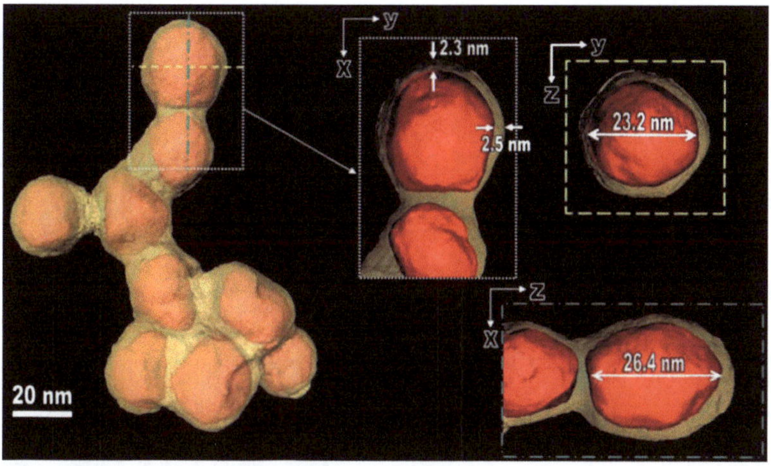

Fig. 19 (*Left*) Tomographic reconstruction of a cluster of Au–Ag core–shell nanoparticles acquired in HAADF-STEM mode. (*Right*) Orthogonal views of the uppermost nanoparticle. The particle has been cut to reveal the Ag shell thickness in cross-section and the Au interior. From Ref. [87], copyright (2008) by Springer Science + Business

which forms the first image of the sample, hence the term 'image corrector'. Resolution in STEM mode is not sensitive to the aberrations of the objective lens, but depends primarily on the size of the electron probe which scans the sample; in this case, the 'probe corrector' compensates for the aberration of the condenser system.

6.1 Monitoring Dynamic Processes at High Resolution with Single Atom Sensitivity by TEM Imaging

To summarize the advantages of an image corrector and a highly coherent electron source, it is useful to compare the contrast transfer function with and without these optical elements (Fig. 20).

By dividing by 10^3 the spherical aberration coefficient of the objective lens, a C_s—corrector makes it possible to attain a point-to-point resolution under 0.06 nm. The record resolution of an electron microscope (0.047 nm) [88, 89] is even smaller than the Bohr radius of the hydrogen atom! Of course, a highly coherent electron source and an outstanding microscope stability are indispensable to reach such a resolution. Indeed by minimizing the damping of the envelope functions, the reduction of both electron energy dispersion and mechanical vibration allow a sub-angstrom information transfer limit. As indicated on Fig. 20b, an information transfer of $1/0.06\ \text{nm}^{-1}$ is documented on the power spectrum of the HRTEM image. Thus high-resolution imaging is not limited any

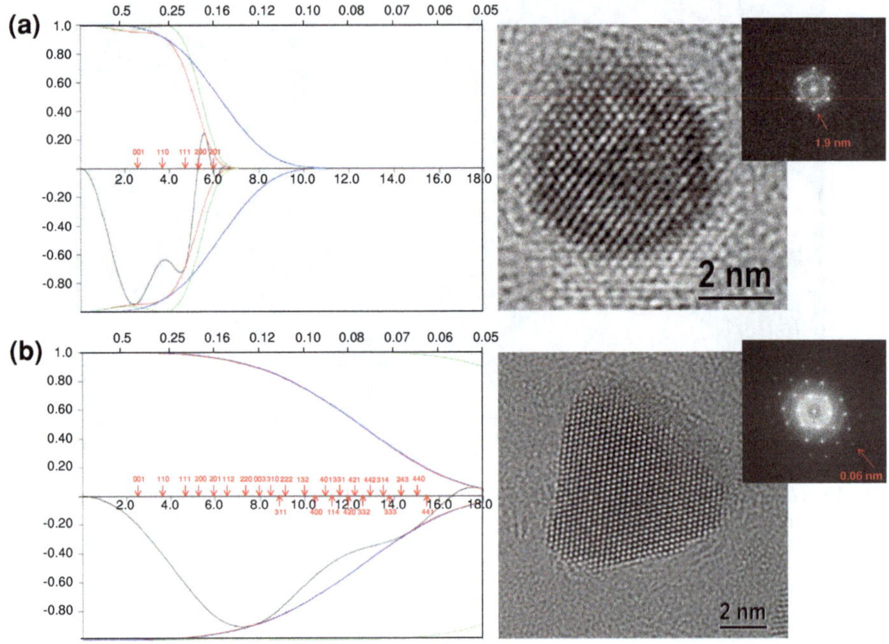

Fig. 20 Total CTF (*black curve*), corresponds to the product of the CTF $T(\nu)$ by the partial spatial coherence (E_s, *green curve*) and the partial temporal coherence (E_c, *blue curve*) envelopes. The *red curve* corresponds to the total envelope function. These curves are computed for an acceleration voltage of 200 kV and the inter-reticular distances from the CoPt system are indicated in *red*. The HRTEM image of a CoPt nanoparticle in [110] zone axis orientation was acquired under the same optical conditions than the computed CTF. The corresponding power spectrum of the HRTEM image is presented in insert. **a** Microscope with no corrector and a conventional FEG, $C_s = 0.5$ mm, $\Delta f = -42$ nm (Scherzer focus), $\delta = 10$ nm. **b** Microscope with a C_s-corrector and a monochromated electron source. C_s = 0,005 mm, Δf = −0.5 nm (Scherzer focus), δ = 1.4 nm

more for the structural study of metal nanoparticles randomly oriented on a substrate: today one can resolve many more inter-reticular distances (Fig. 20b) and thus display the structure of nanoalloys in many more orientations.

Another significant advantage of corrected microscopy for studying nano-objects comes from the fact that the Scherzer focus is very close to 0, so that corrected images do not present delocalization of atomic contrasts outside of the particles (comparison of the Fig. 20a, b). Thus, aberration correctors allow a better visualization of nanoparticles surfaces, which opens many avenues for the characterization of catalytic nanoparticles, for which the structure and composition of the surface are crucial information [90–92].

This resolution capacity and the possibility of displaying interfaces clearly is also very useful to study dynamic processes on an atomic scale, as for example, the coalescences mechanisms which affect the size and shape of nanoparticles. As can be seen by looking at Fig. 21a, b, c, the coalescence of the two gold nanoparticles

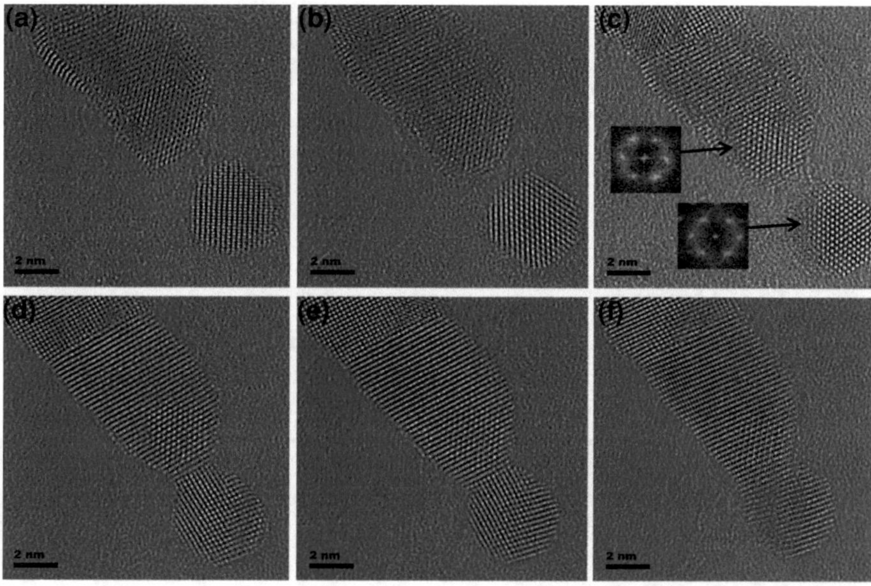

Fig. 21 Beam-induced coalescence of two gold nanoparticles observed in an aberration-corrected microscope (see description in the text)

involves a dynamic diffusion of the spherical particle (i.e. diffusion of the whole cluster) and thus a rotation of the structure. The sub-ångström resolution of modern microscopes make it possible to observe the atomic structure of the particle independently from its zone axis orientation, and to understand that the creation of the inter-diffusion bridge appearing in Fig. 21d is possible only because the spherical particle had the same crystalline orientation as the tip of the elongated particle (FFTs Fig. 21c). Once the inter-diffusion bridge is created, one observes very active atomic diffusion on the surface of the particles, whose purpose is to enlarge the neck between the two clusters. One also observes important deformation phenomena and even the creation of dislocations ensuring the movement of the two particles towards each other. Here, this coalescence process is a beam-induced phenomenon, but the possibility to perform in situ heating studies at the angstrom level using a hot stage in an aberration corrected environment is very encouraging for the direct analysis of gas–solid reaction at high temperature [93–95].

Although the extraordinary resolution of electron microscopes is put forward by their manufacturers, the most important advance of high-resolution imaging is about sensitivity. The detection and the study of single atoms have long been limited by the noise level of images. Thanks to their enhanced mechanical and electrical stability, modern microscopes can override this limitation and directly display single atoms in various materials. As illustrated in Fig. 22, single atom detection is possible almost independently from the atomic number of the analyzed materials: Platinium (Z = 78), germanium (Z = 32) [11], and even carbon

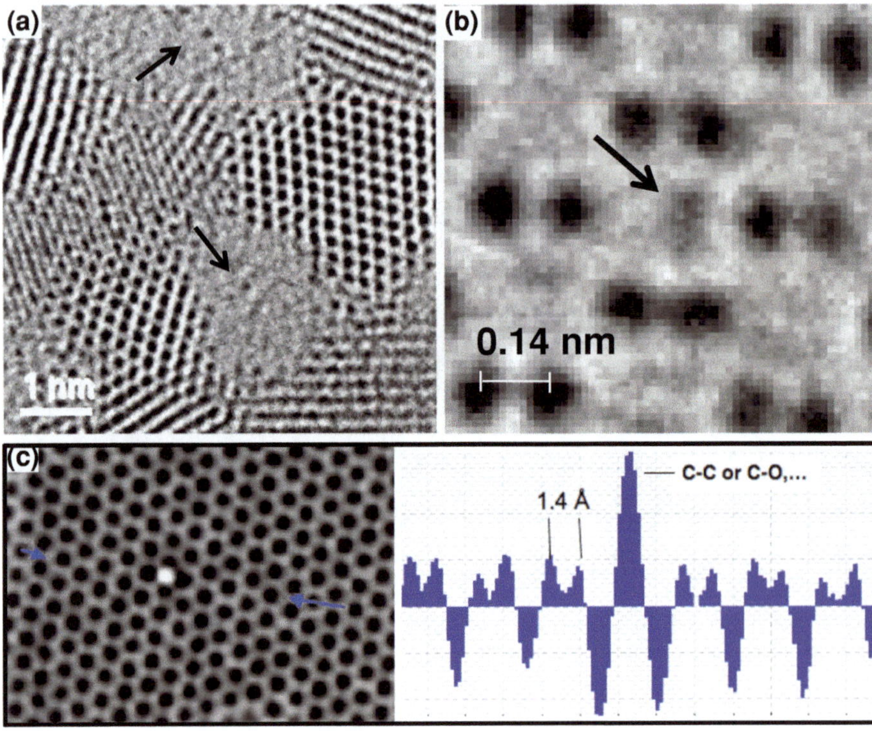

Fig. 22 Single atom analysis with modern microscopes. **a** In a nanostructured Pt thin film fabricated on amorphous carbon.: the *black arrows* indicate a Pt atoms diffusing on the carbon film. **b** In a thin germanium sample: the *black arrows* indicate an interstitial atoms of Ge diffusing into the crystal (from Ref. [11], copyright (2009) by The American Physical Society). **c** Aberration corrected TEM image of a single layer of graphene. The *right panel* shows the intensity profile measured between the two blue arrows on the images. Individual carbon atoms are resolved as white spots 0.14 nm apart. The *brighter white* spot corresponds to a single light atom attached to the graphene sheet. Image simulations confirmed that this ad-atom is either a carbon, a nitrogen or an oxygen atom (results from Ref [96])

(Z = 6) [96]. Combined with image simulation, such images allows for single atom chemical identification. For a few years, many studies have exploited the sensitivity and the good time resolution of HRTEM to observe the movement and the energetically favorable site of individual atoms in real time, on a gold/vacuum interface [89], in a thin Ge sample (Fig. 22b) [11] or on a graphene/vacuum interface [97]. These works demonstrate that aberration-corrected HRTEM is now a cutting edge approach for studying dynamical phenomena in nanomaterial structure with single atom sensitivity, which offers an unprecedented opportunity to confirm theoretical results on atomic structure stability.

One remaining challenge in the study of nanomaterials is the atomic-scale investigation of hybrid system made of a nanoparticle (metallic or nanoalloys) and a shell of soft material (molecule, polymer). Studying of soft/hard material interfaces would enhances the understanding of these nano-systems, which have many applications, in particular in biomedicine. Let us note that aberration correctors were an essential stage for the high-resolution study of materials sensitive to electron beams, like soft materials. In the past, it was necessary to increase the high voltage (to several thousands of kV!!) in order to increase the resolution of the equipment; today it is possible to maintain a resolution in the angstrom range with voltages as low as 80 or 50 kV. Therefore, the performances of modern microscopes allow the observation of single light atoms while resolving the atomic structure of nanocrystal, which open up new way to establish the relationship between the atomic structure and the functionality of hybrid nano-systems [98].

6.2 Analyzing the Atomic Scale-Chemistry of Nanoalloys by STEM Spectroscopy and Imaging

STEM spectroscopy and STEM imaging techniques have also enormously benefitted from the technological advances of electron microscopy for some fifteen years now. One of the main progresses of STEM techniques for the characterization of nanoalloys, is about the spatial resolution of chemical analysis.

Due to their small volume, sub-10 nm nanoparticles generate little inelastic interaction and consequently they generate very poor X-rays signal. The inefficiency in X-ray signal generation is a major limiting factor in the application of EDX spectrum imaging (EDX SI) at the nanoscale.[1] Today, the correction of spherical aberration offers a much better compromise between chemical map resolution (i.e. size of the electron probe) and signal detection. To increase the detection limit, it is necessary to amplify the probe current in order to increase the number of inelastic events. The probe current can be increased by using a larger probe-forming aperture and thus allowing a greater number of the electrons emitted from the source to contribute to the beam. However, using a larger aperture size, which defines the probe convergence angle (α), increases the effect of spherical aberration (C_S) which is proportional to α^3. This trade-off can be overcome to some extent in an aberration-corrected instrument, because, in the absence of C_s, a larger aperture can be used without degrading the resolution. Thus, with a nanometer probe size, the current of the corrected-probe is up to 12 times more important than the one of the uncorrected probe.

The ability to increase the probe current without degrading the instrument's spatial resolution for imaging and analysis allows the chemical mapping of small bimetallic nanoparticles by EDX SI. This technique enriches the functionality of electron microscopes for the characterization of core–shell nanostructures.

---

[1] EDX spectrum imaging is similar in principle to EELS spectrum imaging, but instead of acquiring an EELS spectrum at each pixel (Sect. 4.2.2), one acquires an X-ray spectrum. Similarly, post-acquisition processing of the obtained spectrum imaging data cube allows the production of chemical maps.

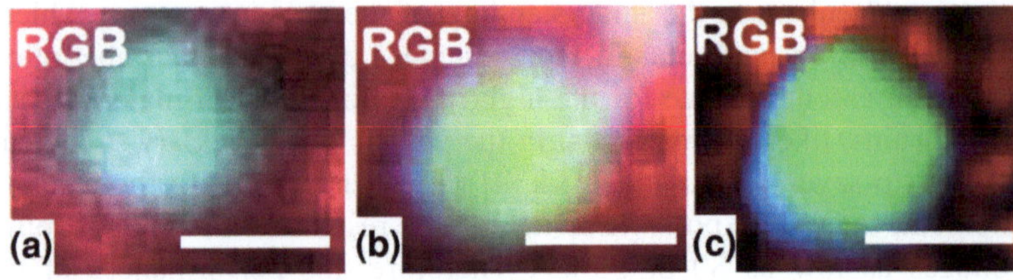

Fig. 23 RGB overlays (*Green*: Au; *Blue*: Pd) for a series of AuPd nanoparticles subjected to different thermal treatments. **a** dried at 120°C, **b** calcined at 200°C, **c** calcined at 400°C. Scale bars corresponds to 30 nm on image a and 20 nm on images *b* and *c*. From Ref. [63], reproduced by permission of the Royal Society of Chemistry

As EDX analysis is more recommended than electron spectroscopy analysis for the characterization of heavy elements, this new way to perform chemical mapping is extremely interesting for the characterization of nanoalloys. As an example, EDX SI has been used to show the surprising behavior of AuPd nanoparticles during annealing process in air [63]. The equilibrium phase diagram of this system, predicts that an FCC solid solution of Au and Pd is formed over the entire compositional range with special ordering compounds existing at the Au_3Pd and $AuPd_3$ compositions. If the chemical map of uncalcined nanoparticles (Fig. 23a) shows expected homogeneous AuPd nanostructures, the analysis of the samples calcined at 200 and 400°C (Fig. 23b, c respectively) clearly show the progressive development of a Pd-rich shell and Au-rich core morphology. From a thermodynamical point of view, the formation of 3 nm nanometer large Pd-shell is also surprising since the surface energies of Au and Pd are 1.50 and 2.05 J m^{-2}, respectively, suggesting that if anything were to migrate to the nanoparticle surface it should be the Au component. This unexpected core–shell morphology is presumably brought about by the preferential formation of Pd–O bonds at the alloy surface since in this 200–400°C temperature range palladium oxidizes more readily than gold.

Conversely, for the same probe current, the probe size is much smaller in the corrected state. This important probe size reduction gives the opportunity to study atomic structures with sub-ångström resolution by using HAADF-STEM imaging. Therefore, high resolution Z-contrast imaging offers the opportunity to quantitatively analyzed columns of atoms, since they can be clearly resolved and their contrast strongly depends on their thickness and composition. However, if this images are more straightforward to interpret than conventional HRTEM imaging, it remains indispensable to compare them with simulated images in order to take account all the optical and physical parameters which influence image contrast (i.e. illumination and collection geometry, thickness and composition of atomic columns…). Combined with image simulations, aberration-corrected HAADF-STEM can then be used to recover the three-dimensional atomic scale structure of monometallic nanoparticles, since in mono-element samples, atomic Z-contrasts are related to the number of atoms in the atomic columns [99]. As illustrated on Fig. 24, this technique is very efficient to study the atomic scale chemistry of

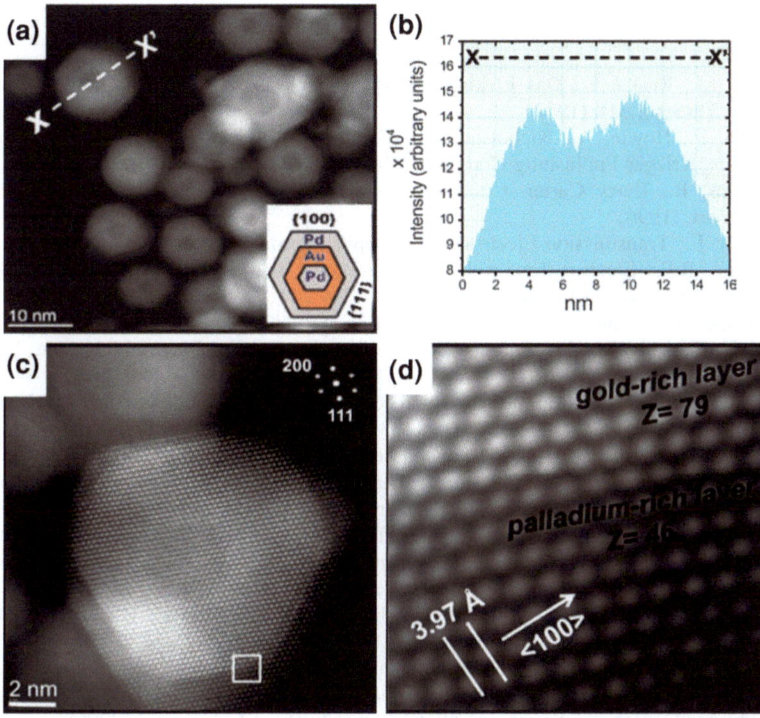

Fig. 24 **a** Conventional STEM HAADF image of Au/Pd nanoparticles with cuboctahedral shape. The contrast is due to a core–shell structure consisting of three layers as sketched in the inset. **b** The experimental intensity profile of a typical HAADF-STEM image shows a lower magnitude on the central portion on the particle (indicated as x–x′) due to the fact that $Z_{Pd} < Z_{Au}$. However, the outer Pd layer is not visible at this magnification. **c** Aberration-corrected high resolution HAADF-STEM image of a cuboctahedral Pd/Au/Pd nanoparticle. The contrast of the three distinct regions can be clearly seen. *Bright dots* represent atomic columns. The inset corresponds to the fast Fourier transform of the nanoparticle. **d** *Enlarged image* of *a* small part of the exterior layer of nanoparticle (indicated as a *white square* in image *c*), exhibiting <110> crystal orientation. From Ref. [100], reproduced by permission of the Royal Society of Chemistry

nanoalloys. In this work [101], the atomic structure of three-layer Pd/Au/Pd nanoparticles was revealed by matching the experimental intensities of atomic columns with simulated images of theoretical models of the three-layer nanoparticles. The outer Pd layer of these complex nanostructures, which is not detect at low magnification (Fig. 24a, b), is easily observed on high resolution HAADF-STEM images (Fig. 24c, d).

These two studies performed on AuPd nanostructures illustrate that both STEM spectroscopy and STEM imaging performed in a C_s-corrected microscope, provide a new level of insight into the characterization of technologically relevant catalysts [101].

References

1. Knoll, M., Ruska, E.: Das Elektronenmikroskop. Zeitschrift für Physik A Hadrons and Nuclei **78**(5–6), 318 (1932)
2. Hirsch, P., Howie, A., Nicholson, R., Whelan, M.J.: Electron Microscroscopy of Thin Crystals. Krieger Publushing Company, Malabar (1965)
3. Willams, B., Barry Carter, C.: Transmission Electron Microscopy, vol. 1–4. Plenum, New York (1996)
4. Reimer, L.: Transmission Electron Microscopy, 4th edn. Springer, Berlin (1997)
5. Egerton, R.F.: Electron Energy-loss Spectroscopy in the Electron Microscope. Plenum Press, New York and London (1986)
6. Kirkland, A.I.: Nanocharacterisation. Royal Society of Chemistry, Cambridge (2007)
7. Shindo, D., Oikawa, T.: Analytical Electron Microscopy for Materials Science. Springer, Tokyo (2002)
8. Wang, Z.L.: Transmission electron microscopy of shape-controlled nanocrystals and their assemblies. J. Phys. Chem. B **104**(6), 1153–1175 (2000)
9. Wang, Z.L.: Characterization of Nanophase Materials. Wiley-VCH, Weinheim (2000)
10. Ayache, J., Beaunier, L., Boumendil, J., Ehret, G., Laub, D.: Sample Preparation Handbook for Transmission Electron Microscopy. Springer, New York (2010)
11. Alloyeau, D., Freitag, B., Dag, S., Wang, L.W., Kisielowski, C.: Atomic-resolution three-dimensional imaging of germanium self-interstitials near a surface: Aberration-corrected transmission electron microscopy. Phys. Rev. B **80**, 014114 (2009)
12. Browning, N.D., Arslan, I., Erni, R., Idrobo, J.C., Ziegler, A., Bradley, J., Dai, Z., Stach, E.A., Bleloch, A.: Monochromators and aberration correctors: Taking EELS to new levels of energy and spatial resolution. In: Brown, P.D., Baker, R., Hamilton, B. (eds.) EMAG-NANO 2005: Imaging, Analysis and Fabrication on the Nanoscale, vol. 26. Journal of Physics Conference Series, pp. 59–64. Iop Publishing Ltd, Bristol (2006)
13. Kimoto, K., Ishizuka, K., Asaka, T., Nagai, T., Matsui, Y.: 0.23 eV energy resolution obtained using a cold field-emission gun and a streak imaging technique. Micron **36**(5), 465–469 (2005)
14. Kimoto, K., Kothleitner, G., Grogger, W., Matsui, Y., Hofer, F.: Advantages of a monochromator for bandgap measurements using electron energy-loss spectroscopy. Micron **36**(2), 185–189 (2005)
15. Konno, M., Suzuki, Y., Inada, H., Nakamura, K.: Application of 80–200 kV aberration corrected dedicated STEM with cold FEG. Electron Microscopy and Analysis Group Conference 2009, vol. 241. Journal of Physics Conference Series. Iop Publishing Ltd, Bristol (2010)
16. Sato, K., Bian, B., Hirotsu, Y.: Fabrication of oriented $L1_0$-FePt and FePd nanoparticles with large coercivity. J. Appl. Phys. **91**, 8516–8518 (2002)
17. Demortiere, A., Petit, C.: CoPt magnetic nanocrystals in the A1/L1(0) transformation. J. Appl. Phys. **109**(8), 9 (2011)
18. Habrioux, A., Vogel, W., Guinel, M., Guetaz, L., Servat, K., Kokoh, B., Alonso-Vante, N.: Structural and electrochemical studies of Au-Pt nanoalloys. Phys. Chem. Chem. Phys. **11**(18), 3573–3579 (2009)
19. Langlois, C., Alloyeau, D., Le Bouar, Y., Loiseau, A., Oikawa, T., Mottet, C., Ricolleau, C.: Growth and structural properties of CuAg and CoPt bimetallic nanoparticles. Faraday Discuss **138**, 375–391 (2008)
20. Penuelas, J., Andreazza-Vignolle, C., Andreazza, P., Ouerghi, A., Bouet, N.: Temperature effect on the ordering and morphology of CoPt nanoparticles. Surf. Sci. **602**(2), 545–551 (2008)
21. Tournus, F., Bardotti, L., Dupuis, V.: Size-dependent morphology of CoPt cluster films on graphite: A route to self-organization. J. Appl. Phys. **109**(11), 4 (2011)

22. Alloyeau, D., Ricolleau, C., Langlois, C., Le Bouar, Y., Loiseau, A.: Flash laser annealing for controlling size and shape of magnetic alloy nanoparticles. Beilstein J. Nanotechnol. **1**, 55–59 (2010)
23. Takahashi, Y.H., Ohkubo, T., Ohnuma, M., Hono, K.: Size effectSize effect on the ordering of FePtFePt granular films. J. Appl. Phys. **93**, 7166 (2003)
24. Desvaux, C., Dumestre, F., Amiens, C., Respaud, M., Lecante, P., Snoeck, E., Fejes, P., Renaud, P., Chaudret, B.: FeCo nanoparticles from an organometallic approach: Synthesis, organisation and physical properties. J. Mater. Chem. **19**(20), 3268–3275 (2009)
25. Alloyeau, D., Langlois, C., Ricolleau, C., Le Bouar, Y., Loiseau, A.: A TEM in situ experiment as a guideline for the synthesis of as-grown ordered CoPtCoPt nanoparticles. Nanotechnology **18**(37), 375301 (2007)
26. Sato, K., Kovács, A., Hirotsu, Y.: Order-disorder transformation in Fe-Pd alloy nanoparticles studied by in situ transmission electron microscopy. Thin Solid Films **519**(10), 3305–3311 (2011)
27. Sellmyer, D.J., Yu, M., Kirby, R.D.: Nanostructured magnetic films for extremely high density recording. Nanostruct. Mater. **12**(5–8), 1021–1026 (1999)
28. Gai, P.L., Kourtakis, K.: Solid-state defect mechanism in vanadyl pyrophosphate catalyst—implications for selective ixidation. Science **267**(5198), 661–663 (1995)
29. Gai, P.L., Boyes, E.D., Helveg, S., Hansen, P.L., Giorgio, S., Henry, C.R.: Atomic-resolution environmental transmission electron microscopy for probing gas-solid reactions in heterogeneous catalysis. MRS Bull. **32**(12), 1044–1050 (2007)
30. Cabie, M., Giorgio, S., Henry, C.R., Axet, M.R., Philippot, K., Chaudret, B.: Direct observation of the reversible changes of the morphology of Pt nanoparticles under gas environment. J. Phys. Chem. C **114**(5), 2160–2163 (2010)
31. Giorgio, S., Cabie, M., Henry, C.R.: Dynamic observations of Au catalysts by environmental electron microscopy. Gold Bull. **41**(2), 167–173 (2008)
32. Hansen, P.L., Wagner, J.B., Helveg, S., Rostrup-Nielsen, J.R., Clauşen, B.S., Topsoe, H.: Atom-resolved imaging of dynamic shape changes in supported copper nanocrystals. Science **295**(5562), 2053–2055 (2002)
33. Gai, P.L.: Development of wet environmental TEM (Wet-ETEM) for in situ studies of liquid-catalyst reactions on the nanoscale. Microsc. Microanal. **8**(1), 21–28 (2002)
34. Williamson, M.J., Tromp, R.M., Vereecken, P.M., Hull, R., Ross, F.M.: Dynamic microscopy of nanoscale cluster growth at the solid-liquid interface. Nat. Mater. **2**(8), 532–536 (2003)
35. Zheng, H.M., Smith, R.K., Jun, Y.W., Kisielowski, C., Dahmen, U., Alivisatos, A.P.: Observation of single colloidal platinum nanocrystal Growth trajectories. Science **324**(5932), 1309–1312 (2009)
36. Ascencio, J.A., Gutiérrez-Wing, C., Espinosa, M.E., Marín, M., Tehuacanero, S., Zorrilla, C., José-Yacamán, M.: Structure determination of small particles by HREM imaging: theory and experiment. Surf. Sci. **396**(1–3), 349–368 (1998)
37. Reyes-Gasga, J., Tehuacanero-Nunez, S., Montejano-Carrizales, J.M., Gao, X.X., Jose-Yacaman, M.: Analysis of the contrast in icosahedral gold nanoparticles. Top. Catal. **46**(1–2), 23–30 (2007)
38. Wang, R., Zhang, H., Farle, M., Kisielowski, C.: Structural stability of icosahedral FePtFePt nanoparticles. Nanoscale **1**(2), 276–279 (2009)
39. Wang, R.M., Dmitrieva, O., Farle, M., Dumpich, G., Ye, H.Q., Poppa, H., Kilaas, R., Kisielowski, C.: Layer resolved structural relaxation at the surface of magnetic FePt icosahedral nanoparticles. Phys. Rev. Lett. **100**(1), 4 (2008)
40. Mejia-Rosales, S.J., Fernandez-Navarro, C., Perez-Tijerina, E., Blom, D.A., Allard, L.F., Jose-Yacaman, M.: On the structure of Au/Pd bimetallic nanoparticles. J. Phys. Chem. C **111**(3), 1256–1260 (2007)
41. Dmitrieva, O., Rellinghaus, B., Kastner, J., Dumpich, G.: Quantitative structure analysis of L1(0)-ordered FePt nanoparticles by HRTEM. J. Cryst. Growth **303**(2), 645–650 (2007)

42. Saha, D.K., Koga, K., Takeo, H.: Stable icosahedral nanoparticles in an as-grown Au–Fe alloy. Eur. Phys. J. D **9**(1–4), 539–542 (1999)
43. Rellinghaus, B., Dmitrieva, O., Stappert, S.: Destabilization of icosahedral structures in FePtFePt multiply twinned particles. J. Cryst. Growth **262**(1–4), 612–619 (2004)
44. Langlois, C., Oikawa, T., Bayle-Guillemaud, P., Ricolleau, C.: Energy-filtered electron microscopy for imaging core–shell nanostructures. J. Nanoparticles Res. **10**(6), 997–1007 (2008)
45. Scherzer, O.: The theoretical resolution limit of the electron microscope. J. Appl. Phys. **20**(1), 20 (1949)
46. Gratias, D., Portier, R.: Time-like perturbation method in high-energy electron-diffraction. Acta Crystallogr. Sect. A **39**, 576–584 (1983). (JUL)
47. Stadelmann, P.A.: EMS—a software package for electron-diffraction analysis and HREM image simulation in materials science. Ultramicroscopy **21**(2), 131–145 (1987)
48. Vandyck, D.: Fast computational procedures for the simulation of structure images in complex or disordered crystals—a new aproach. J. Microsc-Oxf. **119**(MAY), 141–152 (1980)
49. Biskupek, J., Jinschek, J.R., Wiedwald, U., Bendele, M., Han, L., Ziemann, P., Kaiser, U.: Identification of magnetic properties of few nm sized FePt crystalline particles by characterizing the intrinsic atom order using aberration corrected S/TEM. Ultramicroscopy **110**(7), 820–825 (2010)
50. Blanc, N., Tournus, F., Dupuis, V.: Measuring the L1(0) chemical order parameter of a single CoPt nanoparticle smaller than 4 nm. Phys. Rev. B **83**(9), 092403 (2011)
51. Colak, L., Hadjipanayis, G.C.: Evolution of texture and atomic order in annealed sinter-free FePt nanoparticles. Nanotechnology **19**(23), 235703 (2008)
52. Miyazaki, T., Kitakami, O., Okamoto, S., Shimada, Y., Akase, Z., Murakami, Y., Shindo, D., Takahashi, Y.K., Hono, K.: Size effectSize effect on the ordering of $L1_0$ FePt nanoparticles. Phys. Rev. B (Condens Matter Mater. Phys.) **72**(14), 144419 (2005)
53. Sato, K., Hirotsu, Y., Mori, H., Wang, Z., Hirayama, T.: Determination of order parameter of $L1_0$ FePd nanoparticles by electron diffraction. J. Appl. Phys. **97**(8), 084301 (2005)
54. Sato, K., Hirotsu, Y., Mori, H., Wang, Z., Hirayama, T.: Long-range order parameter of single $L1_0$-FePd nanoparticle determined by nanobeam electron diffraction: Particle size dependence of the order parameter. J. Appl. Phys. **98**(2), 024308 (2005)
55. Alloyeau, D., Ricolleau, C., Oikawa, T., Langlois, C., Le Bouar, Y., Loiseau, A.: STEM nanodiffraction technique for structural analysis of CoPt nanoparticles. Ultramicroscopy **108**(7), 656–662 (2008)
56. Ganesh, K.J., Kawasaki, M., Zhou, J.P., Ferreira, P.J.: D-STEM: A parallel electron diffraction technique applied to nanomaterials. Microsc. Microanal. **16**(05), 614–621 (2010)
57. Alloyeau, D., Ricolleau, C., Mottet, C., Oikawa, T., Langlois, C., Le Bouar, Y., Braidy, N., Loiseau, A.: Size and shape effects on the order-disorder phase transition in CoPt nanoparticles. Nat. Mater. **8**(12), 940–946 (2009)
58. Sato, K.: Magnetic nanoparticles: When atoms move around. Nat. Mater. **8**(12), 924–925 (2009)
59. Cliff, G., Lorimer, G.W.: Quantitative analysis of thin specimen. J. Microsc. **1032**, 203–207 (1975)
60. Van Cappellen, E.: The parameterless correction method in X-ray microanalysis. Microsc. Microanal. Microstruct. **1**(1), 1–22 (1990)
61. Alloyeau, D., Prévot, G., Le Bouar, Y., Oikawa, T., Langlois, C., Loiseau, A., Ricolleau, C.: Ostwald ripening in nanoalloys: When thermodynamics drives a size-dependent particle composition. Phys. Rev. Lett. **105**(25), 255901 (2010)
62. Di Vece, M., Bals, S., Verbeeck, J., Lievens, P., Van Tendeloo, G.: Compositional changes of Pd-Au bimetallic nanoclusters upon hydrogenation. Phys. Rev. B **80**(12), 125420 (2009)
63. Herzing, A.A., Watanabe, M., Edwards, J.K., Conte, M., Tang, Z.-R., Hutchings, G.J., Kiely, C.J.: Energy dispersive X-ray spectroscopy of bimetallic nanoparticles in an

aberration corrected scanning transmission electron microscope. Faraday Discuss. **138**, 337–351 (2008)
64. Lifshitz, I.M., Slyozov, V.V.: The kinetics of precipitation from supersaturated solid solutions. J. Phys. Chem. Solids **19**, 35–50 (1961)
65. Che, R., Peng, L.M., Chen, Q., Duan, X.F., Gu, Z.N.: Fe_2O_3 particles encapsulated inside aligned CN_x nanotubes. Appl. Phys. Lett. **82**(19), 3319–3321 (2003)
66. Trasobares, S., López-Haro, M., Kociak, M., March, K., de La Peña, F., Perez-Omil, J.A., Calvino, J.J., Lugg, N.R., D'Alfonso, A.J., Allen, L.J., Colliex, C.: Chemical imaging at atomic resolution as a technique to refine the local structure of nanocrystals. Angew. Chem. Int. Ed. **50**(4), 868–872 (2011)
67. Cho, S.-J., Idrobo, J.-C., Olamit, J., Liu, K., Browning, N.D., Kauzlarich, S.M.: Growth mechanisms and oxidation resistance of gold-Coated iron nanoparticles. Chem. Mater. **17**(12), 3181–3186 (2005)
68. Langlois, C., et al.: HAADF-STEM imaging of CuAg core-shell nanoparticles. J. Phys. Conf. Ser. **241**(1), 012043 (2010)
69. Michel, B., et al.: Mapping surface plasmons at the nanometre scale with an electron beam. Nanotechnology **18**(16), 165505 (2007)
70. Nelayah, J., Gu, L., Sigle, W., Koch, C.T., Pastoriza-Santos, I., Liz-Marzán, L.M., van Aken, P.A.: Direct imaging of surface plasmon resonances on single triangular silver nanoprisms at optical wavelength using low-loss EFTEM imaging. Opt. Lett. **34**(7), 1003–1005 (2009)
71. Nelayah, J., Kociak, M., Stephan, O., Garcia de Abajo, F.J., Tence, M., Henrard, L., Taverna, D., Pastoriza-Santos, I., Liz-Marzan, L.M., Colliex, C.: Mapping surface plasmons on a single metallic nanoparticle. Nat.Phys. **3**(5), 348–353 (2007)
72. N'Gom, M., Li, S., Schatz, G., Erni, R., Agarwal, A., Kotov, N., Norris, T.B.: Electron-beam mapping of plasmon resonances in electromagnetically interacting gold nanorods. Phys. Rev. B **80**(11), 113411 (2009)
73. Cao, L.-F., Xie, D., M-X., G., Park, H.-S., Fujital, T.: Size and shape effects on Curie temperature of ferromagnetic nanoparticles. Trans. Nonferrous Met. Soc. China **17**(6), 1451 (2007)
74. Oldenburg, S.J., Averitt, R.D., Westcott, S.L., Halas, N.J.: Nanoengineering of optical resonances. Chem. Phys. Lett. **288**(2–4), 243–247 (1998)
75. Kim, C., Lee, H.: Shape effectShape effect of Pt nanocrystals on electrocatalytic hydrogenation. Catal. Commun. **11**(1), 7–10 (2009)
76. McEwen, B.F., Marko, M.: The emergence of electron tomography electron tomography as an important tool for investigating cellular ultrastructure. J. Histochem. Cytochem. **49**(5), 553 (2001)
77. Saxton, W.O., Baumeister, W., Hahn, M.: Three-dimensional reconstruction of imperfect two-dimensional crystals. Ultramicroscopy **13**(1–2), 57–70 (1984)
78. Florea, I., Ersen, O., Hirlimann, C., Roiban, L., Deneuve, A., Houlle, M., Janowska, I., Nguyen, P., Pham, C., Pham-Huu, C.: Analytical electron tomography mapping of the SiC pore oxidation at the nanoscale. Nanoscale **2**(12), 2668–2678 (2010)
79. J. Radon: Über die Bestimmung von Funktionen durch ihre Integralwerte längs gewisser Mannigfaltigkeiten. Ber. Verh, K. Sachs. Ges. Wiss. Leipzig, Math. Nat. Kl **69**, 262–277 (1917)
80. Shaikh, T.R., Gao, H., Baxter, W.T., Asturias, F.J., Boisset, N., Leith, A., Frank, J.: SPIDER image processing for single-particle reconstruction of biological macromolecules from electron micrographs. Nat. Protoc. **3**(12), 1941–1974 (2008)
81. Xu, W., Xu, F., Jones, M., Keszthelyi, B., Sedat, J., Agard, D., Mueller, K.: High-performance iterative electron tomography reconstruction with long-object compensation using graphics processing units (GPUs). J. Struct. Biol. **171**(2), 142–153 (2010)
82. Midgley, P.A., Weyland, M., Yates, T.J.V., Arslan, I., Dunin-Borkowski, R.E., Thomas, J.M.: Nanoscale scanning transmission electron tomography. J. Microsc.-Oxf. **223**, 185–190 (2006)

83. Penczek, P., Marko, M., Buttle, K., Frank, J.: Double-tilt electron tomography. Ultramicroscopy **60**(3), 393–410 (1995)
84. Alloyeau, D., Ricolleau, C., Oikawa, T., Langlois, C., Le Bouar, Y., Loiseau, A.: Comparing electron tomography and HRTEM slicing methods as tools to measure the thickness of nanoparticles. Ultramicroscopy **109**(7), 788–796 (2009)
85. Harauz, G., Van Heel, M.: Exact filters for general geometry three dimensional reconstruction. Optik **73**, 146 (1986)
86. Sousa, D., Grigorieff, N.: Ab initio resolution measurement for single particle structures. J. Struct. Biol. **157**(1), 201–210 (2007)
87. Hernandez, J.C., Moreno, M.S., Coronado, E.A., Midgley, P.A.: STEM electron tomography of gold nanostructures European Microscopy Congres, p. 311. Springer, Berlin Heidelberg (2008)
88. Erni, R., Rossell, M.D., Kisielowski, C., Dahmen, U.: Atomic-resolution imaging with a sub-50-pm electron probe. Phys. Rev. Lett. **102**(9), 096101 (2009)
89. Kisielowski, C., Freitag, B., Bischoff, M., van Lin, H., Lazar, S., Knippels, G., Tiemeijer, P., van der Stam, M., von Harrach, S., Stekelenburg, M., Haider, M., Uhlemann, S., Müller, H., Hartel, P., Kabius, B., Miller, D., Petrov, I., Olson, E.A., Donchev, T., Kenik, E.A., Lupini, A.R., Bentley, J., Pennycook, S.J., Anderson, I.M., Minor, A.M., Schmid, A.K., Duden, T., Radmilovic, V., Ramasse, Q.M., Watanabe, M., Erni, R., Stach, E.A., Denes, P., Dahmen, U.: Detection of single atoms and buried defects in three dimensions by aberration-corrected electron microscope with 0.5-Å information limit. Microsc. Microanal. **14**(05), 469–477 (2008)
90. Gontard, L.C., Chang, L.Y., Hetherington, C.J.D., Kirkland, A.I., Ozkaya, D., Dunin-Borkowski, R.E.: Aberration-corrected imaging of active sites on industrial catalyst nanoparticles. Angew. Chem.-Int. Ed. **46**(20), 3683–3685 (2007)
91. Kisielowski, C., Ramasse, Q.M., Hansen, L.P., Brorson, M., Carlsson, A., Molenbroek, A.M., Topsoe, H., Helveg, S.: Imaging MoS(2) nanocatalysts with single-atom sensitivity. Angew. Chem.-Int. Edit. **49**(15), 2708–2710 (2010)
92. Specht, P., Gulotty, R.J., Barton, D., Cieslinski, R., Rozeveld, S., Kang, J.H., Dubon, O.D., Kisielowski, C.: Quantitative contrast evaluation of an industry-style rhodium Nanocatalyst with single atom sensitivity. ChemCatChem. **3**(6), 1034–1037 (2011)
93. Gai, P.L., Boyes, E.D.: Advances in atomic resolution in situ environmental transmission electron microscopy and 1 angstrom aberration corrected in situ electron microscopy. Microsc. Res. Tech. **72**(3), 153–164 (2009)
94. Gai, P.L., Boyes, E.D.: Angstrom analysis with dynamic in situ aberration corrected electron microscopy. Electron Microscopy and Analysis Group Conference 2009, vol. 241. Journal of Physics Conference Series. Iop Publishing Ltd, Bristol (2010)
95. Shiju, N.R., Yoshida, K., Boyes, E.D., Brown, R., Gai, P.L.: Dynamic atomic scale in situ electron microscopy in the development of an efficient heterogeneous catalytic process for pharmaceutical NSAIDS. Catal. Sci. Technol. **1**(3), 413–425 (2011)
96. Kisielowski, C., Specht, P., Alloyeau, D., Erni, R., Ramasse, Q.: Aberration-corrected electron microscopy imaging for nanoelectronics applications. In: Erik, M.S., David, G.S., Rajinder, P.K., Dan, H., Garner, C.M., Robert, M., Alain, C.D. (eds.) pp. 231–240. American Institute of Physics, Merville, USA (2009)
97. Girit, C.O., Meyer, J.C., Erni, R., Rossell, M.D., Kisielowski, C., Yang, L., Park, C.-H., Crommie, M.F., Cohen, M.L., Louie, S.G., Zettl, A.: Graphene at the edge: Stability and dynamics. Science **323**(5922), 1705–1708 (2009)
98. Lee, Z., Jeon, K.J., Dato, A., Erni, R., Richardson, T.J., Frenklach, M., Radmilovic, V.: Direct imaging of soft-hard interfaces enabled by graphene. Nano Lett. **9**(9), 3365–3369 (2009)
99. Li, Z.Y., Young, N.P., Di Vece, M., Palomba, S., Palmer, R.E., Bleloch, A.L., Curley, B.C., Johnston, R.L., Jiang, J., Yuan, J.: Three-dimensional atomic-scale structure of size-selected gold nanoclusters. Nature **451**(7174), 46–48 (2008)

100. Ferrer, D., Blom, D.A., Allard, L.F., Mejia, S., Perez-Tijerina, E., Jose-Yacaman, M.: Atomic structureAtomic structure of three-layer Au/Pd nanoparticles revealed by aberration-corrected scanning transmission electron microscopy. J. Mater. Chem. **18**(21), 2442–2446 (2008)
101. Liu, J.: Advanced electron microscopy of metal–support Interactions in supported metal catalysts. ChemCatChem **3**(6), 934–948 (2011)

Electronic Structure of Nanoalloys: A Guide of Useful Concepts and Tools

Guy Tréglia, Christine Goyhenex, Christine Mottet, Bernard Legrand and Francois Ducastelle

Abstract The aim of this lecture is to give an overlook about methods developed in infinite (bulk) and semi-infinite (surface) metallic materials and some tracks to extend them to finite size systems. In this framework we will first study the effect of bond breaking and dimension lowering on electronic structure, at surfaces of pure metals (surface states, atomic level shifts, reconstructions and relaxations) and in monometallic clusters. Then we will illustrate the influence of chemical ordering on electronic structure (and vice versa) by considering firstly bulk alloys (diagonal *versus* off-diagonal disorder) and then bimetallic surfaces (stress effect induced by either surface segregation or epitaxial growth). These two approaches will then naturally be combined in the peculiar case of nanoalloys. The methods will be developed following two main goals. The first one is to determine local electronic densities of states (LDOS), the knowledge of which is essential to the understanding and the analysis of nano-objects. The second one is to derive from these LDOS energetic models well suited to both the degree of complexity of the systems under study (bulk and surface crystalline structure, chemical ordering, ...) and their implementation in numerical simulations (Molecular Dynamics, Monte

G. Tréglia (✉) · C. Mottet
CINaM – CNRS, Campus de Luminy, Case 913, 13288
Marseille Cedex 9, France
e-mail: treglia@cinam.univ-mrs.fr

C. Goyhenex
IPCMS – CNRS – UDS, 23 rue du Loess, 67034
Strasbourg cedex 2, France

B. Legrand
SRMP – DMN, CEA Saclay, 91191
Gif sur Yvette Cedex, France

F. Ducastelle
LEM – CNRS/ONERA, B.P. 72, 92322
Châtillon cedex, France

D. Alloyeau et al. (eds.), *Nanoalloys*, Engineering Materials,
DOI: 10.1007/978-1-4471-4014-6_5,

Carlo). The different sections of the lecture will be illustrated by examples issued from studies performed on systems which can be considered as archetypal in the nano-alloy community, such as CoPt, CoAu and CuAg.

1 Introduction

The peculiar properties of nanoalloys depend on the *local* electronic structure on the various inequivalent sites resulting from their chemical and morphological structures, which in turn depend on this electronic structure. It is then essential to understand how these structures are coupled to one another and how they vary with the metal species, the concentration and the orientation of the surface for semi-infinite materials and/or the cluster size for finite ones. Modelling these phenomena would indeed allow us to design the best suited binary system for a given property.

The aim of this lecture is to give the tools for characterizing the electronic structure of bimetallic clusters, and to show how they can be used to predict both their atomic and chemical structures. These electronic structure methods extend from ab initio calculations to *semi-phenomenological* models such as *Tight-Binding* approximation for transition metals. Since the most commonly used nanoalloys are made of metals of the end of the transition series, we will put some emphasis on the latter by giving some details on moment and continued fraction methods. We will underline how the electronic structure is modified at surface and cluster sites, first for pure metals and then for bimetallic systems. Then, we will show how the energetics of the system (cohesive energy, surface tension, mixing energy) can be derived from electronic structure by using more or less sophisticated many-body potentials (*SMA: Second Moment Approximation, TBIM: Tight-Binding Ising Model*). This will allow us to get trends as a function of the number of valence electrons for various properties such as the crystalline structure of pure metals, the relaxation or reconstruction of surfaces, the shape of clusters and finally the chemical structure of infinite systems (tendency to ordering or phase separation) and finite ones (surface or site segregation). In turn, we will illustrate the dependence of the local densities of states with respect to the equilibrium (geometrical, chemical) environment defined as above.

2 Concepts and Methods (Pure Bulk Metal)

2.1 Chemical Bonding and Periodic Table

The Hamiltonian of a system with N nuclei located at R, and N_e electrons, located at r with a spin σ, writes in the most general way:

$$H = \sum_{I=1,N} \frac{P_I^2}{2M_I} + \sum_{i=1,N_e} \frac{p_i^2}{2m} + \sum_{i,j} \frac{e^2}{|\mathbf{r}_i - \mathbf{r}_j|} + \sum_{I,J} \frac{Z_I Z_J e^2}{|\mathbf{R}_i - \mathbf{R}_j|} - \sum_{i,I} \frac{Z_I e^2}{|\mathbf{r}_i - \mathbf{R}_j|} \quad (1)$$

This Hamiltonian acts on a many-body wave function $\Phi(x,R)$ which depends on both nuclei (R) and electron ($x \cong (r,\sigma)$) coordinates. Due to the mass difference between the electrons and nuclei, we can decouple their respective movements (adiabatic approximation) which allows one to write the total wave-function as the product of those of the electrons $\psi(x,R)$ and of the nuclei $\chi(R)$: $\Phi(x,R) = \psi(x,R).\chi(R)$. Solving exactly the Schrödinger equation $H\psi = E\psi$ for the electrons is only possible in the simple case of the hydrogen atom with only one proton and one electron. In that case, using spherical coordinates, the solution writes as the product of a radial function and of a spherical harmonic:

$$\psi_{nlm}(r,\theta,\varphi) = R_n^l(r) Y_l^m(\theta,\phi) \quad (2)$$

which involves three quantum numbers n (principal: $n \geq 0$), l (azimuthal: $n \geq l \geq 0$) and m (magnetic: $l \geq m \geq -l$), plus a fourth number for the spin ($s = \pm 1/2$). The energy associated to the function ψ_{nlm} only depends on n ($E_n = -E_0/n^2$, with $E_0 = 13.6$ eV) so that the ground state of the system, which corresponds to the minimal energy, is obtained by filling the respective levels as a function of increasing n. All the electronic states corresponding to the same energy are then labelled by n, even though their properties essentially depend on the value of l, which drives the shape of the orbitals (see Fig. 1a), giving rise to the usual denomination: ns ($l = 0$), np ($l = 1$), nd ($l = 2$), nf ($l = 3$), ... The magnetic number gives the degeneracy of each state (*i.e.* the maximal number of electrons it can contain) which, counting the spin, is: ns^2, np^6, nd^{10}, nf^{14}, ...

In fact, the degeneracy of the different l-levels corresponding to a given n state will then be lifted by introducing interaction between electrons for atoms containing more than one electron, leading to the variation of E_{nl} schematized in Fig. 1b, which directly leads to the classification of all elements within the Mendeleiev classification table. The various types of elements, characterized by the nature of their valence electrons (*s* for simple metals, *sp* for covalent elements, *d* for transition metals and *sd* for noble metals) are illustrated in Fig. 2.

2.2 One electron approximation: band structure (Hartree–Fock, DFT)

In condensed matter, one has to deal with the general problem of N_e electrons moving in the potential V^{ion} of N fixed ions. The Hamiltonian then writes:

$$H = \sum_{i=1}^{N_e} \left(\frac{p_i^2}{2m} + V^{ion}(\mathbf{r}_i) \right) + \frac{1}{2} \sum_{i,j=1}^{N_e} \frac{e^2}{|r_i - r_j|} \quad (3)$$

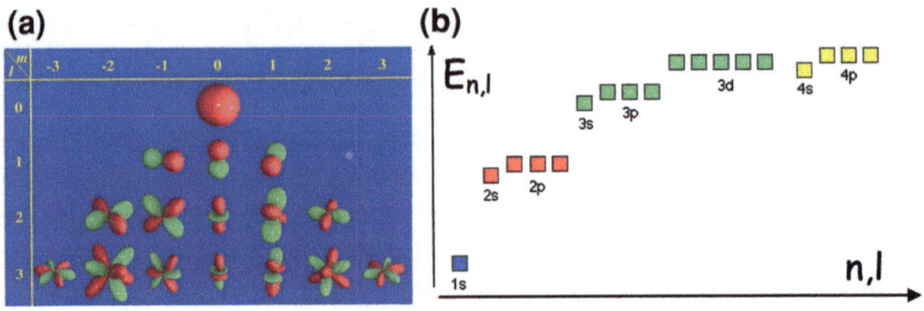

Fig. 1 Schematic orbitals (*left*) for the various (*l,m*) numbers and corresponding level energies (*right*)

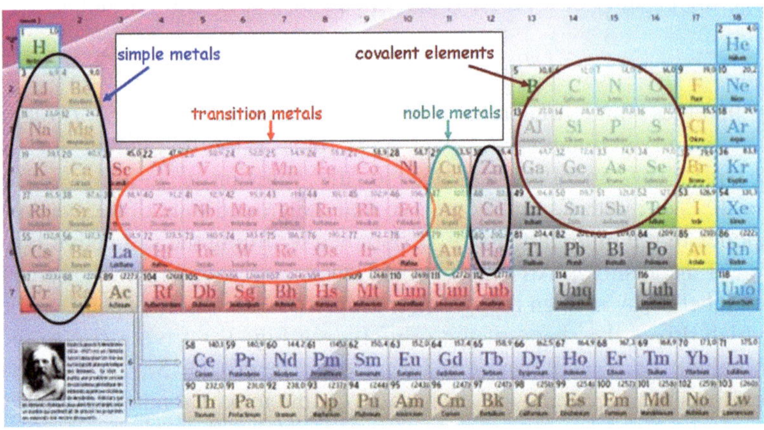

Fig. 2 Mendeleïev periodic classification table

The motions of the electrons are then correlated due to their Coulomb interactions, which is a quantic many-body problem involving at least all valence electrons (external shells). Assuming that a given (single) electron interacts with all the others by means of an effective mean-field $V^{eff}(r)$, Eq. (3) reduces to a "one electron" Hamiltonian, the eigenfunctions and eigenvalues of which are solutions of the Schrödinger equation:

$$\left[\frac{p^2}{2m} + V^{ion}(\mathbf{r}) + V^{eff}(\mathbf{r})\right]\psi_\alpha(\mathbf{r}) = \varepsilon_\alpha\psi_\alpha(\mathbf{r}) \tag{4}$$

The ground state of the system at $T = 0$ K is obtained by stacking the electrons in the lowest energy states available, leading to *N*-body states characterized by the occupation numbers (or Fermi functions) f_α which are defined such as $f_\alpha = 1$ if $\varepsilon_\alpha < E_F$ (E_F being the Fermi level) and $f_\alpha = 0$ otherwise. The spatial density of states is then defined as:

$$n(\mathbf{r}) = \sum_\alpha f_\alpha |\psi_\alpha(\mathbf{r})|^2 \tag{5}$$

Within the Hartree approximation, the effective potential writes:

$$V^{eff}(\mathbf{r}) = V^H(\mathbf{r}) = e^2 \int d\vec{r}' \frac{n(\mathbf{r}')}{|\mathbf{r} - \mathbf{r}'|} \tag{6}$$

The Eq. (4) is solved by an iterative procedure, starting from an initial (guessed) density of states n(r), which allows one to calculate $V^{eff}(r)$ using (6), then to solve the Eq. (4) from which one obtains ψ_α and then $n(r)$ through (5). The procedure is then iterated as long as self-consistency is not achieved.

Unfortunately, in spite of its physical content, this remains an approximation which does not account for the correlated motion of all the electrons. In particular, this Hartree approximation does not account for the Pauli principle and then totally misses the existence of the so-called *exchange and correlation hole* which makes electrons avoiding each other at short distance. This is somewhat corrected in the Hartree–Fock approximation which improves the Hartree potential by including a so-called *exchange* contribution, which damps the Coulomb potential contribution for parallel spins. Unfortunately, this is a rather asymmetric way to treat the electron interactions since all the electrons should avoid one another. Therefore, whereas the electronic correlations are completely neglected in the Hartree scheme, they are treated in a too much asymmetric way in the Hartree-Fock approximation.

A main progress with the Density Functional Theory (DFT), which is the most widely used ab initio method, is that it treats the correlations in a more symmetric way. It is based on the Hohenberg and Kohn theorem [1], which assumes that the ground state energy E_0 of an inhomogeneous interacting electron gas under an external potential V^{ion} can be written as a functional of the charge density $n(r)$, $E_0 = E_0[n(r)]$, which is minimum for the real density of the system. This leads to write $n(r)$ under the same "one electron" form as (5), using wave functions $\psi\alpha$ which are solutions of a Hamiltonian similar to (4), but with now an effective potential:

$$V^{eff}(\mathbf{r}) = V^H(\mathbf{r}) + V^{xc}(\mathbf{r}) \tag{7}$$

which differs from (6) by the introduction of an *exchange–correlation* term V^{xc}, which is the functional derivative of a contribution $E^{xc}[n(r)]$ to $E_0\ [n(r)]$:

$$V^{xc}(\mathbf{r}) = \frac{\partial E^{xc}[n(\mathbf{r})]}{\partial n(\mathbf{r})} \tag{8}$$

All the difficulties are then transferred in this term which has to be approximated. Within the usual Local Density Approximation (LDA), one assumes that E^{xc} is a local functional of $n(r)$, i.e., that it is defined from the knowledge of the density at **r** only.

$$E^{xc}[n(\mathbf{r})] \approx \int n(\mathbf{r})\varepsilon^{xc}[n(\mathbf{r})]d\mathbf{r} \tag{9}$$

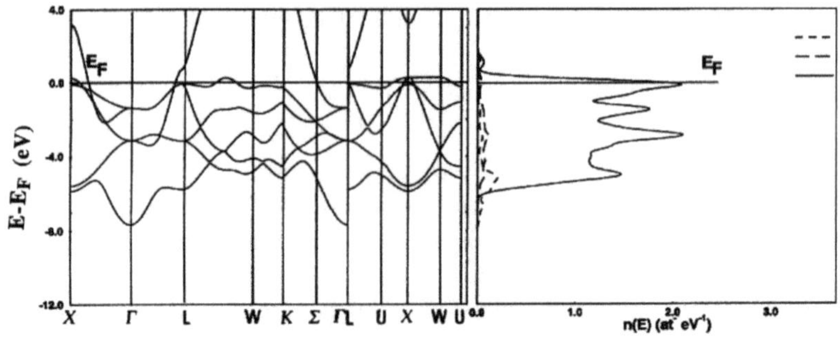

Fig. 3 LMTO calculation of the band structure and density of states of Pd. The full line indicates the *d*-partial LDOS and the *dotted* and *dashed* ones the *s–p* ones. Courtesy of S. Sawaya

$\varepsilon^{xc}(n)$ is the so-called "exchange and correlation" energy of a uniform electron system with density n, which is widely taken as the average value of the exchange potential for a free electron system ($\propto n^{1/3}$), weighted by an empirical factor α (X_α method). When this local approximation fails, due to rapid variations of the density, it can be corrected by introducing corrections linked to the gradient of the density: $\vec{\nabla}n(\vec{r})$ (GGA: Generalized Gradient Approximation). Finally, inserting Eqs. (7) an (8) in the Eq. (4) leads to the well-known Kohn–Sham equations which have to be solved iteratively by using the same self-consistent procedure as already described. From the resulting eigenvalues ε_α one can then access to the electronic band structure, and from that to the density of states following:

$$n(E) = \sum_\alpha \delta(E - \varepsilon_\alpha) = Tr\delta(E - H) \tag{10}$$

where the operator $\delta(E{-}H)$ is defined by: $\delta(E - H).\psi_\alpha = \delta(E - \varepsilon_\alpha).\psi_\alpha$ and the trace (Tr) is performed on the electronic states α. This is illustrated in the case of Pd (Fig. 3).

Even though DFT is a non parameterized method, it requires performing some important choices, in addition to that of the exchange–correlation term (LDA, GGA). The first one is that of the electron potential among a wide variety: full potential (FP), "muffin tin" (MT) potentials (the potential is calculated exactly in spherical regions centred on the nuclei whereas it is taken equal to zero in the interstitial region), atomic sphere approximation (ASA) or pseudopotentials (PP). The latter have been developed to explain how a nearly-free behaviour of electrons could be consistent with a potential V^{ion} which diverges in the ion vicinity. Indeed in this region, their wave functions oscillate rapidly to orthogonalize to the inner shell states, leading to a large kinetic energy which almost compensates the potential energy. One can then define weak pseudopotentials associated to pseudo-nearly free wave functions (e.g. Ashcroft empty core [2]). The second choice is that of the basis which determines the efficiency of the method depending on the system under study. In this framework the plane waves basis provides the simplicity and speed of Fourier development whereas localised orbitals (Gaussians for chemists, or numerical in SIESTA) have the advantage to give a quasi-atomic view (consistent with Tight-Binding approximation, see later). Finally augmented methods (APW) combine the best of these two opposite points of view, by calculating wave functions at fixed energy inside an atomic sphere, which are matched to plane waves outside.

The DFT method not only gives very good results concerning both the band structure and density of states, but also for the lattice parameter, elastic constants and cohesive energies, at least when gradient corrections are taken into account. This is very satisfying since this method is ab initio, *i.e.,* "without parameters" (contrary to more empirical methods which will be developed later), which does not mean that it is "without approximation" as shown above! Nevertheless, this method remains less suited for non periodic systems, in presence of defects, and tedious to use coupled with numerical simulations such as Molecular Dynamics… even though *Car-Parinello* type methods [3] have been developed which take into account simultaneously the movements of ions and electrons. But such methods remain heavy to handle for large systems, which justify developing simpler methods, using semi-empirical potentials suited to the system under study.

2.3 Tight-Binding Approximation and Local Density of States

The Tight-Binding (TB) method [4] starts from isolated atoms with discrete levels, which form energy bands when the atomic wave functions overlap… but not too much! It assumes that any one electron electronic state $\psi(r)$, delocalised in the solid, can be written as a linear combination of atomic orbitals (LCAO) $|n, \lambda\rangle$ where λ labels the orbital at site n: $\psi(r) = \sum_{n,\lambda} a_n^\lambda |n, \lambda\rangle$, which is the more justified as the overlap between the orbitals is weak (*d* states of transition metals, *sp* valence electrons of semi-conductors,…). The corresponding TB Hamiltonian then writes:

$$H = \sum_{n,\lambda} |n, \lambda\rangle(\varepsilon_{\lambda,0} + \alpha_\lambda)\langle n, \lambda| + \sum_{n,m,\lambda,\mu} |n, \lambda\rangle \beta_{nm}^{\lambda\mu} \langle m, \mu| \tag{11}$$

in which $\varepsilon_{\lambda,0}$, α_λ and $\beta_{nm}^{\lambda\mu}$ are respectively the atomic level, crystal field and hopping integrals, the latter being rapidly damped (after 1st or 2nd neighbours) and directly related to the bandwidth. Due to the spherical symmetry of atomic potentials, the $[\beta]$ matrix is diagonal for each l-value in the basis of spherical harmonics with the z-axis along $(m\text{–}n)$, with eigenvalues defined as the integrals σ, π, δ according to the quantum magnetic number $|m| = 0, 1, 2$. This leads to different hopping integrals labelled $ss\sigma$, $pp\sigma$, $pp\pi$, $dd\sigma$, $dd\pi$, $dd\delta$ (see Fig. 4) to which are added integrals coupling two different l: $sp\sigma$, $sd\sigma$, $pd\sigma$, $pd\pi$. In this framework, one can define *d–d* canonical parameters such as: $|dd\sigma| \approx 2\ |dd\pi|$, $dd\delta \approx 0$ [5].

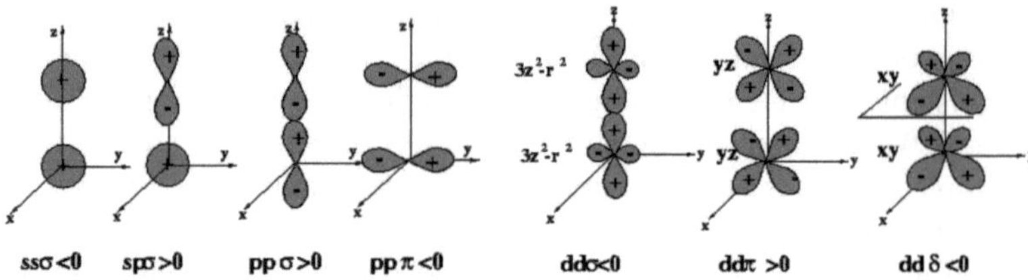

Fig. 4 Schematic *sp–sp* and *d–d* hopping integrals

The essential advantage of the method is that it allows working in the direct space to calculate densities and energies, without resorting to diagonalisation of the Hamiltonian, and then without need for the Bloch theorem. This allows one to deal with non crystalline solids and defects. Indeed, one can derive $n(E)$ from the trace of $\delta(E-H)$ Eq. (10) which can be calculated within any basis, and in particular in the basis of atomic orbitals $|n,\lambda\rangle$. More precisely, using mathematical properties of δ-functions, one can define in a simple way the local density of states (LDOS) at a given site n_0,

$$n_{n_0}(E) = \lim_{\eta \to 0^+} \left[-\frac{\text{Im}}{\pi} \sum_{\lambda} \langle n_0, \lambda | G(E + i\eta) | n_0, \lambda \rangle \right] \tag{12}$$

without resorting to any periodicity condition (n_0 can be a defect site), from the projection of the Green function: $G(z) = (z-H)^{-1}$ on the atomic orbital at site n_0. This projection writes as a continued fraction [6],

$$\langle n_0, \lambda | G(z) | n_0, \lambda \rangle = \cfrac{1}{z - a_1 - \cfrac{b_1^2}{z - a_2 - \cfrac{b_2^2}{z - a_3 - \cfrac{b_3^2}{\cdots\cdots}}}} \tag{13}$$

the coefficients of which can be calculated by two different ways. The first one is to derive them from the knowledge of the p first moments μ_p of $n_{n_0}(E)$:

$$\begin{aligned} \mu_p(n_0) &= \int_{-\infty}^{+\infty} E^p n_{n_0}(E) dE = \sum_{\lambda} \langle n_0, \lambda | H^p | n_0, \lambda \rangle \\ &= \sum_{\lambda, i\mu, j\nu, \ldots} \langle n_0, \lambda | H | i, \mu \rangle \langle i, \mu | H | j, \nu \rangle \ldots \langle ., . | H | n_0, \lambda \rangle \end{aligned} \tag{14}$$

which gives more and more details on the LDOS when p increases, and are obtained by counting closed paths on the lattice [6]. The second way is to calculate them directly by constructing a new basis tridiagonalising H within the so-called *recursion* method [7]. The LDOS is the most precise as the number of calculated coefficients is large, since N pairs of exact coefficients ensure the LDOS to have $2N$ exact moments. The problem is then to terminate the continued fraction. For a

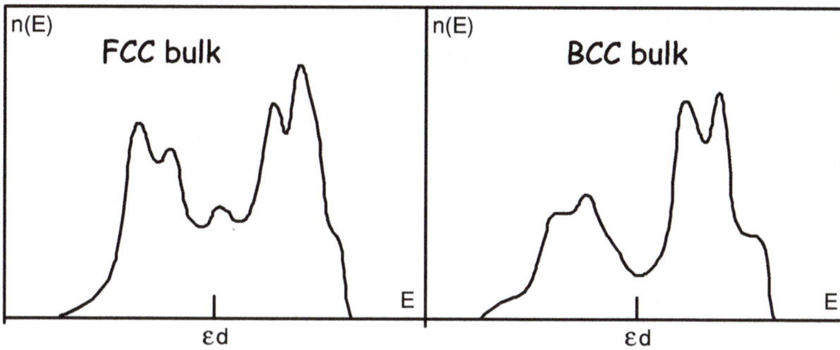

Fig. 5 Typical *d*-LDOS for fcc (*left*) and bcc (*right*) bulk structures

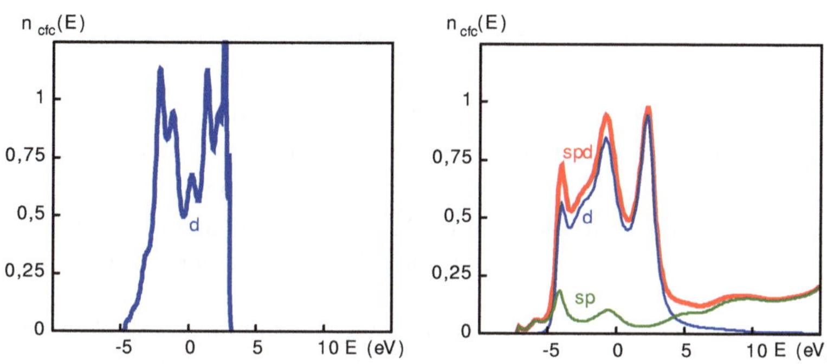

Fig. 6 Influence of *sp-d* hybridization on the fcc LDOS

bulk material, the coefficients converge towards asymptotic values a_∞ and b_∞ which are related to band edges, at least for a band without gap [8], so that they can be fitted to band structure calculations.

In this framework, restricting ourselves to *d*-orbitals as commonly admitted for transition metals and using canonical Slater parameters, one obtains archetypal LDOS for the different crystallographic structures shown in Fig. 5. As can be seen, the fcc LDOS is characterized by a high peak in the upper part which is at the origin of the possible occurrence of magnetism (see Sect. 2.5), whereas the bcc one presents a quasi-gap in the middle of the band, which separates bonding states from anti-bonding ones, which tends to favour strongly this structure for half-filled *d*-band elements. However, it is worth noticing that at least at the end of the transition series, it is necessary to take into account the *s* and *p* valence electrons and their hybridization with the *d* ones to get a density of states in good agreement with that derived from DFT calculations [9]. As can be seen in Fig. 6, this strongly modifies the LDOS.

2.4 Energetics and Link Between Electronic and Crystallographic Structure

As usual in any mean-field approximation, the total band energy of the system is not equal to the sum over the one-electron energies (ε_α), since it counts twice the electron–electron interactions which have to be substracted once. The DFT band energy then writes:

$$E_{b,0}^{LDA} = \sum_\alpha f_\alpha \varepsilon_\alpha - \frac{1}{2}\int d\mathbf{r} V^H(\mathbf{r})n(\mathbf{r}) + \int d\mathbf{r} n(\mathbf{r})\left(\varepsilon^{xc}[n(\mathbf{r})] - \frac{\partial \varepsilon^{xc}[n(\mathbf{r})]}{\partial n(\mathbf{r})}\right) \quad (15)$$

The total energy is then obtained by adding the ion–ion contribution to the band one:

$$E_0 = E_{b,0}^{DFT} + \frac{1}{2}\int d\mathbf{r} V^{ion-ion}(\mathbf{r}) n_{ion}(\mathbf{r}) \quad \text{with } V^{ion-ion}(\mathbf{r}) = \int d\mathbf{r}' \frac{n_{ion}(\mathbf{r}')}{|\mathbf{r}-\mathbf{r}'|} \quad (16)$$

where $n_{ion}(r)$ is the ionic density: $n_{ion}(r) = Z\,\delta(r\text{-}n)$, for Z charges at sites *n*.

Within the TB approximation, one can develop this equation, by introducing the local density of states $n_n(E)$, and the corresponding charge $N_n (= \int^{E_F} n_n(E)dE)$ at site *n* and by assuming charge neutrality ($N_n = Z_n$ ionic charge) [10]. The cohesive energy is then obtained by subtracting the reference of isolated atoms:

$$E_{coh} = \sum_{n,\lambda} \int^{E_F} (E - \varepsilon_{0,\lambda}) n_n^\lambda(E) dE + \frac{1}{2}\sum_{n \neq m} \iint dr dr' \frac{Q_n(r) Q_m(r')}{|r - r'|} \quad (17)$$

The first term is the band energy ($E_{coh,b}$) and the second one the pair interaction ($E_{coh,r}$) between neutral atoms with charge density: $Q_n(r) = Z_n\,\delta(r-n) - N_n(r-n)$. Unfortunately, $E_{coh,r}$ is not sufficient to account for the repulsive part of the energy. Actually, the TB approximation fails to reproduce part of the repulsion at short distance since it does not account for the non-orthogonality of wave functions on different sites and for the compression of *sp* electrons which play an important role before the Coulomb repulsion becomes really efficient.

Therefore, in a first step, we will put some emphasis on properties for which the dependence with *d*-band filling suggests that they are mainly driven by the band term.

This is in particular the case of the quasi-parabolic variation of the cohesive energy but also of the atomic volume and bulk modulus experimentally evidenced for each of the transition series (see Fig. 7). This parabolic behaviour is indeed reproduced by calculating the band term of (17) from the previous LDOS. In fact, the integral depending weakly on details of *n(E)*, this band term can be approximated from a schematic rectangular density of states presenting the same second moment (related to the mean width of the LDOS) as the exact one: $E_{coh,b} = -N_e(10 - N_e)\beta\sqrt{Z}$ where β is an "effective" hopping integral, corresponding to the *Z* first neighbours. Obviously, some features are not well reproduced in this crude approximation, in particular the asymmetry of the trends which requires a

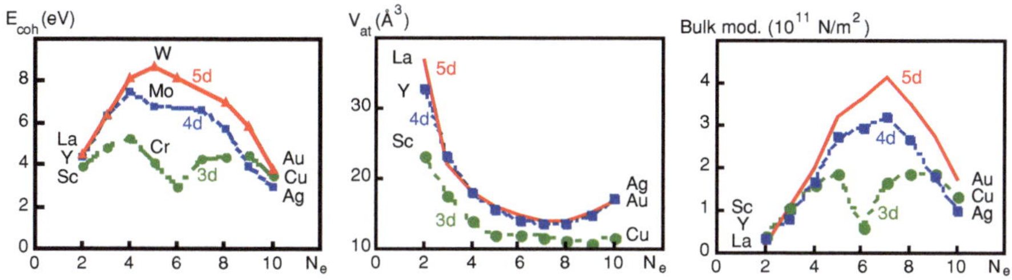

Fig. 7 Experimental variation of the cohesive energy, atomic volume and bulk modulus along transition metal series

Table 1 Crystallographic structure of transition metals

N_e	2	3	4	5	6	7	8	9	10
	Sc	Ti	V	Cr	Mn	Fe	Co	Ni	Cu
	Y	Zr	Nb	Mo	Tc	Ru	Rh	Pd	Ag
	La	Hf	Ta	W	Re	Os	Ir	Pt	Au
	HCP		CC		HCP		FCC		

larger number of exact moments and/or including *sp-d* hybridization to be accounted for. The deep hole of E_{coh} for the first series is attributed to magnetism.

On the other hand, as can be seen in Table 5.1, the crystalline structure of transition metals is clearly related to the band filling. As shown in Fig. 8 the main trends are correctly reproduced by the TB calculation for the bcc/fcc as well as for hcp/fcc systematics (which of course requires to go beyond second moment since hcp and fcc structures are identical up to second neighbours), except for the nearly filled band for which the bcc structure is found instead of fcc. Fortunately, this is corrected if one takes into account *sp-d* hybridization which, as previously mentioned, plays a major role at the end of transition series. In fact, in the case of the preference for hcp or fcc structure, which involves a weak energy balance and an accuracy of the LDOS beyond second moment, the *sp-d* hybridization plays a role on the overall trend, consistently with the corresponding influence on the shape of the LDOS.

As shown in Fig. 8, the *sp*-contribution to the energy balance is small as expected since the overall behaviour is driven by that of the partial *d*-band which significantly differs from the non hybridized *d*-band. In the following, it has to be kept in mind that the energetics only depends on the *d*-band, but once distorted by the *sp-d* hybridization.

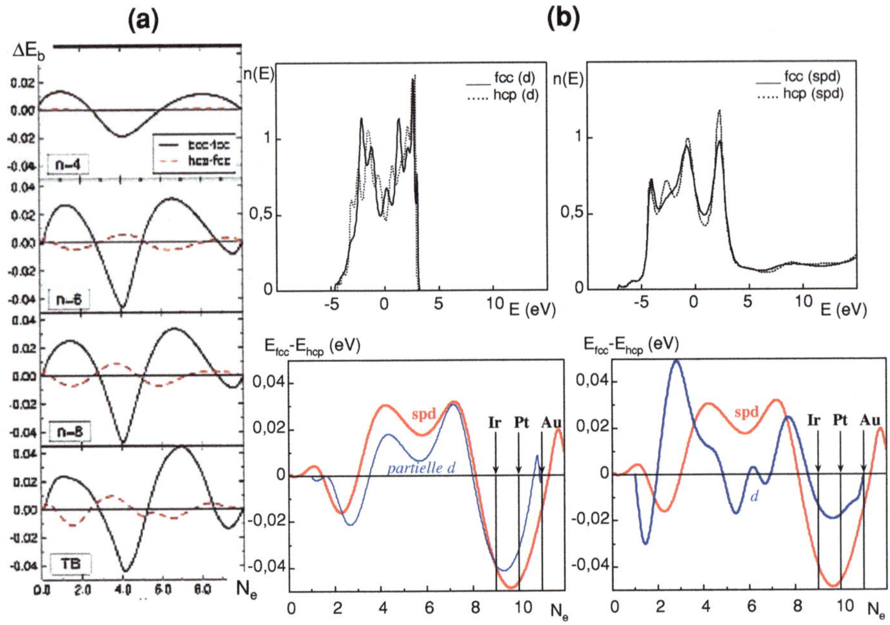

Fig. 8 **a** Stabilities of fcc relative to bcc (*a*) and hcp (*b*) structures in the tight-binding framework (note the two different orders of magnitudes). Effect of *sp-d* hybridization on the latter competition is also shown in (**b**). From Refs. [9–11], copyright (2008), with permission from Elsevier

2.5 Magnetism Within TB Approximation

For a few metallic elements, the energy of the system can be lowered by shifting the two spin bands, inducing different numbers of electrons with *up* and *down* spins ($N_\uparrow$, $N_\downarrow$), and therefore a finite magnetic moment $\mu = N_\uparrow - N_\downarrow$. In the framework of collinear magnetism and in absence of spin–orbit coupling, the *up* and *down* states are decoupled, so that the sub-systems of *up* and *down* electrons can be treated separately, keeping in mind that one must define a single Fermi level (E_F) for both spin directions in order to get the right total *d*-band filling $N_e = N_\uparrow + N_\downarrow$. Thus, in a canonical approach, each spin partial LDOS is obtained from the paramagnetic one $n_0(E)$ by simply shifting its barycentre ε_0 by $\pm\Delta\varepsilon/2$:

$$n_\uparrow(E) = n_0\left(E + \frac{\Delta\varepsilon}{2}\right) \; n_\downarrow(E) = n_0\left(E - \frac{\Delta\varepsilon}{2}\right)$$

In that case, the magnetic moment is simply given by:

$$\mu = N_\uparrow(E_F) - N_\downarrow(E_F) = \int_{E_F - \frac{\Delta\varepsilon}{2}}^{E_F + \frac{\Delta\varepsilon}{2}} n_0(E)dE \qquad \text{(18a)}$$

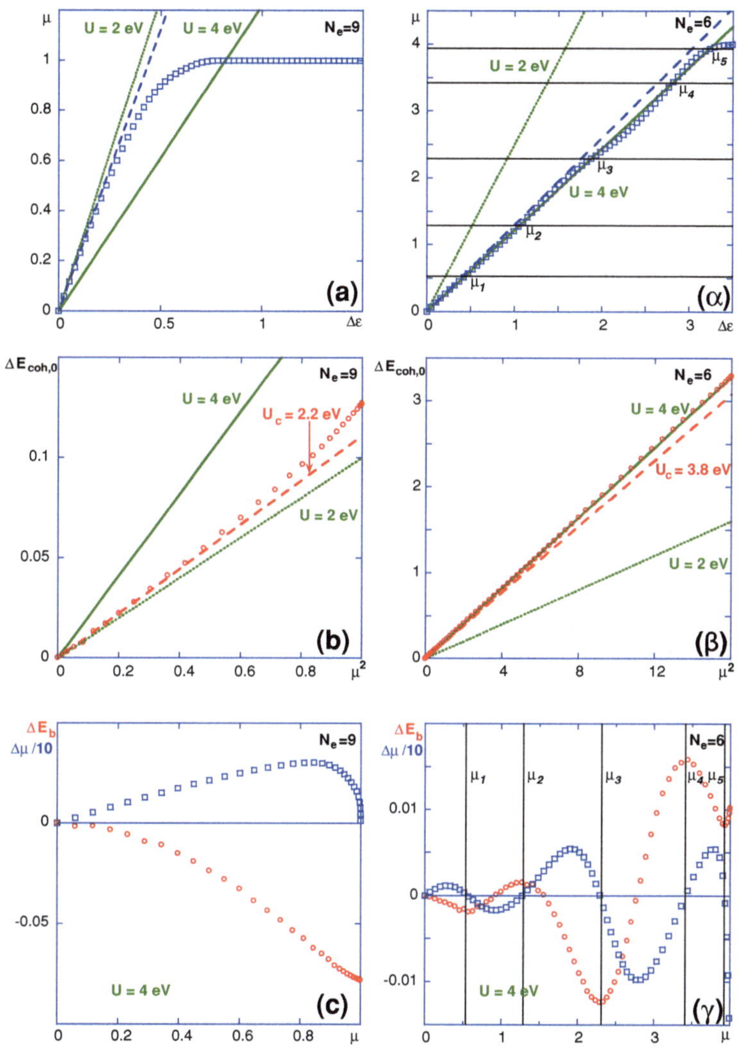

Fig. 9 **a,** ***α*** Self-consistent determination of μ from the crossing point of Eqs. (18a) (*blue dots*) and (18c) (*green lines*) for two values of U. The slope given by Eq. (18b) appears as the dotted blue line **b,** ***β*** μ^2-dependence of band term of the energy ($\Delta E_{coh,0}$, *red dots*) given by Eq. (20b) and of the magnetic term U $\mu^2/20$ for two values of U, *green lines*. The *dotted red* line represents the slope given by $1/4n_0(E_F)$. (c,γ) μ-dependence of $\Delta\mu$ Eq. (19) and ΔE_b Eq. (20b) for U = 4 eV

which in the limit of weak magnetisation (small $\Delta\varepsilon$) reduces to:

$$\mu \cong n_0(E_F)\Delta\varepsilon \tag{18b}$$

This means that the slope at the origin of the μ-curve as a function of $\Delta\varepsilon$ is nothing but the value of the paramagnetic density of states at the Fermi level.

This is illustrated for an almost filled ($N_e = 9$) in Fig. 9a and for an almost half-filled FCC d-band ($N_e = 6$) in Fig. 9α.

On the other hand, linearizing the Hamiltonian leads to a self-consistency relation between the d-level shifts and the magnetic moment through the Coulomb integral U which writes: $\Delta\varepsilon = U\mu/5$, giving an other $\Delta\varepsilon$-variation law for μ:

$$\mu = \frac{5}{U}\Delta\varepsilon \tag{18c}$$

A self-consistent determination of the magnetic moment then requires finding the crossing points of the two curves as a function of $\Delta\varepsilon$ for the actual value of the U parameter. The asymptotic value of μ being that which corresponds to a full *up* d-band filling, the curve given by Eq. (18a) saturates at the value $\mu_s = 10 - N_e$, being then below the line given by Eq. (18b). That means that it can cross the curve given by Eq. (18c) only if the so-called Stoner factor $S_0 = Un_0(E_F)/5$ is larger than unity (Stoner criterion). As can be seen, for a value of U which satisfies this criterion (here $U = 4$ eV) this crossing point only exists for the saturation value ($\mu_s = 1$) for $N_e = 9$, which corresponds to the limit of strong ferromagnetism (full *up* band). At the opposite, different crossing points ($\mu_1, \ldots, \mu_5$) are found below the saturation value ($\mu_s = 4$) for $N_e = 6$ (weak ferromagnetism), in agreement with experiments. This is still clearer in Fig. 9(c, γ), in which is plotted (for $U = 4$ eV) the difference:

$$\Delta\mu = \int_{E_F-\frac{\Delta\varepsilon}{2}}^{E_F+\frac{\Delta\varepsilon}{2}} n_0(E)dE - \frac{5}{U}\Delta\varepsilon \tag{19}$$

In that case, the equilibrium value of μ is the one that which minimizes the band energy E_b:

$$E_b(\mu) = \int^{E_F} En_\uparrow(E)dE + \int^{E_F} En_\downarrow(E)dE - N_e\varepsilon_0 + \frac{5}{4}U\left(\frac{\mu}{5}\right)^2$$

where the last two terms account for the double counting of interactions in the one-electron term. Taking advantage of the self-consistent relation (18c), this energy also writes:

$$E_b(\mu) = E_{coh,0}(N_\uparrow) + E_{coh,0}(N_\downarrow) - \frac{1}{20}U\mu^2 \tag{20a}$$

with: $E_{coh,0}(N_\sigma) = \int^{E_F^\sigma} En_0(E)dE - N_\sigma\varepsilon_0; E_F^\sigma = E_F \pm \frac{\Delta\varepsilon}{2}$ depending on $\sigma = \uparrow, \downarrow$.

The gain (or loss) in energy due to magnetism for a given d-band filling is then given by:

$$\Delta E_b(\mu) = \Delta E_{coh,0}(N_e, \mu) - \frac{1}{20}U\mu^2 \tag{20b}$$

$$\Delta E_{coh,0}(N_e,\mu) = E_{coh,0}(N_\uparrow) + E_{coh,0}(N_\downarrow) - 2E_{coh,0}(N_e)$$

As can be seen, for small values of the magnetic moment, the first term, independent on U, is nothing but the second derivative of the curve $E_{coh,O}(N_e)$ at the considered band filling. From the convexity of this curve, it appears that this band term is positive and then disfavours magnetism whereas the second (magnetic) term is explicitly negative and then favours its occurrence. In the limit of small values of the magnetic moment μ or in the particular case where the LDOS does not vary much around the Fermi level, the equations (20b) can be developed into:

$$\Delta E_b(\mu) \cong \frac{\mu^2}{20}\left(\frac{5}{n_0(E_F)} - U\right) \tag{21}$$

The balance between these two terms then gives the sign of the slope of $\Delta E_b(\mu)$ curve at the origin ($\mu = 0$). This allows to recover the previous Stoner criterion, the critical value of U being $U_c = 5/n_O(E_F)$ (2.2 eV for $N_e = 9$ and 3.8 eV for $N_e = 6$). However, finding the actual value of the magnetic moment requires to go beyond such approximations (small μ or LDOS almost constant around the Fermi level). One sees in Fig. 9(b, β) that the band term $\Delta E_{coh,O}$ indeed deviates from a simple μ^2 behaviour. This is still more apparent in Fig. 9(c, γ) in which we plot ΔE_b given by Eq. (20b). As expected from the self-consistent treatment of μ (Fig. 9(a, α), ΔE_b is decreasing up to the saturation value μ_s for $N_e = 9$ (strong ferromagnetism), whereas one recovers five extrema (with three minima), corresponding to the five crossing points of Fig. 9α for $N_e = 6$ (weak magnetism). In the latter case, the equilibrium value corresponds to the absolute minimum ($\mu_3 = 2.3$). The relative stabilities of the three minima is obviously strongly dependent on the value of U (in the range between U_c and 4.2 eV), or equivalently from variations of the LDOS with respect to its equilibrium shape, due to variation of interatomic distances (*e.g.* epitaxial growth, defects, dilation), which implies that the magnetisation could change under small variations of experimental conditions.

3 Pure Metal Surfaces and Clusters

3.1 Surface LDOS, Charge Self-Consistency and Atomic Level Shifts

If one neglects the crystal field α_λ, two bulk parameters should vary at the surface: first the hopping integrals $\beta_{nm}^{\lambda\mu}$ that we will assume unchanged at the surface (no relaxation), then the effective λ levels ε_λ. Let us first assume that the latter is also unchanged at the surface. In that case, the first effect of bond breaking (ΔZ) is to narrow the LDOS at the surface, due to simple second moment arguments.

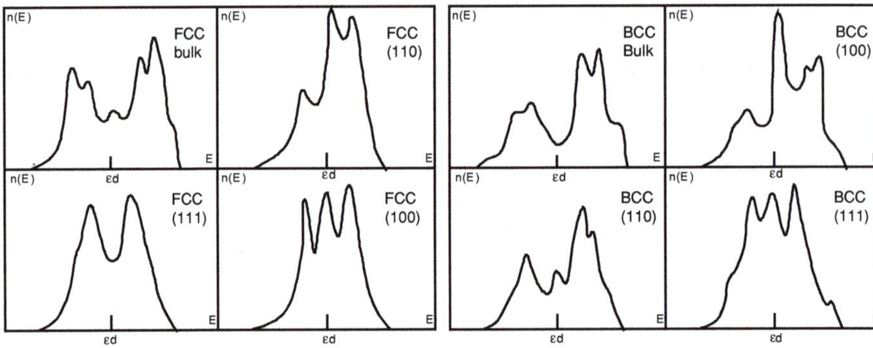

Fig. 10 Surface LDOS for fcc (*left*) and bcc (*right*) low index surfaces

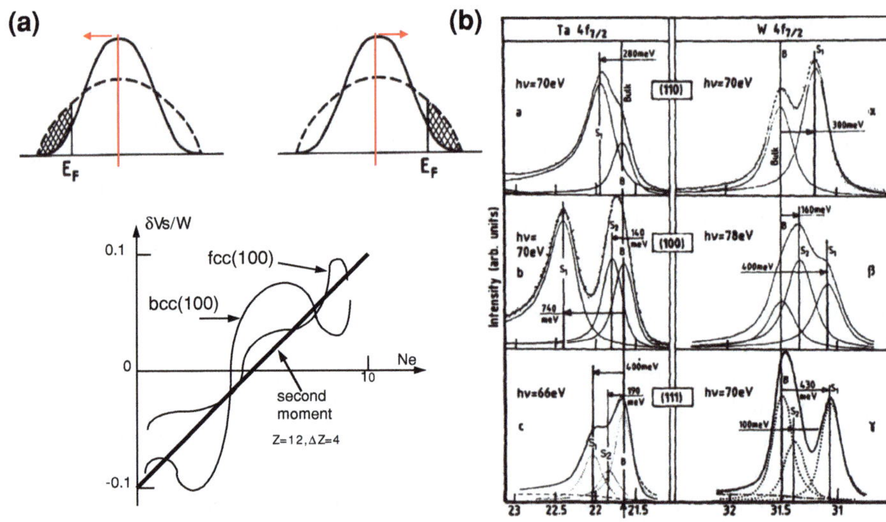

Fig. 11 **a** Variation with the *d* band filling (for realistic and rectangular densities of states) of the surface valence level shifts induced by the local charge neutrality requirement (from Ref. [12]). **b** Experimental surface core level shifts for Ta and W low index surfaces (from Ref. [13], copyright (1985), with permission from Elsevier)

In addition, as can be seen in Fig. 10, the LDOS corresponding to the most open surfaces (namely (110) and (100) for fcc, (100) and (111) for bcc) present quasi surface states, which vanish beyond the surface layer.

However, once the Fermi level is fixed by the bulk density of states, keeping ε_λ unchanged at the surface (Fig. 11a) leads to an unrealistic electronic charge transfer which must obey some self consistent rule, since this charge redistribution modifies the potential and then shifts ε_λ by $\delta\varepsilon_{\lambda,s}$.

This is illustrated in Fig. 11 for a single *d*-band. As can be seen the dependence of $\delta\varepsilon_{d,s}$ with *d*-band filling follows some general trends: it changes sign near the

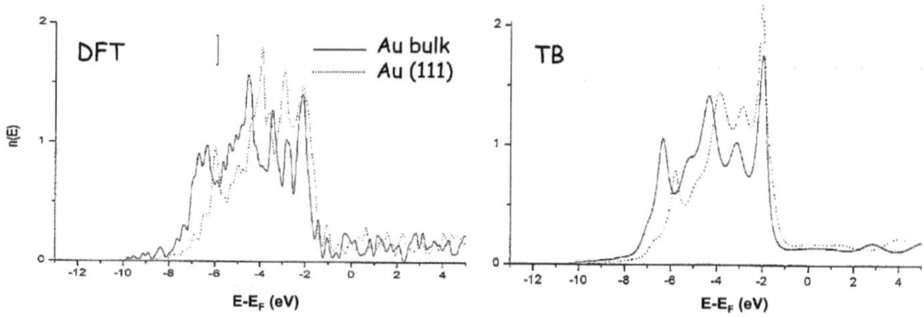

Fig. 12 Au density of states, calculated either by DFT (SIESTA) (*left*) or in the tight-binding framework (*right*) with *sp-d* hybridization with 25 exact couples of coefficients. From Ref. [15]

middle of the series (between W and Ta in the 5d series), its absolute value increases with the number of broken bonds (fcc: $|\delta\varepsilon_{d,110}| > |\delta\varepsilon_{d,100}| > |\delta\varepsilon_{d,111}|$), with a maximum value of about W/10 (W: *d*-band width). Finally, it is worth mentioning that this *d* level shift is almost rigidly followed by the core levels, which is confirmed experimentally by core level spectroscopy [13].

Let us note that this charge neutrality condition at the surface is confirmed by ab initio calculations. More precisely the DFT calculations even show that, when considering *sp-d* hybridization, charge neutrality has to be achieved, not only for each inequivalent site, but also for each orbital [14]. As can be seen in Fig. 12 for the Au(111) surface the resulting LDOS's compare satisfactorily to those of DFT calculations. Moreover, *sp-d* hybridization leads to surface energies in better agreement with experiments [15].

3.2 Relaxations and Reconstructions: Second Moment Approximation

Due to the broken bonds, the surface atoms can undergo displacements with respect to their bulk positions. In all cases, there is at least a vertical relaxation, which is experimentally known to be *inwards* (contraction of the first interlayer distance) for transition metals. In order to model this behaviour, we need not only the band part of the energy but also the repulsive one. Unfortunately as already mentioned, TB does not give such a repulsive part. To go beyond this difficulty, the idea is to build a semi-phenomenological TB model in which the band part, coming from the electronic structure, has a many-body character whereas the repulsive one is a pairwise potential fitted to some physical properties. Subtracting as usual (Sect. 2.4) the contribution due to this correction which is counted twice, the Eq. (17) reduces to:

$$E_{coh} = \sum_{n,\lambda} \int^{E_F} (E - \varepsilon_{0,\lambda}) n_n^{\lambda}(E, \delta\varepsilon_\lambda) dE - \sum_{\lambda=s,p,d} N_0^{\lambda} \delta\varepsilon_\lambda + A \sum_R e^{-p(\frac{R}{R_0}-1)} \quad (22)$$

Since the integral does not depend on details of *n(E)*, the band term can be calculated from a schematic rectangular density of states under the single assumption that it has the same second moment (SMA) as the exact one, which leads to [6, 16]:

$$E_{coh} = -\beta \sqrt{\sum_R e^{-2q(\frac{R}{R_0}-1)}} + A \sum_R e^{-p(\frac{R}{R_0}-1)} \quad (23)$$

where β is an "effective" hopping integral, corresponding to the 1st neighbour distance R_0. In practice, for a given metal, the four parameters (A, β, p, q) are determined by fitting experimental quantities such as the cohesive energy, lattice parameter and elastic constants or the so-called *universal equation* [17]. Note that this SMA-potential is similar to those derived elsewhere within the Embedded Atom Model (EAM [18]) or within the Glue Model [19]. Its main advantage compared to the latter is its physical transparency which clearly shows its limitations… and then its possible improvements (increasing the number of exact moments).

The first success of SMA potentials is to reproduce the *inwards* surface relaxation, which is found to be of the order of a few percents and proportional to the number of broken bonds, contrary to simple pair potential models which predict an *outwards* relaxation [13]. This comes from the stronger decrease with coordination Z of the repulsive term compared to the attractive one ($\sim \sqrt{Z}$). One has then to use such many body potentials to study surface atomic rearrangements.

In some cases, not only vertical but also lateral atomic rearrangements can occur, changing the two-dimensional periodicity and leading to so-called surface reconstructions. In those cases, one observes trends which can be either along the transition series (zig-zag reconstruction of the (100) face of bcc crystals occurring for column VIa but not for V_a) or along a column ((110) missing row reconstruction or pseudo-hexagonal densification of (100) fcc surfaces which only occurs in the 5*d* series). Both trends are well interpreted in the framework of TB calculations. The physical origin of the column sequence for fcc reconstructions, attributed to relativistic effects, can be taken into account within SMA potentials through the increase of the *q* parameter from the 1st series to the 3rd one [20]. On the other hand, understanding the zig-zag reconstruction requires going beyond second moment arguments. Indeed, as illustrated in Fig. 13a, it is due to the broadening of the quasi-atomic surface peak of the local bcc (100) density of states under the lattice distortion, which leads to an energy gain for *d* band filling around 5 (middle of the series) [21]. A similar detailed description of the LDOS is also required to study the possible occurrence of an hcp/fcc staking fault at the surface of late transition elements, in which case one also needs to account for *sp-d* hybridization [9]. The corresponding fault energy is found to be in good agreement with DFT calculations (see Fig. 13b). Its weak value for Au is consistent with the experimental observation of the herringbone reconstruction which leads atoms of the (111) surface layer to occupy both the hcp and the fcc sites.

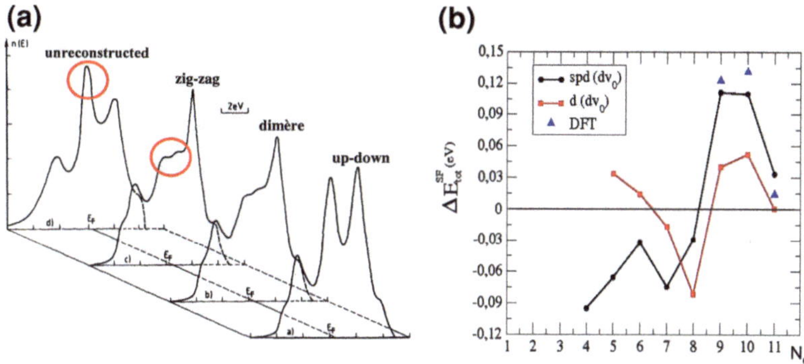

Fig. 13 **a** Influence of the dimer reconstruction on the (100) LDOS of bcc metals (from Ref. [21]). **b** Variation of the surface fcc-hcp stacking fault energy with d band-filling from self-consistent TB and DFT calculations (from Ref. [9], copyright (2008), with permission from Elsevier)

3.3 Monometallic Clusters: LDOS

Obviously, ab initio methods are particularly suited to the study of clusters with very small sizes but become very cumbersome when these sizes reach those which are useful for catalysis purposes (more than 100 atoms). The TB method is then very useful, since it describes the electronic structure in a wide range of sizes, and is able to give reliable site energies. Technically, for a finite cluster, the coefficient $b_n \rightarrow 0$ beyond a given level so that the continued fraction see Eq. (13) is truncated leading to a discrete spectrum for the LDOS of any site. This is illustrated for the central site of a 55 atoms cuboctahedron in Fig. 14.

Here also, when interested in elements of the end of transition series, it is necessary to take into account the *sp-d* hybridization and to perform a self-consistent treatment of the relation between charge and potential. This is achieved as for surfaces by shifting the λ-levels in a different way on each inequivalent site, following the same neutrality rule *per site* and *per orbital* [23]. The variation of these shifts for the different orbitals and sites as a function of their coordination numbers is also plotted in Fig. 15a. Note that theses shifts can be directly related to the activity of the corresponding sites [24], reflecting the high activity of the low coordination sites.

The resulting surface LDOS present a band width which decreases with the site coordination (from facets to edges and vertices) and are significantly modified near the Fermi level depending on the site. Note that, in view of the arguments developed in Sect. 2.5, this band narrowing could induce occurrence of magnetism for clusters of elements which are non magnetic in the bulk. Moreover the cluster symmetry has a strong influence on the density of states. The influence of size and structure is illustrated in Fig. 15b where we plot the total density of states for cuboctahedral and icosahedral (Ih) clusters, by taking the average of the densities

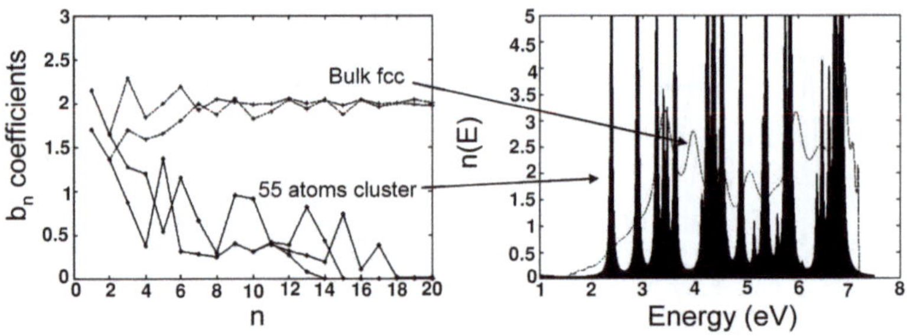

Fig. 14 Variation of the b_n coefficients for a bulk fcc structure and the central site of a 55-atoms cuboctahedron (*left*) and corresponding LDOS (*right*). From Ref. [22]

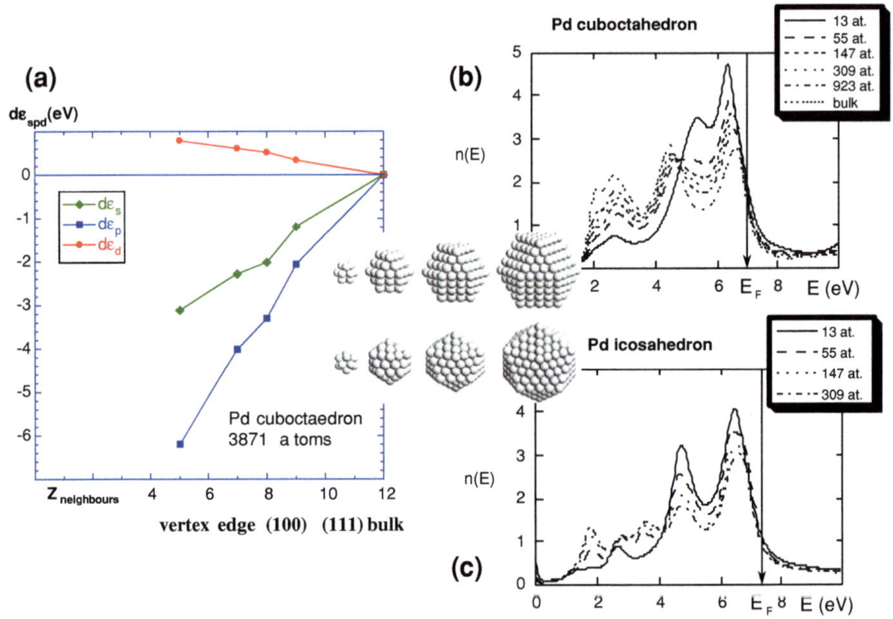

Fig. 15 **a** Core level shifts on the various surface sites of a 3,871 Pd cuboctahedron. **b** Variation with size of the average LDOS for Pd cuboctahedra and icosahedra from self-consistent TB calculations. From Ref. [23], copyright (1996), with permission from Elsevier

on inequivalent sites (vertices, edges, facets and core) weighted by the corresponding number of sites [23].

From energetic calculations performed by Quenched Molecular Dynamics within the SMA model, it appears that the competition between fcc and Ih structures is strongly element dependent. Thus, while Ih Ag clusters are found to be stable until 309 atoms, Au clusters recover their bulk fcc morphology (Wulff polyhedron) beyond 13 atoms. This critical size strongly depends on the local relaxation which differs remarkably between fcc-type and Ih clusters. The latter adopt a very inhomogeneous atomic relaxation profile, in which the contraction of the intershell distance is not limited to the surface shell (as in fcc-type structures) but is also present for the inner shells. More precisely, a considerable core contraction is found in the Ih case, which increases as a function of the cluster size [25]. A spectacular consequence of this contraction of the inner shells for the Ih structure is the existence of a strong compressive pressure in the core which can be relaxed by introducing constitutive vacancies [25]. The stability of these vacancies increasing with the cluster size, a fourfold tetrahedral shaped cavity becomes even more stable than the single one, but beyond the morphological transition to fcc-type structures. Taking into account the stability domain of the Ih relatively to fcc structure, there should then exist a stability range of size for Cu and Ag icosahedra with a central constitutional vacancy but not for Au ones. Let us recall however that SMA potentials are less suited than DFT calculations to model too small clusters. Thus, for 13 atoms, DFT calculations find structures which are neither Ih nor fcc [26].

4 Bulk Alloys A_cB_{1-c}: Link Between Electronic and Chemical Structure

4.1 Influence of Chemical Ordering on LDOS

Extending the Tight-Binding Hamiltonian to the case of a binary alloy A_cB_{1-c} requires to make its parameters depend on chemical configuration $\{p_n^i\}$, p_n^i=1 if site n is occupied by atom of i-species (i = A, B) and p_n^i=0 if not, through the relation:

$$H = H_d + H_{nd} \tag{24}$$

$$H_d = \sum_{n,\lambda} |n, \lambda\rangle \varepsilon_{n\lambda} \langle n, \lambda| \; \varepsilon_n = \sum_{i=A,B} p_n^i \varepsilon_{n\lambda}^i$$

$$H_{nd} = \sum_{n,m,\lambda,\mu} |n, \lambda\rangle \beta_{nm\lambda\mu} \langle m, \mu| \; \beta_{nm\lambda\mu} = \sum_{i,j=A,B} p_n^i p_m^j \beta_{nm\lambda\mu}^{ij}$$

The parameter ε_λ^i which appears in the diagonal contribution H_d is the barycentre of the partial $i\lambda$-LDOS, projected on the λ-orbital of atom of type i, while the off-diagonal contribution H_{nd} involves the hopping integral between the λ-orbital at site n occupied by an atom of type i and μ-orbital at site m occupied by an atom of type j. Thus the modifications undergone when two elements (A,B) are mixed into an A_cB_{1-c} alloy comes from two effects [10]. The first one, which is called *diagonal disorder* effect, is induced by the difference in energy between the barycentres of the valence (essentially d) bands of the A and B pure elements, i.e., the corresponding atomic levels $\varepsilon_{d,0}^A$ and$\varepsilon_{d,0}^B$ (the index 0 refers to the pure bulk value), and it is quantified by the parameter $\delta_{d,0} = \varepsilon_{d,0}^A - \varepsilon_{d,0}^B$. The second

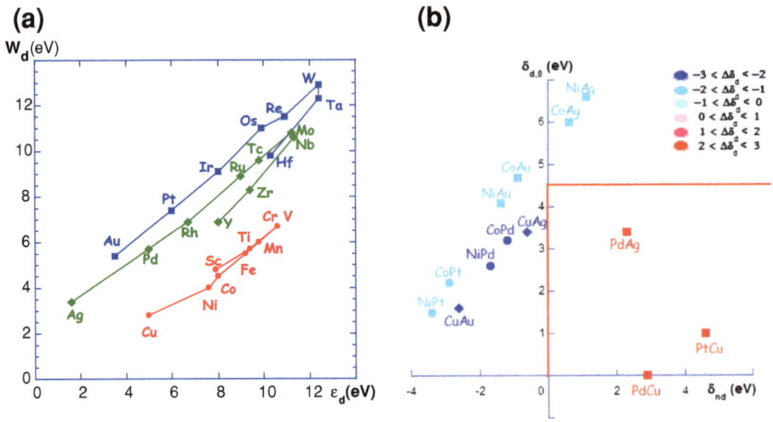

Fig. 16 **a** Correlated variations of the d atomic level $\varepsilon^i_{d,0}$ and effective bandwidth W^i ($\sim -8dd\sigma$) interpolated from ab initio band structure calculations (results from Ref. [27]) along the three transition metal series from which are derived ($\delta_{d,0}$, δ_{nd}) for each alloy. (b) 2D ($\delta_{d,0}$, δ_{nd}) map used to classify the variation $\Delta\delta_d$ of $\delta_{d,0}$ induced by the charge neutrality condition (from Ref. [32], copyright (2011) by The American Physical Society)

one, which is called *off-diagonal disorder* effect, comes from the difference between the hopping integrals for pure A and B elements, and therefore between the pure A and B valence bandwidths W^A and W^B which are directly related (see Sect. 2.3), so that it is quantified by the parameter $\delta_{nd} = W^A - W^B$. The relative magnitude of these two parameters actually drives the redistribution of the electronic states with respect to those of pure elements and therefore induces the new properties of the alloy with respect to those of the two isolated elements.

The values of $\varepsilon^i_{d,0}$ and W^i have been interpolated from ab initio calculations of band structure and tabulated for all transition elements by D. A. Papaconstantopoulos [27]. Their systematic variation along the three transition metal series, i.e., as a function of the d band filling N^i_d, is displayed in Fig. 16. As can be seen from these trends, none of the diagonal and off-diagonal variations seems to prevail, even though it has often been argued that the former prevailed on the latter [10]. In addition both variations are clearly correlated, which means that one can expect both effects to be important or not in the same time for a given system.

Once a given alloy has been characterized by a set of parameters ($\delta_{d,0}$, δ_{nd}), let us detail how to calculate more precisely the density of states for an alloy. In the ordered case, the same methods (continued fraction, recursion) can be used as for pure elements, taking just into account the ordered configuration of A and B atoms to assign the levels ε^i_λ and the hopping parameters $\beta^{ij}_{nm\lambda\mu}$. The situation is more complicated for a disordered system since it requires to calculate the average value, over all configurations, of $n(E)$ and therefore of $G(E)$. In the absence of off-diagonal disorder ($\delta_{nd} = 0$), i.e., assuming that $\beta^{ij}_{nm\lambda\mu} = \beta_{nm\lambda\mu}$, the off-diagonal part of the Hamiltonian (H_{nd}) is the same as for the pure elements, and only the diagonal part (H_d) depends on the chemical configuration through ε_n. If one notes $G_0 = (zI - H_{nd}) - 1$ the Green function for the pure metal, one can calculate the average Green function G, within a mean-field approximation, by introducing an effective local potential $\Sigma(z)$ such as:

$$G(z) = (z - H_0 - \Sigma(z))^{-1} \text{ with } \Sigma(z) \approx \sum_n |n\rangle\sigma(z)\langle n| \quad (25)$$

which means that in the average medium, the levels ε_λ^i are replaced by $\sigma(z)$ at each site. This effective potential can be determined by a self-consistency condition which imposes that fixing the occupancy of a site and then making the average on this site would lead to recover the same potential. This is the *Coherent Potential Approximation* (CPA) [28], which leads to the condition:

$$\sum_{i=A,Bi} c^i t^i = 0 \text{ with } t^i = \frac{\varepsilon^i - \sigma}{1 - \langle n|G(z)|n\rangle(\varepsilon^i - \sigma)} \quad (26)$$

which is self-consistent since the Green function in the disordered state $\langle n|G(z)|n\rangle$ depends on $\sigma(z)$ through the relation: $\langle n|G(z)|n\rangle = G_0(z - \sigma(z))$ which only requires the knowledge of the Green function of the pure element. The alloy densities of states obtained in this way (recursion method for ordered system, CPA and continued fraction for disordered ones) [29] are in good agreement with those obtained by LMTO calculations [30].

If one uses canonical parameters for the pure elements, the electronic structure of the alloy depends on c and $\delta_{d,O}$. In the case of weak diagonal disorder, perturbations at the lowest order lead to a density of states for the disordered alloy which is almost the same as that of the pure metal, but centred on the average level $\bar{\varepsilon} = c\varepsilon^A + (1 - c)\varepsilon^B$ with an average bandwidth $\bar{W} = c\, W^A + (1 - c)\, W^B$. On the contrary, for a strong diagonal disorder, a gap is opened since electronic states for the alloy have to lie between the bounds for the pure metals [10]. One can then analyze schematically the effect of chemical ordering in the following way. In the case of phase separation, the alloy density is the average of those of the pure metals whereas the sub-bands are narrower in the case of perfect order since the number of neighbours of the same type is reduced. For disordered systems, the width is in between but tails are present due to the finite probability of finding pure A and B clusters of any size. Finally, the total LDOS can be decomposed into its partial contributions projected on each element. All these qualitative behaviours are illustrated in Fig. 17.

4.2 Charge Self-Consistency and Atomic Level Shifts

It is worth pointing out that, up to now, a specific alloy has only be defined by the set of values of ($\delta_{d,O}$,δ_{nd}) and implicitly its crystallographic structure, without

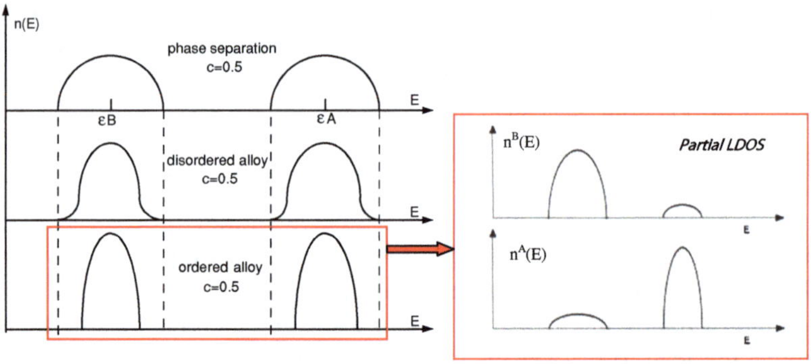

Fig. 17 Schematic variation of the density of states upon alloying for strong diagonal disorder. From Ref. [10]

resorting to the respective number of electrons of each species per orbital N^i_λ. However starting from the values in the pure elements $N^i_{\lambda,0}$ (which are consistent with the respective atomic levels $\varepsilon^i_{\lambda,0}$) while fixing a common Fermi level in the alloy leads to some charge transfer between species and orbitals. This in turn modifies the barycentres of the partial sub-bands ε^i_λ, the procedure having to be iterated up to convergence towards their self-consistent values. Thus, the parameter which actually drives the variation of the electronic structure between the alloy and its pure constituents is not $\delta_{d,0}$ but instead the difference between the barycentres of the two alloy partial sub-bands with respectively A and B characters: $\delta_d = \varepsilon^A_d - \varepsilon^B_d$.

From the experimental point of view, as previously stated for surfaces, it is difficult to identify such a shift of the *d*-bands under alloying due to their dispersion, but it is easier to follow the corresponding shift of the core levels. However, although a lot of work has been made in the case of surface core-level shifts in pure metals [13], only a few have been achieved for alloys. In the latter case, the most documented work in the literature is due to Olovsson et al. [31] who calculated core level binding energy shifts for various disordered alloys within density functional theory (DFT) using the coherent potential approximation (CPA). This allowed them to get good agreement with experimental data, but not a unified physical picture permitting to predict the general behaviour of any alloy.

The main difficulty is then to determine the effective atomic level ε^i_λ for each partial *i*-sub-band and λ-orbital in order to ensure the charge self-consistency. This requires shifting these levels for each orbital λ with respect to those in the bulk by a value $\delta\varepsilon^i_\lambda$ in order to satisfy a given rule on the different band fillings per orbital and species N^i_λ. The local charge neutrality rule *per site* and *per orbital* already justified for surfaces with only *d* electrons [14] detailed in Sect. 3.1 has been extended to a neutrality *per chemical species* in the case of bimetallic compounds [15]. Applying such a rule implies to ensure partial charge neutrality of each (s, p, d) orbital and to find for each one the appropriate band shift $\delta\varepsilon^i_{\lambda,0}$ for each

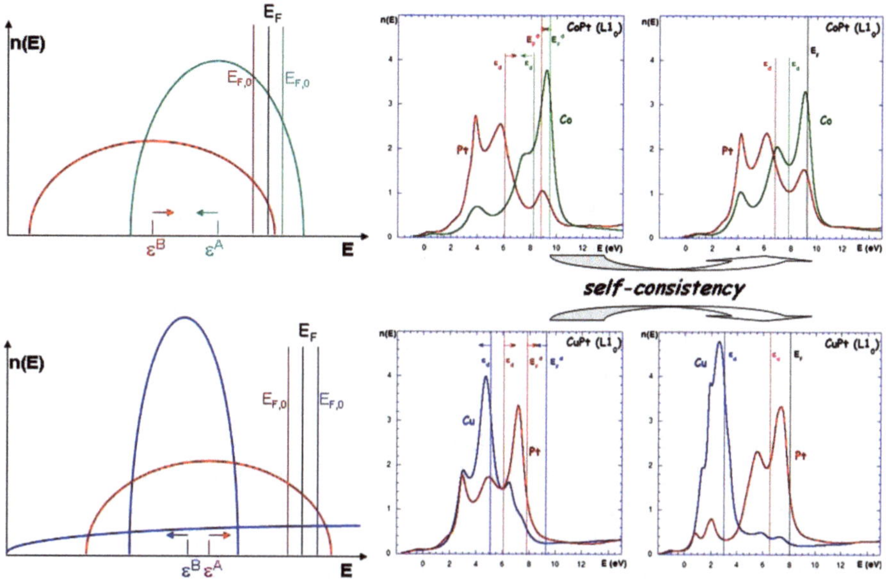

Fig. 18 Schematic (*left*) and TB (*right*) LDOS's in (*up*) CoPt and (*down*) PtCu systems in the $L1_0$ phase. The TB LDOS are presented before and after the charge self-consistency treatment

element. As an example, this procedure has been applied to predict the behaviour of late transition series alloys constituted of one element of the first transition series (Co, Ni, Cu) and the other elements of the second (Pd,Ag) and third (Pt,Au) series, in terms of the single *diagonal* and *off-diagonal disorder* parameters, which have been the subject of X-ray spectroscopy experiments and in addition are extensively studied for their peculiar properties in various applications, and in particular as nanoalloys. Adopting as a general rule that we denote respectively A and B two elements such as $\delta_{d,0} = \varepsilon^A_{d,0} - \varepsilon^B_{d,0} > 0$, the corresponding values of the variations $\Delta\delta_d = \delta_d - \delta_{d,0}$ under self-consistency are displayed in the 2D $(\delta_{d,0}, \delta_{nd})$ mapping of Fig. 16 [32].

As can be seen, almost all the systems present a decrease of the diagonal disorder parameter between 30 and 6%. Only two systems present an increase of this parameter, which fall in a well delimited region of this map ($\delta_{d,0}.\delta_{nd} > 0$, $\delta_{d,0} < \delta_{d,c}$). The widely commonly encountered decrease of the diagonal parameter under self-consistency is indeed expected [33] and can be intuitively understood by considering that, in the absence of self-consistent treatment, the only way to ensure *d* charge neutrality per chemical species in a mixed system should be to consider two different *unphysical* Fermi levels for each sub-band, $E^A_{F,0}$ and $E^B_{F,0}$. In most cases, and in particular when A and B are late transition series elements with unfilled *d*-band, $E^A_{F,0}$ and $E^B_{F,0}$ are ordered in the same way as the corresponding barycentres $\varepsilon^A_{d,0}$ and $\varepsilon^B_{d,0}$. This is illustrated schematically in Fig. 18, and more precisely in the archetypal case of CoPt for which LDOS is represented for a

$L1_0$ phase before any self-consistency. In such a situation the *unphysical* Fermi levels $E^i_{F,0}$ fall into the upper part of the d LDOS's. In view of the rather large order of magnitude of the density of states in this region, $E^i_{F,0}$ weakly varies with d-band filling, and in particular significantly less than the barycentre, so that the epsilon sequence. In these conditions, the effect of self-consistency is obviously to merge the two unphysical Fermi levels into an intermediate single one, which requires opposite shifts of both d levels and then reduces their difference δ_d.

In view of these arguments, the only way to observe the reversal behaviour, i.e. an increase of δ_d under self-consistency, should be to consider an alloy in which one of the two elements is a noble metal, with a full d-band, for which the *unphysical* Fermi level $E^i_{F,0}$ falls into the low density *sp* band, as schematically shown in Fig. 18, in which case it must be strongly shifted to ensure local charge neutrality. In such a situation, as can be guessed from Fig. 16a, except for PdCu system, the barycentre of the noble element sub-band is lower in energy than that of the transition element with unfilled d-band, which implies that, according to our sign convention, A is the transition element and B the noble one. In that case, according to Fig. 16a, one also satisfies $\delta_{nd} > 0$, except for B = Au. Then the corresponding unphysical Fermi levels can be ordered in energy in the opposite way compared to the barycentres of the sub-bands, provided that the latter are not too distant (i.e., that $\delta_{d,0}$ is not too large), and that the narrow band is located below the larger one ($\delta_{nd} > 0$). In other words, one can expect a reversal behaviour as soon as $\delta_{d,0}$ and δ_{nd} have the same sign, provided that the former is not too large. This is indeed what occurs in the PtCu system, as can be seen in the same Fig. 18, for which the only way to achieve self-consistency is to move both levels in opposite directions, leading to an increase of δ_d.

From this analysis one expects to observe different concentration dependence of the d (and core) level lines under alloying in an A_cB_{1-c} alloy. Indeed in the general case (decrease of δ_d), one should observe a symmetrical behaviour of the curves associated to A and B where the level of each element decreases with increasing concentration c, whereas one expects a non symmetrical behaviour in the exceptional case (increase of δ_d), the atomic level of the non noble metal remaining nearly constant on the overall concentration range close to its initial bulk value. This is indeed what occurs as illustrated in Fig. 19 by the variation of the d levels corresponding to each constituent as a function of Pt concentration for the two alloys CoPt and PtCu. In the dilute limit, we show for the matrix only the d levels of the atoms which are first neighbours of the impurity, the other ones keeping their bulk value. The overall behaviour is found in remarkable agreement with the evolution of core level shifts measured in XPS experiments [34] which are recalled in the insets, even though the calculated shifts are found larger than the experimental ones. Note that these curves also allow us to follow the variation with concentration of the diagonal disorder parameter δ_d with respect to the initial value calculated from pure metal data $\delta_{d,0}$ which is found to be very weak in the general case (CoPt) but larger in the exceptional one (PtCu). Finally, the influence of concentration and of the off-

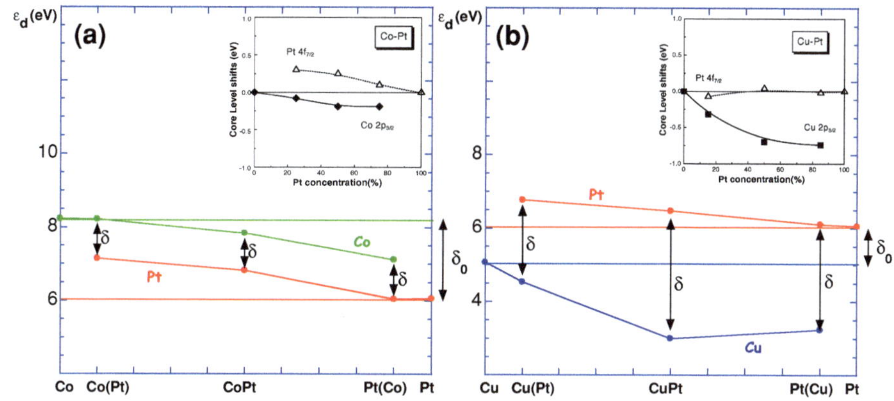

Fig. 19 Calculated shift of the atomic *d* level as a function of the concentration in Pt atoms derived from TB calculations for Co and Pt atoms in CoPt alloy (**a**), and for Cu and Pt atoms in CuPt alloy (**b**) (from Ref. [32], copyright (2011) by The American Physical Society). The experimental absolute core level shifts taken from [34] are shown in the insets

diagonal disorder on the LDOS is illustrated in the Fig. 20 in the particular cases of the CoAu [15] and CoPt systems. As can be seen, it compares satisfactorily to that derived from DFT calculations.

4.3 Rigid Lattice TBIM

As previously stated for the pure metals, the total energy of the alloy, for a given configuration, cannot be described as a sum of pair interactions. Nevertheless, if one neglects (in a first time) the effect of off-diagonal disorder ($\delta_{nd} = 0$) the (small) part of the energy which depends explicitly on the configuration (and which is essential in ordering problems) can be written as a sum of effective pair interactions by developing the energy in a perturbative way with respect to the disordered state [35]:

$$E_{coh}\left(\{p_n^i\}\right) = \bar{E}(c) + \frac{1}{2}\sum_{n,m,i,j} p_n^i p_m^j V_{nm}^{ij} \tag{27}$$

$$V_{nm}^{ij} = -\frac{\text{Im}}{\pi}\int^{E_F} dE t_n^i(E) t_m^j(E) \sum_{\lambda\mu} \bar{G}_{nm}^{\lambda\mu}(E)\bar{G}_{mn}^{\mu\lambda}(E)$$

in which the interatomic Green function: $\bar{G}_{nm}^{\lambda\mu}(E) = \langle n\lambda|\bar{G}(E)|m\mu\rangle$ is calculated using the CPA approximation developed in the previous section. Any energy balance which accounts for changes in the chemical configuration (mixing or ordering energies) will in fact involve the combination:

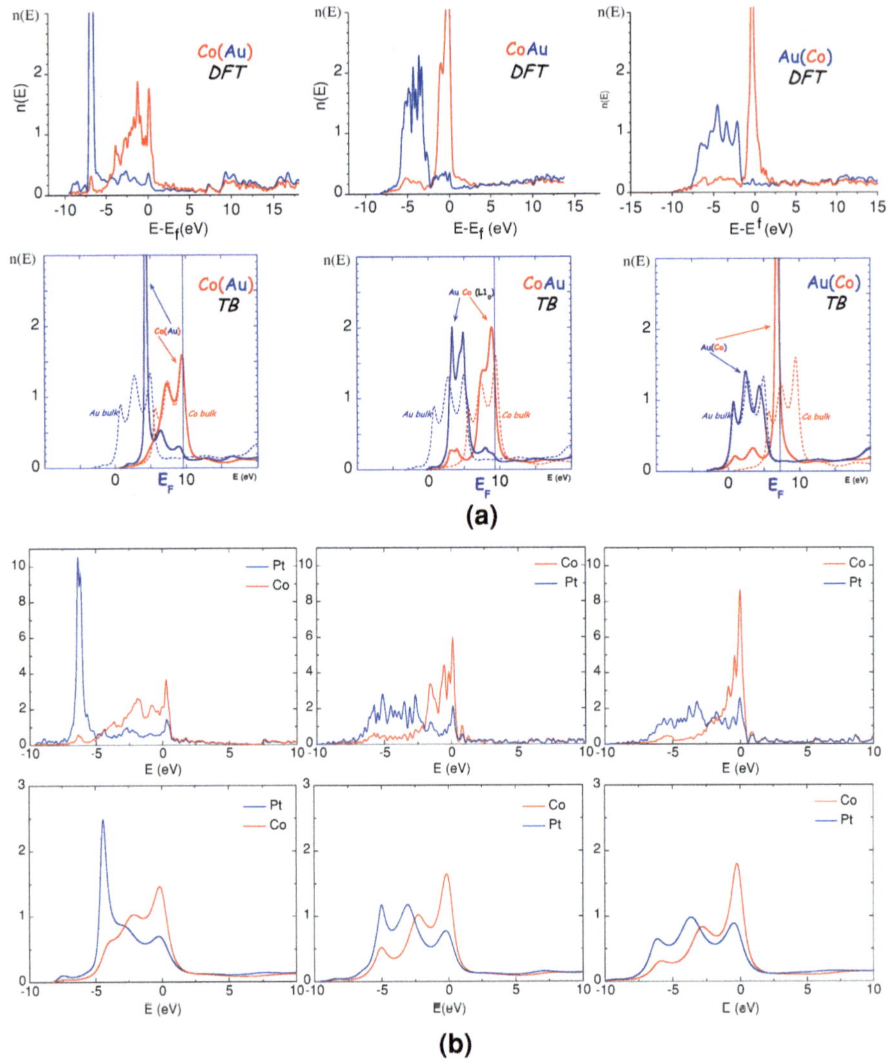

Fig. 20 Variation of the **a** Co-Au (from Ref. [15]) and **b** Co–Pt (courtesy of L. Zosiak) LDOS as a function of concentration from DFT and self-consistent TB calculations

$$V_{nm} = \frac{\left(V_{nm}^{AA} + V_{nm}^{BB} - 2V_{nm}^{AB}\right)}{2} \tag{28}$$

These effective pair interactions (EPI's) have been shown to decrease rapidly with the distance (n-m) (for the fcc structure: $V_1 >> V_2,\ V_3,\ V_4 >> V_5,\ \ldots$). We will therefore limit ourselves to the first neighbour interactions which will be denoted $V = V_1$ in the following. In this framework the sign of V indicates the

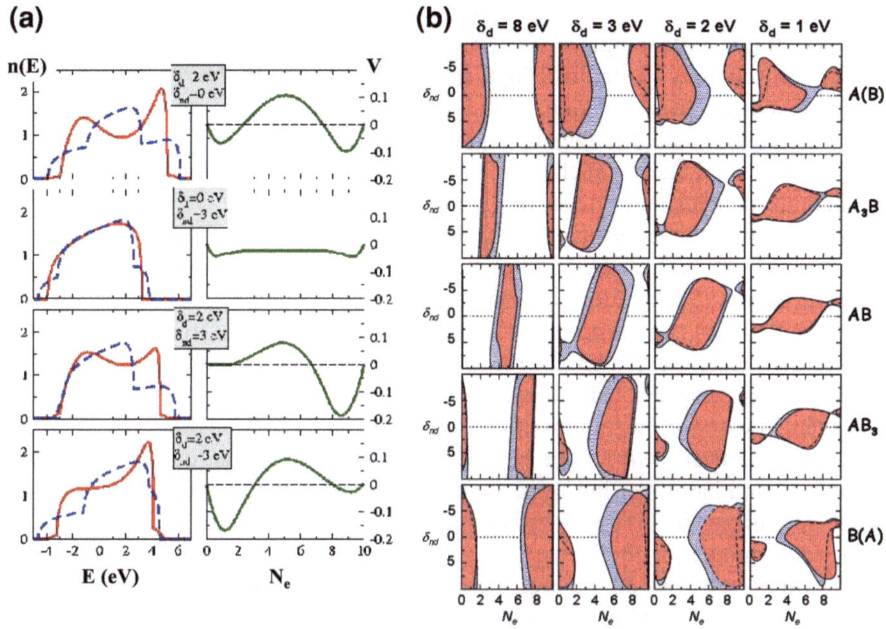

Fig. 21 (courtesy of J. Los): **a** Effects of diagonal (δ_d) and off-diagonal (δ_{nd}) disorder on the *d* band LDOS, *n(E)*, (*left-hand side*) for the *L1₀* (*solid line*) and separated (*dashed line*) phases, and on the band-filling (N_e) dependence of the effective pair interaction *V* (*right-hand side*), with $\bar{W}$ = 8 eV. (b) Ordering (*shaded area*) and demixing (*white areas*) domains in the parameter space spanned by δ_{nd} and N_e for different values of δ_d. From Ref. [36], copyright (2011) by The American Physical Society

tendency of the system to order ($V > 0$) or to phase-separate ($V < 0$). This sign depends (weakly) on bulk concentration (which could change the tendency to order or phase separate in a system in a few cases) and (strongly) on the average *d* band filling $\bar{N}_e = cN_e^A + (1-c)N_e^B$ [35]. In practice, it is possible to calculate *V* either directly from the above formula or indirectly from the expression of the formation energies (per atom) of some ordered phases ($E_{form}(L1_0)$ = -4 V, $E_{form}(L1_2) = -3$ V), or from that of the solution energies $E_{sol} = -12$ V. Thus from the formation energy of the $L1_0$ phase for a realistic value of δ_d, calculated using continued fractions with two exact levels (fourth moment approximation: FMA) and a constant termination, one gets a typical variation of V shown in the Fig. 21a [36]. A comparison with the EPI's calculated with a larger set of exact moments [35] shows that truncating the continued fraction expansion to the second level is sufficient, which confirms the validity of the FMA [36]. As can be seen alloys with a nearly half filled band tend to order whereas those with nearly filled or empty bands tend to phase separate.

The main interest of such a simplified energetic model is that it can be efficiently implemented for lattice Monte-Carlo simulations in order to investigate ordering and segregation phenomena using appropriate values for the EPI's (*see contribution by F. Ducastelle in* Chap. 6). This approach was experimentally validated for a wide range of systems but some exceptions remained, among which, unfortunately, systems of high interest such as CoPt, NiPt or CuAu which are predicted to phase separate although they are archetypal systems known to form ordered phases. It has then be argued that these discrepancies could be due to peculiar neglected effects, such as magnetism or spin–orbit coupling [10], without any definite conclusion up to now.

In addition to these serious drawbacks, a major problem of this Ising model is that it relies on a rigid lattice assumption which makes it completely unsuited to study the effects of relaxation by atomic displacements in the case of a strong size-mismatch between the constituents. In that case, a solution was to extend the semi-empirical interatomic potentials previously developed for pure metal surfaces, among which the previously described Second Moment Approximation (SMA). However, in essence, this approximation only feels the *off-diagonal disorder* and not the *diagonal disorder*, since a second moment calculation, consisting of two-hopping closed paths starting from an atom, does not involve the atomic levels of the neighbouring sites. The SMA is therefore not justified from the point of view of the Ising model based exclusively on the diagonal disorder effect.

The simplest extension that treats alike both diagonal and off-diagonal disorder effects is to use a LDOS based on a fourth moment approximation (FMA). Indeed this procedure allows first to revisit the TB Ising model, up to now limited to the diagonal disorder effect, by introducing off-diagonal disorder, and provides a well founded basis for a new generation of empirical potentials for alloys beyond SMA, based on the FMA, in particular for nanoalloys. Some typical LDOS's and corresponding EPI's for the $L1_0$ phase as a function of the average d band filling N_e and different values of δ_d and δ_{nd} are shown in the same Fig. 21a. The case $\delta_d = 0$ shows that the influence of off-diagonal disorder alone is to favour phase separation for any d band filling. The behaviour of the EPI's in the two limit cases ($\delta_d = 0$ or $\delta_{nd} = 0$) can be easily understood from simple qualitative arguments based on the respective band edges in the ordered and phase separated systems previously given. Finally, one sees that coupling both effects significantly modifies the previous curves by asymmetrising the d band filling dependence, in an opposite way depending on the sign of δ_{nd}, which in particular displaces the range of existence of ordering phases.

To generalize these curves, one can derive 3D maps from these EPI's, which, for a given concentration c, shows the tendency of a system to order or phase separate as a function of δ_d, δ_{nd} (within the same physical ranges as in Fig. 16) and N_e. Sections of these 3D maps for different concentrations and selected values of δ_d are displayed in Fig. 21b. Each section shows the respective domains for the existence of ordered and separated phases as a function of N_e (x-axis) and δ_{nd} (y-axis). As can be seen, the effect of off-diagonal disorder strongly changes the overall trends derived from calculations taking into account diagonal disorder only ($\delta_{nd} = 0$ in the maps). The most spectacular effect in this sense is probably the opening of ordering tendency domains, for reasonable values of the off-diagonal parameter, in the limits of small or large d band fillings for which only phase separation was predicted before. The overall effect of concentration is to shift the ordering domains from larger to lower d band filling from one dilute limit to the other. As a consequence, a given system can reverse its ordering tendency as a function of concentration, in particular for the largest values of δ_d. However, let us keep in mind that the solution energy is very sensitive to atomic relaxations around the impurity which could change the map in the dilute case in presence of strong size-mismatch e.g., in the CuAg case [37].

There remains to see to what extent the ordering behaviour of real systems such as CoPt, NiPt or CuAu which could not be explained by considering only diagonal disorder falls in the right place in the new domains. For this, we need to find the right point in the appropriate map of Fig. 21b, depending on the actual value of the parameters set (δ_d, δ_{nd}, N_e), δ_d (see Fig. 16) and N_e being issued from the self-consistent treatment upon *sp-d* hybridization. One then finds (1,−2.9,8.35) for CoPt which then now falls into an ordering region as it should in the map of Fig. 21b without resorting to other effects such as magnetism or spin–orbit coupling. A similar agreement should be found in the map corresponding to $\delta_d = 0.2$ for NiPt (0.2, −3.4, 8.85). The new maps also allow us to find the right places for alloys made of two noble metals, which slipped through the previous description, and in particular to explain why two systems as close as CuAg (3.4, −0.6, 9.85) and CuAu (1.6, −2.6, 9.75) present two opposite behaviours, phase separation in the former case and ordering in the latter case.

5 Alloy Surfaces and Clusters

5.1 Alloy Surfaces

The LDOS at the surface of an alloy has to combine both bond breaking (as for pure metal surfaces) and alloying (as in bulk systems) effects. In particular, the self-consistency rule per species, site and orbital has to be applied for each inequivalent surface site depending on its occupation by A or B atoms. As shown in Fig. 22, for the CoAu system previously treated in the bulk (Fig. 20) but now in the particular configuration of a Co layer deposited on Au(111), this rule allows one to get LDOS in perfect agreement with DFT calculations.

At the difference of pure metal, the presence of the surface not only introduces atomic but also chemical rearrangements. Indeed, due to broken bonds, the equilibrium concentration at the surface has no reason to be the same as in the bulk, which leads to the phenomenon of surface segregation. The natural way for treating this problem is to extend to the case of surfaces the perturbation treatment of the energy (with respect to configuration fluctuations) previously developed for modelling the ordering processes in the bulk. This leads to the so-called Tight-Binding Ising Model (TBIM) that extends the Eq. (27) into [38]:

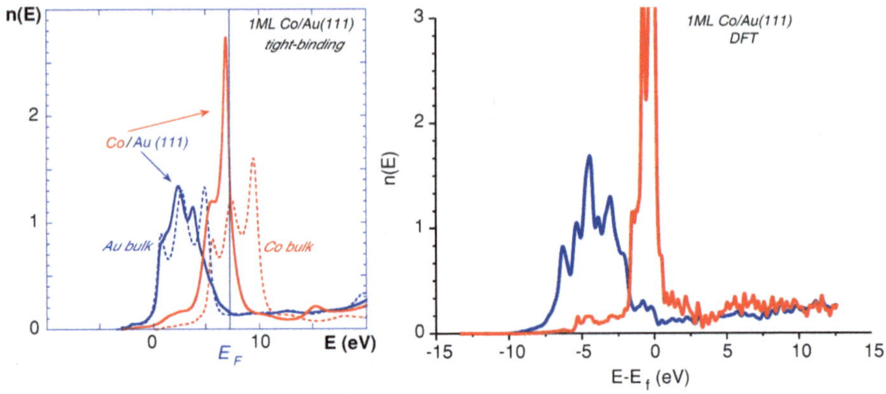

Fig. 22 Co/Au(111) LDOS from self-consistent TB (*left*) and DFT (*right*) calculations. From Ref. [15]

$$H^{TBIM}(\{p_n^i\}) = \sum_{n,i} p_n^i h_n^i + \frac{1}{2} \sum_{n,m,i,j} p_n^i p_m^j V_{nm}^{ij} \tag{29}$$

The main difference with the bulk case is the existence of sites which are no longer equivalent from the geometrical point of view, leading to a local on-site term:

$$h_n^i = \frac{\mathrm{Im}}{\pi} \int^{E_F} dE \sum_\lambda \log\left[1 - \left(\varepsilon^i - \sigma_n\right)\bar{G}_{nn}^{\lambda\lambda}(E)\right] \tag{30}$$

In the simple case of a binary alloy $\left(p_n = p_n^A\right)$, one can determine the concentration profile $\{cp\}$ ($c_p = <p_n>$ for any site n in the pth plane parallel to the surface: $p = 0$) as the one which minimises the free energy. Within mean-field approximation (*see B. Legrand contribution in* Chap. 7), the segregation energy, which accounts for the exchange of a A bulk atom with a B surface one, only involves the double differences: $\Delta h_p = (h_p^A - h_p^B) - (h_{bulk}^A - h_{bulk}^B)$. This local term Δh_p is negligible in the bulk ($p > 0$) and almost identical to the difference in surface energies for $p = 0$ ($\Delta h_0 \sim \tau^A - \tau^B$) [38]. In fact, Δh_0 is the main driving force which leads to the segregation of the element with the lowest surface energy. The EPI's in Eq. (29) are also changed at the surface. More precisely, V Eq. (28) is enhanced by a factor 1.5–2 with respect to its bulk value [38], at least in absence of size-mismatch.

Finally, let us note that, up to now, the derivation of TBIM has been performed on a rigid lattice, which is probably too crude in the case of large size mismatch between the constituents. However, there are two ways to introduce this effect. The first one is to add a third contribution to the segregation energy, $\Delta E_p^{size}(c) \neq 0$ if $p = 0$ (and 1 for open surfaces). $\Delta E_0^{size}(c)$ can be calculated in both dilute limits

($c \rightarrow 0,1$) in the framework of SMA, by determining the four mixed A-B parameters in order that A and B only differ by their size [39]. This leads to a contribution which significantly differs from the one derived from elasticity theory since the latter leads in both limits to the segregation of the impurity, whatever its size. On the contrary, the SMA term is found strongly asymmetric, leading to a segregation of the impurity when it is the largest only (at least for close-packed surfaces). This comes from the anharmonicity of the potential which exhibits a strong asymmetry between tensile and compressive pressures. Size-mismatch can also strongly modify the EPI in the case of phase-separating systems such as CuAg. Indeed, an SMA relaxation of a system containing two impurities shows that bond breaking can reverse the sign of V at the surface, leading to surface ordering in spite of bulk phase separation [37].

This so-called "three effects" rule (cohesive, alloying and size effects) proved to be quantitatively relevant for many different environments (flat or vicinal surfaces, grain boundaries, clusters) in alloys of transition metals with a chemical tendency to either phase separation (CuAg) or to ordering with a low mixing energy, but not for systems with size-mismatch similar to Cu–Ag, but which exhibit a strong tendency to order (CoPt). To elucidate the origin of this disagreement, a second approach has been proposed which couples these three effects (CTEM: Coupled Three Effects Model [40]), based on the systematic study of the permutation enthalpies in the bulk and at the surface as a function of the value of the mixed interaction parameter involved in the TBSMA potential. This allows one to explain both previous observations, disagreement for CoPt and agreement for CuAg, as due to the variation of the EPI's at the surface and by the existence of coupling coefficients between the three effects. More specifically, if one indeed recovers that the surface EPIs are proportional to the bulk ones in the absence of significant size-mismatch, they are found to differ by an additive constant value in the presence of a strong size effect.

5.2 As a Conclusion: Nanoalloys

Determining the electronic structure of nanoalloys within TB approximation needs to combine the features of pure metal clusters and alloy surfaces. This means that one first has to extend the self-consistent neutrality rules per element, site and orbital by shifting the atomic levels differently for vertices, edges, facets depending on their occupation by A or B atoms to ensure the same orbital filling as in their respective bulks (coupling Figs. 15a, 16b and 19). The resulting LDOS per inequivalent surface sites should then also combine the features of those for pure clusters (Fig. 15b) and semi-infinite alloys (Fig. 22).

From the energetic point of view, the coupling between segregation and reconstruction [41] should be particularly important in bimetallic clusters, due to the effects of finite matter (the available quantity of segregant matter could be lower than the quantity of surface sites) and geometrical frustrations (coexistence

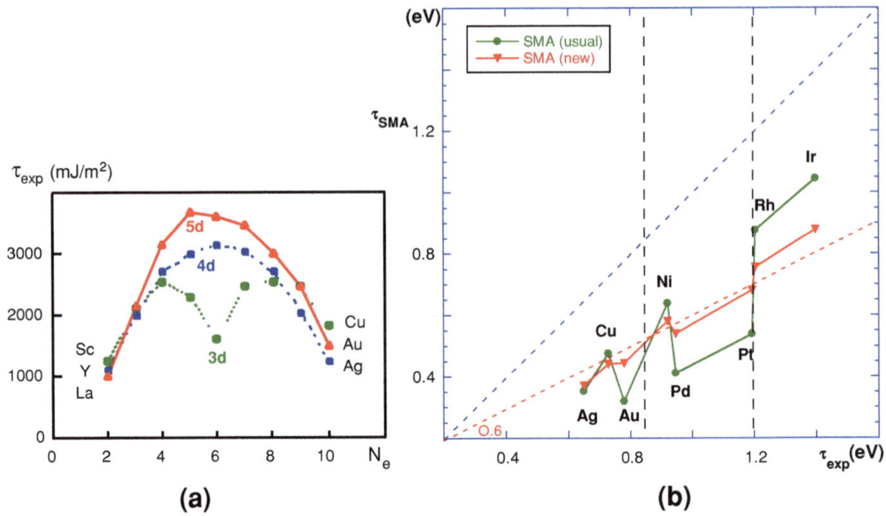

Fig. 23 **a** Experimental variation of the surface energy along the transition metal series. **b** SMA surface energy as a function of experimental one for usual parameters fitting only the cohesive energy (*green line*) and for parameters fitting also a value of surface energy of about 0.6 its actual value

of vertices, edges, facets with different orientations). In practice, this coupling between chemical and atomic structure is now too strong to allow us to separate them. A first attempt to achieve it is to mimic the case of alloy surfaces by combining TBIM/CTEM (segregation and ordering) and SMA (distance dependence of the interatomic potential) approaches, in order to be able to treat various morphologies (fcc polyhedra, bcc dodecahedra, icosahedra) and to study the competition or synergy between bulk ordering and surface segregation. This can be done in both directions, either using TBIM/CTEM on a rigid lattice with effective parameters (local field and EPI's) calculated by SMA relaxation procedure to account for size-mismatch, or reversely by using SMA potentials with parameters ensuring the TBIM/CTEM prescriptions by reproducing mixing energies and difference in surface energies between the constituents. This latter point is not obvious. Actually, as shown in Fig. 23, the usual SMA potential with parameters fitting bulk properties fails not only to reproduce the absolute values of these surface energies (which are found too small by a factor of two) but also, which is more important here, their variation from an element to the other. It is indeed impossible to fit simultaneously both cohesive energies and surface energies with this type of potential. A possible solution could be to accept this global lowering of the surface energies by introducing as an additional constraint in determining the SMA parameters that of reproducing 0.6 τ^i instead of τ^i. As can be seen in Fig. 23, this indeed allows recovering the good sequence from an element to the others. Such new SMA potentials are under development [42].

This coupled approach allowed to evidence the coupling between atomic relaxations and chemical arrangements, in two systems presenting opposite thermodynamics behaviours, CuPd [43] and CuAg [44] which respectively tend to order and to phase separate. Thus in the CuPd case, for small sizes, if the usual sequence of relative stabilities (icosahedron, fcc, and -well above- bcc dodecahedron) was recovered in the disordered state, chemical ordering at low temperature leads to a spectacular reversal in which the bcc structure is stabilized with respect to fcc by chemistry, the icosahedron being destabilized by chemical order. Moreover, a surface induced disorder is observed with respect to inner sites. On the other hand, in the CuAg case, one finds that the segregation hierarchy based on broken-bond arguments (preferential segregation to the vertices, less to edges, and least to facets) is not at all universal and that the segregation driving forces for cuboctahedral and icosahedral nanoalloys may differ, being similar for the vertex and edge sites, but not for the sites of the triangular facets due to dilations of orthoradial distances in the icosahedral structure.

The alternative solution to this mixed approach is to directly couple the chemical and atomic requirements by giving up SMA/TBIM in favour of FMA potentials. Indeed, the previous results on ordering trends in bulk alloys (Fig. 21) not only allow to revisit the TB Ising model by accounting for both diagonal and off-diagonal disorder effects, but also provide a well founded basis for a future extensive use in nanoalloys of FMA interatomic potentials, up to now limited to covalent materials [45] or pure metals [46]. Here also some work is currently done for the archetypal CoPt and CuAg systems.

References

1. Kohn, W., Becke, A.D., Parr, R.G.: Density functional theory of electronic structure. J. Phys. Chem. **100**, 12974–12980 (1996) (and references therein)
2. Ashcroft, N.W.: The Fermi surface of aluminium. Philos. Mag. **8**, 2055–2083 (1963)
3. Car, R., Parrinello, M.: Unified approach for molecular dynamics and density-functional theory. Phys. Rev. Lett. **55**, 2471–2474 (1992)
4. Friedel, J.: Physics of Metals, vol. 1, Cambridge University Press, Cambridge (1978)
5. Ducastelle, F.: Structure électronique des métaux de transition et de leurs alliages. Thèse Orsay (1972)
6. Lambin, P., Gaspard, J.P.: Continued-fraction technique for tight-binding systems: a generalized-moments method. Phys. Rev. B **26**, 4356–4368 (1982)
7. Haydock, R., Heine, V., Kelly, M.J.: Electronic structure based on local atomic environment for tight-binding bands. J. Phys. C **5**, 2845–2858 (1972); C **8**, 2591–2605 (1975)
8. Turchi, P., Ducastelle, F., Tréglia, G.: Band gaps and asymptotic behaviour of continued fraction coefficients. J. Phys. C **15**, 2891–2924 (1982)
9. Jaafar, A., Goyhenex, C., Tréglia, G.: Role of sp-d hybridization in the formation of stacking defects at metal surfaces. Surf. Sci. **602**, 2681–2688 (2008)
10. Ducastelle, F.: Order and Phase Stability in Alloys. North-Holland, Amsterdam (1991)
11. Turchi, P.: Structure électronique et stabilité des alliages de métaux de transition: effets de structure cristalline et d'ordre configurationnel. Thèse Paris (1984)
12. Desjonquères, M.C., Spanjaard D.: Concepts in Surface Physics. Springer, Berlin (1995)

13. Spanjaard, D., Guillot, C., Desjonquères, M.C., Tréglia, G., Lecante, J.: Surface core level spectroscopy of transition metals: a new tool for the determination of their surface structure. Surf. Sci. Rep. **5**, 1–85 (1985)
14. Sawaya, S., Goniakowski, J., Mottet, C., Saúl, A., Tréglia, G.: Charge redistribution at Pd surfaces: ab initio grounds for tight-binding interatomic potentials. Phys. Rev. B **56**, 12161–12166 (1997)
15. Jaafar, A., Goyhenex, C., Tréglia, G.: Rules for tight-binding calculations in bimetallic compounds based on density functional theory: the case of CoAu. J. Phys. Condens. Matter. **22**, 505503 (2010)
16. Rosato, V., Guillopé, M., Legrand, B.: Thermodynamical and structural properties of fcc transition metals using a simple tight-binding model. Philos. Mag. A **59**, 321–336 (1989)
17. Spanjaard, D., Desjonquères, M.C.: Universal features of bonding in metals. Phys. Rev. B **30**, 4822–4827 (1984)
18. Foiles, S.M., Baskes, M.I., Daw, M.S.: Embedded atom method functions for the fcc metals Cu, Ag, Au, Ni, Pd, Pt and their alloys. Phys. Rev. B **33**, 7983–7991 (1986)
19. Garofalo, M., Tosatti, E., Ercolessi, F.: Structure, energetics, and low temperature behaviour of the Au(110) reconstructed surface. Surf. Sci. **188**, 321–326 (1987)
20. Guillopé, M., Legrand, B.: (110) surface stability in noble metals. Surf. Sci. **215**, 577–595 (1989)
21. Legrand, B., Tréglia, G., Desjonquères, M.C., Spanjaard, D.: A "quenched molecular dynamics" approach to the atomic stability of the (100) face of bcc transition metals. J. Phys. C **19**, 4463–4472 (1986)
22. Mottet,C.: Étude par simulation numérique d'agrégats libres mono- et bi-métalliques. Thèse, Université Aix-Marseille II (1997)
23. Mottet, C., Tréglia, G., Legrand, B.: Electronic structure of Pd clusters in the tight-binding approximation: influence of spd-hybridization. Surf. Sci. **352–354**, 675–679 (1996)
24. Hammer, B., Morikawa, Y., Norskov, J.K.: CO Chemisorption at metal surfaces and overlayers. Phys. Rev. Lett. **76**, 2141–2144 (1996)
25. Mottet, C., Tréglia, G., Legrand, B.: New magic numbers in metallic clusters: an unexpected metal dependence. Surf. Sci. **383**, L719–L727 (1997)
26. Wang, L.L., Johnson, D.D.: Density functional study of structural trends for late-transition-metal 13-atom clusters. Phys. Rev. B **75**, 235405 (2007)
27. Papaconstantopoulos, D.A.: Handbook of Electronic Structure of Elemental Solids. Plenum, New York (1986)
28. Velicky, B., Kirkpatrick, S., Ehrenreich, H.: Single-site approximations in the electronic theory of simple binary alloys. Phys. Rev. B **175**, 747–766 (1968)
29. Bieber, A., Ducastelle, F., Gautier, F., Tréglia, G., Turchi, P.: Electronic structure and relative stabilities of L_{12} and DO_{22} ordered structures occurring in transition metal alloys. Solid State Comm. **45**, 585–590 (1983)
30. Kudrnovsky, J., Bose, S.K., Andersen, O.K.: Comparative study of the electronic structure of ordered, partially ordered and disordered phases of the Cu_3Au alloy. Phys. Rev. B **43**, 4613–4621 (1991)
31. Olovsson, W., Göransson, C., Pourovski, L.V., Johansson, B., Abrikosov, I.A.: Core-level shifts in fcc random alloys: a first-principles approach. Phys. Rev. B **72**, 064203 (2005)
32. Goyhenex, C., Tréglia, G.: Unified picture of d band and core level shifts in transition metal alloys. Phys. Rev. B **83**, 075101 (2011)
33. Tréglia, G., Ducastelle, F., Gautier, F.: Generalised perturbation theory in disordered transition metal alloys: application to the self-consistent calculation of ordering energies. J. Phys. F **8**, 1437–1456 (1978)
34. Lee, Y.-S., Lim, K.-Y., Chung, Y.-D., Wang, C.-N., Jeon, Y.: XPS core-level shifts and XANES studies of Cu-Pt and Co-Pt alloys. Surf. Interface Anal. **30**, 475–478 (2000)
35. Bieber, A., Gautier, F., Tréglia, G., Ducastelle, F.: Electronic structure, pairwise interactions and ordering energies in binary fcc transition metal alloys. Solid State Comm. **39**, 149–153 (1981)

36. Los, J., Mottet, C., Tréglia, G., Goyhenex, C.: Ordering trends in transition metal alloys from tight-binding electronic structure calculations. Phys. Rev. B **84**, 180202(R) (2011)
37. Meunier, I., Tréglia, G., Legrand, B.: Surface-induced ordering in phase separation systems: Influence of concentration and orientation. Surf. Sci. **441**, 225–239 (1999)
38. Tréglia, G., Legrand, B., Ducastelle, F.: Segregation and ordering at surfaces of transition metal alloys: the tight-binding ising model. Europhys. Lett. **7**, 575–580 (1988)
39. Tréglia, G., Legrand, B.: Surface-sandwich segregation in PtNi and AgNi alloys: two different physical origins for the same phenomenon. Phys. Rev. B **35**, 4338–4344 (1987)
40. Creuze, J., Braems, I., Berthier, F., Mottet, C., Tréglia, G., Legrand, B.: Model of surface segregation driving forces and their coupling. Phys. Rev. B **78**, 075413 (2008)
41. Tréglia, G., Legrand, B., Ducastelle, F., Saúl, A., Gallis, C., Meunier, I., Mottet, C., Senhaji, A.: Alloy surfaces: Segregation, reconstruction and phase transitions. Comput. Mat. Sci. **15**, 196–235 (1999)
42. Goyhenex, C.: Revised tight-binding second moment potential for transition metal surfaces. Surf. Sci. **606**(3–4), 325–328 (2012)
43. Mottet, C., Tréglia, G., Legrand, B.: Theoretical investigation of chemical and morphological ordering in Pd_cCu_{1-c} clusters. Phys. Rev. B **66**, 045413 (2002)
44. Moreno, V., Creuze, J., Berthier, F., Mottet, C., Tréglia, G., Legrand, B.: Site segregation in size-mismatched nanoalloys: application to Cu–Ag. Surf. Sci. **600**, 5011–5020 (2006)
45. Amara, H., Bichara, C., Ducastelle, F.: Understanding the nucleation mechanisms of carbon nanotubes in catalytic chemical vapor deposition. Phys. Rev. Lett. **100**, 056105 (2008)
46. Los, J.H., Pellenq, J.M.: Determination of the bulk melting temperature of nickel using Monte Carlo simulations: inaccuracy of extrapolation from cluster melting temperatures. Phys. Rev. B **81**, 064112 (2010)

Chemical Order and Disorder in Alloys

François Ducastelle

Abstract Chemical ordering in bulk alloys has been studied for a long time, and microscopic theories are now well developed, both from the electronic structure point of view and from the statistical physics point of view. The practical importance of surface segregation effects has also stimulated many studies to understand the modifications of ordering processes induced by the presence of a surface. The effect of still more reduced dimensionalities as in the case of clusters or nanoparticles has been considered more recently, but now with the progresses of the experimental techniques and of the computation facilities, nanoalloys are objects of growing interest. In this introductory chapter we first recall the main tools of statistical physics applied to the study of order-disorder transitions in bulk alloys: statistical ensembles, Ising model, mean field theory and Landau theory of phase transitions. These tools can be extended to inhomogeneous systems, in the presence of interfaces and surfaces. This is described in a second part. The same tools can be used for nanoalloys, but finite size effects are now involved. Genuine phase transitions no longer exist, but in practice they can still be defined for particles of nanometer size. The transitions are broadened and the ordering temperatures are modified. This is discussed in the last part.

1 Bulk Alloys

Before examining the effect of reduced dimensionalities, we recall the basic features of alloy theory for bulk systems.

F. Ducastelle (✉)
LEM, ONERA-CNRS, BP72, 92322, Châtillon Cedex, France
e-mail: Francois.Ducastelle@onera.fr

D. Alloyeau et al. (eds.), *Nanoalloys*, Engineering Materials,
DOI: 10.1007/978-1-4471-4014-6_6, © Springer-Verlag London 2012

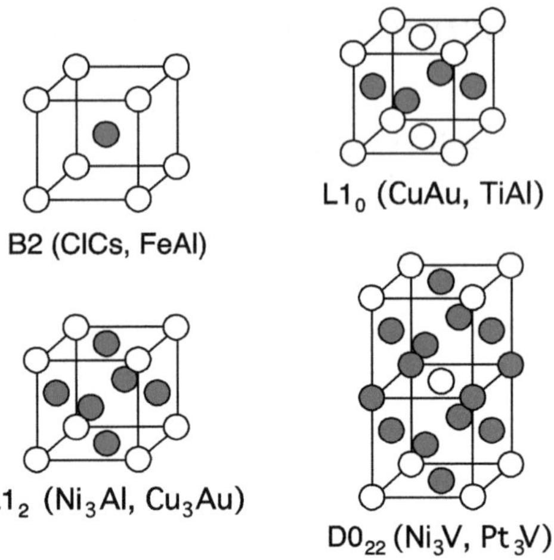

Fig. 1 Typical ordered structures on the underlying BCC and FCC structures

1.1 Order and Phase Separation on a Fixed Lattice

Due to the presence of interatomic interactions, any alloy should eventually order or phase separate at low temperature. This is observed in many systems. When the ordering interactions are strong, the alloy may remain ordered up to the melting point (Ni_3Al for instance). When they are very weak, the solid solution can be the only observable phase since at low temperature atomic diffusion is no longer efficient (Cu–Ni). In the intermediate regime we have a more interesting situation with one or several ordered phases at low temperature which disorder at a critical temperature before melting (Cu–Au, Pd–V, Ni–Fe, etc.). In the simplest case all phases are built on a fixed underlying lattice such as the FCC or BCC lattice, but more frequently, several structures are involved and many compounds display structures which are not observed in elemental metals (Laves phase, σ phase, A15 phases, etc.). A few examples are given in Figs. 1 and 2. In the following we will mainly deal with binary alloys and ordering on a fixed lattice, which is more easy to describe theoretically [1, 2].

At finite temperature, thermodynamics and phase diagrams are involved. We have basically two different cases to consider depending on whether similar atoms have a tendency to attract (homo-atomic interactions) or to repel (hetero-atomic interactions) each other. This leads to two types of phase diagrams. In the case of homo-atomic interactions, we have a tendency to phase segregation with the appearance of a two-phase domain in the concentration-temperature plane (Fig. 3). In the case of ordering, the ordered phase has generally a domain of existence separated from solid solutions by two-phase regimes. Actually, there is no basic difference in principle between the two situations: phase separation can be considered as a special case of ordering. The phase diagrams look different because the order parameter describing phase separation is precisely the concentration: for a fixed concentration, the order parameter is said to be conserved, which is not the case for genuine ordering (see below and Ref. [1]).

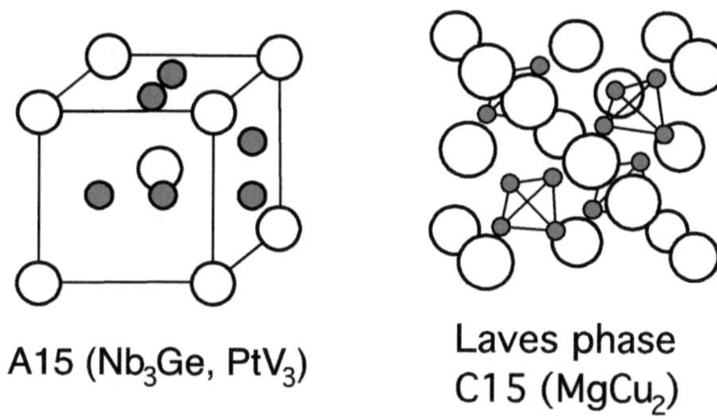

Fig. 2 A15 and C15 ordered compounds

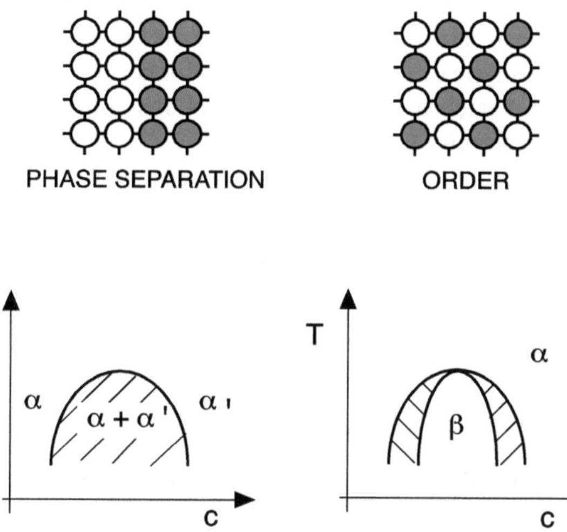

Fig. 3 Phase diagrams in the case of phase separation (*left*) and of ordering (*right*); the α and α' phases are solid solutions of different concentrations; β is an ordered phase. The hatched regions are two-phase domains

1.2 Displacements

We now consider atoms of different sizes. Then, in general, for a given atomic configuration, the atoms do not sit exactly at their reference lattice sites. In a crystalline state however, each atom can still be referred to a specific lattice site $\mathbf{n}$, its actual position being now given by $\mathbf{R}_n = \mathbf{n} + \mathbf{u_n}$, where $\mathbf{u_n}$ is the atomic displacement (see Fig. 4). To be more precise, assume that we start from a pure metal A and that we add at random big B atoms. This has two effects. First, the lattice parameter increases, and then the atoms are displaced from the positions of the mean lattice. In more general situations the symmetry of the lattice may also change with the introduction of impurities and several shear parameters should be introduced to describe the homogeneous deformation of the unit cell. Notice here that, even in the case of ordered structure on a fixed lattice, the symmetry generally decreases upon ordering so that we expect also distortions of the unit cell. For example the $L1_0$ ordered structure (see Fig. 1) is tetragonal with a c/a ratio which generally deviates from unity.

Fig. 4 Displacements of the atoms from the mean lattice positions

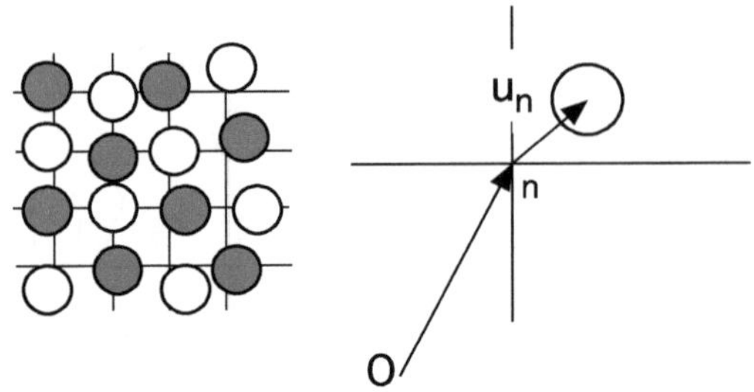

To avoid any ambiguity, it is therefore important in practical cases to specify what is the lattice of reference. Changes of lattice parameters produce displacements proportional to the interatomic distances whereas the displacements from the mean lattice remain finite and are generally small.

This type of description can no longer be used in the case of strongly disordered systems such as liquids or amorphous solids, and more generally in the case of strong structural transformations.

1.3 Configurations, Thermodynamics and Energetics

The first thing to do before developing statistical thermodynamic models at a microscopic (atomic) level is to characterize the atomic configurations of the alloy.

1.3.1 Configurations: Chemistry and Displacements

For a binary alloy, we can assign two degrees of freedom to each site, the so-called occupation number $p_{\mathbf{n}} = 0$ or 1, depending on the nature of the atom (A or B) at site $\mathbf{n}$, and the displacement $\mathbf{u}_{\mathbf{n}}$. More generally, in the case of a multicomponent alloy, we use general occupation numbers $p^i_{\mathbf{n}}, i = A, B, C, \ldots$ In a full thermodynamic theory, chemical and displacement effects are obviously coupled. Since we do not consider here the effect of external stresses, the displacements and internal stresses are induced by the chemical effects. We begin with a discussion of these chemical effects on a fixed lattice.

A "chemical" configuration is then characterized by the set of numbers $p_{\mathbf{n}}$, which in one dimension can be viewed as a binary number. Another convenient convention is to use spin-like variables $\sigma_{\mathbf{n}} = \pm 1$, with $p_{\mathbf{n}} = (1 + \sigma_{\mathbf{n}})/2$.

1.3.2 Thermodynamics

We assume that we have an expression for the energy as a function of the configurations $H(p_{\mathbf{n}}, \mathbf{u}_{\mathbf{n}})$. Using standard statistical physics tools we define the partition function Z:

$$Z = \sum_{p_{\mathbf{n}},\mathbf{u}_{\mathbf{n}}} \exp -(H/k_B T), \tag{1}$$

so that the corresponding free energy F is given by $F(V, T, N^i) = -k_B T \text{ Log } Z$, where V is the volume and N^i is the number of atoms of type i. This is just the standard Helmhotz free energy. If we add a term PV where P is the pressure we get the Gibbs free energy $F(P, T, N^i) = \sum N^i \mu^i$ where μ^i is the chemical potential of species i. The variation of the (Helmhotz) free energy as a function of the independent thermodynamic variables (V, T, N^i) is given by the standard expression $dF = -PdV - SdT + \sum \mu^i dN^i$, where S is the entropy. We will frequently consider the case of a constant vanishing pressure and will handle a simple free energy depending on T and N^i. Assuming furthermore a constant number of atoms $N = \sum_i N^i$, *i.e.* for a binary alloy, $N^B = Nc$, $N^A = N(1-c)$, where c is the concentration in B atoms, we obtain:

$$dF(T, c) = -SdT + (\mu^B - \mu^A)Ndc = -SdT + \mu Ndc, \tag{2}$$

where μ is the chemical potential difference $\mu^B - \mu^A$.

Although the concentration c is frequently the genuine control parameter in the case of alloys, it is frequently more convenient to work within a thermodynamic ensemble where the concentration is no longer fixed whereas the total number of atoms N remains fixed. This is sometimes called a semi-grand canonical ensemble. For simplicity we will keep here the grand canonical denomination. In this ensemble, the concentration is fixed by the chemical potential difference μ and $F(T, \mu)$ is defined from $F(T, \mu) = F(T, c) - \mu Nc$. The passage from one ensemble to the other is made through the use of the functions $c(\mu)$ or $\mu(c)$. The graph $c(\mu)$ is called an isotherm.

1.3.3 Energetics

For alloys described with reference to a fixed lattice, it is usual to define chemical and elastic contributions to the energy of a particular configuration. As mentioned previously, the separation depends in practice on the choice of the mean lattice taken as a reference. Chemical interactions are related to the electronic structure of the alloy where all atoms are assumed to occupy the sites of the mean lattice. In metallic alloys where screening effects are efficient the corresponding interactions are short-ranged. This is discussed in detail in the contribution by G. Tréglia (Chap. 5). As far as the elastic interactions are concerned, a fairly complete treatment is available within the (quasi)-harmonic approximation for the displacements [3, 4]. Let $E_{el}(\{\mathbf{u_n}\})$ be the elastic energy, *i.e.* the part of the energy which depends on the displacements:

$$E_{el}(\{\mathbf{u_n}\}) \simeq E_{el,0} - \sum_{\mathbf{n},\alpha} F_{\mathbf{n}}^{\alpha} u_{\mathbf{n}}^{\alpha} + \frac{1}{2} \sum_{\mathbf{n},\mathbf{m},\alpha,\beta} \phi_{\mathbf{nm}}^{\alpha\beta} u_{\mathbf{n}}^{\alpha} u_{\mathbf{m}}^{\beta}, \tag{3}$$

where the coefficients in this second order expansion are the so-called Kanzaki forces and the forces constants. They are calculated on the lattice of reference. α, β denote vector components. In the simplest approximation, the Kanzaki force at some lattice position $\mathbf{n}$ is related to the occupation of neighbouring sites: $\mathbf{F_n} = \sum_{\mathbf{m}} \mathbf{W_{mn}} p_{\mathbf{m}}$, where the $\mathbf{W}$ are short-ranged interactions. If we replace the force constants by average ones on the mean lattice, we have just a linear coupling between the displacements $\mathbf{u_n}$ and the chemical occupation numbers. This is of course an over-simplification. For example, this implies that the elastic constants are assumed to be independent of the nature of the atoms.

Within the harmonic approximation, it is possible to calculate explicitly the sum over the displacements in the definition of the partition function Z (see Eq. (1)) [2–4]. The result is that the partition reduces to a sum over the occupation numbers of $\exp -(H_{\text{tot}}/k_B T)$, where H_{tot} is the sum of the chemical terms and of an induced elastic contribution which is quadratic with respects to the Kanzaki forces, with a coefficient given by the lattice Green function. This function characterizes the effect of a force at one lattice site on the displacement of another site. It is given by the "inverse" of ϕ considered as a matrix. The result is that the displacements have been eliminated but that we have additional chemical interactions of elastic origin on the fixed lattice of reference. In the simplest scheme described above, these interactions are pair interactions which are generally weak and fairly long-ranged with a typical $1/R^3$ variation with distance R [4].

1.4 Ising Model

Even if in general the total energy cannot be written in terms of pair interactions, the configurational part of this energy can frequently be written in this way (see G. Tréglia's contribution):

$$H(\{p^i_{\mathbf{n}}\}) \simeq E_0 + \sum_{\mathbf{n},i} E^i_{\mathbf{n}} + \frac{1}{2} \sum_{\mathbf{n},\mathbf{m},i,j} V^{ij}_{\mathbf{nm}} p^i_{\mathbf{n}} p^j_{\mathbf{m}}, \tag{4}$$

where E_0 is a constant which may well be the most important contribution. $E^i_{\mathbf{n}}$ is a point contribution, relevant in the case of inhomogeneous systems (surfaces, for example). For bulk alloys it does not depend on $\mathbf{n}$ and the corresponding term can be absorbed in the constant E_0 if the concentrations are fixed. Similarly, in the bulk, the pair interactions only depend on $\mathbf{m} - \mathbf{n}$.

In the following we consider binary alloys and work within the (semi)-grand canonical ensemble:

$$H(\{p_{\mathbf{n}}\}) = \frac{1}{2} \sum_{\mathbf{n},\mathbf{m}} V_{\mathbf{nm}} p_{\mathbf{n}} p_{\mathbf{m}} - \mu \sum_{\mathbf{n}} p_{\mathbf{n}}, \quad \text{with} \quad V_{\mathbf{nm}} = V^{AA}_{\mathbf{nm}} + V^{BB}_{\mathbf{nm}} - 2V^{AB}_{\mathbf{nm}}. \tag{5}$$

This is the so-called lattice-gas model which corresponds to a model where $p_{\mathrm{n}} = 0, 1$ indicates the presence or absence of a particle at site $\mathbf{n}$. Switching to spin-like variables $\sigma_{\mathbf{n}}$, this can also be written:

$$H(\{\sigma_{\mathbf{n}}\}) = -\frac{1}{2}\sum_{\mathbf{n},\mathbf{m}} J_{\mathbf{nm}}\sigma_{\mathbf{n}}\sigma_{\mathbf{m}} - h\sum_{\mathbf{n}}\sigma_{\mathbf{n}}$$
$$J_{\mathbf{nm}} = -V_{\mathbf{nm}}/4; \quad h = -\mu/2 - \sum_{\mathbf{m}} V_{\mathbf{nm}}/4. \tag{6}$$

In this form, this is the familiar Ising model. We see that in the alloy model, the chemical potential μ plays the part of the magnetic field. This analogy is very fruitful. It is somewhat hidden by the fact that the control parameter in the case of magnetism is the magnetic field (intensive thermodynamic variable) whereas in the alloy case it is the concentration, equivalent of the magnetization (extensive variable) which is fixed.

1.4.1 Ground States

For given interactions the first problem to solve is to find the configuration minimizing the energy. Let us consider a few examples of increasing difficulty. Much more can be found elsewhere [1].

Consider first a linear chain with first neighbour interactions V_1. We work at fixed concentration $c = 1/2$ and have therefore to minimize $V_1 \sum_n p_n p_{n+1}$. If $V_1 > 0$, the minimum $H = 0$ is attained for a perfectly ordered chain $ABAB\cdots$. More generally, positive interactions V_{nm} favour ordering. In the magnetic language this is equivalent to a tendency to antiferromagnetism (negative J_{nm}). In the case where V_1 is negative, we have to maximize $\sum_n p_n p_{n+1}$ with $\sum_n p_n = N/2$. The best we can do is to completely separate A and B atoms, which yields the configuration $AAAA\cdots BBBB\cdots$. In the thermodynamic limit, $N \to \infty$, c finite, the "interface" energy between the two sequences is negligible. This is a one-dimensional example of phase separation. This is equivalent to ferromagnetism. Remember that $c = 1/2$ means a zero magnetization. In the presence of an infinitesimal fixed magnetic field the ground state would be a ferromagnet with up or down spins, depending on the sign of the field. The complete solution in the presence of a finite field is shown in Fig. 5.

This discussion applies as well to various lattices with first neighbour interactions, more precisely to the so-called alternating lattices which can be divided into two equivalent interpenetrating sublattices. An example is the BCC lattice which can be divided into two simple cubic lattices. The ground state is the familiar CsCl or $B2$ structure $(V_1 > 0)$ or corresponds to phase segregation $(V_1 < 0)$.

Consider now a triangular lattice. In the case of ordering interactions $(V_1 > 0)$, it is no longer possible to "satisfy" all first neighbour bonds. The best we can do is to have AAB or ABB triangles. This is what is called a frustration effect. It is easy

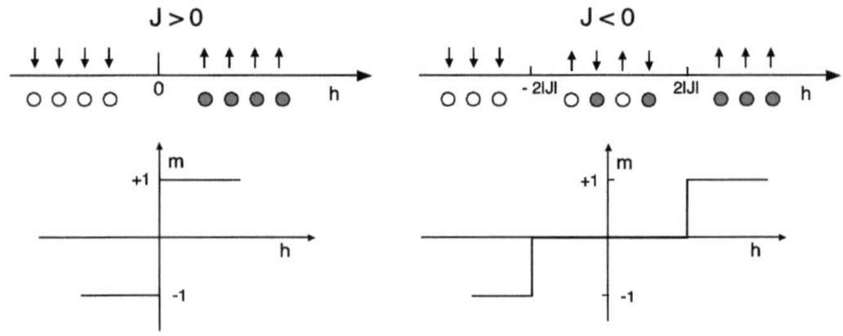

Fig. 5 Ground states of the Ising linear chain; the isotherms $m(h)$, equivalent of the alloy isotherms $c(\mu)$ are also shown

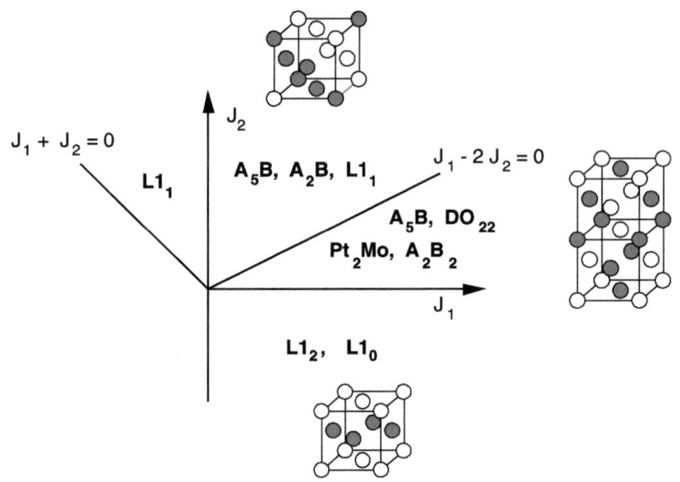

Fig. 6 Ground states of the FCC lattice with first and second neighbour interactions. Descriptions of the structures can be found in different places.Those corresponding to the concentration $c = 1/2$ are shown. Results from Refs. [2, 4, 5]

to realize that there are many ways of filling the lattice with such triangles, all of them being (degenerate) ground states [2]. So, in this case we can solve our problem by considering triangle configurations. In the case of the FCC lattice, tetrahedra produce further frustrations, but it is still possible to obtain fairly easily the ground states (for any concentration or chemical potential) by looking at the tetrahedron configurations. When we increase the range of the interactions the difficulty increases exponentially. In practice, one can say that the ground states are known on any simple lattice for interactions up to second neighbours, but, to take an example, they are not completely known on the FCC lattice for interactions up to fourth neighbours, which is a moderate range. The ground state phase diagram of the FCC lattice with first and second neighbour interactions is shown in Fig. 6.

1.5 Mean Field Theory

In the grand canonical ensemble, μ is fixed, and the concentration is defined as the average $\langle p_{\mathbf{n}} \rangle$ of $p_{\mathbf{n}}$, where the average of any quantity O is defined from $\langle O \rangle = \mathrm{Tr}\rho O$, where the symbol Tr, for trace, means that we sum over all values of the $p_{\mathbf{n}}$, *i.e.* over the total number 2^N of configurations. This is the advantage of working within the grand canonical ensemble: in the canonical ensemble we would have to sum over configurations of fixed concentration.

1.5.1 Simple Mean Field Theory for the Ising Model

Let us first consider the internal energy U,

$$U = \langle H \rangle = \frac{1}{2} \sum_{\mathbf{n},\mathbf{m}} V_{\mathbf{nm}} \langle p_{\mathbf{n}} p_{\mathbf{m}} \rangle. \tag{7}$$

In a first approximation we replace $\langle p_{\mathbf{n}} p_{\mathbf{m}} \rangle$ by $\langle p_{\mathbf{n}} \rangle \langle p_{\mathbf{nm}} \rangle = c_{\mathbf{n}} c_{\mathbf{m}}$ when $\mathbf{n} \neq \mathbf{m}$. Adding the usual entropy term for ideal solutions we obtain an expression for the mean field free energy:

$$F = \frac{1}{2} \sum_{\mathbf{n},\mathbf{m}} V_{\mathbf{nm}} c_{\mathbf{n}} c_{\mathbf{m}} + k_B T \sum_{\mathbf{n}} \{ c_{\mathbf{n}} \ln c_{\mathbf{n}} + (1 - c_{\mathbf{n}}) \ln(1 - c_{\mathbf{n}}) \} - \mu \sum_{\mathbf{n}} c_{\mathbf{n}}. \tag{8}$$

In this expression, the concentrations should be considered as variational parameters. $\partial F / \partial c_n = 0$ then leads to the mean field equations:

$$\sum_{\mathbf{m}} V_{\mathbf{nm}} c_{\mathbf{m}} + k_B T \ln \frac{c_{\mathbf{n}}}{(1 - c_{\mathbf{n}})} - \mu = 0, \tag{9}$$

which can also be written

$$\frac{c_{\mathbf{n}}}{(1 - c_{\mathbf{n}})} = \exp \big((\mu - \sum_{\mathbf{m}} V_{\mathbf{nm}} c_{\mathbf{m}}) / k_B T \big). \tag{10}$$

We see that the effect of the interactions is to replace μ by an effective chemical potential $\mu_n^{eff} = \mu - \sum_{\mathbf{m}} V_{\mathbf{nm}} c_{\mathbf{m}}$. In the magnetic language this is perfectly equivalent to the familiar expression:

$$\langle \sigma_{\mathbf{n}} \rangle = \tanh \beta h_{\mathbf{n}}^{eff}; \quad h_{\mathbf{n}}^{eff} = h + \sum_{\mathbf{m}} J_{\mathbf{nm}} \langle \sigma_{\mathbf{m}} \rangle. \tag{11}$$

For interactions between first neighbours J this yields the usual self-consistent equation for the magnetization $m = \langle \sigma_{\mathbf{n}} \rangle$:

$$m = \tanh \beta (h + ZJm), \tag{12}$$

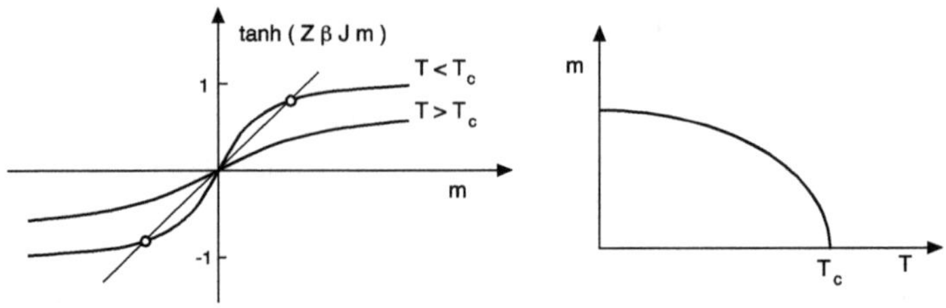

Fig. 7 Graphical solution when $h = 0$ of the self-consistent Eq. (12) (*left*), and variation of the magnetization as a function of temperature (*right*); $T_c = Z\beta J$

Fig. 8 Phase diagrams in the (h, T) and (m, T) planes; the latter can also be viewed as a phase diagram for phase separation if m is replaced by the concentration c

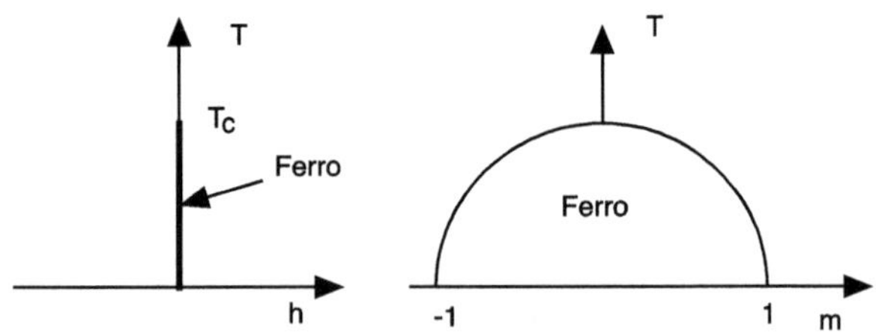

Fig. 9 Isotherms above and below the critical temperature. Notice the discontinuity at the origin below T_c

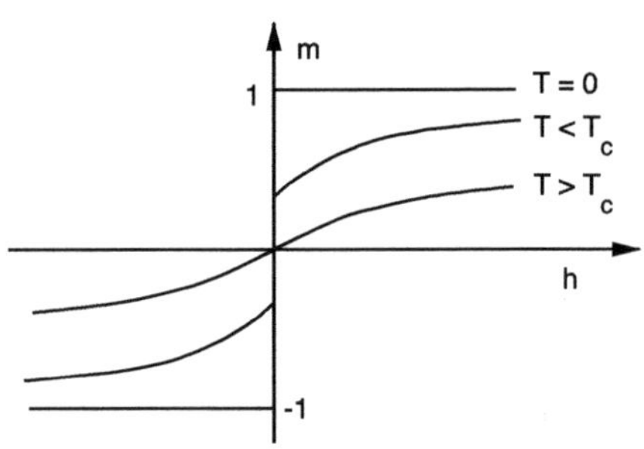

where Z is here the number of first neighbours. A graphical discussion of this equation at vanishing field, $h = 0$, shows that a ferromagnetic solution $m \neq 0$ exists when $\beta ZJ > 1$, *i.e.* when $k_B T < k_B T_c = ZJ$ (see Fig. 7).

1.5.2 Typical Phase Diagrams

The corresponding phase diagrams in the (h, T) and (m, T) planes are shown in Fig. 8. Typical isotherms are shown in Fig. 9. Typical phase diagrams in the case of order-disorder transitions are shown in Fig 10. The transition across the line in the (T, μ) plane can be of second or of first order (see below for details). In the latter case two-phase regions appear as shown in the phase diagram in the (T, c) plane.

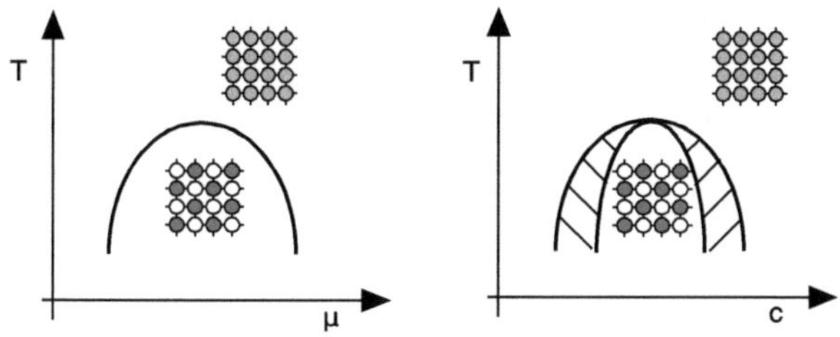

Fig. 10 Typical phase diagrams with order-disorder transitions

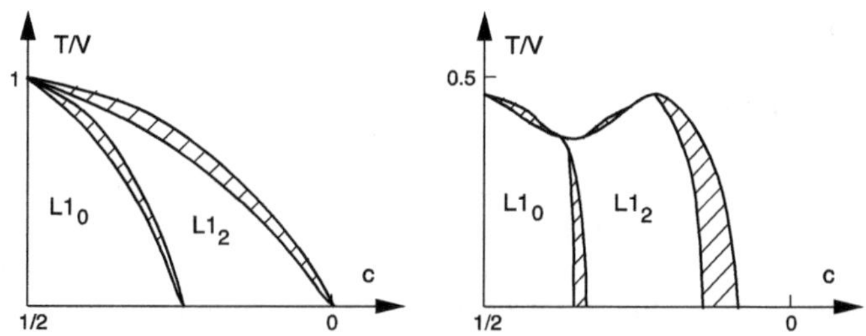

Fig. 11 Mean field phase diagram for the FCC lattice with first neighbour ordering interactions *V* (*left*) and "exact" phase diagram as deduced from CVM and Monte Carlo simulations (right). From Ref. [1], copyright (1991), with permission from Elsevier

1.5.3 Fcc Lattice; Mean Field and Beyond

The phase diagram of the FCC lattice with first neighbour heteroatomic interactions has attracted much attention for a long time when it appeared that the simple meanfield predictions were quite wrong: the topology of the phase diagram is not correct when compared to "exact calculations" and the mean field value for the critical temperature when $c = 1/2$ is too large by a factor of two. This is shown in Fig. 11.

These discrepancies are known to be due to frustration effects and, as a consequence, due to a very bad treatment of short range order and of configurational entropy. Numerous attempts to go beyond the simple mean field theory have been made in the seventies and after. The cluster variation method (CVM) of Kikuchi [1, 5, 6] has been shown to be one of the most efficient method to treat these difficulties. This is a fairly elegant extension of the mean field theory which has the advantage to provide analytical expressions for the free energy. The corresponding mean field equations are however not very simple to solve, and with the progress of computers, Monte Carlo simulations are now very convenient and efficient tools from a practical point of view. To summarize: the usual mean field theory is in general fairly good in the absence of frustration effects. Then the topology of the phase diagrams is correct with overestimations of the critical temperatures of the order of 10–20%. On the order hand, when frustration effects are important, either because of geometrical frustrations or because of the competition between different interactions, errors can be of the order of 100% and the phase diagrams can be qualitatively wrong.

1.6 Order Parameters

At finite temperature, or out of stoichiometry, the ordered phases are not perfectly ordered. Conversely the disordered phases display some local order. Traditionally, the distinction is therefore made between long range order and short range order.

1.6.1 Short Range Order (SRO)

We start from the so-called correlation function $\langle p_\mathbf{o} p_\mathbf{r} \rangle$ which is the average number of *BB* pairs occupying sites **o** and **r**, and for simplicity, assume that we have a single atom per unit cell. Within the standard statistical physics treatment this average is an ensemble average. In an infinite homogeneous system this is equivalent to a spatial average. The above correlation function only depends on **r** and the average is made over all pairs obtained through translations. Quite generally such pair correlation functions can be measured from diffraction experiments. Now when $\mathbf{r} \to \infty$, what happens at sites **o** and site **r** is uncorrelated in a disordered state, and $\langle p_\mathbf{o} p_\mathbf{r} \rangle \to \langle p_\mathbf{o} \rangle \langle p_\mathbf{r} \rangle = c^2$. This behaviour can also be considered as a definition of the disordered state (here a solid solution). The deviation from this behaviour is what is called short range or local order. The so-called Warren-Cowley SRO parameter $\alpha(\mathbf{r})$ is defined from:

$$\langle p_\mathbf{o} p_\mathbf{r} \rangle - \langle p_\mathbf{o} \rangle \langle p_\mathbf{r} \rangle = c(1-c)\alpha(\mathbf{r}). \tag{13}$$

In the disordered state, $\alpha(\mathbf{r}) \to 0$ when $\mathbf{r} \to \infty$. If not, this signals a symmetry breaking and the appearance of long range order.

1.6.2 Long Range Order (LRO)

Let us consider for instance an ordered linear chain *ABABAB*.... The product $p_o p_r$ is equal to $0, 1, 0, 1, \ldots$ or to $0, 0, 0, 0, \ldots$ when $r = 0, 1, 2, 3, \ldots$ depending on the choice of the origin. Taking the average, this gives $\alpha(r) = -1, +1, -1, +1, \ldots$.

In the ordered phase, because of the lowering of symmetry, translation and rotation variants generally appear, each of them being not completely ordered. Separate averages should then be performed in principle for each variant, which is not so easy to do without ambiguity. Formally the selection of definite variants can be achieved by introducing an appropriate conjugate field [1]. Then, one can write $\langle p_\mathbf{n} \rangle = c + \delta c_\mathbf{n}$. In our previous example, δc_n in the fully ordered state takes values

$\pm 1/2$. In general we write $\delta c_n = (-1)^n \phi/2$. In other terms, ϕ is the amplitude of the concentration modulation whose values for a given variant are in the interval (0,1). Changing ϕ into $-\phi$ amounts to exchange translation variants. When $c \neq 1/2$ we write $c_n = c + (-1)^n \phi/2$. Then $|\phi| < 2c$ when $c < 1/2$ or $< 2(1-c)$ when $c > 1/2$. In this precise case, ϕ can also be defined locally through $\phi = |c_{n+1} - c_n|$. Conversely, it is also very convenient to describe ordered states in terms of concentration waves, *i.e.* in reciprocal space. In the linear chain model we can indeed write also $c_n = c + \exp(ikna)\phi/2$, where a is the lattice parameter and $k = \pi/a$ is the wave vector of the modulation, defined up to a vector of the reciprocal lattice.

$L1_0$ ordered structure
In the case of the $L1_0$ structure a similar description can be used since this structure displays alternating A or B plane along the $0z$ direction (see Fig. 1). We can then write:

$$c_{\mathbf{n}} = (c + \phi \exp(i\mathbf{k}_3.\mathbf{n}))/2; \quad \mathbf{k}_3 = \frac{2\pi}{a}[001]. \tag{14}$$

The interest of this description is that it clearly displays the Fourier components of the atomic density which yield Bragg peaks in diffraction experiments. The $L1_0$ reciprocal space is obtained from the FCC reciprocal space by adding superstructure spots at $\mathbf{k}_3$ and equivalent positions. In real space ϕ can also be defined from the difference between the concentrations of successive planes in the z direction. For a given choice of the z axis there are obviously two equivalent ordering possibilities *ABAB*... or *BABA*.... They are called translation variants. There are also three orientational variants. Finally this gives six possible variants.

$L1_2$ ordered structure
Here the ordered structure can be viewed as the superposition of three concentration waves along the three axis:

$$c_{\mathbf{n}} = [c + \phi(\exp(i\mathbf{k}_1.\mathbf{n}) + \exp(i\mathbf{k}_2.\mathbf{n}) + \exp(i\mathbf{k}_3.\mathbf{n}))]/4 \tag{15}$$

$$\text{with} \quad \mathbf{k}_1 = \frac{2\pi}{a}[100], \quad \mathbf{k}_2 = \frac{2\pi}{a}[010], \quad \mathbf{k}_3 = \frac{2\pi}{a}[001]. \tag{16}$$

It is fairly easy to realize that we have here four translation variants. Notice that more complicated cases where the amplitudes of the different modulations are not equal can be defined. Then three different order parameters should be used, but it is more convenient to view them as the components of a vectorial order parameter ϕ.

1.6.3 Order Parameters: A Short Discussion

To summarize, LRO parameters on a fixed lattice can fairly easily be defined provided that the type of order is known and that definite variants can be identified. It is even possible to define "local" LRO parameters if the amplitude $\phi_{\mathbf{n}}$ of the

LRO modulation is slowly varying in space. For example, in the case of the $L1_0$ ordered structure, $\phi_\mathbf{n}$ can be defined from the difference of the concentrations of successive [001] planes.

In the previous discussion, SRO has been defined in the disordered state, *i.e.* when there is no LRO. Actually when LRO is not perfect, SRO parameters can also be defined from the deviations of the pair correlations functions from their values in a completely ordered state, but they are not easy to measure in real as well as in reciprocal space. For bulk crystals SRO in the disordered phase is conveniently measured through diffuse scattering experiments: a fully disordered crystal produces a so-called Laue diffuse scattering, constant in reciprocal space which is modulated in the presence of SRO, the modulation being proportional to the Fourier transform of the SRO parameters. Close to disorder-order transitions this diffuse scattering becomes frequently peaked on specific values which announce the tendency to long range order associated with the corresponding wave vectors. Diffraction, diffuse scattering and related topics are discussed in references [1–4].

Finite size effects will be discussed in more detail later on, but we can already put forward some points. LRO and SRO parameters are well defined for infinite systems, but it is fairly clear that LRO in a nanoparticle whose shape is related to the bulk lattice can perfectly be defined if there is a single variant, or well separated variants. For finite systems different lengths are involved, depending on the observed quantities. For example the Bragg peaks of the underlying lattice have a width in reciprocal space of the order of the inverse of the size of the crystal. Superstructure peaks have a width proportional to the inverse of the size of the ordered domains. Finally SRO peaks have a width related to the so-called correlation length, to be discussed in more detail below.

1.7 Landau Theory of Phase Transitions

Let us start from the high temperature disordered phase in which the (scalar) LRO parameter ϕ, at equilibrium, vanishes. If we impose some finite value of this parameter, it is possible—at least in principle—to define restricted sums within the partition function, over configurations of fixed ϕ. This allows us to define free energies $F(c, T, \phi)$, in the canonical ensemble. The Landau theory of phase transitions is a phenomenological approach based on the general behaviour of this free energy once the LRO parameters have been properly defined. At high temperature the disordered state should be stable, which means that F should be minimum at $\phi = 0$, $\delta F = F(\phi) - F(\phi = 0) \simeq r\phi^2/2$.

1.7.1 Continuous or Second Order Phase Transitions

We assume first that F is an even function of ϕ, which is the case when ϕ and $-\phi$ correspond to different variants. This is in particular the case of $L1_0$. Below some temperature the ordered phase should be stable. At the transition the coefficient r

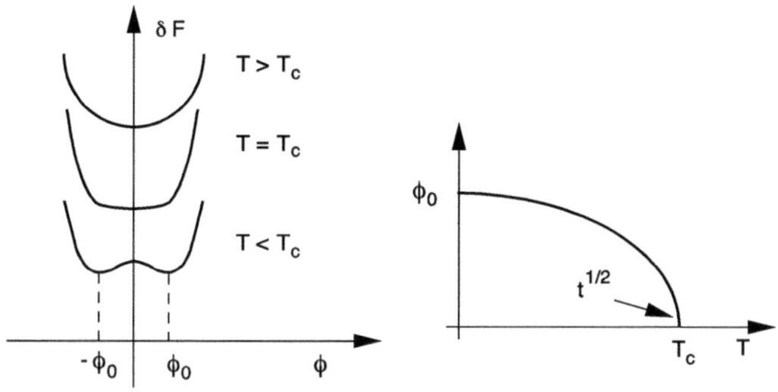

Fig. 12 Typical variation of the Landau free energy in the ϕ^4 model (left) and corresponding variation of the order parameter as a function of temperature (right)

should therefore vanish and should become negative below. The simplest assumption is therefore that $r \sim (T - T_c)$ where T_c is the critical order-disorder temperature. Since the free energy should be bounded from below we must add in the free energy a term of higher order, and we end up with the simplest ϕ^4 model:

$$\delta F = \frac{1}{2} r \phi^2 + \frac{1}{4} u \phi^4; \quad r \propto (T - T_c). \tag{17}$$

This free energy is shown in Fig. 12, as well as the corresponding variation of the order parameter as a function of temperature. We see that the transition is continuous, which implies in particular that there is no domain of coexistence between the two, ordered and disordered, phases: the two-phase domain in Fig. 10 does not exist. In the old Ehrenfest terminology, this type of transition is called a second order phase transition (discontinuities as a function of temperature appear on the derivative of ϕ, *i.e.* on the second derivative of the equilibrium free energy.

1.7.2 First Order Phase Transitions

Actually, in most cases, the order parameter is more complex or, even in the case of scalar order parameter, the symmetry $\phi \to -\phi$ no longer exists, as for example in the case of $L1_2$ ordering. The generic Landau free energy then contains a cubic term:

$$\delta F = \frac{1}{2} r \phi^2 + \frac{1}{3} w \phi^3 + \frac{1}{4} u \phi^4. \tag{18}$$

It is then easy to realize that, when lowering the temperature, a second minimum appears before reaching the temperature where the curvature at the origin vanishes (Fig. 13). Both phases co-exist at $T = T_0$ and there is a temperature range around T_0 where both phases are either stable or metastable, which is the source of hysteresis phenomena. The order parameter is now discontinuous at T_0 and the

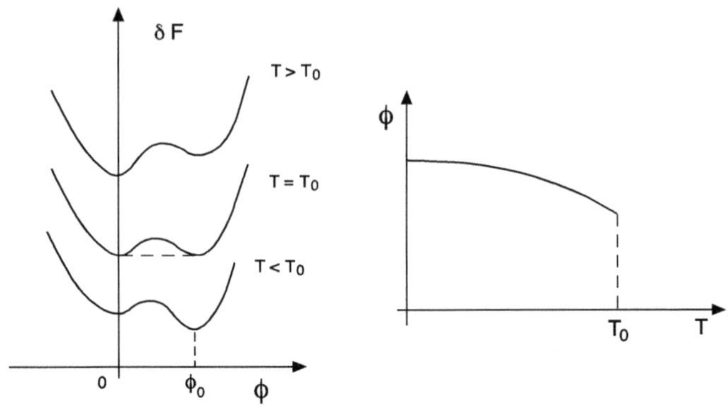

Fig. 13 Typical variation of the Landau free energy in the presence of a ϕ^3 contribution (left) and corresponding variation of the order parameter as a function of temperature (right)

phase diagram displays two-phase domains. First order transitions also occur if the coefficient of the ϕ^4 term is negative, in which case Landau expansion has to be pushed up to sixth order. This is in practice the case of the order-disorder transitions involving the $L1_0$ ordered state as in CuAu, CoPt, FePt, etc. Many physical quantities are then discontinuous at the transition: the internal energy or enthalpy, but also other parameters such as the c/a ratio in the case of $L1_0$. Such parameters are generally called in Landau terminology secondary order parameters. They are coupled to the main order parameter and can be used, once carefully calibrated, to determine experimentally this LRO parameter.

2 Interfaces and Surfaces

There are many cases where ordered alloys melt before disordering. When they are elaborated from the melt such alloys order directly from the liquid state where diffusion is very efficient so that in general a single variant is selected. In the presence of order-disorder transitions in the solid state on the other hand, different variants appear at low temperature, separated by so-called antiphase boundaries (see [7, 8] and references therein).

2.1 Antiphase Boundaries

The simplest antiphase boundaries (APB) are those separating translation variants. In one dimension this corresponds to a defect of type ...*ABABABBABABA*.... In the simplest case where the LRO parameter is scalar and symmetric the two domains in contact have LRO parameters equal to $\pm\phi_0$ far away from the boundary. Within this boundary, ϕ should therefore vary, and we will assume that

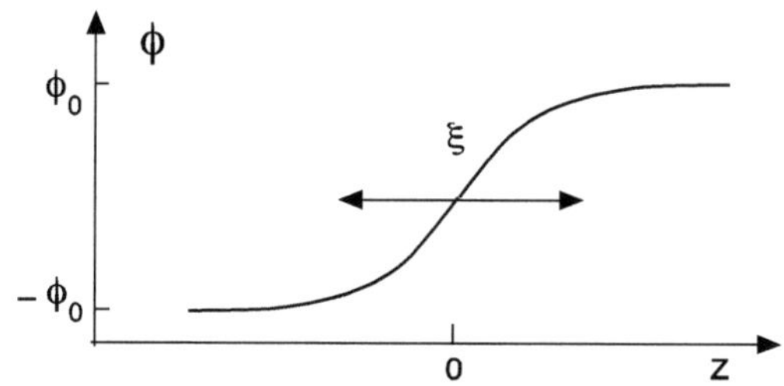

Fig. 14 APB profile $\phi(z)$ in the case of a second order phase transition

it becomes a function $\phi(z)$ of the coordinate z normal to the interface. This continuous approximation is valid provided the APB width is larger than the lattice parameter.

2.1.1 Landau-Ginzburg Theory

Within Landau theory it is natural to define local free energies $F_l(\phi(z))$ related to the value of the order parameter at point z. The total free energy can then be written as an integral over z of this local free energy. But this is not sufficient: varying ϕ has an energy cost because, at a microscopic level, wrong bonds are introduced (a *BB* bond in our one dimensional example). The simplest way to take this effect into account is to make an expansion in terms of derivatives of $\phi(z)$. By symmetry there is no linear term, and finally the simplest phenomenological expression for the free energy writes:

$$F = \int \frac{dz}{L} \left[F_l(\phi(z)) + \frac{1}{2} m \left(\frac{d\phi}{dz} \right)^2 \right], \tag{19}$$

where m is a "stiffness" coefficient characterizing the energy cost of wrong bonds. The free energy is a functional of $\phi(z)$ which describes the profile of the APB. This type of phenomenological theory has been first used by Landau and Ginzburg to describe inhomogeneous superconductors. We have then to minimize F with respect to $\phi(z)$. This is a familiar Euler-Lagrange problem which can easily be solved within a mechanical analogy. Actually if the term involving the derivative $d\phi/dz$ is viewed as a kinetic energy, the quantity to be integrated looks like a Lagrangian provided that $-F_l$ plays the part of the potential energy. Within this analogy, ϕ and z play the role of position and of time, respectively.

Second order phase transition

For a second order transition where the ϕ^4 model applies, the result is [4, 9] (Fig. 14):

$$\phi(z) = \phi_0 \tanh(z/\xi), \quad \text{with} \quad \xi = \sqrt{m/2r}, \tag{20}$$

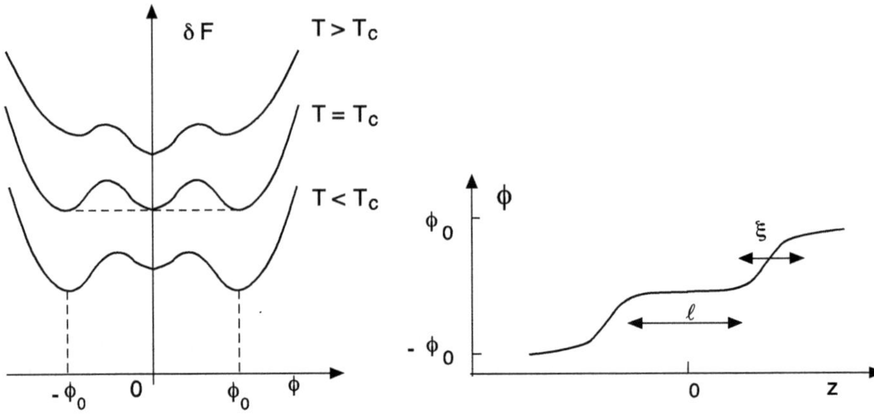

Fig. 15 Free energy in the ϕ^6 model (*left*) and order parameter profile (*right*)

where ξ is the so-called correlation length which therefore depends on temperature as $(T_c - T)^{-1/2}$. This is a typical mean field result. More accurate estimates based on more elaborate theories of phase transitions (renormalization group in particular [10]) predict that the exponent $1/2$ is replaced by a critical exponent ν which, in three dimensions and for the Ising model is about 0.66. We see that the order parameter profile strongly depends on temperature close to T_c: its width diverges whereas the amplitude of variation, equal to ϕ_0, vanishes. The APB broadens and vanishes at T_c.

The correlation length is a very important quantity in the modern theories of phase transitions. It measures the range of the perturbation introduced by a local defect. In the ordered phase considered here, it measures the range of the disordered region induced by the APB. In the disordered phase, one can equivalently define a correlation length measuring the ordered region induced by a local perturbation. In particular the short range order parameter behave as $\alpha(\mathbf{R}) \sim \exp(-R/\xi)$. The correlation lengths in both ordered and disordered phases have not the same meaning, but they behave similarly close to T_c with similar critical exponents ν.

First order phase transitions

In the case of first order transitions, the free energy is different, but the mechanical analogy can still be used. For convenience we consider a symmetric ϕ^6 model in which three minima co-exist close to T_c corresponding to the disordered phase and two variants. As a consequence we see that the order parameter profile presents a plateau at $\phi = 0$, *i.e.* a thin layer of width l develops between the two variants: the APB splits into two order-disorder interfaces. Simple models show that this width diverges according to a logarithmic law: $l \sim \xi \,\mathrm{Log}(T_c - T)/T_c$ (Fig. 15). So, in this model the bulk first order transition has been replaced by a continuous transition where the disordered phase nucleates at the interfaces. This is a wetting phenomenon [11].

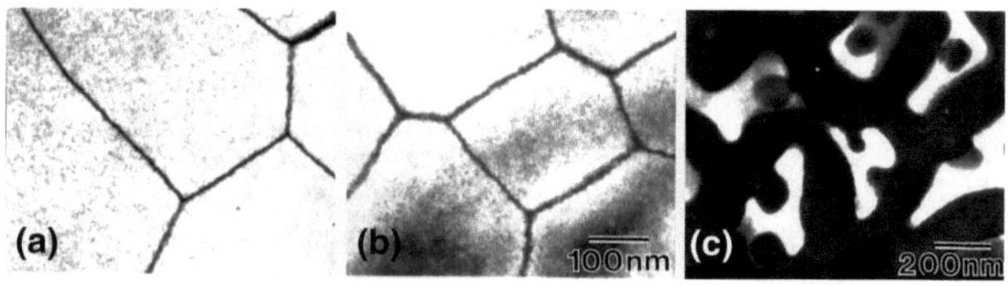

Fig. 16 Wetting of APB in Cu_3Pd and heterogeneous nucleation of the disordered phase. From Ref. [12], copyright (1992) by The American Physical Society

Fig. 17 Wetting of APB in Cu_3Au; **a** dark field image; the numbers 2 and 3 label APB characterized by the vectors $\mathbf{R}_2$ and $\mathbf{R}_3$. **b** High resolution TEM and heterogeneous nucleation of the disordered phase. From Ref. [13], copyright (1994) by Springer Science + Business Media

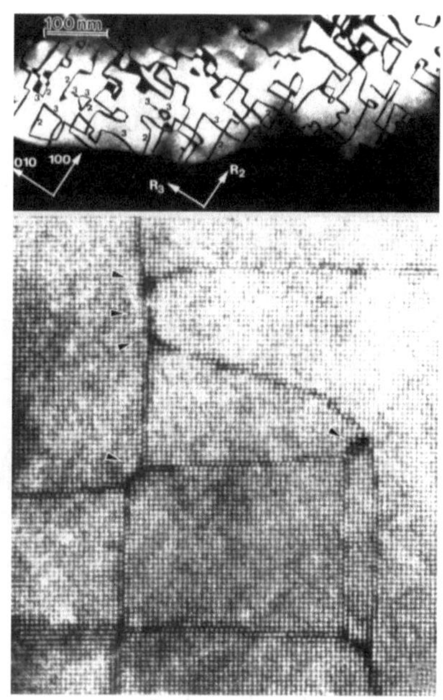

2.1.2 Experimental Observations

First order transitions

Wetting phenomena have been observed in several alloys showing (first order) $L1_2$−disorder transitions; Cu_3Pd, Cu_3Au, Co_3Pt. It turns out that in $L1_2$ there are three types of APB and that in usual dark field transmission electron microscopy (TEM) observations, two over three only are observable. Actually this extinction rule is violated when increasing the temperature, which is the signature of prewetting phenomena, and finally strong wetting and heterogeneous nucleation of the disordered phase on the APB is clearly seen (Fig. 16). In Cu_3Au the APB have a strong tendency to be perpendicular to [100] directions, and wetting is more anisotropic (Fig. 17).

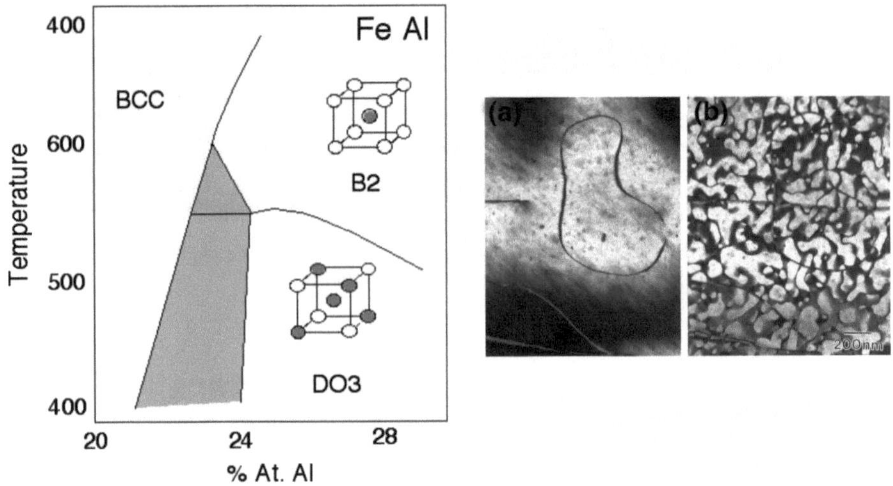

Fig. 18 *Left* phase diagram of the FeAl system. The DO_3-B2 transition is a second order phase transition. *Right* typical dark field images of APB. The big loop (*left*) is an APB separating variants of B2 wheras the smaller ones (*right*) correspond to different variants of DO_3 within B2

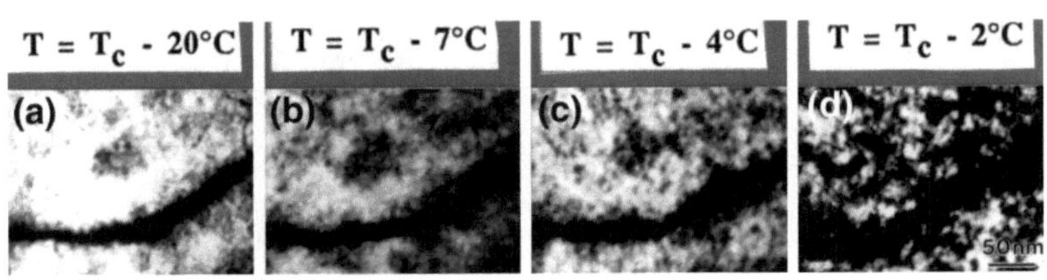

Fig. 19 *Dark field images* showing the critical behaviour of APB in the case of a second order phase transition. Results from Ref. [14]

Second order transitions
Second order phase transitions are less common in alloys but do exist in the FeAl system. FeAl orders according to the B2 structure, but the B2-disorder transition occurs at high temperature and is difficult to observe. On the other hand Fe_3Al, which is also B2 at high temperature transforms into the so-called DO_3 structure (Fig. 18) which can be viewed as a further ordering of the partially disordered B2 structure, and the transition is of second order.

When approaching the critical temperature from below we observe that APB broaden and disappear while critical fluctuations also appear in the bulk (Fig. 19).

2.2 Surfaces

The case of alloy surfaces is treated in detail in B. Legrand's contribution and here we just discuss a few points. As far as the Ising model, written in its magnetic

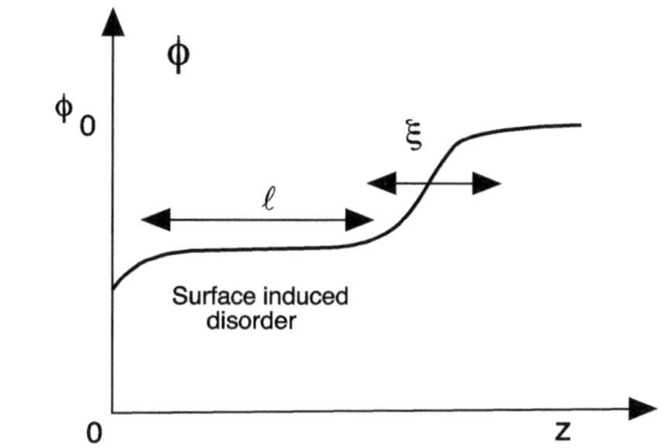

Fig. 20 Surface induced disorder close to the order–disorder transition temperature

formulation, is concerned (see Eq. (6)), the main complication is that all sites are no longer equivalent and there are now relevant point (single site) contributions to the energy, which can be included into site-dependent fields $h_{\mathbf{n}}$. In particular the field at the surface contains a contribution related to the difference between the surface energies of both elements of the alloy. This is still more obvious in a mean field treatment of the Ising model since now the effective field $h_{\mathbf{n}}^{eff} = h + \sum_{\mathbf{m}} J_{\mathbf{nm}} \langle \sigma_{\mathbf{m}} \rangle$ contains a contribution depending on the neighbours of site **n** (see Eq. (11)). As a consequence the local concentrations depend on the plane parallel to the surface where they are defined. Surface segregation is then the rule more than the exception. This is discussed by B. Legrand (Chap. 7).

In the case of ordering processes, such effects also exist but we must also define order parameter profiles, as in the case of APB. Actually the Landau-Ginzburg discussion applies also here, the only difference being that the boundary conditions are not the same. Considering the simplest case, APB are defined by the conditions that $\phi \to \pm 1$ when $z \to \pm\infty$, whereas in the case of surfaces we have to write a more complex boundary condition at the surface. The LRO profile is the same, but the position of this profile with respect to the surface depends on this condition. Depending on the surface field and on the nature of the transition (second order, first order) different profiles can be obtained. In the case of bulk first order transitions, the equivalent of the wetting phenomenon described above is the so-called surface induce disorder where a layer of disordered phase can wet the surface when approaching the phase transition from below (see Fig. 20). In the case of melting this is also called premelting. Such effects have been discussed in detail in the literature, from the theoretical side, but also from the experimental side [11, 15–17]. Surface induced disorder has first been observed in Cu_3Au using grazing incidence X-ray scattering. As mentioned in Sect. 1.6.2 the situation is slightly more complex in that case since the order parameter has three components. In the case of [100] surfaces one should therefore distinguish between longitudinal and transverse components of the order parameter. Similarly, in the case of $L1_0$ different behaviours are expected depending on the orientation of the c axis with respect to the surface.

In the case of surface induced disorder, the order parameter at the surface vanishes continuously while the bulk order parameter shows a discontinuity. It is generally considered as natural that even without wetting the surface has a tendency to be less ordered than the bulk. In a simple mean field theory the critical temperature is more or less proportional to the number of first neighbours, so that a surface completely decoupled from the bulk should order at lower temperatures. Actually this is not necessarily true in the presence of frustration effects. For example Schweika et al. have shown that a [100] surface of the $L1_0$ structure with a c axis along [001] can remain ordered *above* the bulk critical temperature [18]. Surface induced ordering can therefore exist, but it does not seem that experimental evidence of such an effect has been reported.

Surface segregation profiles have also been observed in Cu_3Au above T_c, and typical oscillations reminiscent of the low temperature phase are well observed [19], in agreement with theoretical arguments [20]. Finally the behaviour of surfaces close to second order phase transitions in the case of BCC alloys has also been discussed [21, 22].

3 Nanoalloys

In the case of nanoalloys, size effects are obviously more important still. Several relevant lengths involved in order-disorder phenomena have to be compared with the size of the nanoparticles. Let D be the diameter of the nanoparticle. When D is about a few nanometers, the particle contains 10^3 atoms or more, half of them being at the surface. We therefore expect variations of numerous physical properties for these nanometric sizes.

3.1 Fluctuations

In the disordered state the relevant length, as far as chemical order is concerned, is the correlation length ξ. In the case of second order phase transitions this length diverges at T_c. This poses a problem for finite systems. Actually, it is known that there is, strictly speaking, no phase transition in finite systems. From a mathematical point of view, this is simply due to the fact that phase transitions are associated with singularities ("non analyticities") of thermodynamic quantities: free energy, correlation function etc. For a finite system the number of configurations involved in the partition function is finite so that the partition function is an analytical function of temperature or of external fields. In practice, one can estimate a width of the transition. Since ξ varies as $1/\sqrt{|T - T_c|}$, the width ΔT of the transition is of the order of $1/D^2$ and $\Delta T/T_c$ is of the order of a^2/D^2, where a measures the range of the interactions. Within our simple Ising model, a is of the order of the distance between first neighbours, *i.e.* 0.2–0.3 nm so that $\Delta T/T_c$ is

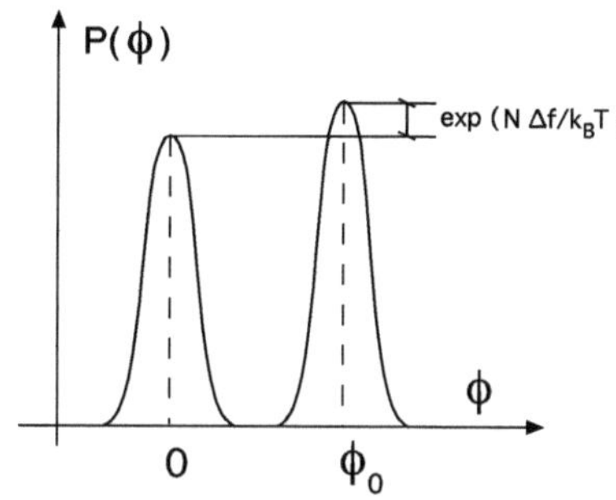

Fig. 21 Probability $P(\phi)$ in the coexistence regime of a first order transition; $\Delta f : f(\phi = 0) - f(\phi_0)$

about 10^{-2} for nanometric particles. This is not negligible but this shows that phase transitions can still be defined safely in most cases.

In the case of first order transition, the argument is different. Instead of having to consider fluctuations close to a single minimum of the free energy—the correlation length is related to the curvature of the free energy at the equilibrium position—we have to compare two different minima (see Fig. 13). Let $Nf_{1,(2)}$ be the free energy of the phases 1 (2) at equilibrium. The weight of each phase is proportional to $\exp(-Nf/k_BT)$. In the limit $N \to \infty$ they are comparable only if $\Delta f = f_2 - f_1$ vanishes. For finite N coexistence exists as long as $\Delta f \sim k_BT/N$. Now, close to the bulk first order transition, $\Delta f \simeq k_B(T - T_c)\Delta s$, where s is the entropy difference per atom between both phases in units of k_B, which is of order unity, and finally coexistence exists when $(T - T_c)/T_c \sim 1/N$, which is negligible for nanometric particles with thousands of atoms but not as soon as the size decreases $(N \sim D^3)$. Then the first order transition is replaced by a continuous transition, but a coexistence regime can still be defined as long as fluctuations for each phase remain small. To be more precise, let us define the probability of having an order parameter $\phi : P(\phi) \propto \exp(-Nf(\phi)/k_BT)$. For a single phase, and large N, $P(\phi)$ has a gaussian shape $P(\phi) \sim \exp(-Nr(\phi - \phi_0)^2/k_BT)$, where ϕ_0 is the equilibrium order parameter at the considered temperature. In the coexistence case $P(\phi)$ is the sum of two gaussians centred on different order parameters (see Fig. 21). The separation between them is significant only if their widths $\sim \sqrt{k_BT/Nr}$ is weak enough. More detailed discussions can be found in many places, for example in Ref. [23, 24].

3.2 Size Effect on the Critical Temperature

Thus, provided the system is not too small (say not below about 10^2 atoms), order-disorder temperatures can be defined, but they are not necessarily equal to the corresponding bulk temperatures. As mentioned earlier, a general belief based on simple mean field arguments is that order-disorder transitions should occur at

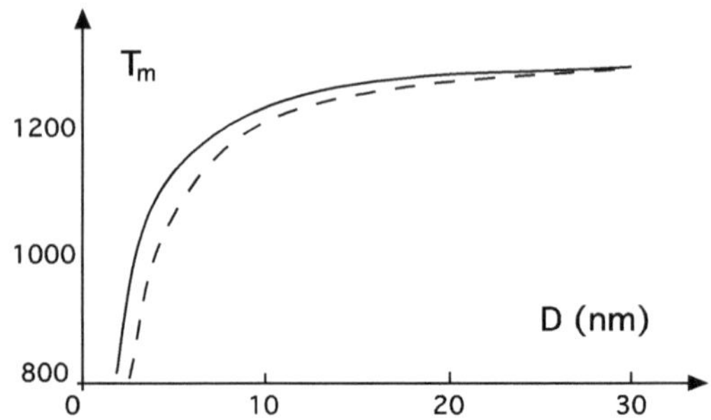

Fig. 22 Melting of small gold particles; *dashed lines* experiments; *solid line* Pawlow theory. Results from Ref. [25]

lower temperatures for small systems. In particular many studies have been devoted to the study of melting of nanoparticles (pure element). There are several theoretical arguments showing that the melting temperature should obey a $1/D$ law (Pawlow law) and this is observed in several systems (Fig. 22). Actually, the theoretical discussion is not so easy. Since surface and interface energies play a role, the melting process depends on the morphology of the two-phase regime and the final formulae depend on the approximations made.

In the case of alloys the problem is more difficult still since, except at congruent points, we have already two-phase equilibrium domains in the bulk. Phase diagrams can certainly be defined in the plane (μ, T), but complications arise in the (c, T) plane. Here again the morphology of the particles in the coexistence regime has to be discussed (core shell configuration for example), a further complication being that the concentration is no longer constant. In particular whereas wetting regimes and two-phase coexistence have been clearly distinguished in bulk systems, it is not obvious that the same can be done for nanoalloys. Then, even the simple case of phase separation requires a detailed treatment. This is discussed in the contribution by B. Legrand (Chap. 7). The case where we are dealing with genuine order-disorder transitions can be more complicated still because of obvious geometrical constraints. On the other hand the concentration is no longer the main LRO parameter and is not directly involved in the phase coexistence process.

3.3 *Order and Disorder in Nanoalloys*

We have seen that, starting from big systems, size effects on thermodynamic properties appear for particle sizes in the range 2–3 nm (about 10^3 atoms). Below 10^2 atoms we are dealing with clusters for which standard thermodynamics hardly applies. It turns out that this interesting intermediate regime can now be studied numerically. Systematic ab initio electronic structure calculations are not yet possible but empirical or semi-empirical methods can be used (see G. Tréglia's contribution, Chap. 5).

Once sufficiently simple schemes have been derived for calculating the energies of the atomic configurations, standard tools can be used to calculate thermodynamic quantities. Mean field methods are very useful tools to analyze the phenomena and are described in detail by B. Legrand in the case of tendencies to phase separation (Chap. 7). In the case of ordering effects it is frequently necessary to go beyond the simplest mean field approximation. But implementing the CVM for inhomogeneous systems is not easy and has not been used systematically up to now to study nanoalloys. Monte Carlo simulations on the other hand are comparatively fairly easy to implement, and a few systems have been studied in some detail recently.

3.3.1 $L1_0$ Nanoalloys

$L1_0$ ordered alloys such as FePt or CoPt are interesting for their magnetic properties. In these compounds the magnetic moment is principally localized on Fe or Co atoms, so that $L1_0$ alloys show a strong magnetic anisotropy due to the succession of two-dimensional magnetic and (almost non-) magnetic planes along the [001] direction. Notice that although the lowering of symmetry, from cubic to tetragonal, implies deviations of the c/a ratio from unity, the main source of magnetic anisotropy has a chemical origin. Such nanoalloys are supposed to be good candidates for magnetic storage devices provided of course size effects do not modify too strongly the ordering process (for recent discussions, see Refs. [26–32]). We review below some recent results based on Monte Carlo simulations.

We describe first the results of Monte Carlo simulations presented recently by Yang et al. concerning ordering in FePt alloys [33]. They used an Ising lattice-gas model with first and second neighbour interactions ($V_2/V_1 = -0.1$) and a surface field to account for surface segregation effects. The $L1_0$-disorder transition is found to be strongly of first order in the bulk with a jump of ϕ at the transition about 0.8 as observed experimentally, and as found in several other alloys [1]. In the case of nanoparticles, they consider truncated octahedra (TO). The sizes considered are in the range 2–5 nm (500–5,000 atoms). When the sizes decreases, the transition becomes continuous and the transition region broadens as expected with a typical decrease of the transition temperature of 15–20% (compared to the bulk value) when varying D from 5.5 to 3.3 nm. This also depends on the strength of the segregation field. The tendency to surface segregation which is of course in competition with the ordering process is also discussed in detail by the authors (Fig. 23). The coexistence regime discussed above is also studied. Actually the authors deduce from their Monte Carlo simulations the probability $P(E)$ for the configurations to have an energy E. Since the internal energy is also discontinuous at a first order transition, $P(E)$ should behave as $P(\phi)$. Yang et al. have checked that $P(E)$ displays the expected bimodal shape shown in Fig. 21 in the bulk.

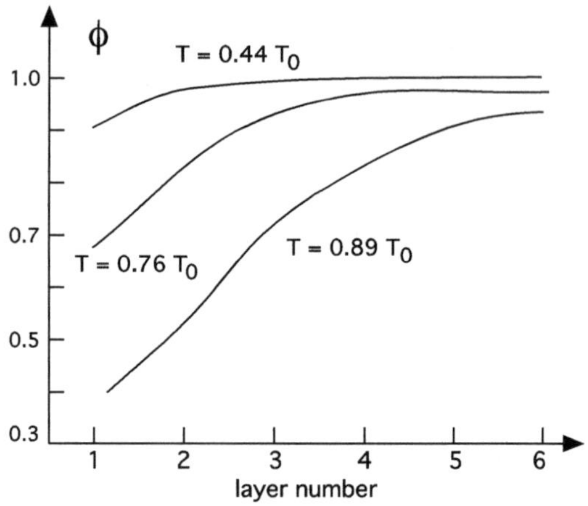

Fig. 23 LRO profile from the centre (*right*) to the surface (*left*) of the nanoparticle of diameter $D = 4.79$ nm. Results from Ref. [33]

For small particles on the other hand this bimodal regime disappears, which is a little bit surprising in view of our previous discussion, but to some extent, consistent with further analyses of the radial LRO profiles. Even well below the critical temperature, the LRO decreases significantly when approaching the surface, which seems to indicate a surface induced disordering (SID) process. These questions of correlations between the concentration or LRO profiles in connection with the nature of the transition can be subtle as discussed in great detail by B. Legrand in the case of a tendency to phase separation (Chap. 7). Even if the situations are not completely similar, it is clear that further studies are necessary here to have a better understanding of size effects in the case of order disorder transitions.

Quite similar results have been obtained by Müller and Albe [34]. More recently, Müller et al. [35] extend their discussion by using a phenomenological continuous interatomic potential instead of an Ising model on a rigid lattice, which allows them to calculate in a consistent way the usual cohesive properties (elastic constants, surface energy, melting temperature, etc.) as well as the quantities characteristic of order-disorder processes. They also find a slight decrease of the ordering temperature when the nanoparticle size decreases, but the effect is weaker than for the Ising model. The LRO profiles have also similar shapes. Finally the structural relaxation related to the deviation from unity of the c/a ratio is not found to be important. The authors conclude that the simplest Ising model is appropriate to study reliably order-disorder effects in these systems. Similar Monte Carlo simulations based on a semi empirical tight-binding potential confirm the lowering of the ordering temperature of about 175 K compared to the bulk ($T_0 = 825°\text{C}$) and also show that this lowering depend on the shape of the nanoparticle, in agreement with experimental results on CoPt [28] (Fig. 24). This lowering has also been found in the case of FePt using potentials deduced from ab initio electronic structure calculations [36]. See also calculations on ordered phases of the AuPd system [37].

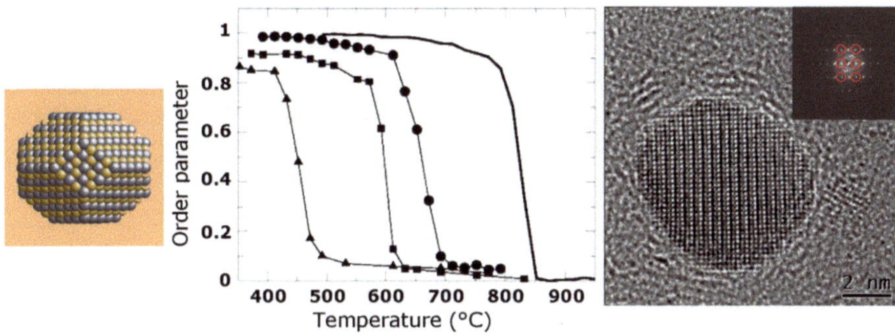

Fig. 24 Long range order parameter of Co-Pt systems from Monte Carlo simulations: bulk (*full line*) and clusters of 1,289 atoms (*circle*), 807 atoms (*square*), 405 atoms (*triangle*). TEM illustration of an ordered CoPt cluster. Results from Ref. [28]

3.3.2 Structural and Ordering Effects

Ordering effects in nanoparticles have been principally be studied using "crystalline" clusters. For example $L1_0$ ordering has generally been studied using truncated octahedra. On the other hand icosahedra or twinned decahedra are generally more stable for small sizes when dealing with pure elements. In the case of alloys, interplay between chemical and structural effects is highly plausible and several studies have been recently devoted to this problem. A detailed discussion can be found in the contribution by P. Andreazza (see also Fig. 25). We just show here the result of theoretical computations based on the tight-binding model mentioned previously. As expected the $L1_0$ TO ordered structure is found to be the stablest structure for sizes of the order of 3 nm (1,000 atoms). On the other hand more complex multitwinned configurations have been studied by Gruner et al. [27], and Dannenberg et al. have recently argued that (111) Pt-enriched surfaces are so favoured in FePt, CoPt or MnPt, that the $L1_1$ ordered structure, which is just a stacking of pure planes along a [111] direction, may well be stable in the case of small particles [31]. This surprising result does not seem to be consistent with the experimental observations (see [28] and the contribution by D. Alloyeau, Chap. 4) and certainly requires further studies.

3.4 Discussion

From the theoretical side, the number of studies dealing with ordering effects in nanoalloys are rapidly growing, and we have only presented here a brief and partial account. It is still difficult to have a broad synthetic view of the main effects involved, but the case of $L1_0$ systems, FePt, CoPt already provides us with some partial conclusions: the ordering temperatures seem to decrease systematically when decreasing the size of the particles, but this is not a big effect: reduction of

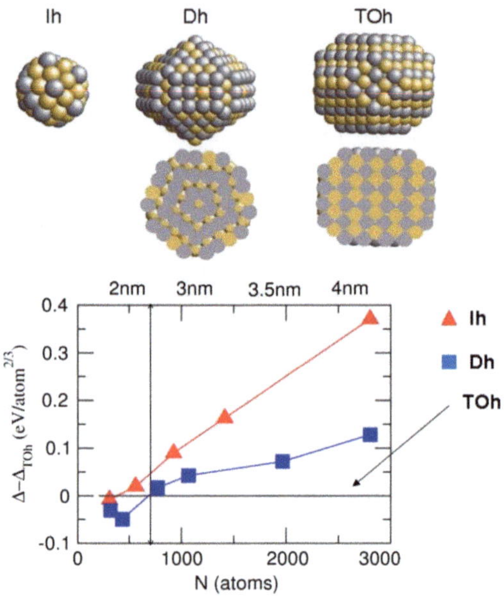

Fig. 25 Phase diagram at 0 K (ground state) of CoPt clusters representing the energy difference as referred to the FCC Truncated Octahedron (TOh) for non periodic structures as the Icosahedron (Ih) and the Decahedron (Dh) of different magic sizes (TOh 201, 314, 405, 807, 1289, 2075, 2951 atoms, Ih 309, 561, 923, 1415, ... and Dh 318, 434, 766, 1067, 2802, ...) at equi-concentration. The energy optimization is performed by Monte Carlo simulations at finite temperature for the chemical ordering including atomic displacements and completed by quenched molecular dynamics simulations for the final atomic relaxation. The structural transition between Dh at small sizes and TOh at large sizes is predicted at about 2–2.5 nm within the tight binding semi-empirical potential used in this study. From Ref. [32], copyright (2010) by The American Physical Society

about 10–15% for particle sizes of a few nm. Segregation effects modify the LRO profiles, and surface induced disorder appears frequently, but this is not a genuine wetting effect with the appearance of well defined interfaces.

These effects should certainly affect the magnetic properties of FePt and CoPt nanoparticles, but it is obviously difficult to measure LRO and magnetic profiles, although important progresses have been made recently. Ordering of nanoparticles of nanometric size is now well established and their magnetic properties are also well characterized [28–30, 38].

References

1. Ducastelle, F.: Order and Phase Stability in Alloys. Elsevier Science, New York (1991)
2. Ducastelle, F.: In: Belin-Ferré, E., Berger, C., Quiquandon, M., Sadoc, A. (eds.) Quasicrystals: Current Topics. World Scientific, Singapore (2000)

3. Krivoglaz, M.A.: X-Ray and Neutron Diffraction in Nonideal Crystals. Springer, Berlin (1996)
4. Khachaturyan, A.: Theory of Structural Transformations in Solids. Wiley, New York (1983)
5. De Fontaine, D.: Solid State Phys. **34**, 73 (1979)
6. Kikuchi, R.: Phys. Rev. **81**, 988 (1951)
7. Ducastelle, F.: In: Morán-López, J.L., Meija Lira, F., Sanchez, J.M. (eds.) Structural and Phase Stability of Alloys, pp. 231. Plenum Press, New York (1992)
8. Loiseau, A.: Curr. Opin. Solid St. Mat. **1**, 369 (1996)
9. Cahn, J.W., Hilliard, J.C.: J. Chem. Phys. **28**, 258 (1958)
10. Domb, C., Green, M. (eds.): Phase Transitions and Critical Phenomena, vol. 6. Academic, London (1976)
11. Dietrich, S.: In: Domb, C., Lebowitz, J.L. (eds.) Phase Transitions and Critical Phenomena, vol. 12, p. 2. Academic, London (1988)
12. Ricolleau, C., Loiseau, A., Ducastelle, F., Caudron, R.: Phys. Rev. Lett. **68**, 3591 (1992)
13. Potez, L., Loiseau, A.: J. Interface Sci. **2**, 91 (1994)
14. Le Floc'h, D., Loiseau, A., Ricolleau, C., Barreteau, C., Caudron, R., Ducastelle, F., Pénisson, J.M.: Phys. Rev. Lett. **81**, 2272 (1998)
15. Lipowsky, S.: Critical Phenomena at Interfaces. Springer, New York (1993)
16. Dosch, H.: Critical Phenomena at Surfaces and Interfaces. Springer, Berlin (1992)
17. Teraoka, Y., Seto, T.: Surf. Sci. **255**, L579 (1991)
18. Schweika, W., Landau, D.P., Binder, K.: Phys. Rev. B **53**, 8937 (1996)
19. Reichert, H., Eng, P.J., Dosch, H., Robinson, I.K.: Phys. Rev. Lett. **74**, 2006 (1995)
20. Mecke, K.R., Dietrich, S.: Phys. Rev. B **52**, 2107 (1995)
21. Mailänder, L., Dosch, H., Peisl, J., Johnson, R.L.: Phys. Rev. Lett. **64**, 2527 (1990)
22. Leidl, R., Diehl, H.W.: Phys. Rev. B **57**, 1908 (1998)
23. Barber, M.N.: In: Domb, C., Lebowitz, J.L. (eds.) Phase Transitions and Critical Phenomena, vol. 8, p. 2. Academic, London (1983)
24. Labastie, P., Calvo, F.: In: Bréchignac, C., Houdy, P., Lahmani, M. (eds.) Nanomaterials and Nanochemistry, p. 55. Springer, Berlin (1988)
25. Borel, J.P.: Surf. Sci. **106**, 1 (1981)
26. Šipr, O., Minár, J., Mankovsky, S, Ebert, H.: Phys. Rev. B **78**, 144403 (2008)
27. Gruner, M.E., Rollmann, G., Entel, P., Farle, M.: Phys. Rev. Lett. **100**, 087203 (2008)
28. Alloyeau, D., Ricolleau, C., Mottet, C., Oikawa, T., Langlois, C., Le Bouar, Y., Braidy, N., Loiseau, A.: Nat. Mater. **8**, 940 (2009)
29. Tournus, F., Tamion, A., Blanc, N., Hannour, A., Bardotti, L., Prével, B., Ohresser, P., Bonet, E., Epicier, T., Dupuis, V.: Phys. Rev. B **77**, 144411 (2008)
30. Tournus, F., Blanc, N., Tamion, A., Hillenkamp, M., Dupuis, V., Magn, J.: Magn. Mater. **323**, 1868 (2011)
31. Dannenberg, A., Gruner, M.E., Hucht, A., Entel, P.: Phys. Rev. B **80**, 245438 (2009)
32. Andreazza, P., Mottet, C., Andreazza-Vignolle, C., Penuelas, J., Tolentino, H.C.N., De Santis, M., Felici, R., Bouet, N.: Phys. Rev. B **82**, 155453 (2010)
33. Yang, B., Asta, M., Mryasov, O., Klemmer, T., Chantrell, R.: Acta Materialia **54**, 4201 (2006)
34. Müller, M., Albe, K.: Phys. Rev. B **72**, 094203 (2005)
35. Müller, M., Erhart, P., Albe, K.: Phys. Rev. B **76**, 155412 (2007)
36. Chepulskii, R.V., Butler, W.H.: Phys. Rev. B **72**, 1 (2005)
37. Atanasov, I., Hou, M.: Surf. Sci. **603**, 2639 (2009)
38. Kovács, A., Sato, K., Lazarov, V.K., Galindo, P.L., Konno, T.J., Hirotsu, Y.: Phys. Rev. Lett. **103**, 115703 (2009)

Segregation and Phase Transitions in Reduced Dimension: From Bulk to Clusters via Surfaces

Jérôme Creuze, Fabienne Berthier and Bernard Legrand

Abstract To describe the thermodynamics of bimetallic clusters, we use approaches that have been successfully employed for bulk alloys and their surfaces. We detail what happens for bulk and surface phase transitions when considering nanoalloys with a tendency to phase separation. A rigid-lattice approach allows us to analyze the behaviour of surface and core shells of the nanoalloys. We discuss the existence of bistabilities (or dynamical equilibrium) which are the analogous of surface and bulk phase transitions in semi-infinite alloys. Such dynamical equilibrium is susceptible to affect the cluster facets in an individual way (individual dynamical equilibrium), whereas the inner shells show a collective bistability (collective dynamical equilibrium). Then, we compare these bistabilities obtained in the semi-grand canonical ensemble with the results obtained in the canonical ensemble, before discussing the relation between experimental conditions and the two thermodynamic ensembles. Finally, a first attempt to establish generalized phase diagram for bimetallic clusters is proposed.

J. Creuze · F. Berthier
ICMMO/LEMHE, Univ. Paris-Sud, 91405 Orsay, France

F. Berthier
CNRS, UMR 8182, 91405 Orsay, France

B. Legrand (✉)
SRMP-DMN, CEA Saclay, 91191, Gif sur Yvette Cedex, France
e-mail: bernard.legrand@cea.fr

D. Alloyeau et al. (eds.), *Nanoalloys*, Engineering Materials,
DOI: 10.1007/978-1-4471-4014-6_7, © Springer-Verlag London 2012

1 Introduction

Bimetallic clusters are used for both their surface properties (catalysis) and their bulk properties (magnetism), these properties being generally modified by the finite nature of the clusters [1, 2]. It is therefore appropriate to take advantage of a knowledge of bulk and surface phase diagrams of alloys to understand the ordering in nanoalloys, while striving to clarify what is specific to clusters. Some simple considerations deduced from the physics of surfaces or thin layers allow one to illustrate the complexity expected in the thermodynamics of nanoalloys.

If surface segregation is generally well characterized and understood in metallic alloys, its dependence with the surface orientation is somewhat subtle [3–6]. A good control of the role of cristallographic orientation on surface segregation is particularly important for clusters, as they exhibit different types of facets, for instance (111) and (100) facets for Wulff polyhedra. For nanoalloys with a tendency to order, exposing several facets, even of equivalent orientation, leads to an additional complexity, which has some similarities with the behaviour of vicinal surfaces in ordered alloys. Consider a stacking of planes parallel to the surface of alternating type, e.g., A/B/A/B... as for the stacking of (100) planes for the $L1_0$ ordered structure. If the surface segregation tends to promote A-segregation and therefore the surface termination of type A, what happens for a vicinal surface exposing (100) terraces? To avoid the unfavorable B-termination, experimental studies [7–9] and theoretical ones [10] reveal an original way to accommodate chemical and geometric constraints. The single steps observed in the disordered state are replaced by double steps in the ordered state, leading to a coupling between the bulk order–disorder transition and a surface structural transition.

Thin films of a given thickness lead to similar problems; thus, in the previous example, if we consider a thin film limited by two (100) surfaces, the chemical compatibility between $L1_0$ ordering and the two (100) surfaces depends on the parity of the number of planes parallel to the surfaces [10, 11]. Geometric constraints are even stronger for a cluster due to the multiplicity of facets. These simple examples illustrate the interest of knowledge of bulk and surfaces phase diagrams to study the thermodynamics of nanoalloys. Furthermore, the study of wetting that may occur at surfaces of semi-infinite alloys when approaching bulk phase transitions [12–15] is of special interest in nanoalloys where surface and inner shells (or "bulk shells") have comparable number of atoms. When compared to bulk and surface behaviour, properties due to the intrinsic characteristics of clusters (finite system, coexistence of many inequivalent sites such as vertices, edges, facets or various concentric shells from the surface toward the core) are then easier to analyze, especially in terms of cluster size.

To determine bulk phase diagram from the theoretical point of view, it is usual to consider the semi grand canonical (S-GC) ensemble where the alloy chemical potential $\Delta\mu$ is fixed. Then, the concentration is controlled by $\Delta\mu$. In Monte Carlo (MC) simulations, it has the advantage to avoid the appearance of interfaces for phase separating systems, contrary to the canonical ensemble when the (fixed) concentration is located in the miscibility gap. For infinite systems, both ensembles are equivalent and allow one to obtain the same bulk phase diagram. For semi-infinite systems, it is much easier to discriminate between surface wetting and phase separation in the S-GC ensemble than in the canonical one and the same conclusion remains true for nanoparticles. However, if the S-GC seems to be more efficient from the theoretical point of view, is it the most relevant ensemble in relation with experiments? Actually, if we consider an isolated nanoparticle with a given nominal concentration, the canonical ensemble is the natural ensemble to study its behaviour. However, experiments often consider a collection of nanoparticles. In this case, if the nanoparticles can exchange atoms with each other while avoiding the coalescence, they can be considered as being in mutual equilibrium with a fixed alloy chemical potential and then, the S-GC ensemble is the pertinent one. The same is true if a mutual equilibrium is reached during the growth of an assembly of nanoparticles before each one can be considered as isolated. Thus, the choice of the pertinent statistical ensemble to compare experimental and theoretical results for nanoalloys requires a detailed analysis of the experimental conditions used to achieve equilibrium provided it is reached.

Given the complexity of the subject, we adopt a rigid-lattice approach and a simple energetic model, which has to be considered as a first step before introducing the role of atomic relaxations in future work. This article is organized as follows: using an Ising model, we recall briefly the main characteristics of bulk phase diagrams for alloys with a tendency to phase separate, using numerical simulations (Monte Carlo) or analytical approaches (mean-field approximation) (Sect. 2). The properties of alloy surfaces (segregation isotherm and surface phase transition) will be recalled in the same frame, as well as the relationship between surface segregation and bulk phase separation via the wetting phenomenon (Sect. 3). Then, the extension of these results to bimetallic clusters is the subject of Sect. 4 and leads to a presentation of a generalized phase diagram for nanoalloys with a tendency to phase separate. This review will be concluded with a discussion about the relevant thermodynamic ensembles for comparison with experiments, as well as some outlooks, particularly on the role of atomic relaxations which is not treated in this work.

2 Bulk Phase Diagram

In this section we just recall the main characteristics of bulk phase diagrams for alloys with a tendency toward phase separation within the Ising model. The Hamiltonian of the system is then written for an alloy A_cB_{1-c} [16]:

$$H = \frac{1}{2} \sum_{i,j} \sum_{n,m \neq n} p_n^i p_m^j V_{nm}^{ij}, \tag{1}$$

where V_{nm}^{ij} is the interaction energy between an atom of type i at a site n and an atom of type j at a site m $(i, j = \text{A, B})$; p_n^i is the occupation factor at site n by an

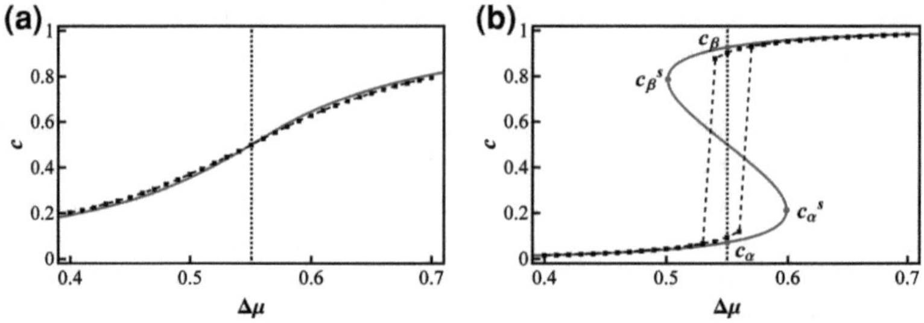

Fig. 1 Isotherm giving the bulk concentration as a function of the chemical potential difference $\Delta\mu$ (in eV) at $T/T_c = 1.7$ (**a**) and $T/T_c = 0.8$ (**b**), where T_c is the MC critical temperature. The dots represent Monte Carlo simulations and the solid line the mean field approximation; the energetic parameters are fitted to Cu–Ag system: $\tau = 46$ meV and $V = -30$ meV. In **b**, (c_α, c_β) are the solubility limits and (c^s_α, c^s_β) are the spinodal limits. The vertical dotted line indicates the critical value of $\Delta\mu$

atom i: it equals 1 if site n is occupied by an atom of type i and 0 otherwise. For the sake of simplicity, we only consider interactions between nearest neighbours and the results will be illustrated for an fcc lattice. A simple way to obtain the phase diagram is to perform Monte Carlo simulations (MC) in the semi-grand canonical ensemble (S-GC), where the number of sites in the simulation box is fixed and the bulk concentration c is controlled by the chemical potential difference between A and B: $\Delta\mu = \mu_A - \mu_B$ [17, 18]. Thus, at sufficiently high temperature, we obtain a one to one relationship between c and $\Delta\mu$, often called bulk isotherm (Fig. 1a). Such an isotherm is also obtained by minimizing the free energy in the mean-field approximation (MFA), which leads to the following relationship [19]:

$$\frac{c}{1-c} = \exp\left(-\frac{\Delta H^{perm}_{bulk} - \Delta\mu}{k_B T}\right), \tag{2}$$

where ΔH^{perm}_{bulk} is the bulk permutation enthalpy, i.e., the enthalpy change when switching a bulk atom from type B to type A. This energy is expressed simply in terms of the energetic parameters of the Ising model [19]:

$$\Delta H^{perm}_{bulk} = Z\,(\tau - V) + 2\,Z\,V\,c, \tag{3}$$

with $V = (V_{AA} + V_{BB} - 2V_{AB})/2$, $\tau = (V_{AA} - V_{BB})/2$ and Z is the bulk coordination number. In alloys with a tendency to phase separate that we consider in this work, the homoatomic pairs are favoured, and therefore V is negative. At high temperature, Fig. 1a shows that the isotherm obtained within MFA is very close to the one deduced from MC simulations.

Below a critical temperature T_c, given in MFA by $T_c^{MFA} = -ZV/2k_B$ [16], the isotherm defined by Eq. (2) is no longer bijective. It presents a van der Waals loop with two metastable parts, the first one between c_α and c^s_α, and the other one between c^s_β and c_β. c^s_α and c^s_β are the spinodal limits bordering the unstable part

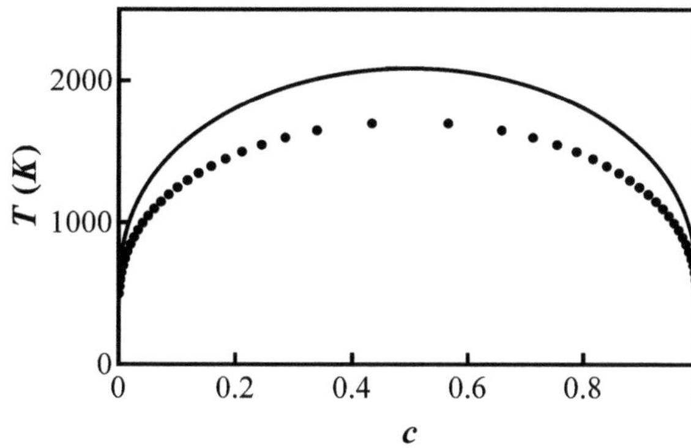

Fig. 2 Comparison of bulk phase diagrams obtained by MC simulations (*dots*) and MFA (continuous line); $V = -30$ meV

$[c_\alpha^s - c_\beta^s]$ of the isotherm [19]. The rule of equal areas is used to obtain the solubility limits c_α and c_β, leading to the determination of the bulk phase diagram (Fig. 2).

Below the critical temperature, MC simulations in the S-GC ensemble produce a concentration jump that may lead to more or less pronounced hysteresis, depending on the size of the box, the length of the simulation and temperature (Fig. 1b). The comparison between the phase diagrams obtained by MFA and MC simulations shows that MFA overestimates the critical temperature (by a factor of 1.22 in the case of the fcc lattice with interactions between nearest neighbours [20]) and the width of the miscibility gap. While the bulk isotherm depends on both τ and V, the phase diagram depends only on V [16]. Note that MC simulations in the canonical ensemble (i.e., with a fixed number of sites and a given nominal concentration) lead to a two-phase state for nominal concentrations within the miscibility gap, the proportion of the two phases of concentration c_α and c_β being given by the lever arm rule [21].

3 Alloy Surfaces

3.1 Isotherms, Driving Forces of Surface Segregation and Concentration Profiles

To describe the driving forces leading to the enrichment of the surface of an alloy by one of its constituents, we use the same mean-field approximation as in the previous paragraph. The semi-infinite alloy A_cB_{1-c} is considered as a stacking of planes parallel to the surface, c_p being the A-concentration of the pth plane ($p = 0$ for the surface plane and $p \to \infty$ for bulk). Minimizing the free energy with respect to all the c_p's in the S-GC ensemble leads to the fundamental equation of segregation [22, 23]:

$$\frac{c_p}{1-c_p} = \exp\left(-\frac{\Delta H_p^{perm} - \Delta\mu}{k_B T}\right), \tag{4}$$

where ΔH_p^{perm} is the permutation enthalpy on the p-plane, i.e., the enthalpy changes when switching an atom of the p-plane from type B to type A. Similarly to Eq. (3), this energy is expressed in terms of the energetic parameters of the Ising model [22, 23]:

$$\Delta H_p^{perm} = Z_p\ (\tau - V) + 2\ V \sum_q Z_{pq} c_q, \tag{5}$$

where Z_{pq} is the coordination number between sites of p- and q-planes and $Z_p = \sum_q Z_{pq}$ is the coordination number of a site of the p-plane. In the case of a semi-infinite alloy, it is possible to combine Eq. (5) with the bulk isotherm (2) to express the concentration of the p-plane as a function of the bulk concentration c [22, 23]:

$$\frac{c_p}{1-c_p} = \frac{c}{1-c} \exp\left(-\frac{\Delta H_p^{seg}}{k_B T}\right), \tag{6}$$

where $\Delta H_p^{seg} = \Delta H_p^{perm} - \Delta H_{bulk}^{perm}$ is the segregation enthalpy in the p-plane, i.e., the enthalpy balance in the exchange of a B atom in the p-plane with an A atom in the bulk. Eqs. (3) and (5) lead to the following expression for ΔH_p^{seg}:

$$\Delta H_p^{seg} = -\Delta Z_p\ (\tau - V) + 2\ V \left(\sum_q Z_{pq} c_q - Zc\right), \tag{7}$$

where $\Delta Z_p = Z - Z_p$ is the number of broken bonds for a site in the p-plane. Equation (7) allows one to analyze the driving forces of segregation by separating the effects of τ and V [24, 25]. Remember that $\tau = (V_{AA} - V_{BB})/2$ is proportional to the difference of cohesive energies between the two elements A and B; more precisely $\tau\ \Delta Z_p = \gamma_A - \gamma_B$, where γ_i is the surface energy of pure element i (i = A or B).

The term proportional to τ, generally known as the effect of surface energy, favours the segregation of the element of lowest surface energy, regardless of the bulk concentration (Fig. 3). This term is all the more important as the surface is "open", i.e., the number of broken bonds ΔZ_0 is high. Note that the possibility to observe a surface almost pure with an element together with a bulk almost pure with the other element, similar to what is often called a "core–shell" configuration for clusters [26–28], is not a prerogative of alloys which tend to phase separate, as it is sometimes mentioned. Thus, Fig. 3c shows that even for an ideal solid solution ($V = 0$), the surface can be pure in A and the bulk almost pure in B if the surface energy effect is sufficiently high or if temperature is sufficiently low.

The term proportional to V, often called the alloying effect, leads to a reversal of the segregating element with the bulk concentration (Fig. 4). Thus, for alloys

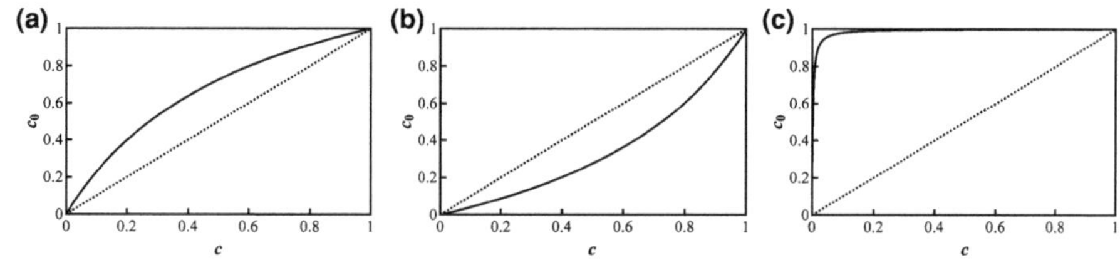

Fig. 3 Segregation isotherm for $V = 0$ and $\tau > 0$ (**a**), $\tau < 0$ (**b**) and a low value of kT/τ (**c**). In **a** and **b**, $|kT/\tau| \approx 4$ and in **c** $kT/\tau \approx 0.65$. The diagonal dotted line indicates the absence of surface segregation

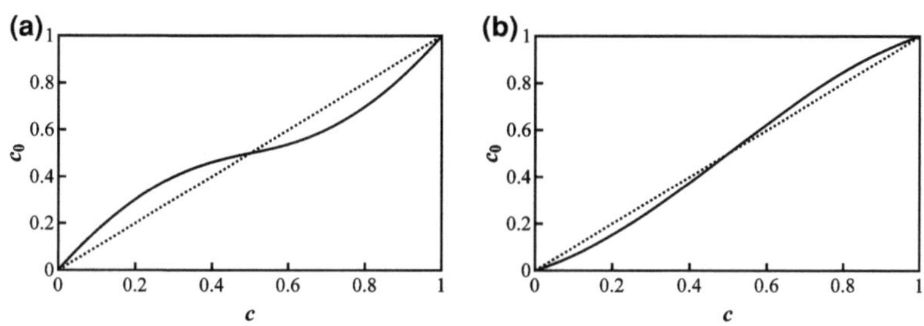

Fig. 4 Segregation isotherm for $\tau = 0$ and $V < 0$ (**a**), $V > 0$ (**b**) with $|kT/V| \approx 6$. The diagonal line indicates the absence of surface segregation

with a tendency toward phase separation as considered here ($V < 0$), this term tends to segregate the minority element (Fig. 4a), whereas it leads to segregation of the majority element for alloys which tend to order ($V > 0$), Fig. 4b.

The sign of V is also driving the nature of the concentration profile near the surface as shown in Fig. 5. A tendency to form homoatomic pairs ($V < 0$) leads to a monotonic profile (Fig. 5a), while a trend to form heteroatomic pairs leads to an oscillating profile (Fig. 5b) [29–31]. All these results derived from a mean-field approach are also obtained with Monte Carlo simulations, in both S-GC-ensemble and canonical ensemble, the ideally infinite bulk then acting as a reservoir with respect to the surface. Finally mention that the "two effects rule" defined by Eq. (7) is generalized into the "three effects" rule, when the size difference between the constituents is taken into account and an additional "elastic" contribution to the segregation enthalpy is obtained after relaxation of atomic positions [24, 25]. Among these three effects, the effect of "surface energy" is usually dominant as shown by maps assessing the relative importance of the three effects [32].

3.2 *Surface Phase Transitions*

In the previous section, bulk and surface were in a disordered state, i.e., forming a solid solution, due to a sufficiently high temperature. This is reflected by the monotonic

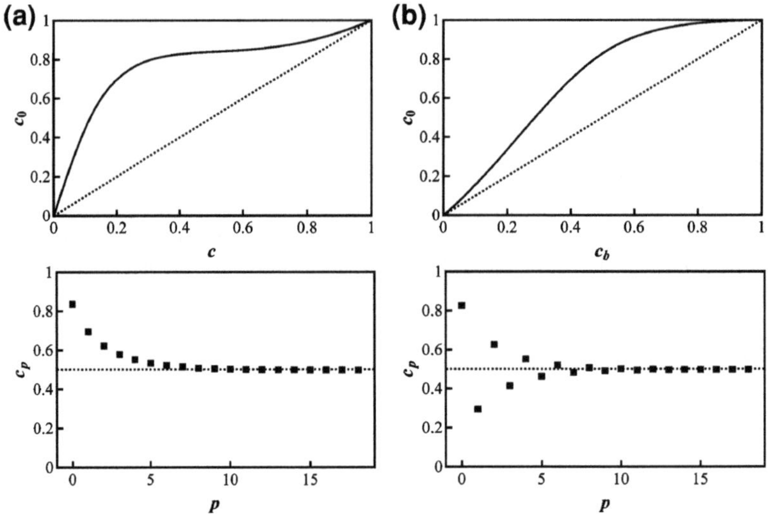

Fig. 5 Segregation isotherm (*upper part*) and concentration profile for equiatomic alloys (*lower part*) with $\tau > 0$, $V < 0$ ($\tau = 46$ meV, $V = -30$ meV and $|kT/V| \approx 6.3$) (**a**) and $\tau > 0$, $V > 0$ ($\tau = 95$ meV, $V = 80$ meV and $kT/V \approx 2.4$) (**b**). In the *upper part*, the *diagonal line* indicates the absence of surface segregation and in the *lower part* the *horizontal line* indicates the bulk concentration

character of segregation isotherm. What happens if the temperature is lowered, the bulk remaining however in the domain of solid solution? To answer this question, we consider a simple "monolayer" model, which is equivalent to suppose that only the surface plane is affected by segregation [19, 33]. Furthermore, we assume that the bulk is very diluted in the segregating species; under such conditions, the problem is reduced to the study of a 2D-system [23]. The mean-field approximation then allows one to determine the critical temperature for the surface plane and the critical chemical potential associated with surface miscibility gap. When comparing these values with those obtained for the bulk, one obtains:

$$T_c^{surf} = -Z^{//}V/2k_B \quad \text{and} \quad \Delta\mu_c^{surf} - \Delta\mu_c^{bulk} = -Z^{\perp}(\tau + V), \tag{8}$$

where $Z^{//}$ and $Z^{\perp}$ are respectively the numbers of bonds (per atom) within the plane and with the upper (or lower) planes for a bulk plane parallel to the surface ($Z = Z^{//} + 2Z^{\perp}$). Recall the analogous relations for the bulk:

$$T_c^{bulk} = -\left(Z^{//} + 2Z^{\perp}\right) V/2k_B \quad \text{and} \quad \Delta\mu_c^{bulk} = \left(Z^{//} + 2Z^{\perp}\right) \tau. \tag{9}$$

The critical temperature being all the more overestimated in MFA as the coordination number is low [16], this leads to an overestimation of the ratio between surface and bulk critical temperatures as shown in Table 1. Equation (8) predicts the effect of surface orientation on the characteristics of the surface

Table 1 Bulk and surface critical temperature determined within the mean-field approximation (MFA) and with Monte Carlo (MC) simulations

	Bulk	(111)	(100)
MFA	T_c^{MFA}	$0.5\ T_c^{MFA}$	$0.33\ T_c^{MFA}$
Monte Carlo	T_c	$0.37\ T_c$	$0.23\ T_c$

For the bulk remember that the ratio of MFA and MC critical temperatures is given by: $T_c^{MFA}/T_c = 1.22$ [20]

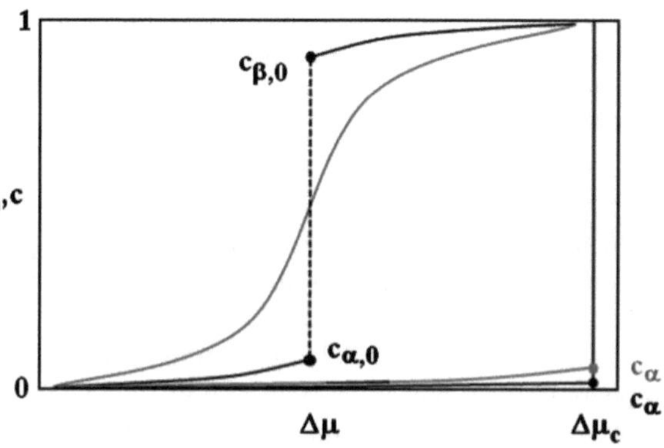

Fig. 6 Schematic behaviour of surface and bulk isotherms at two temperatures, $T_c > T_1 > T_c^{surf}$: *grey lines* and $T_c^{surf} > T_2$: *black lines*. Bulk solubility limits ($c_\alpha(T_n)$ for $n = 1$ and 2) and surface solubility limits ($c_{\alpha,0}\ (T_2)$ and $c_{\beta,0}\ (T_2)$) are indicated

isotherm. As $Z^{//}$ decreases and $Z^{\perp}$ increases when the surface becomes more and more open, this leads to a decrease of the critical surface temperature and an increase of surface segregation (the average position of the surface isotherm being given by $\Delta\mu_c^{surf}$ which corresponds to $c_0 = 0.5$).

3.3 From Surface to Bulk

In this section, we clarify the link between surface segregation and phase separation in alloys with a tendency to phase separate. To this aim, we consider the side of the phase diagram, where this is the solute A (of concentration c) which segregates, and we analyze the superficial behaviour when the bulk concentration c reaches the solubility limit c_α, either by reducing the temperature at constant c (as in experiments) or by increasing c at constant temperature (similar to the isotherm determination). Fig. 6 illustrates schematically the behaviour of the surface isotherm relative to the bulk one when approaching $\Delta\mu_c$ (or c_α), with two different cases:

- either $T_c > T > T_c^{surf}$: the surface isotherm is then continuous;
- either $T_c^{surf} > T$: the surface isotherm then exhibits a jump of concentration between the solubility limits of the surface plane.

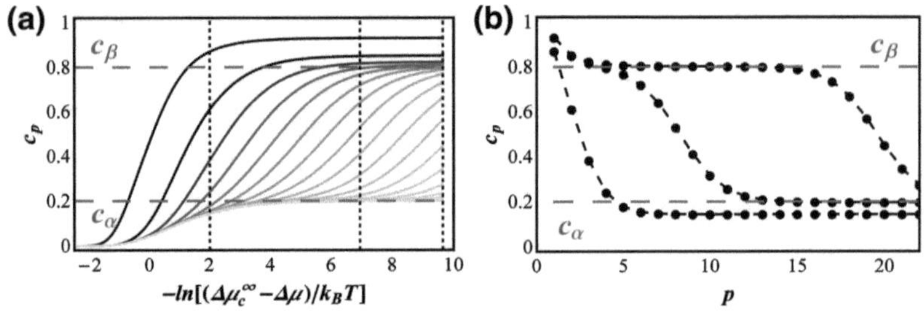

Fig. 7 Schematic illustration of the wetting. Isotherms of the first planes near the surface when approaching $\Delta\mu_c$ (**a**) and corresponding concentration profiles obtained for the three values of $\Delta\mu$ indicated in **a** by the vertical *dotted lines* (**b**). In **a**, the surface isotherm is on the *left* and the innermost ones are on the *right*

When $\Delta\mu$ approaches $\Delta\mu_c$, the number of planes becoming solute-rich may increases gradually, this is the wetting phenomenon [12–15]. Fig. 7a shows the behaviour of the isotherms of the first planes near the surface as a function of $-\ln[(\Delta\mu_c - \Delta\mu)/kT]$. A logarithmic scale is chosen because this behaviour appears only in the immediate vicinity of $\Delta\mu_c$.

The concentration of the planes near the surface then varies from a value very close of the bulk solubility limit c_α (remaining slightly lower due to the fact that the system is still in the single phase part of the bulk phase diagram) to a value very close of c_β. Note that the transition between these two concentrations can be continuous (as shown in Fig. 7a) or discontinuous, depending on the temperature, as does the transition affecting the surface plane (Fig. 6). The concentration profiles obtained for different values of $\Delta\mu$ are shown in Fig. 7b and illustrate some key elements of the wetting, namely:

- a phase initiated at the surface, with a concentration close to c_β and possibly a superficial segregation of A with respect to this concentration;
- an interface between the phases of concentration close to c_β and c_α, that goes gradually away from the surface as $\Delta\mu$ tends to $\Delta\mu_c$;
- a bulk phase of concentration close to c_α (undersaturated solid solution).

The thickness of the wetting layer is a balance between two opposite forces:

- a repulsive force between the surface and the c_α/c_β interface due to the interaction between superficial and interfacial concentration profiles; this leads to a repulsive force depending exponentially on the distance d between surface and interface [16, 34, 35];
- an attractive force due to the free energy cost associated with the presence near the surface of the phase of concentration close to c_β, which is only metastable in the bulk. This force is proportional to d and to $(\Delta\mu_c - \Delta\mu)$, and is at the origin of the logarithmic divergence of the thickness d of the wetting layer with $(\Delta\mu_c - \Delta\mu)$ [16, 34, 35].

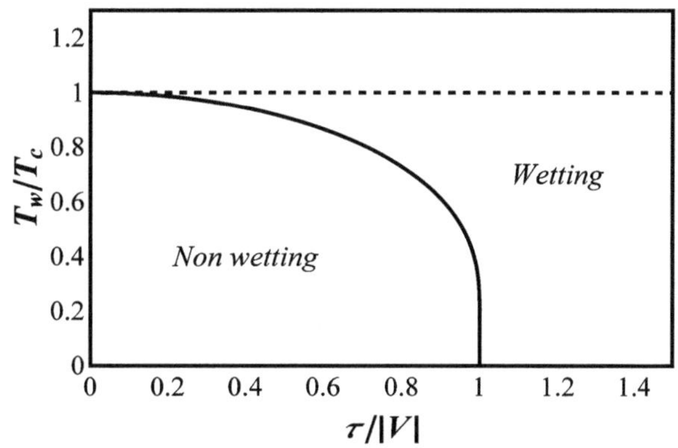

Fig. 8 Domain of existence of the wetting regime as a function of temperature and energetic parameters within the MFA

So far, we merely describe the characteristics of wetting without specifying the conditions of its existence. To stabilize the wetting, it is easy to show that the surface concentration c_0 has to be higher than c_β. In that case, the concentration profile can be "attracted" by the plateau at c_β, this evocative explanation being based on the phase portraits approach [36–39]. The condition $c_0(T) > c_\beta(T)$ can be expressed simply in MFA [36, 40], which allows one to draw the domain of existence of the wetting regime as a function of temperature and energetic parameters (Fig. 8). Note that wetting can only occur if segregation of the minority element is quite strong, which can be the case only on one side of the phase diagram. The above results, obtained by MFA, are also found qualitatively in Monte Carlo simulations, that can also treat the effects of roughness of the c_α/c_β interface, and so the competition between roughening and wetting transition [16].

To prepare the future discussion on the distinction between segregation and phase separation in bimetallic clusters on the one hand and on the characterization of these two phenomena depending on the thermodynamic ensemble on the other hand, we summarize the results obtained for alloy surfaces in systems with $V < 0$ as follows:

- Surface segregation is defined in the single phase state of bulk (i.e., solid solution) and the results are identical in the canonical ensemble and the S-GC one, even when the surface has a phase transition of first order (with a surface miscibility gap). In this case, indeed, the bulk acts as a reservoir for the surface in the canonical ensemble, ensuring equivalence between this ensemble and the S-GC one and preventing the appearance of a superficial two-phases configuration at thermodynamic equilibrium.
- If A has a strong tendency to surface segregate, a wetting in the solid solution B(A) may be observed when approaching the solubility limit if the conditions presented in Fig. 8 are fulfilled. The bulk being in a single phase state, the S-GC ensemble and the canonical ensemble lead again to the same results.
- When the bulk concentration exceeds the solubility limit in the canonical ensemble, a two-phases state is observed in this ensemble: the phase A(B) covers the surface if surface segregation of A is favoured and the ratio of the two

phases is determined by the rule of the lever arm. The S-GC ensemble leads to a first order transition with a jump between two states for the bulk isotherm, i.e., between the B(A) solid solution and the A(B) one. Recall that A surface segregation, and even more the wetting of B(A) bulk solid solution by the A(B) surface phase, prevents any metastability when the bulk isotherm is explored by increasing $\Delta\mu$, i.e., with increasing concentration of A. The segregated surface, or the wetting phase, then serves as a nucleus for the emergence of the A-rich phase. The reverse is not true: a metastability can be observed when the isotherm is explored by decreasing $\Delta\mu$, i.e., when concentration of A decreases. This shows that the hysteresis affecting the bulk isotherm in the presence of a surface is asymmetrical with respect to the true isotherm. This is very similar to what is observed for the melting in the presence or in the absence of a surface [41].

4 Bimetallic Clusters

4.1 The Description by Type of Site

As for bulk or surfaces, the lattice approach for clusters requires to fix a crystal structure. This limitation is more severe than for the bulk, because of the multiplicity of possible structures for the clusters with respect to the bulk [42–45]. Thus, depending on their size and the temperature, monometallic clusters of fcc metals can adopt icosahedral structure, decahedron or truncated octahedron (Wulff polyhedron), whereas cuboctahedral structure is often only metastable [42–45]. Possible structures for bimetallic clusters are still more diverse, making even more convenient the use of a lattice formalism to understand the purely chemical effects before turning to the coupling between chemistry and structure. To be consistent with bulk and surface results obtained for the fcc structure, the results for clusters are shown on a structure based on the fcc lattice too, i.e. the cuboctahedron. This structure has geometrical characteristics simpler than the truncated octahedron (which is also based on fcc lattice) [45] but is only metastable due to a too large proportion of sites belonging to (100) facets relatively to (111) facets. This may induce some changes on the relative position of F(111) and F(100) isotherms between both structures, but this does not change the main conclusions presented below [46, 47].

Figure 9a illustrates a cuboctahedron of 309 atoms with a complete surface shell, corresponding to what is sometimes called a "magic number" because of the additional stability of these clusters. This consists of 4 concentric shells around the central atom; hence the name of cuboctahedron of order 5 (or Cubo5) if the central atom is counted as a shell [45]. To characterize the distribution of both components in this bimetallic cluster, we may define a concentration for each concentric

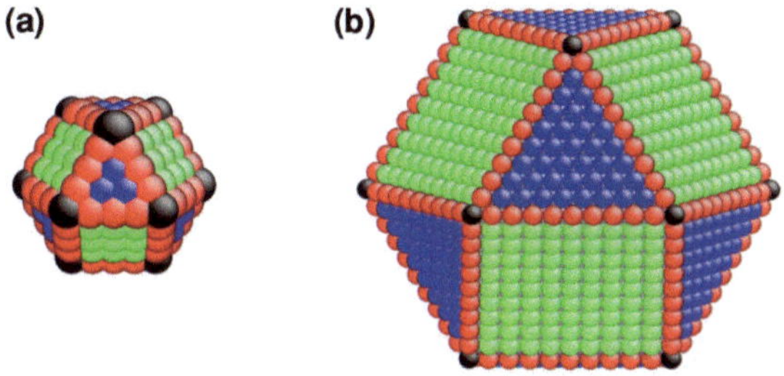

Fig. 9 Cuboctahedron of order 5 (**a**) and of order 11 (**b**); vertices edges (100) facets and (111) facets are illustrated. Copyright (2006), American Institute of Physics

shell, similar to the concentration profile for a surface for which a concentration for each plane parallel to the surface is defined. Nevertheless, Fig. 9a shows that it is necessary to have a more detailed description than for the surface, because the surface of the cluster is composed of vertices (V), edges (E), (111) and (100) facets (F). We therefore define concentrations by type of site i, $c_i = N_i^A/N_i$ where N_i^A is the number of A atoms located on sites of type i and N_i is the number of sites of type i, i = vertices, edges, (100) facets and (111) facets. Note, however, that this definition based on coordination numbers comes to group not strictly equivalent sites such as edge sites neighbours of a vertex with those of the centre of the edges, or facet sites neighbours of an edge with those in the centre of the facet. For the inner shells, it may be sufficient as a first step to define a concentration for each shell, with the possibility of detailing again the vertex, edge, and facets sites within each shell if necessary. In Fig. 9b, we show also the cuboctahedron of order 11 (Cubo11) which has 3871 atoms and will be used in the following to illustrate more clearly some characteristics of the concentration profile.

4.2 Isotherm: The Different Modes of Representation

Before discussing in more detail the segregation isotherms in clusters, let us describe the different forms of representation. Note that the results shown in this section are obtained with a set of energetic parameters suitable for the Cu–Ag system. This system is representative of alloys with a strong tendency to phase separate and with a large difference between surface energies of the constituents that leads to a strong superficial segregation of one component (here Ag). However, remember that the large difference of size between Ag and Cu atoms is not taken into account in the present rigid lattice approach, preventing a direct comparison with experiments [48] or simulations allowing atomic relaxations on this system [49, 50].

In the S-GC ensemble, the most natural representation is to show the concentration (for instance in Ag) of the different shells as a function of $\Delta\mu = (\mu_{Ag} - \mu_{Cu})$, Fig. 10a. To be closer to experimental results, which are usually obtained by controlling the nominal concentration c_{nom}, it is possible to plot these isotherms as a

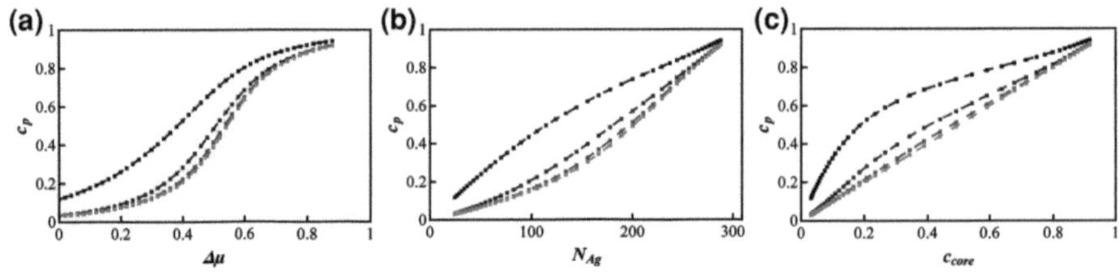

Fig. 10 Different modes of representation for the isotherms of the concentric shells of Cubo5 (309 sites) as a function of: $\Delta\mu$ (**a**), N_{Ag} (**b**) and c_{core} (**c**). These isotherms are obtained in the S-GC ensemble at high temperature ($T/T_c = 1.7$). For each graph, the surface isotherm is on the *left* and the isotherm for the innermost shell is on the *right*. $\Delta\mu$ is represented in eV

function of $N_{Ag} = Nc_{nom} = \sum_p N_p c_p$, where c_p is the concentration of the *pth* shell and N_p is the number of sites in this shell, N being the total number of sites ($N = \sum_p N_p$) (Fig. 10b). Finally it is possible to represent the isotherms as a function of the concentration of the innermost shell c_{core}, allowing the link between segregation in the clusters and at the surface of semi-infinite alloys; c_{core} then plays for clusters the role of the bulk concentration c for semi-infinite alloys (Fig. 10c). Note that the analogy between c_{core} and the bulk concentration c is valid for the rigid lattice formalism used here but may be not relevant in presence of large strain in the core, such as in the icosahedral structure [51, 52].

4.3 Isotherm: The V/E/F Hierarchy

The driving forces for segregation being proportional to the number of broken bonds ΔZ in a lattice formalism (*cf.* Sect. 3.1), it is expected that the vertices are the most segregated sites, followed by the edges, the (100) facets and finally the (111) facets. Fig. 11a shows that this is indeed the case for the surface shell, the segregation hierarchy V/E/F(100)/F(111) being clearly observed. Such a hierarchy, although much attenuated, is still discernible on the second shell (Fig. 11b). The origin of this hierarchy is no longer the number of broken bonds (all atoms in this shell have 12 nearest neighbours as in the bulk), but the number of bonds (per atom) between this shell and the surface shell, which is also equal to ΔZ. Because the homoatomic bonds are favoured in an alloy with a tendency to phase separate as the one considered here, the Ag enrichment in the surface shell promotes the Ag enrichment in the second shell [29, 31]. More precisely, this enrichment for the sites of the second shell is much stronger as these sites have more bonds with surface atoms. When going further toward the core, we observe that the isotherms for the different sites of the third shell are almost indistinguishable (Fig. 11c). Note that these results may be modified when atomic relaxations are taken into account, an additional driving force for segregation of the smallest atom being present in the second shell [49].

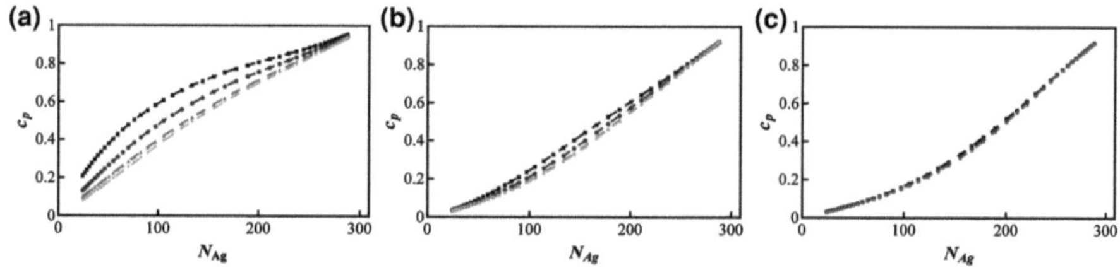

Fig. 11 Isotherms obtained in the S-GC ensemble for the various types of site [V, E, F(100) and F(111)] for the first three shells of Cubo5 at high temperature ($T/T_c = 1.7$) as a function of N_{Ag}. For each shell [surface shell (**a**), second shell (**b**) and third shell (**c**)], the isotherms from left to right correspond to V, E, F(100) and F(111) sites

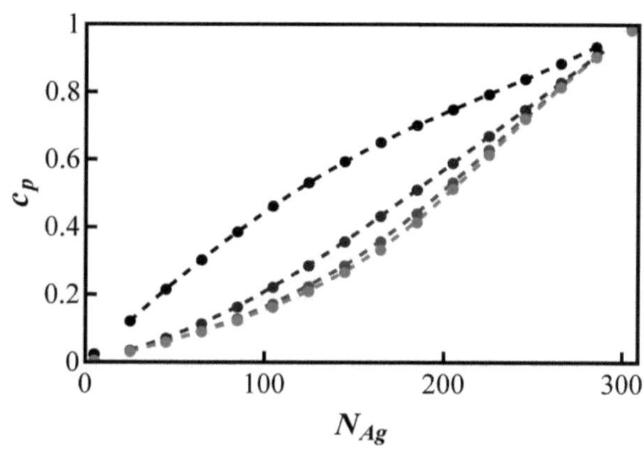

Fig. 12 Comparison between canonical (*full symbols*) and S-GC (*dashed lines*) isotherms for each shell of Cubo5 within the N_{Ag}-representation at high temperature ($T/T_c = 1.7$). The surface isotherm is on the *left* and the isotherm for the innermost shell is on the *right*

4.4 Isotherm: Change with Temperature

The results shown in Figs. 10 and 11 are related to a temperature above the bulk critical temperature T_c. The isotherms obtained in the S-GC and canonical ensembles are then indistinguishable as shown in Fig. 12, where the N_{Ag}-representation allows one an immediate comparison between both ensembles.

Let us now consider a temperature lower than T_c but higher than the critical temperature of the (111) surface, itself higher than the critical temperature of the (100) surface, cf. Table 1. Fig. 13a shows that the isotherms obtained in the S-GC and canonical ensembles are similar in a given range in N_{Ag} (stage I), then they differ for higher values of N_{Ag} (stage II), the differences being more pronounced for the shells close to the core. Finally, the isotherms become again similar in a very restricted range of N_{Ag} (stage III).

To analyze this behaviour, it is very instructive to consider the $\Delta\mu$-representation for the isotherm obtained in the S-GC ensemble (Fig. 13b). There is a quasi-vertical regime, where the Ag concentration of the inner shells change from a low value to a high value, the magnitude of this sudden change in concentration being

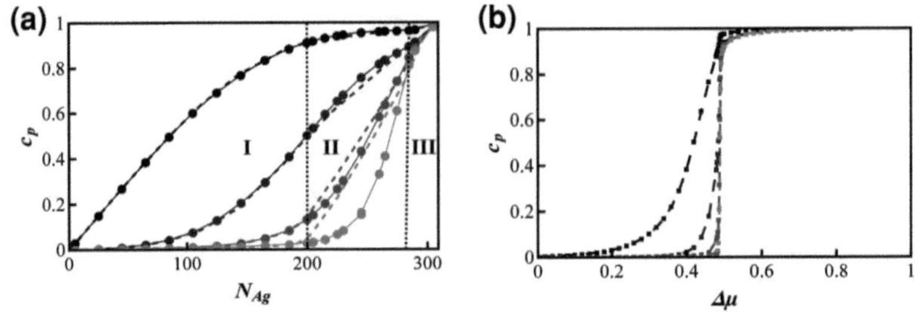

Fig. 13 Isotherms of each shell for Cubo5 at intermediate temperature ($T/T_c = 0.6$): comparison between canonical (*continuous lines* and *full symbols*) and S-GC (*dashed lines*) isotherms within the N_{Ag}-representation (**a**) and S-GC isotherms within the $\Delta\mu$-representation (**b**). The surface isotherm is on the *left* and the isotherm for the innermost shell is on the *right*. In **a**, the three different stages are indicated (see text). $\Delta\mu$ is represented in eV

all the stronger that the shell is closer to the core. Note two important features of these isotherms:

- although almost vertical, these isotherms are continuous. Actually, provided that the $\Delta\mu$-step adopted to calculate the isotherm is sufficiently low, the isotherms are continuous and reversible;
- the almost vertical part of the isotherms, when expressed in terms of $\Delta\mu$, corresponds precisely to the range in N_{Ag} where isotherms (in the N_{Ag}-representation) differ between the S-GC and canonical ensembles (stage II).

4.5 Collective Bistability of the Inner Layers

In order to analyze the behaviour of the cluster in the almost vertical part of the isotherms, we show in Fig. 14a the variation of the concentrations of the different shells as a function of the number of MC steps for a value of $\Delta\mu$ located in the middle of this regime. If the surface concentration has a usual behaviour with small amplitude fluctuations around the average value, it is quite different for the other shells. Their concentrations oscillate between two values. One value, close to 1, is common to all shells and is characteristic of an Ag-rich state. The other depends on the shell and is even closer to 0 as one approaches the core. This bistability, sometimes called dynamic equilibrium (DE), is a *collective* phenomenon affecting all shells of the cluster as evidenced by the simultaneity of the concentration switching for the different shells as a function of the number of MC steps (Fig. 14a). Another way to characterize this bistability is to consider the configurational densities of states (CDOS) for each shell p [40, 47]. Thus $n(c_p)\,dc_p$ is the number of states for which the concentration is between c_p and $c_p + dc_p$. Bistability manifests itself by a bimodality of CDOS (Fig. 14b). It should be noted, however, that the CDOS representation is not able to assess the collective aspect of the bistability in the inner shells.

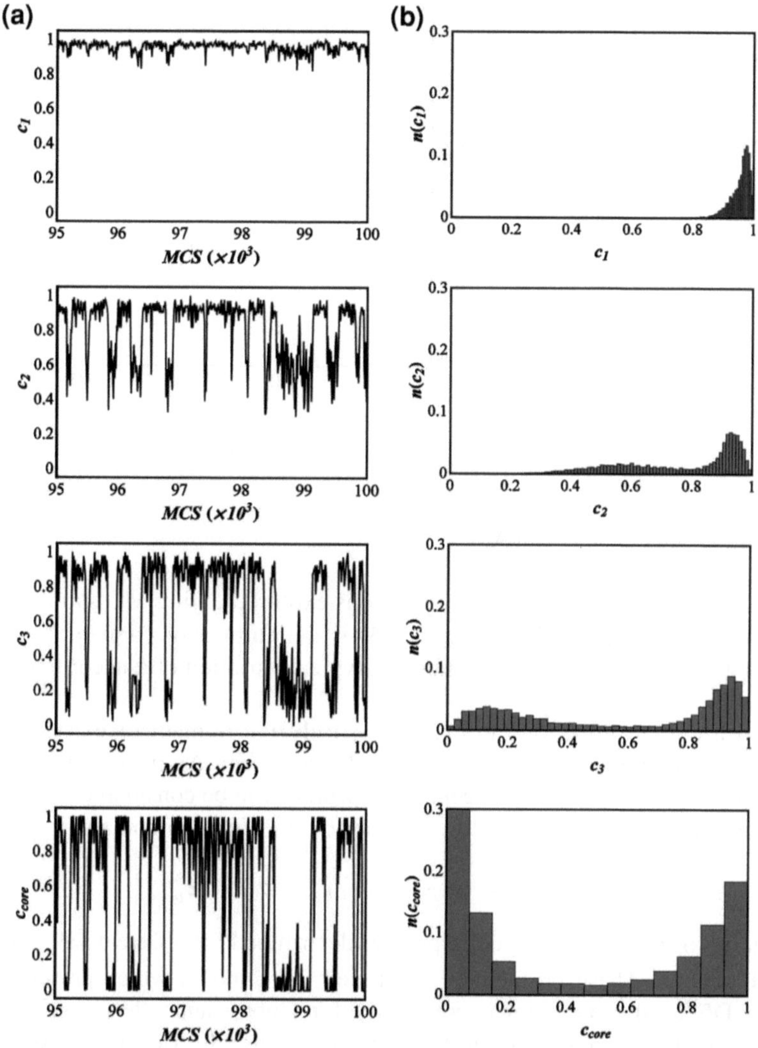

Fig. 14 Evolution of the concentration for the various shells of Cubo5 at intermediate temperature ($T/T_c = 0.6$) in the S-GC ensemble as a function of the number of MC steps (MCS) in the DE regime ($\Delta\mu = 490$ meV) (**a**) and corresponding CDOS (**b**). The results for shells 1–4 are shown from *left* to *right*

After detailing the behaviour of the cluster in the S-GC ensemble for a value of $\Delta\mu$ located in the almost vertical part of the isotherms of the inner shells, we investigate the influence of a change of $\Delta\mu$ in this regime.

The first effect is a change for the relative weight of the two states. To highlight this, it is possible to determine the weight of a state, for instance the Ag-rich one, using the parameter α defined as follows [40]:

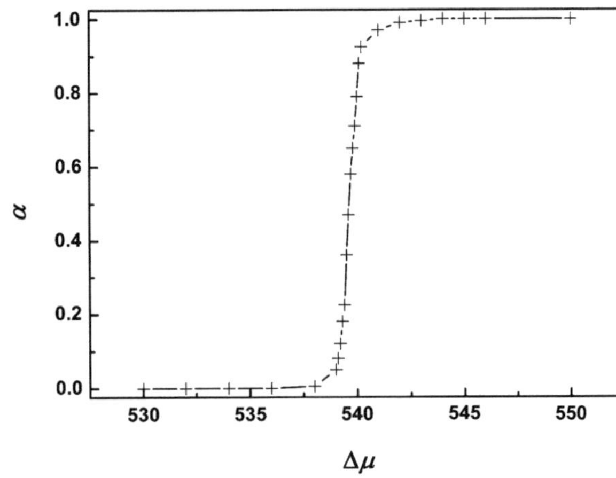

Fig. 15 Variation of the weight of the Ag-rich state as a function of $\Delta\mu$ (in meV) in the DE regime at intermediate temperature for Cubo11 ($T/T_c = 0.88$)

$$\alpha(\Delta\mu) = \frac{N^{config}(c_{core} > 0.5)}{N^{config}}, \tag{10}$$

where N^{config} is the number of configurations explored during MC simulation for a given value of $\Delta\mu$ and $N^{config}(c_{core} > 0.5)$ is the number of configurations corresponding to $c_{core} > 0.5$ (Ag-rich state). A typical variation of α according to $\Delta\mu$ is shown Fig. 15.

Such a curve allows one to define precisely the domain of existence of the bistability and shows that α varies almost linearly with $\Delta\mu$ in the whole range of bistability. In this range, the concentration profiles can be considered as a weighted average of the two critical profiles, c_p^- and c_p^+, limiting the domain of bistability:

$$c_p(\Delta\mu) = (1 - \alpha(\Delta\mu))\, c_p^- + \alpha(\Delta\mu)\, c_p^+. \tag{11}$$

Figure 16 shows the c_p^- and c_p^+ profiles determined for different values of $\Delta\mu$ in the DE regime. These profiles are illustrated for Cubo11, which presents a very similar DE as the one shown previously for Cubo5 and allows one an easier analysis of the concentration profiles. If the c_p^+ profile, rich in the segregating element, is independent of $\Delta\mu$, as assumed in the formula (11) (Fig. 16a), there is a significant variation of the c_p^- profile with $\Delta\mu$ (Fig. 16b). This variation is due to the very first stage of wetting (Sect. 3.3). We will see later why wetting does not develop, whereas it is observed for the semi-infinite alloys and should lead to a move of the Ag/Cu interface toward the centre of the cluster as $\Delta\mu$ increases.

In the canonical ensemble, what is the configuration of the cluster in stage II? For the sake of clarity, we consider again the Cubo11. Fig. 17 shows a progressive movement of the interface separating the Ag-rich outer shells from the Cu-rich inner shells when N_{Ag} varies. Nevertheless, using only the canonical ensemble does not allow one to discriminate stage II from the other two stages, while the S-GC ensemble provides a clear definition of this stage via the presence of

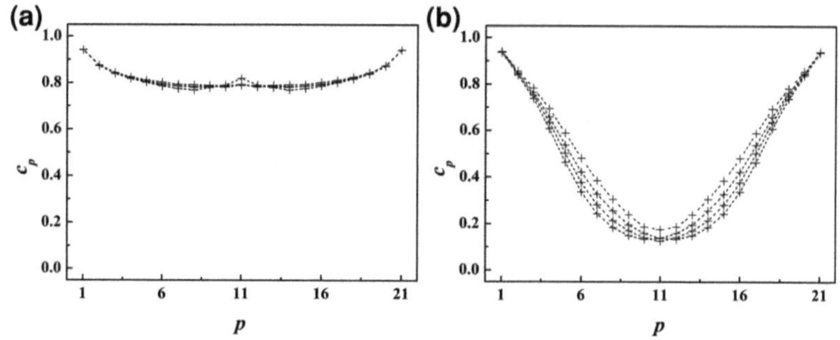

Fig. 16 Concentration profiles for the Ag-rich state c_p^+ (**a**) and for the Cu-rich one c_p^- (**b**) for Cubo11 at intermediate temperature ($T/T_c = 0.88$) for different values of $\Delta\mu$ in the DE regime. The concentration profiles are shown along a diameter of the cluster ($p = 1$ and $p = 21$ correspond to the surface, $p = 2$ and $p = 20$ to the first shell under the surface and $p = 11$ to the central site

Fig. 17 Comparison between canonical (*continuous lines* and *full symbols*) and S-GC (*dashed lines*) isotherms for each shell of Cubo11 at intermediate temperature ($T/T_c = 0.6$) within the N_{Ag}-representation. The surface isotherm is on the *left* and the isotherm for the innermost shell is on the *right*. The three different stages are indicated (see text)

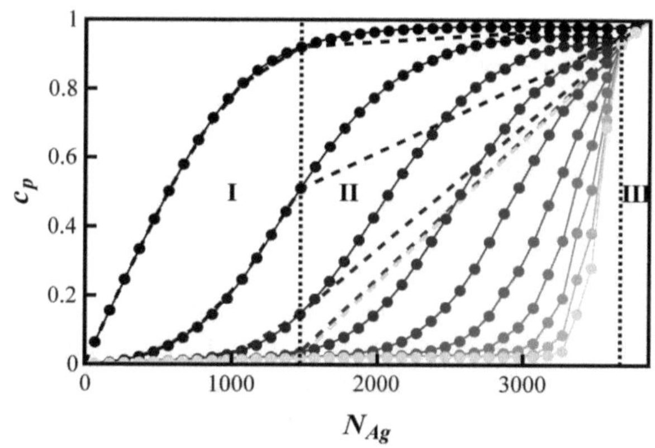

bistability (Fig. 14). Recall that stage II is also characterized by the differentiation between canonical and S-GC isotherms.

If we adopt the concepts and semantics of infinite and semi-infinite alloys, stage I corresponds to a single phase state (i.e., solid solution very dilute in Ag) with a strong Ag surface segregation leading to a pure Ag surface shell and a second shell with a concentration close to 0.5. Stage II corresponds to a two-phase state in the canonical ensemble, the outermost layers being almost pure in Ag and the innermost layers being almost pure in Cu. Finally, in stage III, there is again a single-phase state (i.e., solid solution rich in Ag), in which Ag surface segregation is hardly perceptible.

In the solid solution domains (stages I and III), S-GC and canonical ensembles lead to the same isotherms as expected. In stage II, the S-GC ensemble leads to a first order transition for an infinite volume with a thermodynamic bistability only for the critical value $\Delta\mu_c$ (remember, however, the existence of metastable states and the possibility of observing an hysteresis over a range around $\Delta\mu_c$). Due to the finite size of the cluster, the first order transition for an infinite volume is replaced within a narrow range in $\Delta\mu$ by a dynamic equilibrium between two states, one rich in Cu and affected by a strong Ag surface segregation, the other one rich in Ag [53, 54]. These two states are the parallel of the saturated solid solutions limiting the miscibility gap of the bulk phase diagram.

To mark the difference between single-and two-phase states in bimetallic clusters, we propose to use the term "core–shell" for the configurations corresponding to a solid solution with very low A concentration in the inner shells (i.e., the "core"), but with strong segregation of A to the surface and eventually in the underlying layer (i.e., the "shell"). We reserve the name "cherry stone" for the two-phase regime, corresponding to an outer region rich in A (i.e., the "flesh of the cherry"), whose thickness can reach almost the radius of the cluster and an inner region rich in B ("the stone of the cherry").

It remains to explain why wetting described for semi-infinite alloys (Sect. 3.3) does not appear in the bimetallic cluster studied here. First, recall that the wetting appears in the single phase state of the semi-infinite alloy and should not be confused with a two-phase state within the miscibility gap. This point becomes important in the case of clusters, for which the shape of isotherms obtained in the canonical ensemble could suggest a wetting phenomenon (Fig. 17 compared with Fig. 7). However, we have seen that the concentration increase of successive layers in the cluster in the canonical ensemble is a characteristic of a two-phase state and not of the wetting. In the S-GC ensemble, where the dynamic equilibrium allows one to distinguish unambiguously wetting and phase separation, bistability is not preceded (in $\Delta\mu$) by a wetting regime. The reason comes from the elimination of the Ag/Cu interface which is present in the Cu-rich phase (it separates the Ag-rich surface from the Cu-rich inner shells) and disappears in the Ag rich phase. The free energy gain associated with the disappearance of this interface leads to a decrease of the critical value of $\Delta\mu$ in the cluster (defined, for instance, as the middle of the range of bistability in $\Delta\mu$) compared to the critical value $\Delta\mu_c$ for an infinite volume. The wetting regime appearing in the semi-infinite system for values very close to $\Delta\mu_c$, the decrease of $\Delta\mu_c$ in the cluster removes the range where wetting exists. Only the change in the c_p^- concentration profile with $\Delta\mu$ (Fig. 16b) shows the proximity of a wetting regime. A mean-field approach allows one to show that the maximum thickness of the wetting layer in a cluster increases as ln $(N^{1/3})$ [40]. This excludes the possibility to observe wetting layer of large thickness in clusters of few thousands to several millions atoms.

4.6 Individual Bistability Per Facet

After discussing the connection between the miscibility gap in an infinite alloy and dynamic equilibrium (S-GC ensemble) or the "cherry-stone" configuration (canonical ensemble) in a cluster, let us consider the behaviour of the surface of a cluster when the temperature drops below the critical temperature of surfaces of semi-infinite alloys.

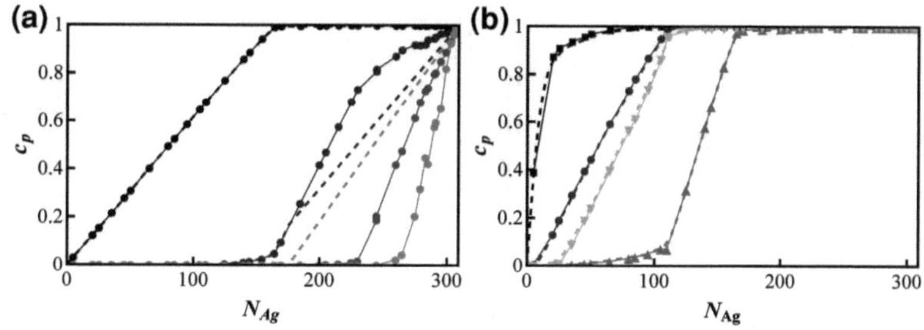

Fig. 18 Isotherms for each shell (**a**) and for the different sites of the surface (**b**) of Cubo5 at low temperature ($T/T_c = 0.18$) in the canonical (*continuous lines* and *full symbols*) and S-GC (*dashed lines*) ensembles within the N_{Ag}-representation. In **a** the surface isotherm is on the *left* and the isotherm for the innermost shell is on the *right*; in **b** from *left* to *right* V, E, F(111) and F(100) isotherms

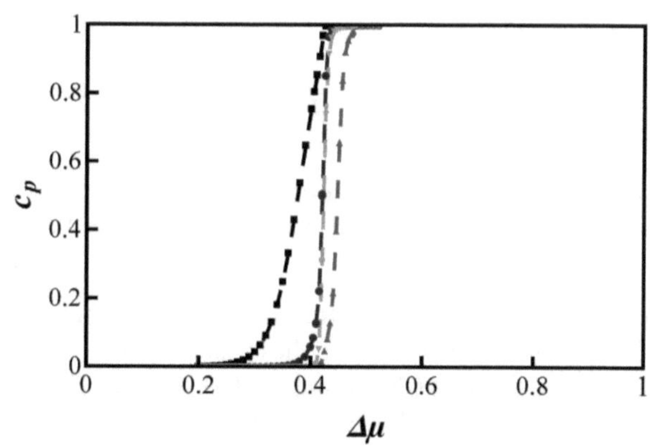

Fig. 19 Isotherms for the different sites of the surface in Cubo5 at low temperature ($T / T_c = 0.18$) in the S-GC ensemble within the $\Delta\mu$-representation. From *left* to *right* V, E, F(111) and F(100) isotherms. $\Delta\mu$ is represented in eV

Figure 18a shows the isotherms obtained at $T/T_c \approx 0.18$, i.e., a temperature below the critical temperature of (111) and (100) surfaces, see Table 1. The isotherms of the different shells are quite similar to those observed at higher temperatures ($T/T_c \approx 0.6$, Fig. 13), the differences in stage II between canonical and S-GC isotherms for the internal layers, however, being more pronounced. On the other hand, both thermodynamic ensembles lead to a similar surface isotherm. Then, it does not allow one to correlate the existence of a surface miscibility gap with a difference between S-GC and canonical surface isotherms, as this is the case for inner layers. To analyze the behaviour of the surface shell, Fig. 18b details the various components of the surface isotherm, i.e., the isotherms of the different sites (vertices, edges, (100) and (111) facets) within the N_{Ag}-representation.

There is no difference between both thermodynamic ensembles and once more again this is the $\Delta\mu$-representation of the S-GC results which is the most informative (Fig. 19). If the vertices isotherm is steadily increasing with $\Delta\mu$, the isotherms for edges and (111) facets, which are almost superimposed, have an abrupt

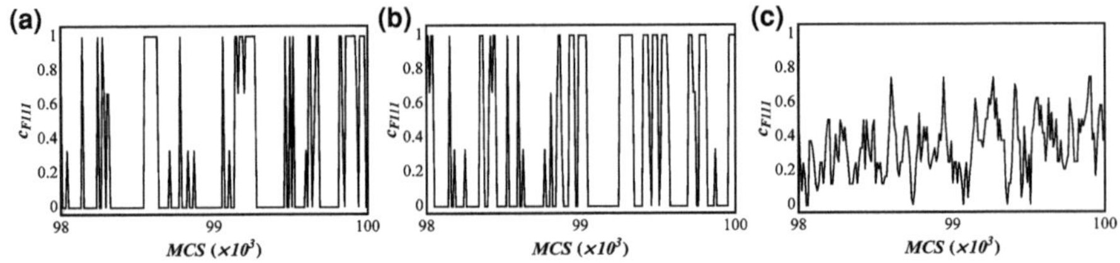

Fig. 20 Evolution of the concentration of two different (111) facets in the DE regime (**a, b**) and of the average of the concentrations of all (111) facets (**c**) in Cubo5 at low temperature ($T / T_c = 0.18$) as a function of the number of MC steps (MCS) in the S-GC ensemble

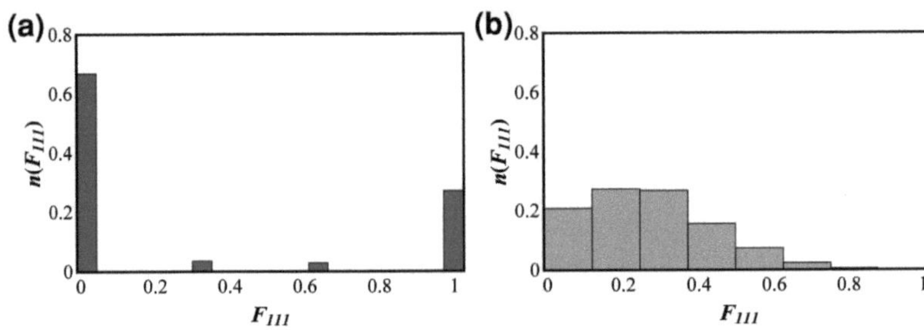

Fig. 21 CDOS for the concentration of a given (111) facet (**a**) and for the concentration of all (111) facets (**b**) for Cubo5 at low temperature ($T/T_c = 0.18$) in the S-GC ensemble

part while remaining continuous. A similar behaviour is observed for the (100) facets. This abrupt part of these isotherms is very similar to the one described above for the inner layers in the DE regime (Fig. 13b).

Indeed, if one follows the evolution of the concentration of a (111) facet as a function of the number of MC steps for a given value of $\Delta\mu$ in the almost vertical part of the isotherm, there are oscillations between two states: a Cu rich state, the other being almost pure in Ag (Fig. 20a, b).

These oscillations lead to the observation of bimodal CDOS for each (111) facet (Fig. 21a). Nevertheless, there is a remarkable difference between this bistability affecting the (111) facets of the cluster and the one observed for the inner layers (Fig. 14). While the inner layers collectively switch from one state to the other, (111) facets oscillate independently of each other as evidenced by the asynchronous oscillations of the two (111) facets shown Fig. 20a, b. Another way to highlight the individual nature of the bistability of the facets is to monitor the concentration of all (111) facets and not of a given (111) facet. Fig. 20c shows the absence of oscillations of large amplitude in this case, resulting in monomodal CDOS for the concentration averaged over all (111) facets (Fig. 21b).

The same behaviour is observed for the (100) facets, a dynamic equilibrium for each (100) facet leading to instantaneous configurations such as those shown in Fig. 22. A remarkable consequence of this bistability for each facet is the coexistence within a cluster of (100) [or (111)] facets rich in silver and others

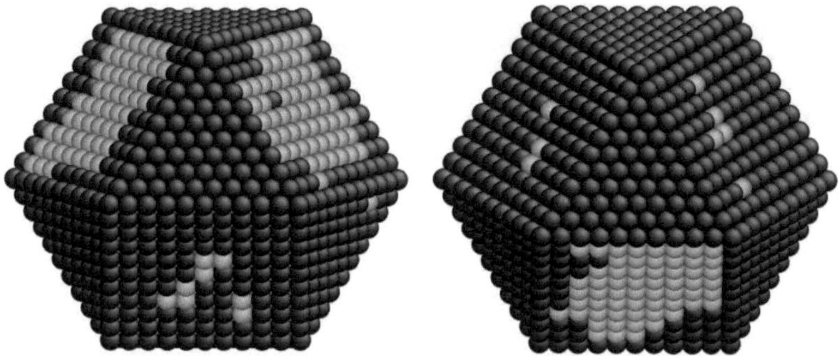

Fig. 22 Instantaneous configurations in the DE regime of (100) facets for Cubo11 at low temperature ($T/T_c = 0.18$) in the S-GC ensemble; Cu atoms are *light* and Ag ones are in *dark*

copper-rich. For clusters which present a catalytic interest, this result may have important consequences because, concerning reactivity properties, it is not at all equivalent to have an equiatomic concentration on all facets of a given orientation or to have half these facets of pure A and the other half pure B.

Moreover, the individual character (i.e., by facet) of the dynamic equilibrium affecting the facets of a given type (i.e., the (100) facets or the (111) ones) is the cause of the superimposition of surface isotherms between S-GC and canonical ensembles. Indeed, in the canonical ensemble, there is a similar bistability for each facet, all facets of the same type acting as a mutual reservoir. Thus, when a facet switches from one state to the other one, another facet switches in an opposite way, which permits the nominal concentration to remain constant.

These results obtained within a rigid-lattice formalism show the richness of the thermodynamics of bimetallic clusters. Thus, in the canonical ensemble, all facets of the same type [(100) or (111)] form a realization of the multi-objects canonical ensemble, in which these objects serve themselves as a mutual reservoir. Multi-objects canonical ensemble and S-GC ensemble lead to the same isotherms, even in the presence of bistability.

The behaviour of internal layers is different because of the strong coupling between them. The collective bistability observed in the S-GC ensemble cannot take place in the canonical ensemble due to the constraint on the nominal concentration. The isotherms in the canonical ensemble are then characteristic of a progressive move of the Ag/Cu interface toward the core of the cluster when the nominal concentration in Ag increases. This progressive movement of the interface then leads to decouple the behaviour of the various internal layers. Therefore, in the canonical ensemble and at sufficiently low temperature, the facets of an inner layer may behave in a similar way as the surface facets, i.e., a dynamic equilibrium by facet can be observed during the rise of the isotherm of this inner layer. This is illustrated by the instantaneous configuration shown in Fig. 23a, where one sees the coexistence of pure Ag and pure Cu (111) facets in each shell n (with $2 \leq n \leq 6$) when the shell has an equiatomic concentration, i.e. when the Ag/Cu interface is localized on this shell. A shift toward a more collective behaviour is observed for the

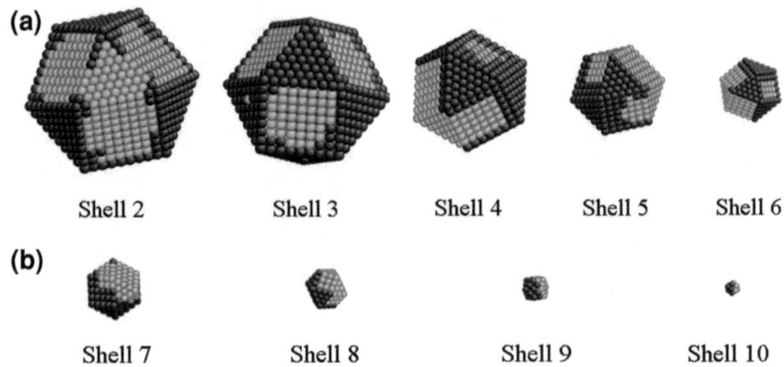

Fig. 23 Instantaneous configurations for shells 2–6 (**a**) and for shells 7–10 (**b**) in the canonical ensemble for Cubo11 at low temperature ($T/T_c = 0.18$). Cu atoms are *light* and Ag ones are *dark*. Each shell is represented when the Ag/Cu interface is located on this shell, i.e. when its concentration is about 0.5 (see Fig. 17)

innermost shells (shells 7–10), for which a Janus-like configuration [48, 55] is stabilized with one hemisphere Ag-rich, the other one being Cu-rich (Fig. 23b). This result illustrates the complexity associated with individual or collective behaviour of the different entities forming clusters, i.e., concentric shells, and within these shells, vertices, edges and facets of different types.

In particular, the individual character of the bistability observed for each facet of the surface, and possibly for the facets of the shells under the surface in the canonical ensemble, can be questioned in the case of longer range interactions leading to a coupling between facets or in the presence of a structural transition related to surface segregation of one element [50].

4.7 Toward Phase Diagram for Nanoalloys

Bistability domains, for both the surface facets and the inner shells, are equivalent of the two-phase domains for infinite or semi-infinite alloys. The aim of a phase diagram for nanoalloys is therefore to define these bistability domains obtained in the S-GC ensemble. Different representations can be chosen according to the intended use, as for the various modes of representation for the segregation isotherms (Sect. 4.2).

We present the phase diagram of Cubo5 in two different forms:

- the first one is the classical representation of phase diagrams, which are plotted usually as a function of bulk or surface concentration and T. Its generalization for the cluster is to plot the bistability domain for each shell or for each type of sites as a function of c_p, the concentration of the pth shell or of the p-type of sites (Fig. 24a);

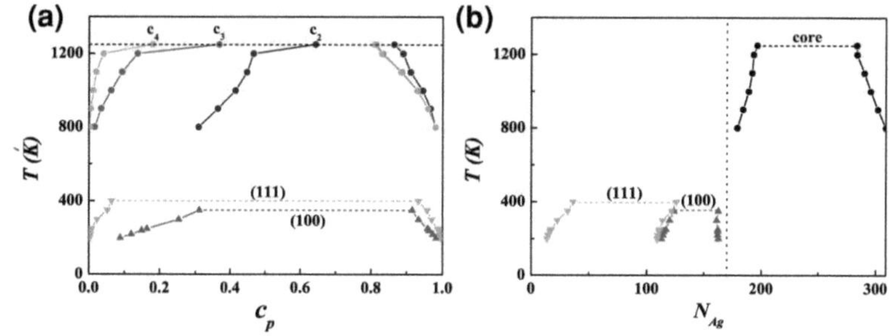

Fig. 24 Phase diagram of Cubo5 (309 sites) obtained in the S-GC ensemble within the c_p - representation (**a**) and within the N_{Ag} - representation (**b**)

- The second one is expressed in terms of the nominal concentration c_{nom} (or, equivalently, N_{Ag}) (Fig. 24b). This representation is more appropriate for comparison with experiments, where c_{nom} is usually the relevant control parameter.

The c_p-representation allows one a detailed analysis of the two states limiting the domain of bistability. Thus, for the inner layers, Fig. 24a clearly shows the strong influence of the segregation profile on the Cu-rich state (the limit of bistability for the outermost shells is richer in silver), while it has almost no effect on the Ag-rich state (the limit of bistability is almost identical for all shells). A similar effect is observed when considering the (100) facets, the segregation profile starting from the edges being responsible for the asymmetry of the bistability limits for these facets (Fig. 24a). Such a phenomenon does not occur for the (111) facets, because the edges bordering a (111) facet and the facet itself behave as a single object (Fig. 19).

The N_{Ag}-representation illustrates the domains of nominal concentration, in which MC simulations predict:

- for the inner shells: a collective dynamic equilibrium (or bistability) in an experimental realization of the S-GC ensemble, or a two-phase state of "cherry stone" type in the canonical ensemble. Emphasize again that this two-phase state is clearly distinguishable from the surface segregation in a single phase only when using simulations in the S-GC ensemble;
- for the surface shell, a dynamic equilibrium (or bistability) which affects individually each facet, both for the (111) and (100) orientation, in the S-GC ensemble as well as in the canonical ensemble.

The right side limit of the bistability domain for the (111) facets may seem surprising, since it leads to a reentrant phenomenon (Fig. 24b). At fixed N_{Ag} and increasing temperature, it shows indeed a zone of bistability (or DE) between classical configurations (i.e., a single state) at lower and higher temperature. This phenomenon comes from the overlap between isotherms of (111) and (100) facets, which is more or less important depending on temperature. Thus, when the isotherm of the (100) facets comes closer to the isotherm of the (111) facets in

Table 2 Comparison of critical temperatures (in K) for bulk and surfaces of infinite or semi-infinite alloys and for the Cubo5 (core and facets) obtained by MC simulations. The critical temperatures of bulk (respectively core) and (111) and (100) surfaces (resp. facets) are indicated

T_c (K)	Bulk	F(111)	F(100)
(semi-) infinite alloy	1,711	637	395
Cuboctahedron of order 5	1,250	400	350

the bistability domain of the latter, this leads to an increase of the right side limit of this domain in the N_{Ag}-representation (Fig. 24b), a feature that does not appear in the c_p-representation (Fig. 24a).

The maximum temperature limiting the bistability domain (equivalent to the critical temperature limiting the miscibility gap for the infinite or semi-infinite alloys) is difficult to determine accurately. Indeed, in the vicinity of T_c, the CDOS's do not conserve their bimodal aspect and they have an almost constant value on a large range of c_p. It is therefore difficult to deduce precisely the concentrations limiting the bistability domain when approaching T_c. However, the bulk phase diagram and even more the surface one being very flat near T_c, we choose to limit the bistability domain by a horizontal dashed line corresponding to a temperature where CDOS's lost a well defined bimodal aspect.

The critical temperatures shown in Fig. 24 indicate a significant lowering of T_c for the core and the (111) facets when comparing with their infinite (bulk) or semi-infinite ((111) surface) counterpart (Table 2). The decrease is less pronounced for the (100) facets because of their larger size in the cuboctahedral structure (Table 2). This lowering of T_c, both for the core and for the facets, is due to the segregation profile occurring between the outer part of these objects (the edges for the facet, or the surface for the core) and their center. Thus, it is expected that the level of segregation as well as the spatial extension of the segregation profile are correlated with the lowering of the critical temperature [19, 56].

These results form a first attempt to draw a phase diagram for a nanoalloy with a given size for a system with a tendency to phase separate. They are obtained in a very restrictive framework, namely a rigid-lattice approach that neglects any effect of atomic relaxation and does not permit a local structural change (e.g. at the surface) or a global one, i.e., affecting the whole cluster, such structural changes may depend on concentration or temperature. However, this very simplified framework allows one to establish a methodology for the study of nanoalloys phase diagrams. In particular, it shows the necessity to complete the studies performed in the canonical ensemble, the most common so far, through simulations in the S-GC ensemble that distinguish clearly between surface segregation regime (with "core–shell" configuration) and phase separation (with "cherry stone" or "Janus" configurations).

Conclusion and Outlooks

Studies of phase diagrams of bimetallic clusters have much to gain by relying on knowledge about bulk and surface phase diagrams. We have restricted the contents of this chapter to the case of alloys that tend toward phase separation, but the issues (surface segregation, surface and bulk phase transition, wetting when approaching bulk phase transition) can easily be generalized to the case of alloys with a tendency to order.

A feature of clusters is the coexistence of several facets of equivalent orientation, which has some similarities with the vicinal surfaces exhibiting terraces of the same orientation separated by steps. In the case of A–B alloys with a tendency to phase separate, this multiplicity of facets of equivalent orientation can lead to coexistence of almost pure A facets and almost pure B ones, both in the canonical ensembles and in the S-GC ensemble. In alloys with a tendency to order, the behaviour may be more complex because of the possible competition between chemical order and segregation and of the possible frustration of chemical order between facets. Again, an analogy with the vicinal surfaces in alloys with a tendency to order can be successful, illustrating the diversity of the modes allowing to relax the frustration when chemical order occurs. Recall in this context the existence of a "simple step/double step" transition predicted for vicinal surfaces in ordering alloys [10] and observed experimentally [7–9].

If the analogy between clusters and vicinal surfaces can be successful, the one with the thin layers is very fruitful too. This is especially true for the analysis of the conditions of existence for wetting. For both thin film and clusters, the gain in free energy due to the possibility of annihilation of the wetting phase/core phase interface has the effect of shifting the critical chemical potential and then of preventing the wetting regime. If the S-GC ensemble is perfectly suited to highlight this point, the canonical ensemble can be confusing because it does not allow to easily distinguish the single phase state with surface segregation (which may go up to the well-known "core–shell" configuration) from the two-phase state (of "cherry stone" type), which itself may be confused with a wetting configuration.

The choice of the thermodynamic ensemble for the simulations naturally raises the question of comparison with experiments. This is particularly crucial in the bistability regime for the core layers predicted by the simulations in the S-GC ensemble. Experimentally, it is unlikely that a given aggregate can oscillate between two states, one rich in A, the other rich in B, because of kinetic limitations on the mass transport required. Moreover, even if this mass transport was effective (e.g., by exchange with atoms of other clusters), then probably it would lead to coalescence between clusters and therefore modify their size, which prevents comparison with predictions for a given size. It seems more promising to seek confirmation of bistability regime of core layers in the analysis of a population of clusters. If each of the cluster is not able to oscillate between two states due to kinetic limitations mentioned above, the bistability can result in a bimodal distribution in nominal concentration for clusters of a given size if they had time to equilibrate with the environment during their growth. If such equilibrium did not have time to occur and if the clusters can be considered as isolated thereafter, the canonical ensemble description becomes relevant and clusters whose nominal concentration is in the miscibility gap adopt a two-phase configuration of "cherry stone" or "Janus" type.

The problem is different for the bistability of the facets of the surface, because even if the kinetic limitations do not allow every facet of a cluster to switch between two states, the observation of a bimodal distribution for the facet concentration of a cluster, even isolated, may be possible, as evidenced by the agreement between simulations in the S-GC ensemble and in the canonical one.

If a first phase diagram is proposed in this work, it remains to determine the influence of the cluster size on this phase diagram, especially the dependence of the critical temperature and the limits of bistability with the number of atoms. A study of the linear chain within de MFA shows that the size dependence of the phase diagram is very different whether the size of the chain is more or less than the correlation length of the system [40]; similar behaviour is also expected for clusters.

If the lattice approach has proven successful, it nevertheless constitutes only a first step in establishing realistic phase diagram in nanoalloys. Thus, taking into account the coupling between chemistry and structure through local atomic relaxations or global structural changes, appears especially important in the case of systems with high lattice mismatch. It is important too for cluster structures that give rise to high internal stress such as the icosahedral structure [51, 52]. First examples of a chemical bistability coupled to a structural bistability have been observed [50] and, again, the alloy surfaces are a useful reference for analyzing the behaviour of clusters. In particular, this leads to determine whether the reconstruction or superstructures appearing in the surface alloy due to segregation [57, 58] are also observed on the facets of the clusters. Moreover, we can expect to distinguish different behaviours depending on the compatibility between the facets size and the superstructure unit cell. This may lead to new "magic" numbers, in analogy with what is observed on vicinal surfaces when the widths of the terraces are compatible with the unit cell of the superstructures of the flat surfaces [59, 60].

Finally, mention that the lattice approach can incorporate the role of relaxations in a variation of the energetic parameters as a function of the geometrical and chemical environment, as shown for surfaces and grain boundaries [24]. In particular, in the presence of a strong size mismatch, the variation of the alloying effective pair interactions at the surface can lead to a reversal of their sign relative to the bulk [61, 62]. This has been verified for the surfaces of semi-infinite alloys [63, 64] but remains an open question for the bimetallic clusters.

Acknowledgments It is a pleasure to acknowledge the many collaborators and colleagues who have significantly contributed to this work: Virginie Moreno, Florence Lequien, Laure Delfour, Mariem Lamloum and Mohamed Briki. We are indebted to Isabelle Braems, Christine Mottet, Guy Tréglia, François Ducastelle, Christian Ricolleau and Cyril Langlois for many stimulating and useful discussions throughout the course of this research.

References

1. Sinfelt, J.H.: Bimetallic Catalysts: Discoveries Concepts and Applications. Wiley, New York (1983)
2. Ferrando, R., Jellinek, J., Johnston, R.L.: Nanoalloys: from theory to applications of alloy clusters and nanoparticles. Chem. Rev. **108**, 845–910 (2008)

3. Gauthier, Y., Baudoing, R.: Low energy electron diffraction from alloy surfaces. In: Dowben, P.A., Miller, A. (eds.) Surface Segregation and Related Phenomena, pp. 169–206. CRC, Boca Raton (1990)
4. Khoutami, A., Legrand, B., Tréglia, G.: On a surprising anisotropy of surface segregation in CuPt alloys. Surf. Sci. **287–288**, 851–856 (1993)
5. Senhaji, A., Tréglia, G., Legrand, B.: On the various terminations occurring in CuPt-ordered alloys: the TBIM approach. Surf. Sci. **307–309**, 440–444 (1994)
6. Gallis, C., Legrand, B., Saúl, A., Tréglia, G., Hecquet, P., Salanon, B.: Competition or synergy between surface segregation and bulk ordering: the Cu-Pd system. Surf. Sci. **352–354**, 588–591 (1996)
7. Barbier, L., Goapper, S., Salanon, B., Caudron, R., Loiseau, A., Alvarez, J., Ferrer, S., Torrelles, X.: X-Ray observation of a chemical order driven morphological transition on the surface of an A_3B type alloy. Phys. Rev. Lett. **78**, 3003–3006 (1997)
8. Le Goff, E., Barbier, L., Goapper, S., Salanon, B.: Step pairing and bulk chemical order at alloy vicinal surfaces. Surf. Sci. **466**, 73–88 (2000)
9. Piccolo, L., Barbier, L.: Chemical order driven morphology of a vicinal surface of Fe_3Al (111): an He diffraction and STM study. Surf. Sci. **505**, 271–284 (2002)
10. Senhaji, A.: Etude théorique du dépôt de couches minces métalliques: de la structure électronique à la cinétique. Université Paris XI-Orsay, Thèse (1993)
11. Gallis, C.: Couplage entre ordre de volume et ségrégation de surface:des surfaces d'alliages aux alliages de surface. Université Paris VI, Thèse (1997)
12. Dietrich, S., Domb, C., Lebowitz, J. (eds.): Phase Transitions and Critical Phenomena, Vol. 12. Academic, London (1983)
13. Lipowsky, R.: Critical Phenomena at Surfaces and Interfaces. Springer Tracts in Modern Physics, p. 127. Springer, Berlin (1993)
14. Teraoka, Y., Seto, T.: Surface segregation and bulk phase separation in segregating alloys. Surf. Sci. **255**, 209–218 (1991)
15. Saúl, A., Legrand, B., Tréglia, G.: Link between the surface wetting in Cu(Ag) and the layer-by-layer dissolution mode of a thick Ag deposit on a Cu substrate. Surf. Sci. **331–333**, 805–810 (1995)
16. Ducastelle, F.: Order and Phase Stability in Alloys. North-Holland, Amsterdam (1991)
17. Binder, K.: The Monte CarloMonte Carlo Method in Condensed Matter Physics. Springer, Berlin (1995)
18. Foiles, S.M.: Calculation of the surface segregation of Ni-Cu alloys with the use of the embedded-atom method. Phys. Rev. B **32**, 7685–7693 (1985)
19. Creuze, J.: Des surfaces aux joints de grains: la ségrégation dans tous ses états. Université Paris XI-Orsay, Thèse (2000)
20. Fisher, M.E.: The theory of equilibrium critical phenomena. Rep. Prog. Phys. **30**, 615–730 (1967)
21. Philibert, J., Vignes, A., Bréchet, Y., Combrade, P.: Métallurgie: du minerai au matériau. Masson, paris (1998)
22. Tréglia, G., Legrand, B., Ducastelle, F., Saúl, A., Gallis, C., Meunier, I., Mottet, C., Senhaji, A.: Alloy surfaces: Segregation, reconstruction and phase transitions. Comput. Mat. Sci. **15**, 196–235 (1999)
23. Creuze, J.: Atomic-scale modelling of intergranular segregation: The case of alloys with strong size-effect. Defect Diffus Forum **203–205**, 3–36 (2002)
24. Berthier, F., Legrand, B., Tréglia, G.: How to compare superficial and intergranular segregation? A new analysis within the mixed SMA-TBIM approach. Acta Mater. **47**, 2705–2715 (1999)
25. Creuze, J., Braems, I., Berthier, F., Mottet, C., Tréglia, G., Legrand, B.: Model of surface segregation driving forces and their coupling. Phys. Rev. B **78**, 075413 (2008)
26. Portales, H., Saviot, L., Duval, E., Gaudry, M., Cottancin, E., Pellarin, M., Lermé, J., Broyer, M.: Resonant Raman scattering by quadrupolar vibrations of Ni-Ag core-shell nanoparticles. Phys. Rev. B **65**, 165422 (2002)

27. Cazayous, M., Langlois, C., Oikawa, T., Ricolleau, C., Sacuto, A.: Cu–Ag core-shell nanoparticles: A direct correlation between micro-Raman and electron microscopy. Phys. Rev. B **73**, 113402 (2006)
28. Ciuculescu, D., Amiens, C., Respaud, M., Falqui, A., Lecante, P., Benfield, R.E., Jiang, L., Fauth, K., Chaudret, B.: One-pot synthesis of core-shell FeRh nanoparticles. Chem. Mat. **19**, 4624–4626 (2007)
29. Tersoff, J.: Oscillatory segregation at a metal alloy surface: Relation to ordered bulk phases. Phys. Rev. B **42**, 10965–10968 (1990)
30. Mecke, K.R., Dietrich, S.: Segregation profiles in Cu_3Au above the order-disorder transition. Phys. Rev. B **52**, 2107–2116 (1995)
31. Kaddouri, S.: Etude du comportement asymptotique des profils de ségrégation superficielle. LEMHE, Université Paris XI-Orsay, Rapport de stage (2001)
32. Roussel, J.-M., Tréglia, G., Legrand, B.: Surface segregation maps derived from tight-binding ising model. Solid State Phenom. **172–174**, 1008–1015 (2011)
33. Williams, F.L., Nason, D.: Binary alloy surface compositions from bulk alloy thermodynamic data. Surf. Sci. **45**, 377–408 (1974)
34. Ducastelle, F., Ricolleau, C., Loiseau, A.: Order and disorder at interfaces. Mat. Sci. Forum **155–156**, 367–386 (1994)
35. Le Bouar, Y., Loiseau, A., Finel, A., Ducastelle, F.: Wetting behavior in the Co–Pt system. Phys. Rev. B **61**, 3317–3326 (2000)
36. Pandit, R., Wortis, M.: Surfaces and interfaces of lattice models: mean-field theory as an area-preserving map. Phys. Rev. B **25**, 3226–3241 (1982)
37. Pandit, R., Schick, M., Wortis, M.: Systematics of multilayer adsorption phenomena on attractive substrates. Phys. Rev. B **26**, 5112–5140 (1982)
38. Tréglia, G., Legrand, B., Maugain, P.: Surface segregation in CuNi and AgNi alloys formulated as an area-preserving map. Surf. Sci. **225**, 319–330 (1990)
39. Ducastelle, F., Legrand, B., Tréglia, G.: Surface segregation in transition metal alloys: from electronic structure to phase portraits. Prog. Theor. Phys., suppl **101**, 159–180 (1990)
40. Lequien, F.: Thermodynamique des agrégats bimétalliques: surface, volume et effet de taille finie. Université Paris XI-Orsay, Thèse (2008)
41. Kofman, R., Cheyssac, P., Aouaj, A., Lereah, Y., Deutscher, G., Ben-David, T., Penisson, J.-M., Bourret, A.: Surface melting enhanced by curvature effects. Surf. Sci. **303**, 231–246 (1994)
42. Baletto, F., Ferrando, R.: Structural properties of nanoclusters: energetic, thermodynamic, and kinetic effects. Rev. Mod. Phys. **77**, 371–423 (2005)
43. Baletto, F., Ferrando, R., Fortunelli, A., Montalenti, F., Mottet, C.: Crossover among structural motifs in transition and noble-metal clusters. J. Chem. Phys. **116**, 3856–3863 (2002)
44. Mottet, C., Goniakowski, J., Baletto, F., Ferrando, R., Tréglia, G.: Modeling free and supported metallic nanoclusters: structure and dynamics. Phase Transitions **77**, 101–113 (2004)
45. Mottet, C.: Etude par simulation numérique d'agrégats libres mono- et bi-métalliques. Université Aix-Marseille II, Thèse (1997)
46. Lequien, F., Creuze, J., Berthier, F., Legrand, B.: Superficial segregation in nanoparticles: from facets to infinite surfaces. J. Chem. Phys. **125**, 094707 (2006)
47. Lequien, F., Creuze, J., Berthier, F., Legrand, B.: Dynamical equilibrium in nanoalloys. Faraday Discuss. **138**, 105–117 (2008)
48. Langlois, C., Alloyeau, D., Le Bouar, Y., Loiseau, A., Oikawa, T., Mottet, C., Ricolleau, C.: Growth and structural properties of CuAgCuAg and CoPt bimetallic nanoparticles. Faraday Discuss. **138**, 375–391 (2008)
49. Baletto, F., Mottet, C., Ferrando, R.: Growth of three-shell onionlike bimetallic nanoparticles. Phys. Rev. Lett. **90**, 135504 (2003)
50. Delfour, L., Creuze, J., Legrand, B.: Exotic behavior of the outer shell of bimetallic nanoalloys. Phys. Rev. Lett. **103**, 205701 (2009)

51. Mottet, C., Tréglia, G., Legrand, B.: New magic numbers in metallic clusters: an unexpected metal dependence. Surf. Sci. **383**, L719–L727 (1997)
52. Moreno, V., Creuze, J., Berthier, F., Mottet, C., Tréglia, G., Legrand, B.: Site segregation in size-mismatched nanoalloys: application to Cu–Ag. Surf. Sci. **600**, 5011–5020 (2006)
53. Lequien, F., Creuze, J., Berthier, F., Braems, I., Legrand, B.: Superficial segregation, wetting, and dynamical equilibrium in bimetallic clusters: a Monte Carlo study. Phys. Rev. B **78**, 075414 (2008)
54. Gulminelli, F.: Phase coexistence in nuclei. Ann. Phys. Fr. **29**(6), 1–121 (2004)
55. Langlois, C., Oikawa, T., Bayle-Guillemaud, P., Ricolleau, C.: Energy-filtered electron microscopy for imaging core-shell nanostructures. J. Nanoparticles Res **10**, 997–1007 (2008)
56. Berthier, F., Creuze, J., Tétot, R., Legrand, B.: Multilayer properties of superficial and intergranular segregation isotherms: a mean-field approach. Phys. Rev. B **65**, 195413 (2002)
57. Creuze, J., Berthier, F., Tétot, R., Legrand, B.: Phase transition induced by superficial segregation: the respective role of the size mismatch and of the chemistry. Surf. Sci. **491**, 1–16 (2001)
58. Tétot, R., Berthier, F., Creuze, J., Meunier, I., Tréglia, G., Legrand, B.: Cu–Ag (111) polymorphism induced by segregation and advacancies. Phys. Rev. Lett. **91**, 176103 (2003)
59. Sotto, M., Bouillard, J.-C.: Thermally induced transitions on clean stepped gold surfaces through LEED. Surf. Sci. **214**, 97–110 (1989)
60. Cohen, C., L'Hoir, A., Moulin, J., Schmaus, D., Sotto, M., Domange, J.-L., Bouillard, J.-C.: Study of atomic relaxations on clean and oxygen covered (100), (410) and (510) copper surfaces by channeling. Surf. Sci. **339**, 41–56 (1995)
61. Meunier, I., Tréglia, G., Legrand, B.: Surface-induced ordering in phase separation systems: Influence of concentration and orientation. Surf. Sci. **441**, 225–239 (1999)
62. Braems, I., Creuze, J., Berthier, F., Tétot, R., Legrand, B.: Effect of a size mismatch on bulk and surface alloy interactions: the illustrative example of the Cu–Ag system. Surf. Sci. **602**, 1903–1915 (2008)
63. Sprunger, P.T., Laegsgaard, E., Besenbacher, F.: Growth of Ag on Cu (100) studied by STM: From surface alloying to Ag superstructures. Phys. Rev. B **54**, 8163–8171 (1996)
64. Christensen, A., Ruban, A.V., Stoltze, P., Jacobsen, K.W., Skriver, H.L., Norskov, J.K., Besenbacher, F.: Phase diagram for surface alloys. Phys. Rev. B **56**, 5822–5834 (1997)

Computational Methods for Predicting the Structures of Nanoalloys

Riccardo Ferrando

Abstract Determining the geometric structure and chemical ordering of alloy nanoparticles is a crucial step for understanding and tailoring their properties. Here we review the methods for exploring the energy landscape of nanoalloys in order to find the most stable structural motifs and chemical ordering patterns. These methods are known under the name of global optimization, and range from simulated annealing, to genetic algorithms and basin hopping algorithms. The thermodynamics of the melting transition and kinetic effects in the growth of gas-phase nanoalloys are also discussed. For all topics, specific examples are presented.

1 Introduction

From the point of view of theory and simulation, the determination of the structures of alloy nanoparticles is quite challenging. The equilibrium structure of a gas-phase nanoalloy A_mB_n, with $m + n = N$, depends in general on its size, composition, and temperature. For nanoalloys in contact with an environment, such as a substrate or a solvent, structure can strongly depend on the interactions with the environment. Moreover, depending on the method by which a nanoalloy is grown, non-equilibrium structures can be produced. For these reasons, a great variety of nanoalloy structures is found in experiments [1–3], and the explanation of their origin is often difficult.

R. Ferrando (✉)
Dipartimento di Fisica dell'Università di Genova,
via Dodecaneso 33, 16146 Genova, Italy
e-mail: ferrando@fisica.unige.it

D. Alloyeau et al. (eds.), *Nanoalloys*, Engineering Materials,
DOI: 10.1007/978-1-4471-4014-6_8,

In order to tackle the problem of structure determination, specific computational methods have been developed. The aim of this chapter is to review these methods, showing also their applications to specific examples. After describing in general which kinds of *structural motifs* can be found for nanoalloys (Sect. 2), we first focus on the algorithms for searching their low-energy structures (Sect. 3). These algorithms, which have been impressively developed in recent years [2–4], are known as *global optimization* methods. Then, we consider the simulation methods for taking into account finite-temperature effects and studying phase transitions (Sect. 4). Finally, we focus on the simulations of nanoalloy growth, which are well suited to single out how kinetics can affect the structures that are actually produced in experiments (Sect. 5).

2 Geometric Motifs and Atomic Ordering Patterns in Nanoalloys

The structure of a nanoalloy can be described in terms of its *geometric motif* and of its *chemical ordering* pattern. The former describes the spatial arrangement of the atoms without reference to their chemical species. Chemical ordering, on the other hand, describes the pattern in which the atoms are arranged within the geometric structure.

Geometric motifs can be crystalline, when the nanoparticle can be seen as a fragment of a bulk crystal, such as the truncated octahedron of the face-centered cubic crystal lattice. Non-crystalline motifs are often in the form of polyhedra that are not space-filling. High-symmetry non-crystalline structures are icosahedra, decahedra, and polyicosahedra (see Fig. 1). Decahedra and icosahedra are made of five and twenty tetrahedra, respectively. The decahedron can be seen also as made of two pentagonal pyramids sharing a common basis. However, in most cases, decahedra are more energetically favourable in a specific truncated form, the Marks decahedron [5], which is more compact than the complete pentagonal bipyramid. The icosahedron and the decahedron have six and one fivefold rotation axes, respectively.

Chemical ordering can assume a variety of patterns. Some of them, such as the random mixing and the ordered alloy patterns, reproduce at the nanoscale atomic arrangements that can be found in infinite bulk crystals. However, there are typical nanoalloy patterns that are more easily related to what is found in semi-infinite alloys. For example, core-shell patterns are possible, in which one atomic element constitutes the shell and the other element constitutes the core. This nanoalloy arrangement bears some analogy with a semi-infinite alloy with tendency to phase separation and to segregation of one element at the surface. Multishell and Janus arrangements are also possible. In multishell nanoalloys, three or more concentric shells are present. In a Janus nanoparticle, the two halves of the nanoparticle present clearly different compositions and are separated by a sharp interface.

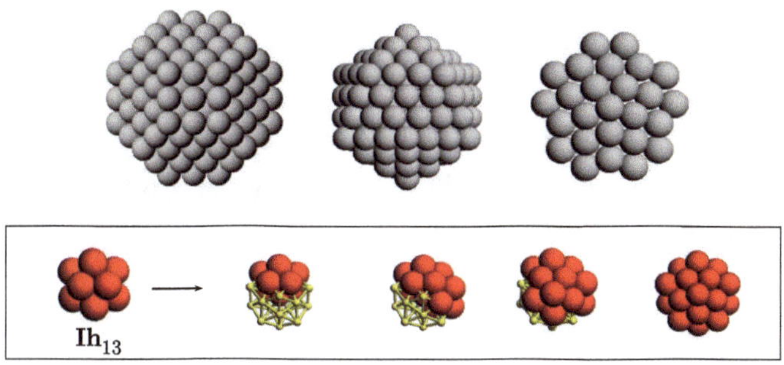

Fig. 1 *Top row* Crystalline fcc truncated octahedral, Marks decahedral and icosahedral structures. *Bottom row* a polyicosahedron which is built up by packing together sever elementary icosahedra of 13 atoms which share some atoms. The resulting structure has 34 atoms. From Ref. [2], copyright (2008) American Chemical Society

Nanoalloy structures that are intermediate between core-shell and Janus arrangements are also possible, with a shell only partially incorporating the core. Other patterns are possible. For example, a nanoparticles can be randomly intermixed in its interior, but with a surface enrichment of an atomic species, so that its chemical ordering has some features of both the random mixing and the core-shell arrangement. This is analogous to the case of a semi-infinite solid solution in which one element has a tendency to surface segregation. A schematic representation of some of these patterns is given in Fig. 2.

This variety of geometric motifs and chemical ordering patterns has been found in a number of recent simulation works [6–20], not to mention a large body of experimental results (see Ref. [2] for a summary).

3 Exploring the Energy Landscape of Nanoalloys by Global Optimization Searches: Methods and Specific Examples

From the qualitative description of possible nanoalloys structural motifs of Sect. 2 it can be easily understood that the study of the energy landscape of nanoalloys is a complex task. Exploring the energy landscape of a nanoparticle [21] amounts to studying its potential energy E, which is a function of all the coordinates $\{\mathbf{r}_i\}$ of the atoms of the nanoparticle. The function $E(\{\mathbf{r}_i\})$ is usually known as *potential energy surface* (PES). In particular, one is interested in finding the low-lying minima on the PES, because they have a larger statistical weight at equilibrium. In principle, the lowest energy minimum should be located, because it gives the equilibrium configuration of the nanoparticle when temperature $T \to 0$.

The problem of finding the lowest minimum of a function of many variables, such as the PES, is known as *global optimization*. The complexity of the problem is due to the large number of minima that a PES can present. In fact, a simple

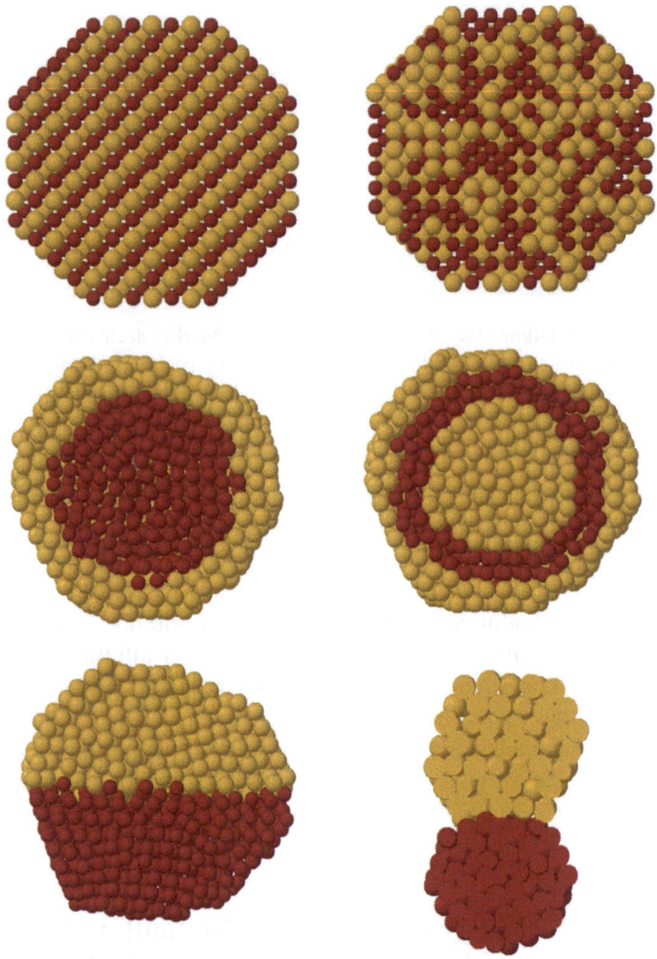

Fig. 2 Schematic representation of typical chemical ordering patterns. Cluster cross sections are given in all snapshots, in order to show the internal arrangements of atoms. *Top row* phase ordered (*left*) and randomly intermixed patterns. These patterns can be found in both bulk alloys and nanoalloys. *Middle row* core-shell and three-shell patterns. *Bottom row* phase-separated (Janus) particles. Real nanoparticles can present ordering patterns that share some features of different patterns: for example, core-shell nanoparticles with a randomly mixed core and surface shell which is enriched by one atomic element. From Ref. [2], copyright (2008) American Chemical Society

argument shows that the number of local minima on the PES increases exponentially with nanoparticle size [22–25]. Let us divide the nanoparticle into p equivalent subsystems of N atoms each. Assuming that every subsystem has independent stable configurations, we find:

$$n_{min}(pN) = [n_{min}(N)]^p, \tag{1}$$

whose solution is an exponential

$$n_{min}(N) = e^{\alpha N}, \tag{2}$$

where the constant α depends on the system

This argument can be applied to any nanoparticle which is large enough to be subdivided into approximately independent subsystems. In any case, the number of local minima in a nanoparticle is expected to increase explosively with its size. For example, in a single-element Lennard-Jones nanoparticles of 40 atoms, the number of local minima is estimated to be as large as 10^{40} [26].

In nanoalloys, a further degree of complexity comes into play. In fact, in a single-element nanoparticle, local minima which differ by simply exchanging the coordinates of two atoms are completely equivalent. In a nanoalloy, this is not true anymore, and leads to the concept of *homotop*, which has been introduced by Jellinek and Krissinel [27]. For a nanoalloy of given size and composition, homotops are isomers sharing the same geometric structure (neglecting local relaxations) but different chemical ordering. In a $A_m B_n$ nanoalloy, the number of homotops is given by

$$N_{homotops} = \frac{(n+m)!}{n!m!}. \tag{3}$$

Even though many of these homotops can be symmetry equivalent, $N_{homotops}$ is generally a huge number.

From these considerations one may deduce that there is no hope of solving the problem of finding the global minimum for PES of interest in the field of nanoalloys (and of nanoparticles in general). In fact, the only way to rigorously demonstrate that a given local minimum is indeed the global minimum is to compare it with *all* local minima on the PES. This is clearly impossible due to the huge number of minima. For this reason we note that all structures that will be denoted as "global minimum" (GM) in the following have to be intended as a *putative* global minima.

However, there are some features of nanoparticle PES that render global optimization feasible and useful. In fact, nanoparticle PES are usually organized in *funnels* [21]. These funnels are regions of the PES in which the pathways to their absolute minimum involves sequences of monotonically decreasing minima separated by low energy barriers. Within a given funnel, the search for the absolute minimum can be relatively simple. Minima belonging to the same funnel usually present similar structures. We can for example single out icosahedral funnels, decahedral funnels etc.. on the same PES. A very well-known example is the Lennard-Jones particle of 38 atoms [21, 28], whose PES presents a wide icosahedral funnel, in which the local minima are (often defective) fragments of the icosahedron of 55 atoms, and a narrow fcc funnel, which contains the GM.

When the PES is organized in funnels, its global optimization reduces to the problem of finding all of them. Then, within each funnel, the search for its absolute minimum (or of low-lying minima at least) might be relatively simple.

One difficulty in practical cases is that funnels are usually separated by huge energy barriers, so that it is quite common that a search procedure remains trapped in one of them. Another difficulty is that, when the nanoparticle size is large, even exploring a single funnel may become cumbersome. This is especially true in nanoalloys, because of the problem of the large number of homotops. However, one should keep in mind that, for large nanoparticles, the location of the GM itself becomes less important. In fact, in a large nanoparticle (say, from a few hundred atoms on), it is likely that there are several similar structures, differing only because of localized defects, that are close in energy and practically indistinguishable from the experimental point of view.

The same kind of consideration applies also to the homotops. It is likely that in a nanoalloy consisting of several hundred atoms, there will be a large number of homotops within a small energy range above the global minimum. This is especially true if the atomic species in the nanoalloy present a tendency to random mixing. In this situation, the location of the truly lowest homotop is an extremely cumbersome (or even impossible) task. But this is not truly necessary, because the real important objective is to determine the typical low-energy chemical ordering pattern, to which many different but similar homotops may belong.

In summary, a good global optimization algorithm should be able to reach quickly the minima at the bottom of a given funnel and to explore different funnels without being trapped forever in the initial one. Several types of global optimization algorithms have been developed and applied to nanoparticles and nanoalloys. A description of the most widely used algorithms is found in Sects. 3.1–3.4 (see also [25]).

Most global optimization searches are applied to approximate models of nanoparticle PES, such as those given by atom-atom potentials. In fact, full *ab initio* global optimization is still too cumbersome, being limited to systems of a few ten atoms and no more. For this reason, it is essential that a global optimization algorithm is able to explore different funnels in order to produce a diversity of structures that can serve as a database [2, 3] for further local relaxation by means of more sophisticated approaches, such as Density Functional Theory (DFT).

3.1 Simulated Annealing

In simulated annealing (SA), the nanoparticle is first equilibrated at high temperature and then cooled by either a Monte Carlo or a molecular-dynamics procedure. It can be shown that, if temperature decreases logarithmically with time, SA brings the system to its potential-energy minimum [29]. Since both Monte Carlo and molecular dynamics are in principle able to sample correctly the equilibrium free energy of a system, this means that SA should reach the GM if simulation time is long enough.

In practical cases (for which simulation time cannot be increased at will), SA suffers from two serious drawbacks. It is a quite time-consuming algorithm in reaching the low-energy part of a given funnel. Moreover, it can be easily get trapped in the initial funnel, being thus unable to locate the GM of multiple-funnel PES. For these reasons, SA is nowadays less commonly used in the global minimization of nanoparticles. Applications to nanoalloys are practically absent.

However, an advantage of SA must be mentioned. When SA is implemented with a molecular-dynamics cooling procedure, it becomes a cooling procedure which closely resembles a real physical cooling process. Therefore, besides producing low-energy structures, SA can provide precious information on the cooling process itself. The global optimization algorithms that will be discussed in the following are likely to be more efficient in global optimization, but they do not produce physical trajectories in the nanoparticle configuration space, and therefore their use is limited to the search of low-energy configurations.

3.2 Basin Hopping

Basin hopping [30] is the most well-known algorithm which uses thermodynamic equilibrium sampling for the global optimization of the PES. In basin hopping, the PES is subjected to an appropriate transformation whose purpose is to maximize the probability of sampling low-energy minima at equilibrium, allowing at the same time a fast exploration of significant portion of the PES itself.

In BH the global minimum is searched for by transforming the original PES $E(\{\mathbf{r}\})$ into a staircase function $\tilde{E}(\{\mathbf{r}\})$. This transformation is simply achieved by applying to each nanoparticle structure a local minimization procedure, and associating to a given point in configuration space the energy of its closest local minimum. $\tilde{E}(\{\mathbf{r}\})$ is sampled at equilibrium at a given temperature T by a Metropolis Monte Carlo algorithm. At each step, to an initial local minimum of energy E_1, a *move* is applied to produce a new configuration, which, in general, is not a local minimum. This new configuration is locally minimized obtaining the energy E_2 of a (hopefully) new local minimum. Energies E_2 and E_1 are compared. If $\Delta E = E_2 - E_1 \leq 0$ the move is accepted, and the initial configuration is substituted by the new one. If $\Delta E > 0$, the move is accepted with probability $\exp[-\Delta E/(k_B T)]$, where k_B is the Boltzmann constant.

The advantage of applying local minimization in searches of low-energy configurations is apparent from Fig. 3. In fact, this transformation eliminates the barriers between different minima to the maximum possible extent. The advantage can be seen also from the point of view of equilibrium sampling. In fact, points in configuration space which belong to the basin of low-energy minima are lowered by larger amount by the transformation, so that they become more likely in an equilibrium sampling.

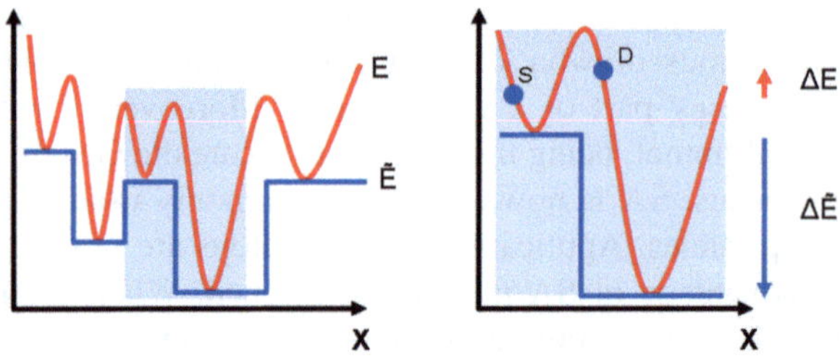

Fig. 3 Schematic one-dimensional representation of the transformation of a PES (*red line*) into a staircase function (*blue line*). From the right panel, it is evident that this transformation favours those points in configuration space that belong to basins of low-energy minima. Before the transformation, point D has a higher energy than point S. After the transformation the opposite is true, with an exponential gain in the probability of sampling point D in an equilibrium simulation. From Ref. [25]

Several different kinds of moves can be applied for generating new structures at each step. A general requirement of a move is that it should be strong enough so that the system can escape from the basin of its present local minimum and generate a truly new configuration. On the other hand, if the move is too strong, the new configuration is likely to be very high in energy. A compromise between these conflicting needs must be found, in a way which depends on the system under study and on the specific kind of move. Typical moves used in BH simulations are [25]

- *Shake move*. Each atom is displaced from its present position within a sphere of radius r_s (or within a cube). Typical values of r_s are close to half of the nearest-neighbour distance between atoms. Variants of the shake move displace only a part of the atoms of the cluster or displace surface atoms by larger amounts than inner atoms.
- *Shell move*. It is designed to improve the arrangement of the cluster surface. A single surface atom is displaced to a random position within a spherical shell which roughly corresponds to the external atomic layer of the cluster.
- *Brownian move*. A short molecular-dynamics (or Langevin) simulation is run at high temperature T_B. Compared to the shake move, this procedure is less likely to produce unphysical configurations, allowing at the same time quite strong rearrangements of the cluster.
- *Exchange move*. The positions of two atoms of different species are exchanged. This is very important for optimizing chemical ordering in nanoalloys. Variants of this move include tailored exchanges [20], for example involving only surface atoms of species A and inner atoms of species B.

The simulation temperature T is simply a parameter that can be tuned in order to achieve the best efficiency of the algorithm. The optimal temperature may however depend on the kind of move. In fact, for the shake move, typical values of T for metallic nanoparticles and nanoalloys are above 1,000 K. The same applies to the Brownian move. On the other hand, the exchange move requires lower temperatures (as low as 100–300 K), especially for nanoalloys with intermixing tendency. Higher temperatures would cause a very high acceptance rate leading to high-energy homotops. This is due to the fact that the typical energy separation between homotops is usually smaller than the separation between geometrically different isomers. Since in basin hopping one has a single temperature, it is often more efficient to proceed as follows [3] when optimizing nanoalloys. First, simulations with either shake or Brownian moves are made, without any exchange. Then, the lowest-energy motifs of the different structures are used as starting configurations of low-temperature simulations with exchanges only. To this purpose, it is useful to classify the simulation output by means of an order parameter which is able to distinguish different geometrical structures [25, 31]. The use of order parameters in global optimization searches will be discussed in more detail in Sect. 3.4.

3.3 Genetic Algorithms

Genetic algorithms (GA) have been used in a large variety of fields, from chemistry to physics, economy, computer science, and more. These algorithms mimic a biological evolutionary process, in which a population of individuals evolves generation after generation, trying to optimize a fitness parameter. In the cases of our interest, the individuals are clusters and the fitness parameter is their potential energy.

From an initial population (the parents) with a number P of individuals, new individuals (the sons) are produced by means of mutation and mating operations. In a mutation operation, a single parent is modified in some way (for example by means of the moves described in Sect. 3.2) to produce a son. In a mating operation, two parents produce a single son. For example, two parent clusters generate a son by a cut-and-splice operation in which two halves of different clusters are joined together. When P sons are obtained, a new generation is produced by selecting, among $2P$ sons and parents, the P individuals having the best fitness. Often this condition on the choice of the best individuals is relaxed by some probabilistic rule. Also in genetic algorithms local minimization is applied to all structures, so that the fitness is evaluated on local minimum configurations. This strongly improves the efficiency of the method.

Genetic algorithms have been used by several groups for single-element nanoparticles (see for example [32–36]) and nanoalloys [6, 8–10, 14]. An excellent review about genetic algorithms with applications to nanoalloys is found in Ref. [4].

3.4 Other Methods

3.4.1 Parallel Excitable Walkers (PEW) Algorithm

The transformation of the PES into a staircase function eliminates all energy barriers between minima belonging to a descending sequence. For this reason, BH is extremely efficient for finding the bottom of a given funnel, especially at low temperatures. However, barriers between different funnels are essentially not altered by the transformation, so that BH runs can easily remain trapped in the funnel to which its initial configuration belongs. To avoid trapping, high simulation temperatures should be used, but this may deteriorate the efficiency of sampling of the funnel bottom.

An approach that tries to combine efficiency of BH in sampling the low-energy parts of funnels with an increased probability of exploring several different funnels is the parallel excitable walkers (PEW) algorithm [31]. In the PEW algorithm, n_w Monte Carlo walkers perform BH searches of the PES. These walkers interact with each other in the following way. An order parameter p is defined for each point $\mathbf{X} = (\mathbf{r}_1, \ldots, \mathbf{r}_N)$ of the PES. A neighbouring relation between walkers is defined in the order parameter space. Given a distance δ, walkers a and b are neighbours if they satisfy

$$|p(\mathbf{X}_a) - p(\mathbf{X}_b)| \leq \delta. \tag{4}$$

At each step of the simulation, one walker is randomly chosen. If this walker has no neighbours, its move is either accepted or refused according to the usual Metropolis criterion applied to ΔE as in standard BH. If this walker has at least one neighbour, the Metropolis criterion is applied to $\Delta E - E^*$, where E^* is the excitation energy. This amounts to increasing the energy of the initial configuration of the walker by the quantity E^*. Walkers with neighbours have thus much larger probability of having their moves accepted, so that they are likely to increase the distances from their neighbours. This procedure has proven to be efficient in exploring multiple-funnel PES [31], in both single-element nanoparticles and nanoalloys. Typical values of E^* are in the range of 0.5 eV for transition metal nanoparticles. The distance δ is chosen in such a way that $2n_w\delta$ is about half of the variation range of p. The use of the excitation energy allows to employ low simulation temperatures in the range 100–500 K. Therefore, walkers with no neighbours are efficient in arriving at the bottom of their funnels.

The efficiency of the PEW algorithm depends on the choice of the order parameter. An order parameter is good if it associates well separated values to structures belonging to different funnels. In the case of nanoalloys (and of nanoparticles in general) parameters deriving from the Common Neighbour Analysis [37] have proven to be effective, since they are able to distinguish cristalline structures, decahedra, icosahedra and polyicosahedra. Also parameters measuring the degree of intermixing, such as the percentage of heterogeneous nearest-neighbour bonds, can be quite useful.

A good feature of the PEW algorithm is that it is robust against a bad choice of the order parameter (typically, a choice of an order parameter which associates the same range of values to different structural motifs). In fact, if the order parameter is good, PEW can achieve relevant improvements over pure BH. With a bad choice of the order parameter, the performance of PEW is as good as that of BH [25, 31].

3.4.2 Basin Hopping with Memory: HISTO Algorithm

Another approach which may improve basin hopping takes into account the history of the simulation. In fact, while searching the PES for lower and lower minima, it seems useless to reexplore those regions which have been already visited. The HISTO algorithm uses an order parameter to take into account the memory of already visited places. This algorithm is quite similar to the Energy Lanscape Paving (ELP) algorithm of Hansmann and Wille [38] and bears some resemblance also with the metadynamics approach of Laio and Parrinello [39]. The main difference with respect to the ELP is that the HISTO algorithm samples the transformed PES after local minimization.

In the HISTO algorithm, a normalized histogram H is constructed step by step reporting the frequencies of visited minima in different intervals of the order parameter space. The quantity $\tilde{E}$ is defined by

$$\tilde{E} = E + wH(p), \tag{5}$$

where w is a positive weight and $H(p)$ is the height of the histogram for the value p of the order parameter which is associated to E. The Metropolis criterion is applied to $\Delta\tilde{E}$:

$$\Delta\tilde{E} = E_2 - E_1 + w[H(p_2) - H(p_1)]. \tag{6}$$

If the order parameter interval of minimum 1 has been more frequently visited than the interval of minimum 2, $\Delta\tilde{E} < \Delta E$, so that the memory term favours the move to configuration 2. On the contrary, if $H(p_2) > H(p_1)$, the move is hindered.

The HISTO algorithm has been applied to several models of nanoparticles and nanoalloys [25, 31] and compared to BH and PEW. It turns out that HISTO is more sensitive to the choice of the order parameter. If the choice is good, improvements over BH can be more spectacular than in PEW. If the choice is not good, the performance of HISTO easily deteriorates. For example, if the order parameter assigns overlapping intervals to structures belonging to different funnels, to avoid revisiting order parameter regions which have been already explored can lead to a drastically incomplete sampling of the relevant PES.

3.4.3 Minima Hopping (MH) Algorithm

The minima hopping (MH) algorithm has been developed by Goedecker [40] and originally applied to single-element nanoparticles. Recently, it has been applied also to binary Lennard-Jones particles [41], showing thus its potentiality for an efficient search of nanoalloy global minima. At variance with BH, PEW and HISTO, MH is not based on thermodynamics (this feature is shared also by genetic algorithms). MH takes into account the history of the simulation, but in a different way compared to HISTO. In fact, while HISTO uses a coarse-grained description of configuration space by means of an order parameter, MH makes an explicit list of the minima that have been already visited. Then a strategy that limits repeated visits is implemented, in such a way that it does not penalize crossings through important transition basins. This is achieved by making violent escape moves out of the current basin if this basin has already been visited. The efficiency of MH depends strongly on the type of elementary moves that are used. Moves that find low-barrier escape-paths out of the present minimum generally lead into low energy minima [40].

3.5 Applications to Specific Systems

In this Section we present global optimization results about different nanoalloys. These results are obtained by global optimization searches within an atomistic potential that has been widely used in modeling nanoparticles and nanoalloys, i.e. the potential derived within the second-moment approximation (SMA) to the tight-binding model (see Chap. 2 of this book for a description of the SMA potential). Global optimization has been made by means of genetic, BH and PEW algorithms. In most cases, the results of global optimization have been checked by selecting the lowest isomers of each structural motif and relaxing them locally at the DFT level. In DFT calculations, the Perdew-Burke-Ernzerhof (PBE) exchange-correlation functional [42] has been used, unless otherwise specified.

The systems treated in the following are chosen because they are representative of quite different behaviours. Ag–Cu and Ag–Ni present a wide miscibility gap in the bulk, so that they may be expected not to mix in nanoalloys. Both systems are characterized by a large size mismatch (more than 10 %), with Ag being the "large" atom. Ag has also a lower surface energy, so that it should segregate to the cluster surface. On the other hand, Pt–Co tends to mix and to form ordered phases in bulk systems. Finally, Ag–Pd is a mixing system in the bulk, but it forms random solid solutions instead of ordered phases. Since Ag has a lower surface energy than Pd, a certain degree of Ag surface segregation is expected in nanoalloys.

3.5.1 Ag–Cu, Ag–Ni and Co–Ag

These systems have been intensely studied in recent years in both theory and experiments as prototypical phase-separating nanoalloys [8, 9, 13, 17, 43–47].

The structures of Ag–Cu and Ag–Ni in the size range below 50 atoms are dominated by core-shell polyicosahedral structures [8, 9, 13]. These polyicosahedra are made up of interpenetrating elementary icosahedra of 13 atoms (see Fig. 1), with Ag atoms occupying surface sites. In the most stable structures, the core shell arrangement is perfect, with all Ag atoms being at the surface and all Cu atoms being in the inner part of the cluster, so that the external Ag shell is of

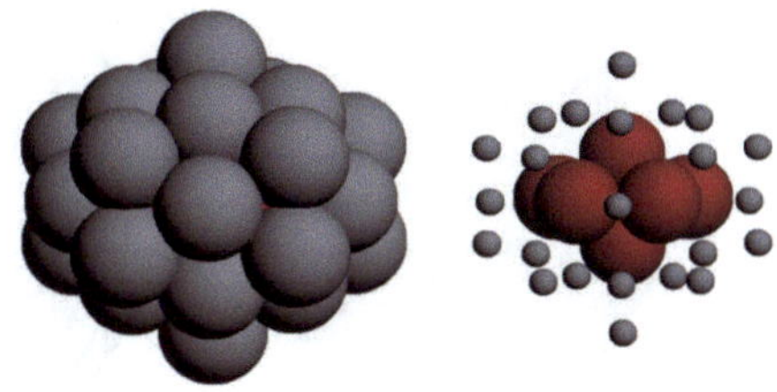

Fig. 4 Global minimum structure of $Ag_{27}Cu_7$ and of $Ag_{27}Ni_7$. In this perfect high-symmetry core-shell structure, all Ag atoms (in *grey*) occupy surface sites, while Cu (or Ni) atoms are in the inner part of the cluster. In the right panel, Ag atoms are represented by small spheres in order to show the cluster core

monoatomic thickness. An example of a very stable structure, whose stability has been checked both at the atomistic potential and at the DFT level, is shown in Fig. 4.

The special stability of core-shell polyicosahedra is justified by simple physical arguments. In fact, polyicosahedra are very compact structures, presenting a very high number of nearest-neighbour bonds for a given size. However, in single-element clusters, these bonds are highly strained, since surface bonds are expanded and internal bonds are compressed. For transition and noble-metal clusters, this is unfavourable, because the bond-orded/bond-length correlation would favour the opposite, i.e. surface bonds shorter than inner bonds. This problem can be avoided in nanoalloys by substituting the inner atoms of the single-element polyicosahedron by atoms of a different species, whose atomic size is smaller, and whose surface energy is higher. In this way, strain is released and the structure strongly gains in stability. This applies perfectly to Ag–Cu and Ag–Ni. Stability may be enhanced also by electronic shell closure effects. For example, for Ag–Cu, size 34 is magic for the spherical jellium model [48]. This implies that the structure of Fig. 4 is of special stability because of both geometric and electronic effects. For other sizes, like size 40 in Ag–Cu, electronic effects may lead to favour other structures than those predicted by the SMA model [13]. These structures are core-shell and still belong to the polyicosahedral family, being incomplete polyicosahedra.

For larger sizes, interesting core-shell icosahedral structures are found. The common Mackay icosahedra [49, 50] (see Fig. 1 for an example) are not favourable because of the atomic size mismatch in core-shell Ag–Cu and Ag–Ni clusters. In fact, the external Ag shell would be too dense, with an enormous strain. For this reason, another form of the icosahedron, which has an external anti-Mackay shell [51] is preferred. The difference between the Mackay and anti-Mackay shells is shown in Fig. 5. Note for example that for covering an icosahedral core of 13 atoms, 42 atoms are necessary in the Mackay shell and 32 atoms in the anti-Mackay shell, which is thus considerably less dense.

When cluster size increases (above 500 atoms in Ag–Cu), a transformation of the anti-Mackay external shell is energetically favourable (see Fig. 5) in such a

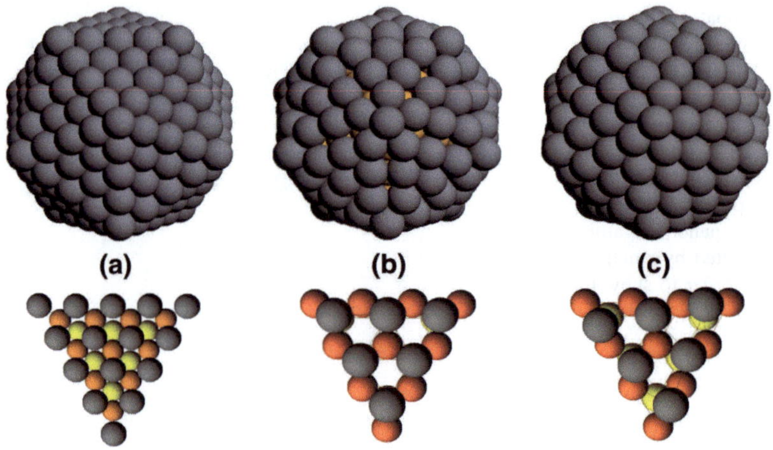

Fig. 5 *Top row* **a** Mackay icosahedron of 309 atoms, for composition $Ag_{162}Cu_{147}$. An external single Ag shell covers a Cu icosahedral core. **b** Anti-Mackay icosahedron of 279 atoms, for composition $Ag_{132}Cu_{147}$. Its Cu core is the same as in the Mackay icosahedron. **c** Chiral icosahedron of 279 atoms, for composition $Ag_{132}Cu_{147}$. Its outer shell is obtained by rotating all triangular Ag islands of the anti-Mackay shell by the same angle so that all mirror symmetries are broken. Neglecting local relaxations, the Cu core preserves the achiral icosahedral symmetry of the previous clusters, while the Ag shell assumes a clearly different structure. Bottom row: The stacking of the three outer atomic shells is shown for the clusters of the top row. Ag atoms are represented by larger spheres (in *grey*). Two layers of copper atoms are shown. The lowest layer is in *yellow* (*lighter grey*). The outer Ag shell is in fcc-like and hcp-like stacking for the Mackay and anti-Mackay clusters, respectively. In the anti-Mackay clusters, the Cu atoms of the third layer are covered by the Ag atoms of the external shell. In the chiral cluster, the Ag *triangles* are rotated so that atoms are displaced from their anti-Mackay sites. From Ref. [20], copyright (2010) American Chemical Society

way that the external shell becomes chiral [20] by a concerted rotation of triangular units. This transformation increases the number of nearest-neighbour bonds in the Ag shell, at the expenses of a somewhat worse matching between the Ag shell and the Cu core. The driving forces for the chiral transformation are of quite general character. Therefore that transformation has been observed, at the DFT level, also in other systems such as Ag-Co and Au-Ni [20].

For compositions that are richer in Ag so that a perfect core-shell structure (with a shell of monoatomic thickness) cannot be formed, another set of interesting structures has been found, in Ag–Co besides Ag–Cu and Ag–Ni [17, 19]. Ag–Co presents the same features about size mismatch and surface energy as Ag–Cu and Ag–Ni. These are core-shell structures where the inner core is placed in strongly asymmetric position, being however still completely covered by silver (see Fig. 6). These particles can be denoted as quasi-Janus particles. The asymmetric position of the core helps the structure to release its strain.

Available experimental data on Ag–Ni and Ag–Cu nanoalloys [43–45] are consistent with the theoretical predictions, at least on a qualitative level. In fact,

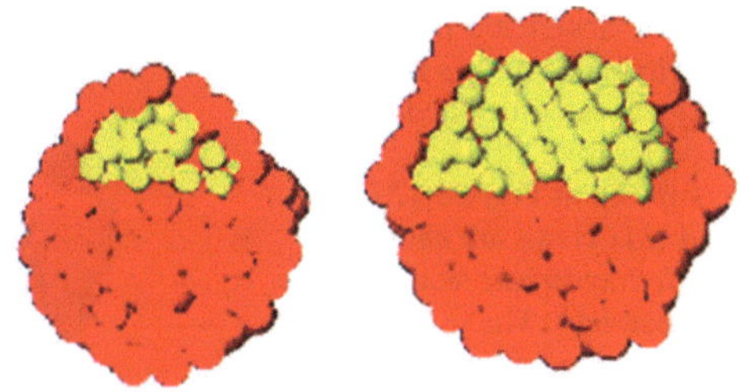

Fig. 6 Cross-sections of quasi-Janus Ag–Co nanoparticles. The same kind of structures are favourable also in Ag–Cu and Ag–Ni. From Ref. [19], copyright (2010) American Chemical Society

the experiments show a transition from core-shell structures (with a centered core) for small sizes (as those considered in the simulations) to Janus-like particles for large sizes. In Janus-like particles, the core is placed asymmetrically and possibly only partially covered by Ag.

3.5.2 Pt–Co

Pt–Co nanoalloys have been intensely studied from the experimental point of view due to their promising properties as nanomagnets [52–55].

Pt–Co is a system which forms ordered phase in the bulk [56]. For example, for 50–50% composition, the $L1_0$ phase forms at low temperatures, being however stable up to 1,100 K. This phase has a tetragonally distorted fcc crystal structure and alternates (001) planes of Co and Pt. Ordered phases form also around 25–75% and 75–25% compositions. In the following we consider mostly compositions close to 50–50%.

For sizes below 100 atoms, global optimization results within the SMA atomistic potential single out polyicosahedral structures as the lowest in energy [57, 58]. These results have been checked also at the DFT-PBE level [58] for size 38, which is a magic size for the truncated octahedron and thus should be favourable for the $L1_0$ bulk phase. The DFT calculations show that polyicosahedral structures are much lower in energy than the $L1_0$ truncated octahedron, by more than 1 eV. This polyicosahedral structure is however different from the one shown in Fig. 4, as can be seen in Fig. 7, in which composition $Pt_{20}Co_{18}$ is considered. The Pt–Co polyicosahedron is made of six interpenetrating elementary icosahedra that are arranged in a hexagonal ring. This structure will be referred to as *sixfold pancake* in the following. The driving force for the formation of this structure is mostly related to the fact that in this polyicosahedral structure is quite compact allowing at the same time a large number of mixed Pt–Co bonds.

For sizes larger than 100 atoms, SMA global optimization calculations only have been made [57]. These calculations show a prevalence of decahedral global minima in the interval 100–500 atoms. A $L1_0$ ordering is reproduced inside each of the five tetrahedra composing the decahedron. From size 500 on, tetragonally distorted fcc clusters with $L1_0$ pattern begin to compete with decahedra, and are expected to prevail above $\sim 1{,}000$ atoms. Icosahedral clusters are in practice never

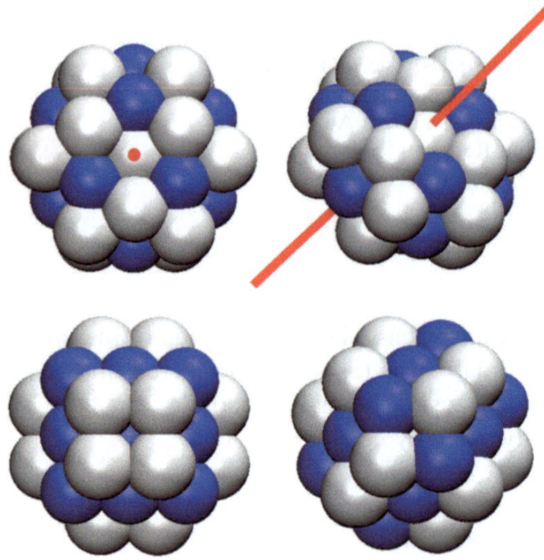

Fig. 7 Schematic pictures of polyicosahedral (*top row*) and $L1_0$ truncated octahedron (*bottom row*) structures for composition $Pt_{20}Co_{18}$. *Top* and *side views* are shown on the *left* and *right* side, respectively. The polyicosahedral cluster is a fragment of a Frank–Kasper phase [59]. Its disclination line in the is indicated. The two atoms along this line have 13 first neighbours. Cobalt atoms are displayed in blue, and platinum atoms are in dark grey. From Ref. [58], copyright (2010) American Chemical Society

prevailing. These results are qualitatively in agreement with the experimental data in Ref. [54] in which mostly decahedral and fcc clusters are formed in the range between 2 and 4 nm of diameter, without evidence in favour of icosahedra.

3.5.3 Ag–Pd

Ag and Pd are fully miscible in the bulk phase, without evidence of the formation of ordered phases [56]. Ag–Pd nanoparticles have been produced by epitaxial vapor deposition onto thin alumina films [60]. These particles were shown to present intermixed chemical ordering but with some silver segregation at the surface.

Global-optimization studies within the SMA potential [10] for size 38 atoms and varying composition have shown a prevalence of polyicosahedral sixfold pancake structures for silver-rich compositions and of fcc truncated octahedra from 50–50% composition on. At variance with the case of Pt–Co, the stability of sixfold pancake structures has not been confirmed at the DFT level [11, 61]. For this reason, a reparametrization [61] of the SMA potential has been developed in such a way that its results are in much better agreement with the DFT data for both the ordering of isomers in clusters of different sizes and compositions and for the energetics of Pd impurities in icosahedral Ag clusters [62]. The weaker stability of

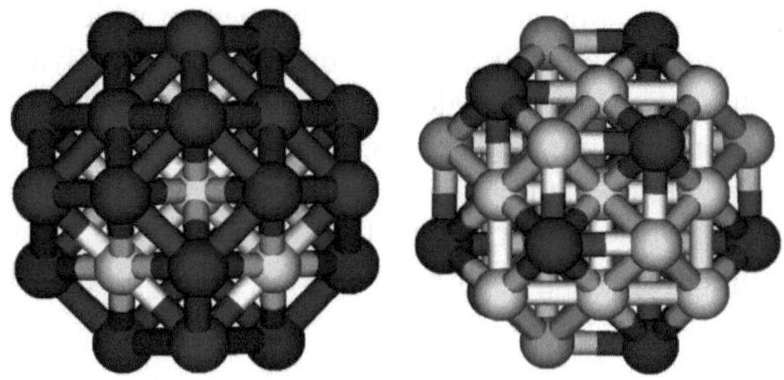

Fig. 8 The most stable 38-atom TO with compositions $Ag_{29}Pd_9$ (*left*) and Ag_9Pd_{29} (*right*). Pd (Ag) atoms are in white (*gray*). From Ref. [61], copyright (2010), American Insitute of Physics

the sixfold pancake with respect to the Pt–Co case can be attributed to the weaker tendency towards mixed bonds and to the smaller size mismatch of Ag–Pd.

Within this reparametrization, three compositions have been considered [61], Ag_9Pd_{29}, $Ag_{19}Pd_{19}$ and $Ag_{29}Pd_9$, in the global optimization searches. These compositions are as close as possible to 1:3, 1:1 and 3:1. For size 38, truncated-octahedral structures have been found for the Pd-rich and the intermediate compositions, and an icosahedral structure for the Ag-rich composition. For size 60, icosahedral structures were found for all three compositions. For size 100, only composition $Ag_{50}Pd_{50}$ was analyzed, finding a decahedral structure as the global minimum. In this case, a remarkable agreement between SMA and DFT calculations on the ordering of several isomers was obtained.

In Ag–Pd, the tendency to form mixed bonds is rather strong, as can be seen by analyzing the most favourable chemical ordering in the truncated octahedra of 38 atoms (see Fig. 8), but not strong enough to induce ordering in alternating Ag and Pd planes. Pd atoms are preferentially placed in the inner part of the cluster or at the center of hexagonal facets. In general, several homotops are however within a small fraction of eV from the global minimum.

In general, Ag–Pd global minima present a strong Ag enrichment of the surface, and some Pd enrichment of the subsurface layer. In total, this results in an Ag enrichment of the external part of the cluster, while the cluster core is intermixed with some local ordering. This chemical ordering pattern agrees with the experimental observations [60].

4 Thermodynamic Effects: The Melting Transition in Nanoalloys

The results shown in the previous sections were related to the search of the lowest-energy configurations of nanoalloys, which is representative of their equilibrium structure in the limit $T \rightarrow 0$. At finite temperature, the nanoalloy has a non-negligible probability of sampling other basins that the basin of the global minimum. When temperature is high enough, the probability of finding the nanoalloy in its global minimum structure becomes negligible, so that a phase change takes place. The simulation of phase changes, of which melting is the most well-known, is made by two different methods, molecular dynamics (MD) and Monte Carlo (MC). We note that global optimization searches serve as a basis for correctly starting melting simulations, because they give the correct structures to start with at low temperatures. In the following, we concentrate on the melting transition. After a brief description of the methods, we present two examples of melting simulations in nanoalloys. A review about melting in nanoalloys can be found in Ref. [2].

4.1 Molecular-dynamics and Monte Carlo Simulations

In MD simulations, Newton equations of motion are solved step by step by discretizing time [63]. Even though the equations of motion are classical, the underlying force field may either derive from a fully quantum calculation at each step (see for example the simulation of the melting of binary alkali clusters in Refs. [64, 65]) or from an atomistic potential, as in most cases [27, 66–68].

Melting is usually studied by calculating the *caloric curve*, which reports the average internal energy E as a function of T. Its derivative gives the thermal capacity. The caloric curve is obtained by starting at low temperatures from the appropriate cluster structure, and slowly increasing the temperature by a thermostat. In the melting region, $E(T)$ presents a smooth step, which gets sharper for increasing cluster size, to approach an abrupt jump in the bulk limit. Examples of caloric curves of nanoalloys are shown in Fig. 9.

Phase transitions in nanoalloys can be studied also by means of Monte Carlo simulations [63]. There are several examples in the literature [17, 47, 57, 69, 70], using either canonical or (semi)grand-canonical schemes. The advantage of Monte Carlo is that sophisticated sampling techniques are more easily implemented, such as parallel tempering [70]. On the other hand, MD simulations offer a more physical description of the melting process, which gives information also on its kinetics and on real transformation mechanisms.

4.2 Single-Impurity Effects on the Melting of Nanoparticles

A clear example showing the sensitivity of the melting transition to changes in composition of the nanoalloy is given by the case of icosahedral Ag clusters with a single impurity [71].

For Ni and Cu impurities, the preferred position in Ag icosahedra is the central site. In fact, both Ni and Cu atoms are of smaller size and more cohesive than Ag. Since the icosahedral central atom in pure Ag clusters is quite strongly compressed, substituting it by a smaller impurity causes a release of the strain. The resulting cluster turns out to be clearly more stable from the thermodynamic point of view that the pure cluster. This effect is relevant for icosahedra of quite large

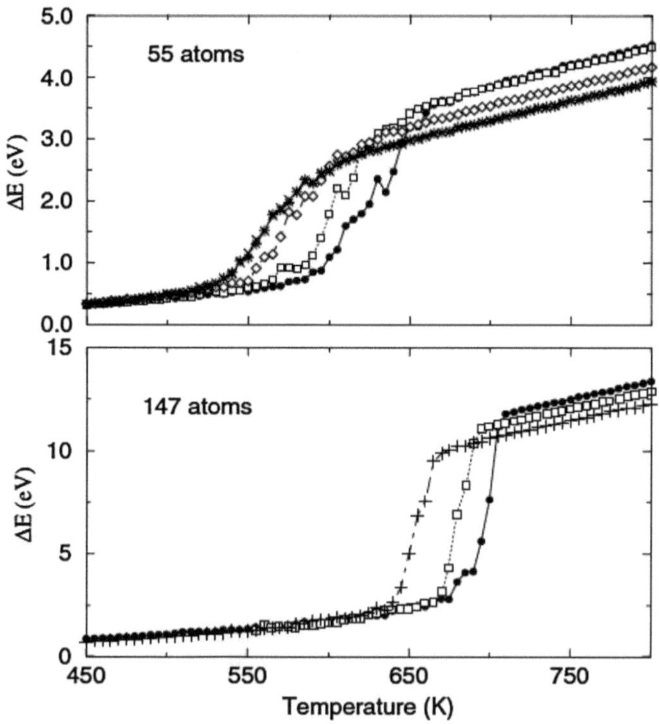

Fig. 9 Caloric curves of the melting of silver icosahedra containing a single impurity obtained by Molecular Dynamics simulations. The quantity plotted is $\Delta E = E(T) - E(0) - 3(N-1)k_BT$, where $E(T)$ is the average cluster internal energy, $E(0)$ is the minimum energy at 0 K, and $3(N-1)k_BT$ is the harmonic part of the energy. In the upper panel, size 55 is considered and all systems (pure Ag, Ag–Cu, Ag–Ni, Ag–Pd and Ag–Au) are shown. In the lower panel, size 147 is considered and only pure Ag, Ag–Cu, and Ag–Ni are shown. Crosses refer to pure Ag, solid circles to Ag–Ni, squares to Ag–Cu, diamonds to Ag–Pd, and asterisks to Ag–Au. From Ref. [71], copyright (2005) by The American Physical Society

size. In fact, as can be seen in Fig. 9, a single Ni impurity causes an increase of more than 50 K of the melting temperature of Ag icosahedra of 55 and 147 atoms. This effect decreases with cluster size, but it is still evident in 561-atom icosahedra. The effect of a Cu impurity is analogous. On the other hand, Pd and Au impurities cause a negligible change in the melting temperature.

These different behaviours are clearly related to the different degrees of decrease of the atomic stress on the central site of the icosahedron caused by different impurities, as shown in Fig. 10. Ni and Cu impurities are by far the most effective in decreasing the atomic stress. We note that it is conceivable that for large clusters, a more important increase of the melting temperature could be obtained by putting several impurities, like an icosahedral 13-atom core.

Interesting points that have not been investigated in this system (but these points will be dealt with in the next example) are about the melting mechanism,

Fig. 10 Atomic stress in Ag icosahedra as a function of size N for different impurities. Crosses refer to pure Ag, solid circles to Ag–Ni, squares to Ag–Cu, diamonds to Ag–Pd, and asterisks to Ag–Au. From Ref. [71], copyright (2005) by The American Physical Society

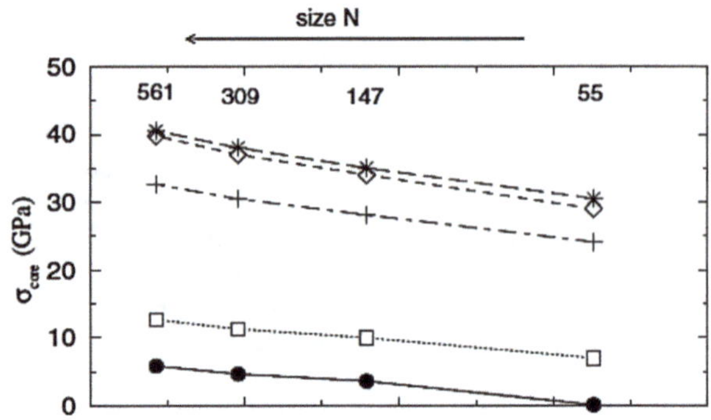

Fig. 11 Anti-Mackay icosahedral structure of $Ag_{32}Ni_{13}$. From Ref. [68], copyright (2008) by The American Physical Society

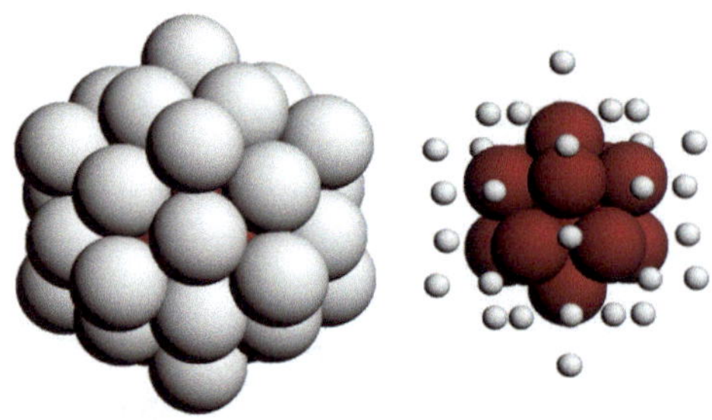

i.e. if melting starts at the surface and then propagates to the inner part of the cluster or it involves directly the whole structure at once.

4.3 Melting of Core-Shell Nanoalloys

Nanoalloys are very interesting systems for studying surface melting phenomena. These are quite clear for core shell structures, as it has been shown in simulations of Cu–Ni, Ag–Co and Pd–Pt [66, 67, 69]. The driving force for surface melting, i.e. the lower coordination of surface atoms, is enhanced in core-shell systems by the fact that the shell metal has a lower melting temperature than the core metal. A clear example of such a phenomenon is given by the melting of $Ag_{32}Ni_{13}$ modelled by the SMA atomistic potential and studied by MD simulations [68]. The global minimum structure in this case is given by a core-shell anti-Mackay icosahedron (see Fig. 11), which contains a 13-atom Ni icosahedral core plus an anti-Mackay Ag shell of monoatomic thickness.

Surface melting in $Ag_{32}Ni_{13}$ is clearly demonstrated by the results of Fig. 12 which reports the probability that the whole structure is in the basin of its global minimum as a function of temperature, together with the probability of finding the Ni core in its original icosahedral configuration. These probabilities are obtained by running MD simulations at each temperature of the curve and periodically

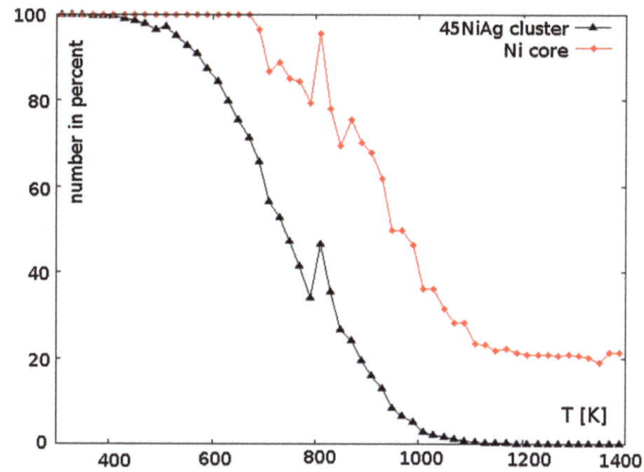

Fig. 12 The black triangles represent the probability that $Ag_{32}Ni_{13}$ is in the basin of its global minimum. The red dots represent the probability that the Ni core is in the basin of the 13-atom icosahedral structure. The former probability drops at significantly lower temperature (by almost 200 K) than the latter. From Ref. [68], copyright (2008) by The American Physical Society

checking whether the whole cluster, and its core separately, are in their lowest-energy configuration by making a local minimization.

The simulations show that there is a rather wide temperature interval in which the cluster is not in its global minimum anymore but its Ni core preserves the original low-temperature structure. This means that only the outer shell has changed its structure, but not the core. One may operationally define the melting temperature as the temperature at which the cluster has a probability of less than 80% of being in its global minimum. In this way, one finds that the outer shell melts at 670 K. On the other hand, by applying the same criterion to the core, one finds that the core melts at 850 K [68]. The choice of a different melting threshold than 80% would simply shift the values of these melting temperatures, preserving however a significant gap between the shell and the core melting temperatures.

5 Kinetic Effects: Growth of Gas-Phase Nanoalloys

The actual nanoalloy structures that are observed in experiments can reflect more the kinetics of their formation process that their thermodynamic equilibrium structures. For this reason, it is important to develop simulation methods that are suited to the study of growth phenomena in nanoalloys. In the following we concentrate on the formation in gas phase, briefly describing the simulation methods and showing a few relevant examples.

5.1 Growth and Coalescence Simulations

The most used simulation method for the formation of nanoparticles is MD. In fact, MD simulations can reproduce a much more realistic growth kinetics than

Monte Carlo because they solve the equations of motion. This is especially important in nanoparticles, where the variety of possible structures (including non-crystalline structures) does not allow to use the coarse-grained lattice models that are appropriate in the simulation of bulk crystal growth [72]. The main limitation of MD is the time scale that can be actually simulated. Since growth phenomena in gas phase nanoalloys of 2-3 nm size can be on the scale from a few microseconds to several milliseconds, ab-initio MD is out of question. Classical MD with atomistic potentials is much more feasible, since it can reach the scale of several microseconds [73, 74], which is not far from the relevant experimental scale.

We can distingush two kinds of MD simulations of formation processes: growth simulations, in which atoms are added one by one on a preexisting seed, and coalescence simulations [7, 16, 75], in which two preformed clusters collide and form a single aggregate [12], which subsequently undergoes rearrangement processes. Both formation processes can produce metastable structures. In several cases, these metastable structures can have sufficiently long lifetimes to be observed on experimental time scales. In the following we focus on growth simulations.

5.2 Formation of Core-Shell and Multi-Shell Nanoalloys

The growth of shells of atomic element B over a core of atomic element A has been studied by molecular dynamics simulations for Ag–Cu, Ag–Ni and Ag–Pd nanoalloys [7, 75]. In these systems, Ag atoms tend, to some degree, to segregate to the cluster surface, because of the lower surface energy of this element. Tendency to segregation is very strong for Ag–Cu and Ag–Ni, because these systems present a wide miscibility gap in the bulk alloy, whereas it is somewhat weaker in Pd, because Ag–Pd makes solid solutions in the bulk [56]. Moreover, the size mismatch effect is much stronger in Ag–Cu and Ag–Ni than in Ag–Pd.

Two types of deposition processes have been considered

- the *direct deposition*, in which element B is the surface-segregating element, i.e. Ag is deposited either on Cu or Pd cores
- the *inverse deposition*, in which element A is the surface-segregating element, i.e. Cu (or Ni or Pd) is deposited on Ag.

Depending on the geometric structure of the initial core, on the type of deposition process, and on temperature, different final structures are produced on a given growth time scale.

Let us consider first the direct deposition [75]. In this case, Ag atoms have been deposited on Cu or Pd cores at rates in the range of one atom each 2.1–7 ns. Temperatures have been chosen in the range 300–600 K. Icosahedral and truncated-octahedral cores have been considered. For all temperatures, and for both core types, well-defined Ag shells of monoatomic thickness have been obtained, as can be seen in the example reported in Fig. 13. The difference is that for Ag–Cu,

Fig. 13 Deposition of Ag atoms on an initial Cu icosahedral core of 147 atoms. In the third snapshot, an almost perfect Ag shell of monoatomic thickness is formed. This shell is an anti-Mackay shell (see the structure in Fig. 5b), which is the equilibrium structure of Ag–Cu for these sizes, compositions and temperatures [20]. This is an example of growth close to the equilibrium. From Ref. [75], copyright (2002) by The American Physical Society

the external shells with the smallest number of defects are produced at high temperatures, while for Ag–Pd the best external shells are produced at lower temperatures. This behaviour is explained by the tendency of Ag to mix with Pd, so that a higher temperature causes diffusion of Ag inside the Pd core, and a few Pd atoms appearing at the surface, resulting thus in an external shell which is no more of pure Ag. On the contrary, Ag and Cu do not show any tendency to intermixing in the temperature range of the simulations, so that high temperatures simply allow a better rearrangement of the external pure Ag shell. We note that in some cases, as the one shown in Fig. 13, the growth sequence is closely reproducing a sequence of equilibrium structures.

The case of inverse deposition leads to a wider variety of results [7], which depend on temperature and also on the geometric structure of the initial core. Icosahedral cores of 147 atoms and truncated octahedral cores of 201 atoms have been considered.

In the deposition of Ni and Cu on Ag icosahedral cores, deposited atoms readily diffuse towards the cluster centre, thus forming the same kind of core-shell structure that is found for direct deposition. The deposition of Pd on the Ag icosahedral core causes the transformation of the latter into a decahedral structure.

However, the most interesting results are found for deposition on the truncated-octahedral core (see Fig. 14) for temperatures up to about 500 K. In this case, deposited atoms quickly enter the cluster, but instead of diffusing towards the cluster centre, they stop in subsurface position, i.e. just one layer below the cluster surface. This causes the formation of three-shell structures, made of an Ag core, an intermediate layer of either Cu, Ni or Pd, and of an external Ag shell of monoatomic thickness. These three-shell structures are metastable, because they transform into core-shell structures upon annealing.

The formation of three-shell structures has been rationalized in terms of the energetics of single impurities inside truncated octahedral clusters [7]. In fact, the most favourable sites for impurities are subsurface sites. Deposited atoms are therefore likely to stop there. On the contrary, in icosahedral Ag clusters, the most favourable site for a single impurity is the central site [7, 61]. Three-shell clusters have been experimentally observed in Au-Pd clusters [76].

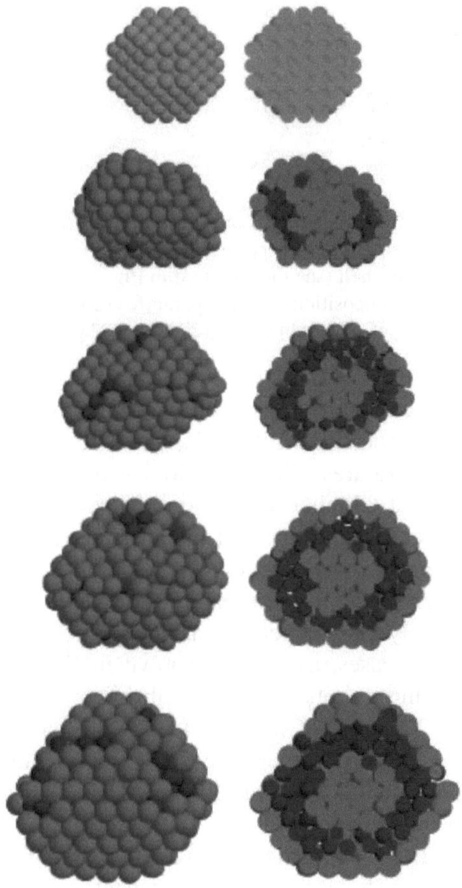

Fig. 14 Deposition of Cu atoms on an initial Ag truncated-octahedral core of 201 atoms. In the right column, cross sections of the clusters are given to show their internal structure. Three-shell onionlike clusters are formed. From Ref. [7], copyright (2003) by The American Physical Society

6 Conclusions

Nanoalloys can assume a great variety of shapes and chemical ordering patterns. Studying their complex energy landscape requires the development of *ad hoc* methods. Efficient global optimization tools have been recently developed so that the low-energy part of the landscape can be efficiently sampled for sizes up to a few hundred atoms in the framework of atomistic models. These searches can serve as a basis for further refinements by ab initio methods.

The actual structure of nanoalloys can however be different from its global minimum, because of finite-temperature effects or because of kinetic trapping phenomena. Also these phenomena can be studied profitably at the computational level, by means of Monte Carlo and molecular dynamics simulations.

However, much work is still to be done, in refining atomistic models, improving the efficiency of search algorithms and of finite-temperature sampling, in order to achieve a better matching with the complexity of real experiments.

References

1. Nanoalloys—from Theory to Applications, Faraday Discuss. **138** (2003)
2. Ferrando, R., Jellinek, J., Johnston, R.L.: Nanoalloys: from theory to applications of alloy clusters and nanoparticles. Chem. Rev. **108**, 845 (2008)
3. Ferrando, R., Fortunelli, A., Johnston, R.L.: Searching for the optimum structures of alloy nanoclusters. Phys. Chem. Chem. Phys. **10**, 640 (2008)
4. Johnston, R.L.: Evolving better nanoparticles: genetic algorithms for optimising cluster geometries. Dalton Trans. **22**, 4193 (2003)
5. Marks, L.D.: Experimental studies of small-particle structures. Rep. Prog. Phys. **57**, 603 (1994)
6. Darby, S., Mortimer-Jones, T.V., Johnston, R.L., Roberts, C.: Theoretical study of Cu–Au nanoalloy clusters using a genetic algorithm. J. Chem. Phys. **116**, 1536 (2002)
7. Baletto, F., Mottet, C., Ferrando, R.: Growth of three-shell onionlike bimetallic nanoparticles. Phys. Rev. Lett. **90**, 135504 (2003)
8. Rossi, G., Rapallo, A., Mottet, C., Fortunelli, A., Baletto, F., Ferrando, R.: Magic polyicosahedral core-shell clusters. Phys. Rev. Lett. **93**, 105503 (2004)
9. Rapallo, A., Rossi, G., Ferrando, R., Fortunelli, A., Curley, B.C., Lloyd, L.D., Tarbuck, G.M., Johnston, R.L.: Global optimization of bimetallic cluster structures. I. size-mismatched Ag–Cu, Ag–Ni, and Au-Cu systems. J. Chem. Phys. **122**, 194308 (2005)
10. Rossi, G., Rapallo, A., Ferrando, R., Fortunelli, A., Curley, B.C., Lloyd, L.D., Johnston, R.L.: Global optimization of bimetallic cluster structures. II. size-matched Ag–Pd, Ag–Au, and Pd–Pt systems. J. Chem. Phys. **122**, 194309 (2005)
11. Ferrando, R., Fortunelli, A., Rossi, G.: Quantum effects on the structure of pure and binary metallic nanoclusters. Phys. Rev. B **72**, 085449 (2005)
12. Mariscal, M.M., Dassie, S.A., Leiva, E.P.M.: Collision as a way of forming bimetallic nanoclusters of various structures and chemical compositions. J. Chem. Phys. **123**, 184505 (2005)
13. Barcaro, G., Fortunelli, A., Rossi, G., Nita, F., Ferrando, R.: Electronic and structural shell closure in AgCu and AuCu nanoclusters. J. Phys. Chem. B **110**, 23197 (2006)
14. Paz-Borbón, L.O., Johnston, R.L., Barcaro, G., Fortunelli, A.: A mixed structural motif in 34-atom Pd–Pt clusters. J. Phys. Chem. C **111**, 2936 (2007)
15. Paz-Borbón, L-O., Johnston, R.L., Barcaro, G., Fortunelli, A.: Structural motifs, mixing, and segregation effects in 38-atom binary clusters. J. Chem. Phys. **128**, 134517 (2008)
16. Toai, T.J., Rossi, G., Ferrando, R.: Global optimisation and growth simulation of AuCu clusters. Faraday Discuss. **138**, 49 (2008)
17. Calvo, F., Cottancin, E., Broyer, M.: Segregation, core alloying, and shape transitions in bimetallic nanoclusters: Monte Carlo simulations. Phys. Rev. B **77**, 121406 (2008)
18. Tran, D.T., Johnston, R.L.: Theoretical study of Cu(38-n)Au(n) clusters using a combined empirical potential-density functional approach. Phys. Chem. Chem. Phys. **11**, 10340 (2009)
19. Parsina, I., Baletto, F.: Tailoring the Structural Motif of Ag–Co Nanoalloys: Core/Shell versus Janus-like. J. Phys. Chem. C **114**, 1504 (2010)
20. Bochicchio, D., Ferrando, R.: Size-Dependent Transition to High-Symmetry Chiral Structures in AgCu, Ag–Co, Ag–Ni, and Au–Ni Nanoalloys. Nano Lett. **10**, 4211 (2010)
21. Wales, D.J.: Energy Landscapes. Cambridge University Press, Cambridge (2003)
22. Stillinger, F.H., Weber, T.A.: Hidden structure in liquids. Phys. Rev. A **25**, 978 (1982)
23. Stillinger, F.H., Weber, T.A.: Packing structures and transitions in liquids and solids. Science **225**, 983 (1984)
24. Stillinger, F.H.: Exponential multiplicity of inherent structures. Phys. Rev. E **59**, 48 (1999)
25. Rossi, G., Ferrando, R.: Searching for low-energy structures of nanoparticles: a comparison of different methods and algorithms. J. Phys. Cond. Mat. **9**, 084208 (2009)
26. Leary, R.H.: Global optimization on funneling landscapes. J. Global Opt. **18**, 367 (2000)

27. Jellinek, J., Krissinel, E.B.: Theory of Atomic and Molecular Clusters. Springer, Berlin (1999)
28. Doye, J.P.K., Miller, M.A., Wales, D.J.: The double-funnel energy landscape of the 38-atom Lennard-Jones cluster. J. Chem. Phys. **110**, 6896 (1999)
29. Wales, D.J., Scheraga, H.A.: Global optimization of clusters, crystals, and biomolecules. Science **285**, 1368 (1999)
30. Wales, D.J., Doye, J.P.K: Global optimization by basin-hopping and the lowest energy structures of Lennard-Jones clusters containing up to 110 atoms. J. Phys. Chem. A **101**, 5111 (1997)
31. Rossi, G., Ferrando, R.: Global optimization by excitable walkers. Chem. Phys. Lett. **423**, 17 (2006)
32. Deaven, D.M., Ho, K.M.: Molecular-geometry optimization by a genetic algorithm. Phys. Rev. Lett. **75**, 288 (1995)
33. Hartke, B.: Global geometry optimization of clusters using a genetic algorithm. J. Chem. Phys. **97**, 9973 (1993)
34. Hartke, B.: Global geometry optimization of clusters using a growth strategy optimized by a genetic algorithm. Chem. Phys. Lett. **240**, 560 (1995)
35. Garzon, I.L., Michaelian, K., Beltran, M.R., Posada-Amarillas, A., Ordejon, P., Artacho, E., Sanchez-Portal, D., Soler, J.M.: Lowest energy structures of gold nanoclusters. Phys. Rev. Lett. **81**, 1600 (1998)
36. Michaelian, K., Rendón, N., Garzón, I.L.: Structure and energetics of Ni, Ag, and Au nanoclusters. Phys. Rev. B. **60**, 2000 (2000)
37. Faken, D., Jónsson, H.: Systematic analysis of local atomic structure combined with 3D computer graphics. Comput. Mater. Sci. **2**, 279 (1994)
38. Hansmann, U.H.E., Wille, L.T.: Global optimization by energy landscape paving. Phys. Rev. Lett. **88**, 068105 (2002)
39. Laio, A., Parrinello, M.: Escaping free-energy minima. Proc. Natl. Acad. Sci. U S A **99**, 12562 (2002)
40. Goedecker, S.: Minima hopping: An efficient search method for the global minimum of the potential energy surface of complex molecular systems. J. Chem. Phys. **120**, 9911 (2004)
41. Sicher, M., Mohr, S., Goedecker, S.: Efficient moves for global geometry optimization methods and their application to binary systems. J. Chem. Phys. **134**, 044106 (2011)
42. Perdew, J.P., Burke, K., Ernzerhof, M.: Generalized gradient approximation made simple. Phys. Rev. Lett. **77**, 3865 (1996)
43. Gaudry, M., Cottancin, E., Pellarin, M., Lermé, J., Arnaud, L., Huntzinger, J.R., Vialle, J.M., Broyer, M., Rousset, J.L., Treilleux, M., Mélinon, P.: Size and composition dependence in the optical properties of mixed (transition metal/noble metal) embedded clusters. Phys. Rev. B **67**, 155409 (2003)
44. Langlois, C.T., Oikawa, T., Bayle-Guillemaud, P., Ricolleau, C.: Energy-filtered electron microscopy for imaging core-shell nanostructures. J. Nanopart. Res. **10**, 997 (2008)
45. Langlois, C., Alloyeau, D., Bouar, Y.L., Loiseau, A., Oikawa, T., Mottet, C., Ricolleau, C.: Growth and structural properties of Ag–Cu and Pt–Co bimetallic nanoparticles. Faraday Discuss. **138**, 375 (2008)
46. Lequien, F., Creuze, J., Berthier, F., Braems, I., Legrand, B.: Superficial segregation, wetting, and dynamical equilibrium in bimetallic clusters: a Monte Carlo study. Phys. Rev. B **78**, 075414 (2008)
47. Delfour, L., Creuze, J., Legrand, B.: Exotic behavior of the outer shell of bimetallic nanoalloys. Phys. Rev. Lett. **103**, 205701 (2009)
48. Brack, M.: The physics of simple metal-clusters—Self-consistent jellium model and semiclassical approaches. Rev. Mod. Phys. **65**, 677 (1993)
49. Mackay, A.L.: A dense non-crystallographic packing of equal spheres. Acta Crystallogr. **15**, 916 (1962)

50. Baletto, F., Ferrando, R.: Structural properties of nanoclusters: energetic, thermodynamic, and kinetic effects. Rev. Mod. Phys. **77**, 371 (2005)
51. Harris, I.A., Kidwell, L.S., Northby, J.A.: Icosahedral structure of large charged argon clusters. Phys. Rev. Lett. **53**, 2390 (1984)
52. Penuelas, J., Andreazza, P., Andreazza-Vignolle, C., Tolentino, H.C.N., Santis, M.D., Mottet, C.: Controlling structure and morphology of Pt–Co nanoparticles through dynamical or static coalescence effects. Phys. Rev. Lett. **100**, 115502 (2008)
53. Alloyeau, D., Ricolleau, C., Mottet, C., Oikawa, T., Langlois, C., Bouar, Y.L., Braidy, N., Loiseau, A.: Size and shape effects on the order-disorder phase transition in Pt–Co nanoparticles. Nature Mater. **8**, 940 (2009)
54. Demortiere, A., Petit, C.: First synthesis by liquid-liquid phase transfer of magnetic Co(x)Pt(100-x) nanoalloys. Langmuir **24**, 2792 (2007)
55. Alloyeau, D., Prévot, C., LeBouar, Y., Oikawa, T., Langlois, C., Loiseau, A., Ricolleau, C.: Ostwald Ripening in Nanoalloys: When Thermodynamics Drives a Size-Dependent Particle Composition. Phys. Rev. Lett. **105**, 255901 (2010)
56. Hultgren, R., Desai, P.D., Hawkins, D.T., Gleiser, M., Kelley, K.K.: Values of the thermodynamic properties of binary alloys. Jossey-Bass Publishers, Berkeley (1981)
57. Rossi, G., Ferrando, R., Mottet, C.: Structure and chemical ordering in Pt–Co nanoalloys. Faraday Discuss. **138**, 193 (2008)
58. Barcaro, G., Ferrando, R., Fortunelli, A., Rossi, G.: Exotic Supported Pt–Co Nanostructures: from Clusters to Wires. J. Phys. Chem. Lett. **1**, 111 (2010)
59. Frank, F.C., Kasper, J.S.: Complex alloy structures regarded as sphere packings. I. definitions and basic principles. Acta Crystallogr. **11**, 184 (1958)
60. Khan, N.A., Uhl, A., Shaikhutdinov, S., Freund, H.J.: Alumina supported model Pd-Ag catalysts: a combined STM, XPS, TPD and IRAS study. Surf. Sci. **600**, 1849 (2006)
61. Negreiros, F.R., Barcaro, G., Kuntová, Z., Rossi, G., Ferrando, R., Fortunelli, A.: Structures of gas-phase Ag–Pd nanoclusters: a computational study. J. Chem. Phys. **132**, 234703 (2010)
62. Kim, H.Y., Kim, H.G., Tyu, J.H., Lee, H.M.: Preferential segregation of Pd atoms in the Ag–Pd bimetallic cluster: density functional theory and molecular dynamics simulation. Phys. Rev. B **75**, 212105 (2007)
63. Frenkel, D., Smit, B.: Understanding Molecular Simulation. Academic Press, San Diego (1996)
64. Aguado, A., López, J.M.: Melting-like transition in a ternary alkali nanoalloy: $Li_{13}Na_{30}Cs_{12}$. J. Chem. Theory Comput. **1**, 299 (2005)
65. Aguado, A., López, J.M.: Molecular dynamics simulations of the meltinglike transition in $Li_{13}Na_{42}$ and $Na_{13}Cs_{42}$ clusters. Phys. Rev. B **71**, 075415 (2005)
66. Huang, S.P., Balbuena, P.B.: Melting of bimetallic Cu–Ni nanoclusters. J. Phys. Chem. B **106**, 7225 (2002)
67. Sankaranarayanan, S.K.R.S., Bhethanabotla, V.R., Joseph, B.: Molecular dynamics simulation study of the melting of Pd–Pt nanoclusters. Phys. Rev. B **71**, 195415 (2005)
68. Kuntová, Z., Rossi, G., Ferrando, R.: Melting of core-shell Ag–Ni and Ag–Co nanoclusters studied via molecular dynamics simulations. Phys. Rev. B **77**, 205431 (2008)
69. van Hoof, T., Hou, M.: Structural and thermodynamic properties of Ag–Co nanoclusters. Phys. Rev. B **72**, 115434 (2005)
70. Calvo, F.: Solid-solution precursor to melting in onion-ring Pd–Pt nanoclusters: a case of second-order-like phase change? Faraday Discuss. **138**, 75 (2008)
71. Mottet, C., Rossi, G., Baletto, F., Ferrando, R.: Single impurity effect on the melting of nanoclusters. Phys. Rev. Lett. **95**, 035501 (2005)
72. Newman, M.E.J., Barkema, G.T.: Monte Carlo Methods in Statistical Physics.. Clarendon Press, Oxford (1999)
73. Baletto, F., Mottet, C., Ferrando, R.: Reentrant morphology transition in the growth of free silver nanoclusters. Phys. Rev. Lett. **84**, 5544 (2000)

74. Baletto, F., Mottet, C., Ferrando, R.: Microscopic mechanisms of the growth of metastable silver icosahedra. Phys. Rev. B **63**, 155408 (2001)
75. Baletto, F., Mottet, C., Ferrando, R.: Growth simulations of silver shells on copper and palladium nanoclusters. Phys. Rev. B **66**, 155420 (2002)
76. Ferrer, D., Torres-Castro, A., Gao, X., Sepulveda-Guzman, S., Ortiz-Mendez, U., Jose-Yacaman, M.: Three-layer core/shell structure in Au–Pd bimetallic nanoparticles. Nano Letters **7**, 1701 (2007)

Magnetism of Low-Dimension Alloys

Véronique Pierron-Bohnes, Alexandre Tamion, Florent Tournus and Véronique Dupuis

Abstract The magnetic properties of metals are very sensitive to the size of the objects, to their organization when non-isolated, as well as to their chemical content and order when alloyed. In this chapter, first we go through the different magnetic energies in competition and the different length scales of importance in the magnetic behaviour and configurations of objects. Different magnetic properties of materials are then described: magnetic ordering and Curie temperature, magnetization reversal and hysteresis loops, the different contributions of the magnetic anisotropy, the thermal stability and superparamagnetism, and finally the effects of nanometric size and of alloying and chemical order on magnetic properties… We thus describe different magnetic properties in pure metals, as well as their modifications due to the effects of surface vicinity, size reduction, interface presence, and alloying. Finally the most classical experimental means will be scanned to allow the reader an enlightened choice when confronted with the magnetism of nanoalloys. This subject is very wide and several books have been recently written on the magnetism of nano-objects [1, 2], the scope of this chapter is thus in no case to be exhaustive, but to give to the reader a first idea of the effects of size and alloying on magnetism of nanoalloys. The results in nanoalloys are today quite scarce because the subject is new and very difficult.

V. Pierron-Bohnes (✉)
IPCMS, CNRS-UDS, 23 rue du Loess, BP 43 67034 Strasbourg Cdx 2, France
e-mail: vero@ipcms.u-strasbg.fr

A. Tamion · F. Tournus · V. Dupuis
LPMCN UMR 5586, CNRS, Univ. Lyon 1, 6 rue Ada Byron,
69622 Villeurbanne Cedex, France
e-mail: Veronique.Dupuis@univ-lyon1.fr

D. Alloyeau et al. (eds.), *Nanoalloys*, Engineering Materials,
DOI: 10.1007/978-1-4471-4014-6_9, © Springer-Verlag London 2012

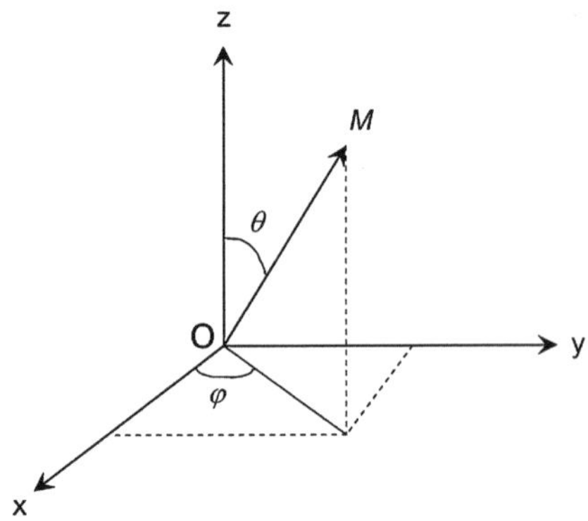

Fig. 1 θ represents the angle between the magnetization and the z axis and φ the angle between the magnetization projection in the xOy plane and the x axis

Definition and Units of Magnetic Quantities

In this prologue we define the magnetic quantities, their units and the usual axis system (Fig. 1) for fine particles used in this chapter.

Quantity	SI unit	Definition
H	$A.m^{-1}$	Magnetic field
M	$A.m^{-1}$	Magnetization
$\boldsymbol{\mu}$	$A.m^2$	Magnetic moment with: $\boldsymbol{\mu} = V\,\mathbf{M}$
K_n	$J.m^{-3}$	Magnetic anisotropy constant
E	J	Energy
F	$J.m^{-3}$	Energy density
T	K	Temperature
χ	none	Magnetic susceptibility

1 The Different Contributions to the Magnetic Energy: Zeeman, Exchange, Anisotropy, Dipolar Energies

In a magnetic metal, the magnetic configuration is the result of the competition of four energies. In this paragraph, we recall their expressions at the atomic level and at the "micromagnetic" (about 10 nm) scale.

The Zeeman energy is due to the interaction of the atomic moments $\boldsymbol{\mu}_j$ or the local magnetization **M**, sum of magnetic moments per volume unit, with the external field **H**. It writes: $E_{zee} = -\mu_0 \sum_j \boldsymbol{\mu}_j.\mathbf{H}$ at the atomic scale and the energy density is $F_{zee} = -\mu_0\mathbf{M}.\mathbf{H}$ at the mesoscopic scale. This energy density is minimized when the moment or the magnetization is aligned with the external magnetic field. The order of magnitude of this energy density depends on the magnetization and the external field: in MJ/Tm^3 unit $\mu_0 M \approx 1.3$ in Co, 1.75 in Fe, 0.53 in CoPt, 0.70 in FePd and 0.77 in FePt.

When the atomic moments are close enough to interact (overlap of the electronic orbitals), the atomic moments are coupled. The simplest way to describe this coupling is to write the exchange energy as the sum of the exchange energies between the atomic moments ($\boldsymbol{\mu}_i$) of nearest neighbour pairs (Heisenberg model):

$$E_{exc} = -\frac{1}{2}\sum_{i,j\ nn\ pairs} J_{ij}\,\boldsymbol{\mu}_i \cdot \boldsymbol{\mu}_j. \quad (1)$$

J_{ij} are the exchange coefficients, positive for a ferromagnetic coupling and negative, at least for some *i,j* couples, for an antiferromagnetic coupling. In the micromagnetic description, it can be written as: $F_{exch} = A\nabla^2\mathbf{M}(\mathbf{r}) + Cst$, where $\mathbf{M}(\mathbf{r})$ is the magnetization at $\mathbf{r}$ point.[1] This energy prevents the magnetization to vary rapidly inside a ferromagnetic material. It is minimal when all moments are parallel within the sample if $J_{ij} > 0$ for nearest neighbours. The order of magnitude of this energy density is 10^{+3} MJ/m^3, equivalent to about 1,000 K (the order of magnitude of the Curie temperatures).

The third energy is the magnetic anisotropy energy.[2] It contains different contributions. The first contribution is the magneto-crystalline anisotropy energy. It is due to the spin orbit coupling, which is present on all magnetic atoms, but is large mainly in *4d/5d* transition metals and rare earths. The electronic orbital moments depend on the lattice symmetry due to the overlap of neighbour-atom orbitals. For *3d* elements the orbital moment is small in condensed matter because the crystal field is non spherical and the magnetization is mainly carried by the spin moments. However thanks to the spin–orbit coupling, the total moment (orbital + spin) will present easy and difficult directions. The symmetry of the lattice fixes the number of easy directions. For example, in hexagonal compact cobalt, the *c*-axis is the only easy direction and the symmetry is uniaxial with noticeable anisotropy energy density (0.45 MJ/m^3), whereas in the body-centered cubic structure, iron presents a small anisotropy energy density (0.04 MJ/m^3) and the symmetry is cubic (<100> easy axes) [3]. The anisotropy energy has other terms due to any symmetry breaking around the atoms; surface, interface, and strain contributions are the most commonly observed.

The simplest expression of a uniaxial anisotropy energy is:

$$E_{mca} = -k_U{\cdot}\mu_0 \sum_j (\boldsymbol{\mu}_j \cdot \mathbf{u}_j)^2 \quad (2)$$

where $\mathbf{u}_i$ is the unit vector of the easy direction and k_U an anisotropy atomic constant without units (containing the different terms: magneto-crystalline, surface, interface, strain…). At the mesoscopic scale, F_{mca} writes: $F_{mca}/V =$

---

[1] If $\boldsymbol{\mu}_i = \bar{\mu} - \Delta\boldsymbol{\mu}$ and $\boldsymbol{\mu}_j = \bar{\mu} + \Delta\boldsymbol{\mu}$, we get $\boldsymbol{\mu}_i \cdot \boldsymbol{\mu}_j = \bar{\mu}^2 - \Delta\mu^2 = \bar{\mu}^2 - 4\nabla^2\mu/d^2$, with d the distance between i and j sites. Thus $A = bJ\mu^2a^2$ with b a factor depending on the lattice type, a the lattice parameter, and J the nearest neighbor exchange constant assumed to be identical for all neighbor pairs.

[2] A system presents anisotropy if it is directionally dependent, as opposed to isotropy, which implies identical properties in all directions.

$-k_U \cos^2(\theta_{MK})$ where θ_{MK} is the angle between the magnetization and the anisotropy axis (Fig. 1). This energy is minimal when the magnetization is parallel to an easy axis; there are at least 2 solutions (both directions). The order of magnitude of this energy density ranges between 10^{+3} and 10^{+7} J/m^3 *i.e.* 1 mK and 10 K. A field of more than 10 T is thus needed to saturate the most anisotropic systems (as the $L1_0$ ordered[3] alloys FePt and CoPt) in their difficult magnetization directions.

The last and most complicated energy is the magneto-static interaction or dipolar energy [4]. It is due to the interaction of the local magnetization with the field created by the other parts of the sample. It writes:

$$E_{dip} = -\frac{\mu_0}{4\pi}\sum_{i,j}\frac{\left[3\left(\boldsymbol{\mu}_i \cdot \mathbf{u}_{ij}\right)\mathbf{u}_{ij} - \boldsymbol{\mu}_i\right]}{r_{ij}^3} \cdot \boldsymbol{\mu}_j \tag{3}$$

where $\mathbf{r}_{ij}$ is the vector linking i and j sites and $\mathbf{u}_{ij} = \mathbf{r}_{ij}/r_{ij}$. At the macroscopic scale, the corresponding energy density can be written as due to a local field: $F_{dip} = \frac{\mu_0}{2} M H_d$ where H_d is the so-called demagnetizing field. It can be calculated from the magnetization distribution using for example the Biot-Savart law, and mainly originates from $\mathbf{M} \times \mathbf{n}$ at surfaces ($\mathbf{n}$ is the unit vector normal to the surface) and $\nabla \times \mathbf{M}$ magnetization in the bulk. It is thus strongly related to the sample shape. The order of magnitude of this energy density is $F_{dip} = ½\ \mu_0 M^2 = 10^{+6}$ J/m^3 equivalent to about 1 K.

2 Magnetization Configurations

The magnetic configuration of an object is determined by the minimization of the total magnetic energy, the complexity of which induces the existence of many different micromagnetic[4] configurations. Different length scales can be defined to take into account the competition of the different energy contributions. In this paragraph, we will only consider systems in a zero external magnetic field. The state will be thus the virgin state which has the lowest total energy. We will not consider remanent states, which strongly depend on the magnetic history of the sample.

For the smallest size of anisotropic nanoparticles (NPs), the magnetic state is a single domain (Fig. 2a): the magnetic moments are parallel in the whole particle which behaves as a macrospin. Simplest models describing particles assemblies (as Stoner–Wohlfarth model, see Sect. 4.2) consider that the individual particles behave like macrospins.

---

[3] $L1_0$ is a AB superstructure formed from the fcc lattice where one (001) plane over 2 is occupied by pure A and the other plane by pure B.

[4] This term is historical and denotes the magnetic configurations below the micron scale, mesomagnetic or nanomagnetic would be more appropriate in the case of nano-objects.

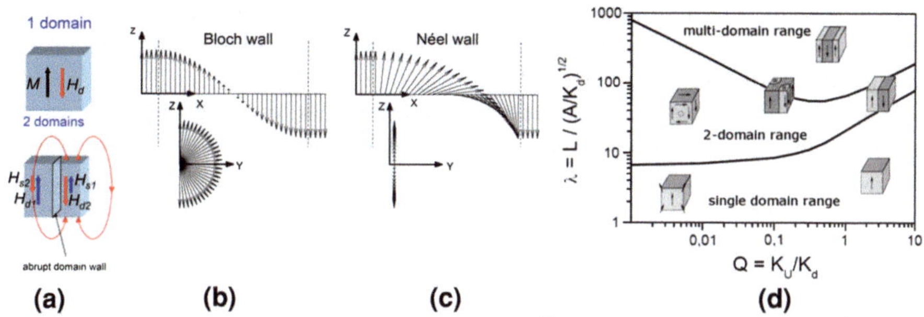

Fig. 2 **a** *Up* single domain object: magnetization $\vec{M}$ and demagnetizing field $\vec{H}_d$. *Down* 2-domain object: idem and self-field. **b**, **c** Bloch and Néel walls seen from side and face: x direction is perpendicular to the wall. **d** Stability domains in cubes of side L. For high anisotropy, domain walls are stable whereas for soft materials other configurations (vortex, flower…) appear. Adapted from Refs. [5, 6]

When the size increases, the dipolar energy becomes high compared to the exchange energy and the anisotropy energy and it becomes more favourable to form a domain wall. The magnetization turns progressively (Fig. 2) with the rotation axis either in the wall plane (Bloch wall) or perpendicular to it (Néel wall). The energy per wall area unit is $\varepsilon_{DW} = \pi\sqrt{A \cdot K_U}$.

Two characteristic lengths can be defined. The first one is $\delta_0 = \pi(A/K_U)^{1/2}$, named the width of domain wall,[5] and reflects the competition between the exchange interactions and the magnetic anisotropy. The typical value of δ_0 is a few hundreds nm in soft materials (small K_U) and a few nm in hard materials (large K_U).

The second one, $l_{exch} = (A/\mu_0\ M_S^2)^{1/2}$, is the exchange length on which the exchange interactions between atomic moments dominate the effects of demagnetizing field. One can also define the quality factor $Q = (K_U/\mu_0\ M_S^2)^{1/2}$ as the competition between the anisotropy and the demagnetizing field (for hard magnets: $Q \gg 1$, while for soft magnets: $Q \ll 1$).

For cubic (quasi-spherical) particles, their L size (R radius) has to be compared to $d_{sd} = 2\pi Q l_{exch}$ ($R_0 = 36\ Q\ l_{exch}$). Below this limit the particle is single domain, otherwise other configurations occur: vortex, flower states (Fig. 2d)… The energies of the different configurations have been calculated in 1946 by Kittel [3] and the stable configurations have been observed experimentally in films, discs [7–11], wires [12, 13] and nanoparticles [5].

For quasi-spherical particles, their R radius has also to be compared to $R_{coh} = (24A/\mu_0\ M_S^2)^{1/2}$. If $R < R_{coh}$, there is a coherent magnetic moment reversal, by keeping all the moments parallel to each other. Otherwise, "curling" effects can be observed for larger particles where $R > R_{coh}$, with vortex core nucleation to reverse the magnetization of the particle [14].

---

[5] If $n = \delta_{DW}/a$ is the number of atomic planes in δ_{DW}, the optimum value of n verifies: $\Delta E_{exch} = A\pi^2/2na = \Delta E_{mca} = naK_U/2$.

In alloys, the configurations of single objects are modified due to the fact that the anisotropy is changed and can be up to 10 times higher than in pure metals, but the global stability diagrams are valid.

When many objects are close to each other, an additional energy due to the magnetostatic interaction between the different objects (dipolar energy F_{dip}) changes the behaviour without and with a magnetic field (hysteresis curves). Using micromagnetic simulations, it has been shown [15] that the reversals of discs are not independent for centre-to-centre distances smaller than 3ϕ, where ϕ is the disc diameter. Yet this calculation was done in the case of vortex states which are not strongly coupled. For NP's on a surface, this corresponds to coverage smaller than typically 1/12. For some synthesis methods, such a low coverage may be difficult or impossible to obtain experimentally: in these cases, interactions between particles have to be taken into account [16–18] to understand the behaviour of NP assemblies. On the other hand, the current models for the analysis of magnetic curves are based on the hypothesis of negligible inter-particle interactions (see Sect. 8).

The magnetic configuration of a system comes from the minimization of the total energy. For a high enough external field the equilibrium state is the saturated state with the magnetization parallel to the applied field everywhere, but at lower field the configuration is more complicated due to the complex variations of the different contributions.

3 Magnetic Ordering and Curie Temperature

When the atomic moments are coupled, a long range magnetic order appears at low temperature. The simplest cases are ferromagnetic coupling (all moments are parallel within a magnetic domain below the Curie temperature) and antiferromagnetic coupling (below the Néel temperature, the neighbour moments are anti-aligned in a direction—defined by the AF vector). For example, in Cr, the AF vector is (½ 0 ½) whereas the moments in the (100) planes are ferromagnetically coupled (Fig. 12).

When dealing with the Heisenberg Hamiltonian (see Eq. (1), Sect. 1) of a ferromagnetic system using a mean field model, a long range magnetic order is found below the Curie temperature: $T_C = ZJ\mu^2$ with μ the atomic moment, J the exchange coefficient, and Z the number of nearest neighbours. All moments are aligned at $T = 0$ K and approaching the Curie temperature, the average moment per volume unit (magnetisation M) decreases and disappears at T_C due to the competition between the thermal and the exchange energies (see Fig. 3). Everything happens as if each atomic moment is within a mean field due to all other moments, which is known as the molecular field: $\mathbf{H}_{mol} = \lambda\mathbf{M}$ (λ is a constant). This field is added to the external field, if any.

The magnetization temperature dependence is given by $M = N\mu\,\mathcal{F}(\mu_0\mu H/k_BT)$, where H is the total effective field $H = H_{ext} + H_{mol}$, N the number of atomic

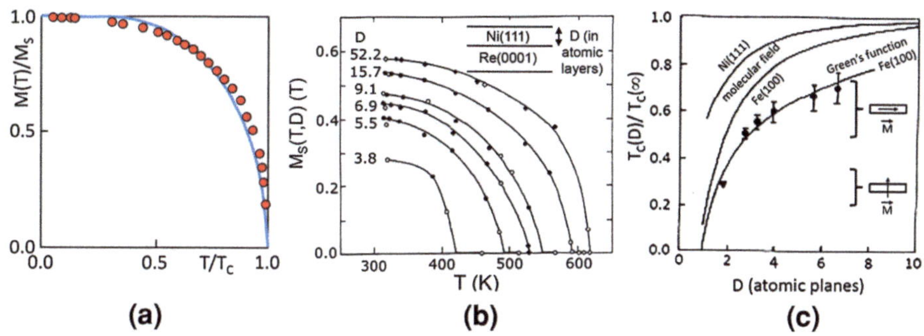

Fig. 3 **a** Variation of the normalized magnetization approaching the Curie temperature (line: Heisenberg model in mean field approximation) (results from [19, 20, 30]). **b** Temperature dependence of magnetization for thin Ni layers deposited on Re(0001) (from Ref. [21], copyright Elsevier (1993)). **c** Curie temperature variation with the number of atomic layers in $Ni_{48}Fe_{52}$/Cu(111) (from Ref. [21], copyright Elsevier (1993))

moments per volume unit, and $\mathcal{F}(x)$ is $\mathcal{L}(x)$, the Langevin function, in the classical limit (no quantization, valid for moments varying continuously):

$$\mathcal{L}(x) = (e^x + e^{-x})/(e^x - e^{-x}) - 1/x = \coth(x) - 1/x, \tag{4}$$

with $x = \mu_0 \mu H/k_B T$. In quantum physics, $\mathcal{F}(x)$ has different forms depending on the $\mathcal{L}$ moment length: the Brillouin function, $\mathcal{B}_{1/2}(x) = (e^x - e^{-x})/(e^x + e^{-x}) = \mathrm{th}(x)$, is valid for moments quantized[6] with 2 possible values ($-1/2$ and $1/2$).

When coupled, without external magnetic field, magnetic moments are in equilibrium at T in the molecular field, the magnetization is thus the solution of: $M = N\mu\ \mathcal{F}(\mu_o \mu \lambda M/k_B T)$. The $M(T)$ variation is plotted as a continuous line on Fig. 3a (classical limit).

The presence of a surface modifies the Curie temperature because the moment and the exchange coupling are different for surface atoms. For thick layers this effect is small, but in thin films and nano-objects, it could be evidenced. Figure 3b shows experimental results in thin Ni films on a non-magnetic Re substrate. In this case, the misfit between the epitaxied magnetic layer and the substrate induces a variation of strain when the thickness varies. The thickness dependence of the Curie temperature is thus amplified by this strain effect. In order to isolate the thickness variation of T_C and compare it to different calculation results (lines on Fig. 3c), Gradmann et al. [21] have deposited on Cu(111) a FeNi alloy, adjusting the composition in order to have the same lattice parameter in the layer and in the substrate. The experimental result is compared to the result with Green's function theory for Fe(100) and molecular field theory of Ni(111) and Fe(100). The former theory appears to describe very well the experimental behaviour.

[6] In quantum mechanics, the projection of the magnetic moment along the external field direction is quantized with 2I + 1 values between –I and +I where I is the norm of the moment.

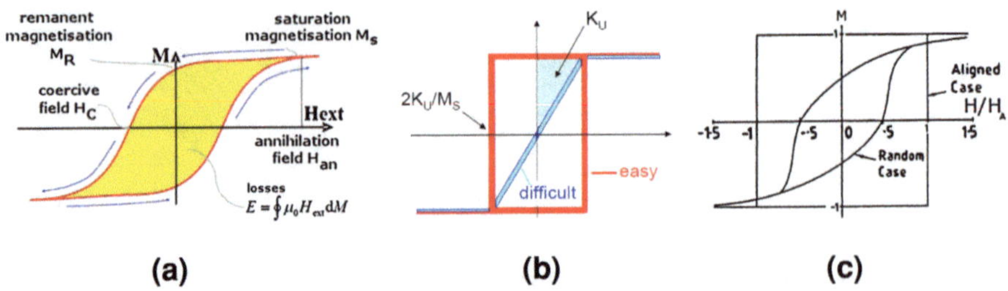

Fig. 4 General form of a magnetic loop (**a**), theoretical form along easy and difficult axis in the case of an anisotropic nano-object with the definition of the anisotropy energy (**b**) and for isolated identical particles randomly oriented or not (Stoner–Wohlfarth model) (**c**)

Due to the combination of the increase of the magnetic moments and the modification of the Curie temperature, the magnetization of a nano-object at a given temperature is difficult to predict as both effects often induce opposite tendencies.

4 Magnetization Reversal and Hysteresis Loops

4.1 Hysteresis Loops

The hysteresis loop of a system is characterized by its remanence (M_R/M_S on Fig. 4a) and its coercive field (H_C in Fig. 4a). The anisotropy of the system can be measured when comparing the hysteresis loop measured along the different directions. The easy axis presents the most square-shaped loop (largest remanence and coercive field) whereas the difficult axes may present a larger saturation field (for macrospins, the saturation field is the same in all directions). The anisotropy energy can be evaluated on single crystals from the surface between the curves along the easy axis (average field of both branches) and the difficult axis (grey/blue on line surface on Fig. 4b).

The hysteresis loops of nano-objects are generally wider than in bulk because there are no magnetic domain walls. The reversal of the magnetization cannot occur by the nucleation and growth of a reversed domain through the propagation of domain walls as in bulk samples. This induces a larger value of the coercive field: whereas in the bulk, the switching field distribution is related to the distribution of pining defects [22, 23] (impurities, inclusions, grain boundaries etc.), in single domain nanoparticles, the coercive field is linked to the activation energy for reversal of the magnetization of the whole particle. While for individual objects [24] the loop is square-shaped (100% remanence; black/red on line curve on Fig. 4b) when the external field is applied along the easy axis and the remanence is 0% along the difficult axes (grey/blue on line curve on Fig. 4b), for a NP assembly with random easy axis directions, the remanence is 50%, as calculated within the Stoner–Wohlfarth model (*cf.* Sect. 4.2, Fig. 4c). However, for an assembly of

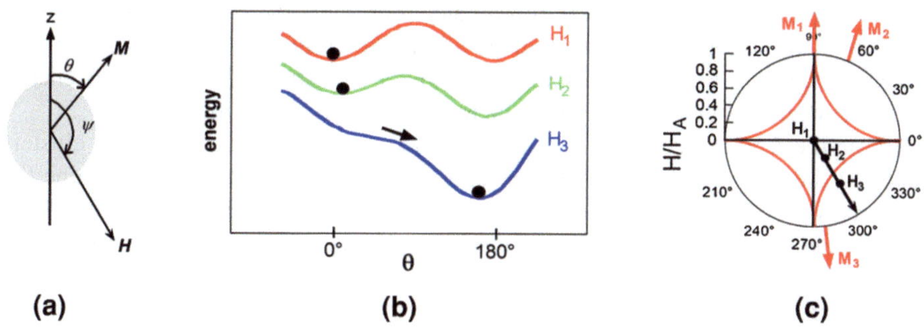

Fig. 5 Axis system (**a**), energy density (**b**) and Stoner–Wohlfarth astroid (**c**) (the magnetization switching occurs for $H_3 > H_{sw}$)

nanoparticles, the distribution of object size, morphology, orientation, atom occupation, etc. often results in a hysteresis loop having a more complicated shape.

4.2 The Stoner–Wohlfarth (S.W.) Model

The S.W. model [25] is widely used in the magnetic clusters community. Indeed this model is very simple and allows determining the expression of the switching field (H_{sw}) of a magnetic particle. The switching field represents the field to apply in order to switch the cluster magnetization from a metastable state to a more stable state. This model is based on several hypotheses:

- The cluster is single domain and the rotation of the magnetic moments is coherent.
- The cluster has a uniaxial anisotropy. The easy axis is the z axis (Fig. 5a).
- The temperature is equal to 0 K.

Due to the axial symmetry of the system, the energy density is simply expressed as: $F = K_{eff} \cdot \sin^2 \theta - \mu_0 M.H.\cos(\psi - \theta)$, sum of the anisotropy and Zeeman energy densities (Fig. 5b) where ψ represents the angle between the applied magnetic field and the easy axis (Fig. 5a). The physical meaning of the effective anisotropy constant (K_{eff}) is explained in details in Sect. 5.2.

The switching of the magnetization occurs when $\frac{\partial F}{\partial \theta} = \frac{\partial^2 F}{\partial \theta^2} = 0$, *i.e.* when the local minimum of a metastable state disappears. If $\psi = \pm\pi/2$ or $\pm\pi$ the switching field is $H_{sw} = H_A = \frac{2.K_{eff}}{\mu_0 Ms}$ also called the anisotropy field (H_A), an important magnetic quantity because it is directly proportional to the anisotropy magnetic constant. For other values of ψ a slightly more complicated calculation leads to:

$$H_{sw}(\psi) = \frac{H_A}{(\sin^{2/3}\psi + \cos^{2/3}\psi)^{3/2}} \qquad (5)$$

This equation plotted in polar coordinates represents the so-called S.W. astroid (Fig. 5c), which represents the magnetization switching field as a function of the applied magnetic field orientation. Inside the S.W. astroid there are two energy minima whereas outside the astroid there is one energy minimum.

From the S.W. model and assuming that there is no interaction between clusters and that the easy axis of each cluster is randomly oriented, we can determine the hysteresis loop of an assembly (*cf.* Fig. 4c). The coercive field is therefore $H_C \approx 0.49 H_A$ and the M_R/M_S ratio is equal to 0.5. As shown in the expression of the switching field, the S.W. astroid and the hysteresis loop do not depend on the cluster size at 0 K. More details about experimental hysteresis loops interpretations using the S.W. model are presented in Sect. 8.1.

4.3 Magnetization Dynamics and Manipulation

The goal of this section is to give a non-exhaustive example of application to study the cluster switching field. For some applications the static magnetic field used to switch the magnetization is difficult to apply. The goal is therefore to decrease the switching field using a radio-frequency (RF) field pulse (precession assisted switching) or a spin polarized current (polarized current assisted switching) or an increase of the temperature (thermally assisted switching). Here we focus on the precession assisted switching which is directly reliable to the astroid study.

In magnetic nano-clusters the problem of thermal stability of the magnetization state (superparamagnetism, see Sect. 6) can be pushed down to smaller particle sizes by increasing the magnetic anisotropy. High fields are then needed to reverse the magnetization, which are difficult to achieve in current devices. C. Thirion et al. have shown that a constant applied field, well below the switching field, combined with a radio-frequency field pulse can reverse the magnetization of a nanoparticle. The efficiency of this method has been demonstrated on a 20 nm-diameter cobalt particle by using the microSQUID technique [26]. Using the same procedure Raufast et al. have performed the switching of 3 nm-diameter cobalt cluster magnetization (Fig. 6a) [27].

In the same way switching the magnetization of a magnetic bit by injection of a spin-polarized current offers the possibility for the development of innovative high-density data storage technologies. Krause et al. [28] have shown how individual superparamagnetic iron nanoislands with typical sizes of 100 atoms can be addressed and locally switched using a magnetic scanning probe tip. They have demonstrated that current-induced magnetization reversal across a vacuum barrier combined with the ultimate resolution of spin-polarized scanning tunneling microscopy allows to separate and quantify three fundamental contributions involved in magnetization switching (*i.e.* current-induced spin torque, heating the island by the tunneling current, and Oersted field effects), thereby providing an improved understanding of the switching mechanism.

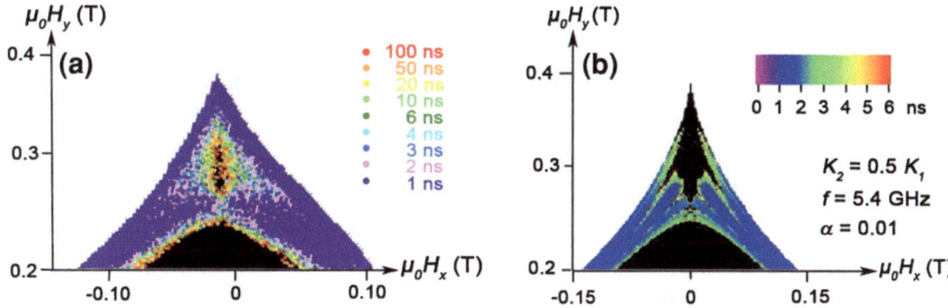

Fig. 6 **a** Switching field map of a 3 nm Co cluster using a radio frequency pulse. **b** Numerical results from integration of the Landau–Lifschitz–Gilbert equation. From Ref. [27], copyright (2008) reprinted with permission from IEEE

The dynamical switching of the magnetization can be simulated using the Landau–Lifschitz–Gilbert equation (Fig. 6b). This equation is expressed as:

$$\frac{\partial \mathbf{M}}{\partial t} = -\gamma_0 \left[\mathbf{M} \times \mathbf{H}_{eff}\right] + \frac{\alpha}{M_S}\left[\mathbf{M} \times \frac{\partial \mathbf{M}}{\partial t}\right] \tag{6}$$

where M_S represents the magnetization at saturation, H_{eff} the effective applied magnetic field, γ_0 the gyromagnetic ratio and α the dumping coefficient. Many studies, in different field of research, try to determine the dumping coefficient which is a very important term in general physics.

5 Magnetic Anisotropy Contributions

5.1 Magneto-Crystalline Uniaxial Anisotropy

The magnetocrystalline energy is related to the symmetry of the lattice structure. The simplest case is the uniaxial anisotropy. Theoretically, in the case of the hexagonal symmetry the anisotropy density energy can be written as:

$$F_{ani} = K_1 \sin^2\theta + K_2 \sin^4\theta + K_3 \sin^6\theta + K_4 \sin^6\theta \cos 6\varphi + \ldots \tag{7}$$

The angles are defined in Fig. 1. In a quadratic symmetry the anisotropy density energy can be written as:

$$F_{ani} = K_1 \sin^2\theta + K_2 \sin^4\theta + K_3 \sin^4\theta \cos 4\varphi + \ldots \tag{8}$$

In these two symmetries the order of the first term is 2 and depends only on θ the angle between the magnetization and the axis with the highest symmetry (order 6 and 4 respectively). In the majority of magnetic systems the other terms are negligible (see Fig. 9) and we express the anisotropy density energy with the first term only:

$$F_{ani} = K_1 \sin^2 \theta \tag{9}$$

Generally this term is called "magneto-crystalline uniaxial anisotropy".

5.2 Shape Anisotropy

In this paragraph we consider, as in the S.W. model, that the rotation of the magnetic moments is coherent *i.e.* the magnetization is constant. For the simplest shapes (ellipsoids, cylinders, parallelepipeds), the dipolar energy density can be written as

$$F_{dip} = \frac{\mu_0}{2}\left(\bar{\bar{N}}_d \mathbf{M}\right) \cdot \mathbf{M}$$

where $\bar{\bar{N}}_d$, the demagnetizing tensor, is related to the shape of the sample. This tensor is diagonal when expressed within the symmetry axes of the system. For example, in the case of a revolution ellipsoid the energy density can be written as:

$$F_{dip} = \frac{\mu_0.M^2}{2}(N_a \cos^2 \theta + N_b \sin^2 \theta) \tag{10}$$

with N_a and N_b the demagnetizing factors respectively parallel and perpendicular to the z axis. In an ellipsoid (Fig. 7a), a plate (Fig. 7b) or a cylinder (Fig. 7c), the shape anisotropy energy is uniaxial. If we suppose that the magneto-crystalline uniaxial anisotropy is parallel to z, we can express the total density energy as:

$$F_{tot} = F_{ani} + F_{dip} = K_{eff} \sin^2 \theta \quad \text{with } K_{eff} = K_1 + \frac{\mu_0 M^2}{2}(N_b - N_a) \tag{11}$$

where K_{eff} is the effective magnetic anisotropy constant already used in the S.W. model (see Sect. 4.2). K_{eff} is widely used in the magnetic clusters community.

5.3 Surface Anisotropy

This energy is intrinsic to the crystal but also depends on the nanometric size of the object through the elastic and surface energies, which will be described later. In alloys the magneto-crystalline energy strongly depends on alloying and on chemical order (*cf.* Sect. 7.3).

The same effect of symmetry breaking in the pair distribution (distribution of pair directions for AA, BB, and AB pairs) gives origin to the surface anisotropy. In the magnetic layers this surface (interface in the case of multilayers) anisotropy can be determined by measuring series with different thickness. Indeed the surface anisotropy can be expressed as a uniaxial anisotropy [31, 32]: $F_S = \frac{K_s}{t} \sin^2 \theta$ with t the layer thickness, which leads to a new expression of the anisotropy constant:

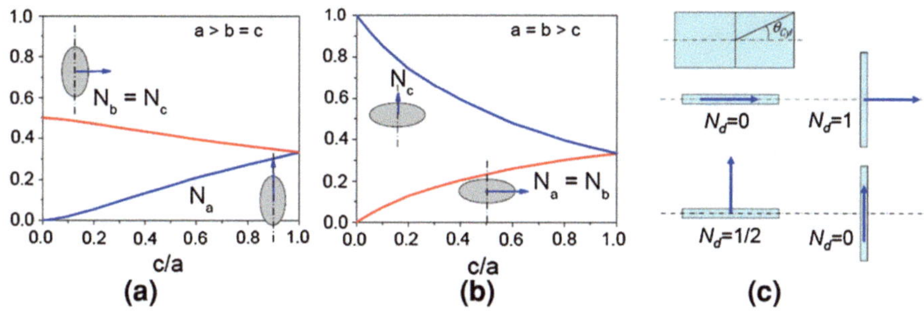

Fig. 7 Demagnetizing factor in ellipsoids (**a**) along *a, b, c* axes and in cylinders (**b**) (results from Refs. [29, 30]). In the case of thin films (**c**), F_{dip} equals $\mu_0 M^2/2$ $(N_d = 1)$ and 0 $(N_d = 0)$ respectively for perpendicular and in plane magnetization; in the case of a needle (**c**), F_{dip} equals $\mu_0 M^2/4$ $(N_d = ½)$ for *M* perpendicular to the needle and 0 for *M* parallel to it

$K = K_{eff} + K_s/t$. When plotting *K.t* as a function of *t*, the curve is a straight line with a slope equal to the sum of the magneto-crystalline volume and the shape anisotropy (K_{eff}), and crossing the ordinate axis at a value corresponding to the surface/interface anisotropy constant (K_s) (Fig. 8b). A departure from this linear behaviour occurs in the case of a biaxial stress, which gives rise to a term of the same form in $cos^2\theta$ (Fig. 8c) due to the piezomagnetic effect.

In nanoclusters the surface anisotropy can be taken into account following the same argument [36]. The anisotropy energy can therefore be written as:

$$E_{ani} = K_V.V \sin^2\theta + K_s.S \sin^2\theta \tag{12}$$

which leads to an energy barrier of:

$$\Delta E_{ani} = K_V.V + K_s.S \tag{13}$$

with *V* and *S* the volume and the surface of the clusters. The surface and interface anisotropies can be described at the first order by the empirical Néel model [37] that attributes to each pair type a contribution parameterized on Legendre's polynomials, the different parameters being determined from magnetoelastic and magnetocrytalline constants in similar systems. This model can be successfully applied to estimate the anisotropy of nano-objects and nano-alloys.

5.4 *Biaxial Anisotropy*

In nanoclusters it is often necessary to take into account a biaxial anisotropy. Jamet et al. [38, 39] have shown, using the Néel model, that in non ideal truncated octahedra there is a biaxial contribution to the anisotropy. In the case of a biaxial anisotropy the energy density is described as:

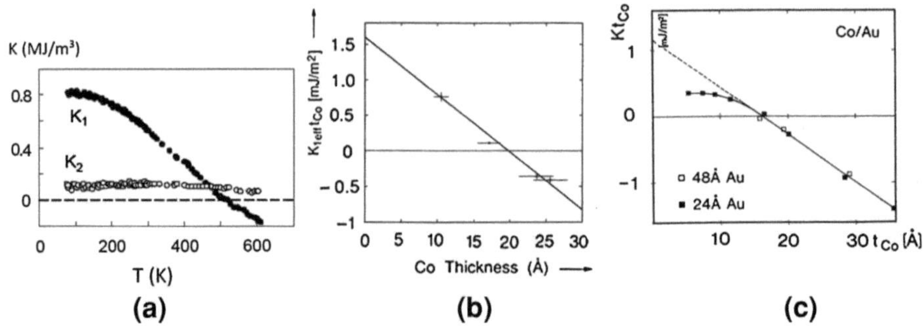

Fig. 8 **a** Temperature variation of the first two anisotropy energies in *hcp* cobalt (from Ref. [33], copyright (1984), with permission from Elsevier): At high T the basal plane is the easy axis whereas it is the 6-fold axis at low T. **b** Dependence of $K.\,t_{Co}$ on t_{Co} in Co/Pd(111) films (from Ref. [34], copyright (1991), with permission from Elsevier). At high t_{Co}, we have: $K.\,t_{Co} = t_{Co}.\,K_V + K_{s(Co)} + K_{I(Co/Pd)}$. **c** Dependence of $K.\,t_{Co}$ on t_{Co} in $(Au/Co)_N$/mica multilayers (from Ref. [35], copyright (1991), with permission from Elsevier). At high t_{Co}, we have: $K.\,t_{Co} = t_{Co}.\,K_V + (2N-1)\,K_{I(Co/Au)} + K_{I(Co/mica)}$

$$F_{ani} = K_1.m_z^2 + K_2.m_y^2 = K_1 \cos^2\theta + K_2 \sin^2\theta \cos^2\varphi \tag{14}$$

the angles are defined in Fig. 1. K_1 is the first order term and can be directly related to K_{eff}, whereas K_2 is the fourth order term. With $K_1 < 0$ and $K_2 > 0$ this energy density expression leads to an easy axis along z and a hard axis along x. Thiaville [40] has generalized the S.W. model in three dimensions using a geometrical approach. This method allows reproducing the switching field of a nanocluster in three dimensions. For example, for a Co nanoparticle (NP) analyzed by micro-SQUID (see Sect. 9.4), Jamet et al. [38] have successfully used this extended S.W. model to adjust the angular dependence of the switching field in three dimensions (Fig. 9) and to determine K_1 and K_2.

However for the majority of nanoclusters, experiments show that the simple expression of the energy density $F = K_{eff} \sin^2\theta$ is a good approximation and leads to a very nice agreement between theory and experimental data.

6 Superparamagnetism

6.1 Blocking Temperature

For a nanoparticle of volume V, considered as a uniaxial macrospin, the energy barrier that must be overcome to switch the magnetic moment is the magnetic anisotropy energy (MAE). It can be written as $\Delta E_{ani} = K_{eff}V$, where K_{eff} is the magnetic anisotropy constant described in Sect. 5. At a given temperature, and even without any applied field, a spontaneous magnetization reversal has a

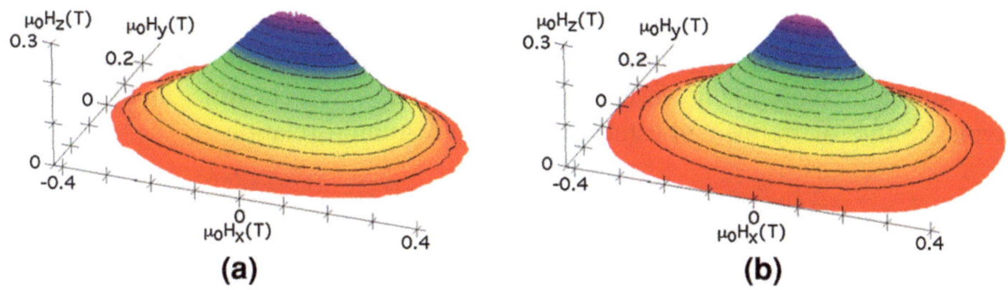

Fig. 9 Three-dimensional switching field distribution of a single cobalt cluster: experimental (**a**) and theoretical (**b**) switching field distributions considering second- and fourth-order terms in the anisotropy energy. From Ref. [38], copyright (2001) by The American Physical Society

non-zero probability to occur. As early shown by Néel [41] and further analyzed by Brown [42], the macrospin relaxation time τ between the two energy minima (two opposite orientations of the magnetic moment along the easy magnetization axis) verifies:

$$\tau = \tau_0 \exp\left(\frac{K_{eff} V}{k_B T}\right) \quad (15)$$

Equivalently, we can write for the relaxation frequency:

$$\nu = \nu_0 \exp\left(\frac{-K_{eff} V}{k_B T}\right) \quad (16)$$

where ν_0 roughly corresponds to the frequency of Larmor precession and has a typical value of several GHz. Note that the exponential term involved in the former expression implies an extremely strong variation of the relaxation frequency with either the particle volume (through the MAE) or the temperature. Since the magnetic anisotropy energy of a particle is almost proportional to its volume (K_{eff} also depends on V in very small NPs), in the case of nanoparticles the relaxation time can reach very low values, smaller than the experimental observation time, even for quite low temperatures: this leads to a regime called *superparamagnetism* [43], which is a particularity of small ferromagnetic particles. In this regime, even if they always bear a permanent magnetic moment, the particles do not display any net magnetization when no external magnetic field is applied. Particles behave as 'superatoms' similarly to the magnetic atoms of a paramagnetic material, with fast thermal fluctuations, and the magnetic response of an assembly appears to be identical to that of a paramagnet.

The temperature T_B for which the relaxation time is equal to the measurement time τ_{meas} is called the *blocking temperature* [43]. This temperature corresponds to the onset of the superparamagnetic regime. We can then write:

$$T_B = \frac{K_{eff} V}{k_B \ln(\nu_0 \tau_{meas})} \quad \text{or} \quad K_{eff} V \approx \varepsilon k_B T_B \quad \text{with } \varepsilon = \ln(\nu_0 \tau_{meas}) \quad (17)$$

It is important to note that its value depends on the type of experiment which is performed (experiments at different timescales will thus not detect the same blocking temperature). If we consider a 100 s measurement time, and given the typical value of ν_0, we can establish the rule-of-thumb relation

$$K_{eff} V \approx 25 k_B T_B (\varepsilon \approx 25). \tag{18}$$

For temperatures below T_B, the particles are in the so-called *blocked* regime, meaning that their individual magnetic moments remain in a fixed orientation (no spontaneous magnetization switching). Of course, the crossover between the blocked and superparamagnetic regimes is gradual and T_B does in fact not correspond to a sudden 'transition' between the two extreme regimes. However, given the expression of the relaxation time, its evolution with temperature is very abrupt and in a first approximation it may be possible and useful to consider that there is indeed an 'abrupt change' at T_B.

Experimentally, the superparamagnetic regime manifests itself as a vanishing coercivity (and remanent magnetic moment) for temperatures higher than T_B, while below T_B open hysteresis loops can be observed. In addition, the blocked to superparamagnetic crossover corresponds to an irreversibility of the low field susceptibility measurements as a function of temperature: this can be observed by using the so-called zero-field cooled/field cooled (ZFC/FC) protocol which is a standard way to characterize magnetic nanoparticles.

6.2 Zero-Field Cooled and Field Cooled Susceptibility Curves

ZFC/FC curves constitute a widespread experimental technique to investigate the magnetic properties of nanoparticles. They are low field susceptibility curves measured as a function of temperature, following a particular procedure:

(i) starting from a high enough temperature (room temperature in most cases) where the system is superparamagnetic, the sample is cooled down to low temperature (a few K), where particles are in the blocked regime, without applying any magnetic field,
(ii) a small measurement field is applied and the induced moment is measured while the temperature is swept up to room temperature, where the superparamagnetic regime is reached again,
(iii) the applied field is kept and the induced moment is measured while the temperature is swept down to the lowest temperature (FC_{down}) and up again (FC_{up}) with the same rate as the ZFC measurement. The final curve is made of two 'branches' corresponding respectively to the zero-field and field cooled parts, that are superimposed when all particles in the sample are in the superparamagnetic regime (the magnetic moment has then its equilibrium value). The separation between the ZFC and FC curves and the existence of a peak in the ZFC curve is a signature of superparamagnetism (but more generally it appears in the case of a slow relaxation that depends on H).

It must be noted that, owing to the superparamagnetic relaxation, the two curves (FC_{down} and FC_{up}) are not exactly identical in the temperature range where the blocking occurs. From a practical viewpoint it must be noted also that it is often difficult to obtain equal rates of warming and cooling, which define the "experimental time". Therefore, for comparison with the ZFC curve (for example to get qualitative information on the anisotropy and size distributions), it is preferable to use the FC_{up} curve, with the exactly same rate of temperature change, sign and amplitude. Actually it is a good practice to perform the two measurements of the FC curves. Indeed a significant difference between the two curves, above the blocking temperature can also indicate an experimental artifact due to time constants of the temperature control system and the sample temperature being different from the displayed sensor temperature.

Figure 10a shows the theoretical ZFC/FC curve that would be obtained for an assembly of particles (uniaxial macrospin, randomly oriented) having a single size and thus a single MAE. This curve has been calculated within a 'progressive crossover model' which has been used to establish a simple analytical expression from the particles dynamic linear response [44]. The crossover between the two regimes occurs on a very narrow range of temperature, near T_B. Note however that T_B does neither correspond exactly to the ZFC inflection point, nor to the ZFC maximum.

In the case of an assembly of particles with different sizes, and consequently different MAE, there are several T_B so that the crossover between the blocked and superparamagnetic regimes occurs at various temperatures in a given range: instead of a steep rise of the ZFC curve, we observe a peak having a width directly related to the dispersion of MAE among the particles (see Fig. 10b, c). The temperature T_{max} of the ZFC maximum cannot be called the 'blocking temperature' of the sample. T_{max} is strongly dependent on the particle size distribution, with no simple rule enabling to quantitatively infer the magnetic anisotropy constant from its value [45]. In particular, it must be kept in mind that T_{max} is not the blocking temperature of the particles having the mean or median diameter and that the rule-of-thumb $25k_BT_{max}$ method should not be used to determine a 'mean' anisotropy energy.

By analyzing ZFC/FC curves, it is possible to precisely characterize the properties of an assembly of magnetic nanoparticles, in particular the magnetic size distribution and the magnetic anisotropy constant. Fitting procedures of experimental curves will be presented in Sect. 8.2. Note that in order to reliably determine the intrinsic properties of particles from measurements on an assembly, it is necessary to ensure that interparticle interactions are negligible; otherwise the response of the system can be altered in a complicated way.

7 Magnetic Properties of Nanoalloys

The magnetic properties of metals and alloys are the consequence of several characteristics of the system: the magnetic moment, the exchange and anisotropy energies, which depend on the system size, composition and chemical order.

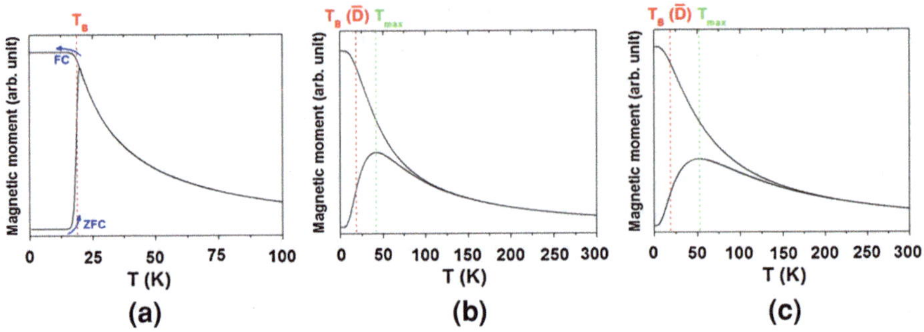

Fig. 10 **a** Theoretical ZFC/FC curves, computed with the 'progressive crossover model', for particles of the same size ($D = 4$ nm), with $K_{eff} = 200$ kJ/m^3 (results from Refs. [44, 45]). The blocking temperature T_B, which is around 19 K, is indicated in the graph. **b**, **c** Theoretical ZFC/FC curves for an assembly of particles having the same mean diameter and the same K_{eff} as in (**a**), but with a lognormal size distribution. The curves in (**b**) correspond to a dispersion parameter of the lognormal equal to 0.2, while it is 0.25 for the curves in (**c**). The temperature T_{max} of the ZFC maximum is compared to the blocking temperature of the mean particle size

In a first part, we will describe this dependence and the effect on the magnetic equilibrium state: the ordering temperature and the hysteresis loops.

7.1 Magnetic Moments

In isolated atoms, the origin of the atomic magnetic moments is the unpaired electrons. In metals, the hybridation of the orbitals induces the formation of electron bands with electron density of states $n(E)$. Electron bands spontaneously split into up and down spins even in the absence of an external magnetic field when the relative gain in exchange interaction is larger than the loss in kinetic energy: *i.e.* when the Stoner criterion $J.n(E_f) > 1$, where J is the exchange coefficient, is fulfilled where E_f is the Fermi level (the highest energy of electrons at 0 K). The net magnetic moment is then related to the difference of electron numbers in majority (up) and minority (down) bands.

When all moments are aligned in an external field, the saturation magnetization is attained. The saturation magnetization of a nano-object may be very different from that of the bulk. As a matter of facts, the atomic magnetic moment is sensitive to [19–21]:

- the local crystalline symmetry (changes the densities of states),
- strains and relaxations near a surface (via magneto-striction),
- the number of neighbours: the presence of a surface changes the magnetic moment due to the change of bond number; in metals, the conduction-band width typically varies as $N_{at}^{1/2}$ where N_{at} is the number of atoms; a surface presence induces a narrowing of the conduction bands, inducing a higher

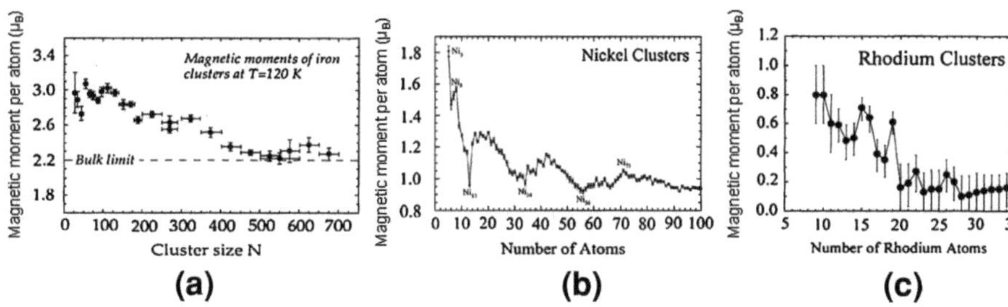

Fig. 11 Variation of magnetic moments per atom with number of atoms in cluster. **a** Fe at T = 120 K (bulk value 2.20 μ_B) (from Ref. [46], copyright (1993) by The American Physical Society). **b** Ni extrapolated to 0 K (bulk value 0.61 μ_B) (from Ref. [51], copyright (1996) by The American Physical Society). **c** Rh extrapolated to 0 K (bulk value 0 μ_B) (from Ref. [58], Copyright (1994) by The American Physical Society)

electronic density at the Fermi level, inducing a change of the band splitting and a change of the electron densities in both bands responsible of the net magnetic moment. Generally an enhancement of the magnetic moment is observed.

For example in Fe, this effect is very strong as both bands are partially filled—Fe is a weak itinerant ferromagnet. Experimental results in clusters [46, 47] and thin layers [21] show an increase of the Fe moment from 2.2 μ_B in the bulk up to 3 μ_B in clusters (Fig. 11a) with less than 100 atoms and 2.5 μ_B in a W(110)\Fe(110)\Ag thin layer. Ab initio calculations in the case of thin layers have been performed with full-potential linearized augmented-plane-wave total-energy method, first without [48] and later including spin–orbit coupling [49]. They have confirmed the experimental increase with a moment as high as 3 μ_B for a free Fe layer. For small clusters up to 6 atoms, recent calculations [50] were performed using the Density Functional Theory (DFT), finding an atomic moment up to 3.3 μ_B in the Fe_6 octahedron-like clusters. Some experimental results are also available in very small Ni clusters [51]. The effect of surface atoms could be clearly evidenced through minimum values for the most packed (minimizing surface) clusters (Fig. 11b).

The same phenomenon has been predicted in antiferromagnetic systems (Fig. 12). At the Cr surface, the Cr-moment continuously increases when approaching the surface [52] (as calculated using a Hubbard tight binding Hamiltonian and 12 levels in the continuous fraction for the *d* orbitals), independently to the alternation due to the antiferromagnetic superstructure. This could explain many experimental results obtained using different spectroscopies.

In nano-alloys, the hybridization effect between different elements is added to the different effects already described for nano-objects. The norm of the moment on one atom can depend on the occupation of the neighbour sites, either decreasing or increasing compared to the pure element. In 3*d* metals, the magnetic moment increases when the average electron number increases up to the half-full *d* band (5 electrons in the *d* band) and decreases for more than half-full *d* band (Slater plot, Fig. 13a). A typical example is FeCo alloys: the progressive filling of the *d* band of

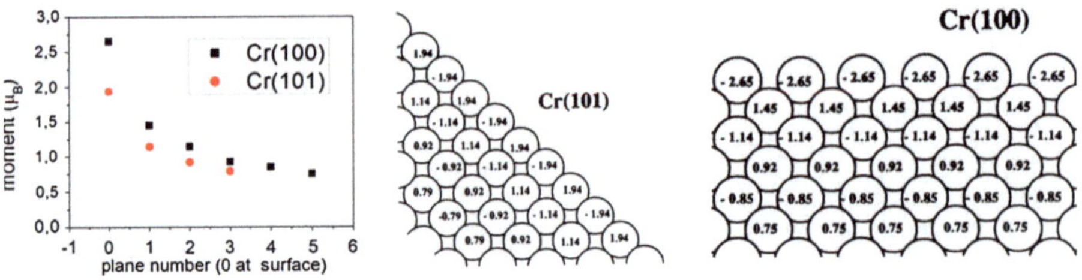

Fig. 12 Cr-moment variation as a function of its distance to the surface and moment configurations near both (101) and (001) surfaces. From Ref. [52], copyright (1994) by The American Physical Society

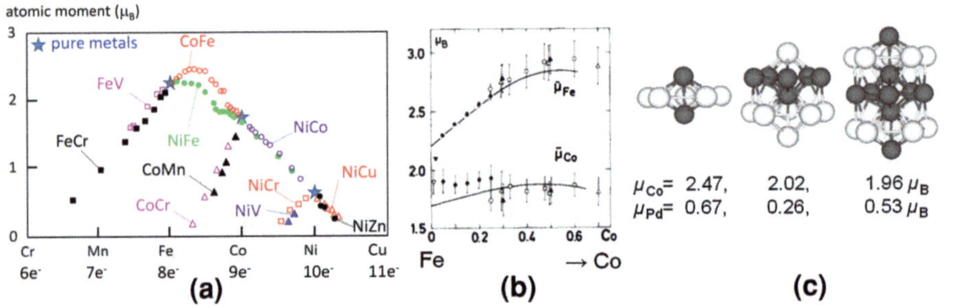

Fig. 13 **a** Average atomic moments of binary alloys of *3d* transition metals (results from Ref. [19]). **b** Variation of the atomic moments on Fe and Co in FeCo calculated in CPA (line) and measured using polarized neutron diffraction (from Ref. [54], copyright (1989), with permission from Elsevier and Ref. [55]). **c** Layer-like order for Co_2Pd_5, Co_6Pd_7 and Co_9Pd_{10} clusters (from Ref. [64], copyright (2006) by The American Physical Society)

Fe shifts the bands and increases Fe moments in the alloy, whereas Co moments remain almost constant (Fig. 13b). These results have been recently confirmed by magnetic dichro (XMCD) measurements in FeCo alloys and multilayers [53].

When an element is close to be magnetic (like Pt or Pd), magnetic moments can be induced by a surface or by alloying. In bulk alloys, by mixing a non magnetic and a magnetic metal, a magnetic moment can be induced on the non magnetic metal through the hybridization of the electronic orbitals. This leads to the very interesting properties of MP alloys with M = *3d* transition metal (as Fe, Co) and P = *4d* or *5d* metal (as Pt, Pd). P presents a strong spin–orbit coupling and zero or negligible moment as pure metal. When alloyed with M (with a large magnetic moment), the coupling between P and M moments is at the origin of the high magnetocrystalline anisotropy of these alloys, which makes their high scientific interest nowadays. The moment induced on Pt in a CoPt $L1_0$ thin film [56] has been found equal to 0.35 μ_B with an unchanged 1.76 μ_B moment on Co as determined through XMCD measurements and local spin-density approximation calculations (all-electron fully relativistic and spin-polarized full-potential muffin-tin orbital method).

Some magnetic moments can even appear when mixing two non ferromagnetic *d* metal in bulk or small clusters: for example a magnetic moment has been evidenced experimentally using XMCD [57] on V atoms, as impurities in Cu (this nevertheless does not give rise to ferromagnetism as the V moments are not enough coupled) and on Rh in clusters with less than 20 atoms (up to $0.8 \pm 0.2\ \mu_B$ measured through Stern–Gerlach experiment [58], see Fig. 11), and confirmed by calculations [59].

In nanoalloys (alloy clusters), both phenomena are combined: enhancement due to the cut bonds and change due to hybridization with other species orbitals. There are few published results. Experimentally, using XMCD in disordered Co_xPt_{100-x} NPs prepared by vapour deposition, Imperia et al. [60] found a linear relation between the Pt amount and the orbital to spin moment ratio (μ_l/μ_s) of Co. Tournus et al. [61] have shown for 3 nm CoPt nanoparticles prepared by vapour deposition, that the orbital Co moment increases in chemically ordered NPs compared to disordered NPs and bulk. Similarly, in FePt 6.3 nm size NPs, Antoniak et al. [62] observed a Fe moment of $2.48(25)\mu_B$ in disordered and $2.59(26)\mu_B$ in ordered phase, and a Pt moment ($0.41(2)\mu_B$ in both cases), whereas in bulk ordered FePt a $2.8(1)\mu_B$ moment was measured on Fe using neutron diffraction [63], in agreement with some DFT calculations ($2.9\mu_B$). The measured moment is thus equal within the error bars in NPs and bulk and slightly larger in the ordered phase.

For Co-Pd clusters, DFT calculations (Fig. 13c) have been done by Aguilera-Granja et al. [64] using the SIESTA code comparing with a combined tight binding model (Gupta potential—second moment). The moment values are strongly dependent on the symmetry and concentration of the cluster. Pd moment reaches $0.67\mu_B$ and Co moment $2.47\mu_B$ in the smallest Co_2Pd_5 cluster. Of course, the different configurations of a cluster, with the same atom numbers, give rise to different relaxations and hence influence the magnetic moment length.

7.2 Interplay Between Chemical and Magnetic Orders

The Curie temperature is also sensitive to alloying; on the one hand, because J_{ij} depends on the atoms present on *i* and *j* sites and, on the other hand, due to the change of the average moment. The Curie temperature in disordered alloys often has a linearly decreasing variation when adding a non magnetic element in a magnetic element, as long as the crystallographic structure remains the same. Less predictable variation laws can be observed, for example when mixing two itinerant magnetic elements. In FeCo, FeNi and CoNi, T_C presents a maximum in γ-Fe-Ni and varies monotonically in α-Fe-Co and γ-CoNi, whereas the average moment has a maximum in FeCo and is almost linear in FeNi and CoNi (Fig. 14). Complete electronic structure and coupling calculations are then needed to understand these features.

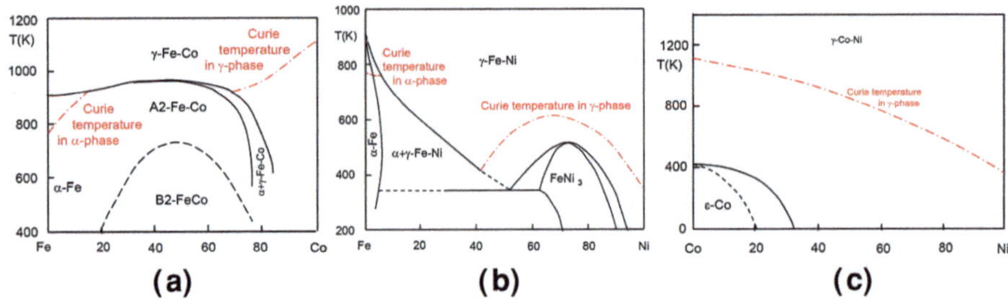

Fig. 14 Phase diagram and Curie temperature of Fe-Co, Fe–Ni and Co–Ni systems. Results from Refs. [65, 66]

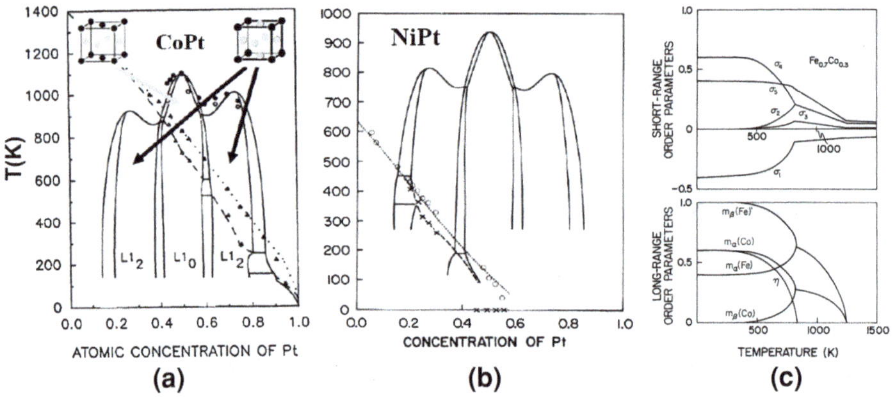

Fig. 15 **a** Co-Pt (results from Ref. [69–71]) and **b** NiPt (from Ref. [72], copyright (1985) by The American Physical Society) phase diagrams showing the strong decrease of Curie temperature in ordered $L1_0$ and $L1_2$ phases and the sensitivity of chemical long range order (LRO) to magnetic LRO. **c** Variation of the chemical and magnetic order parameters in FeCo: η is the chemical LRO parameter, σ_1 is the nearest neighbour chemical short range order parameter, m_i are the average magnetisations on the different sublattices, $\sigma_{2\text{-}5}$ are the nearest neighbour magnetic short range order for Fe-Co, Fe–Fe, Co–Co and Co-Fe ($T_{Curie} \sim 1250K$; $T_{bcc\text{-}B_2} \sim 800K$) (from Ref. [68], copyright (1985) by The American Physical Society)

In alloys, the magnetic coupling is strongly dependent on the chemical ordering (distribution of the different species on the lattice). This induces a concentration dependence of the Curie temperature in the disordered alloys and a sensitivity of the Curie temperature on the chemical order (at long range and at short range) in the ordering alloys. This effect has been experimentally observed in several systems; it is particularly strong in CoPt and NiPt alloys (Fig. 15). It can be satisfactorily described theoretically using a generalized Ising Hamiltonian containing both chemical and magnetic interactions [67, 68]. Chemical and magnetic orders are coupled because the magnetic coupling depends on the site occupation:

$$H = \frac{1}{2}\sum_{I,J,n,m} \varepsilon^{IJ}_{nm} p^I_n p^J_m - \sum_{I,n} \mu^I p^I_n + \frac{1}{2}\sum_{I,J,n,m} J^{IJ}_{nm} p^I_n p^J_m \mathbf{m}^I_n \cdot \mathbf{m}^J_m - \sum_{I,n} p^I_n \mathbf{H} \cdot \mathbf{m}^I_n \quad (19)$$

ε^{IJ}_{nm} is the chemical interaction between I and J atoms placed on n and m sites. J^{IJ}_{nm} is the exchange interaction between $\mathbf{m}^I_n$ moment (spin operator) of I atom placed on n site and $\mathbf{m}^J_m$ moment of J atom placed on m site. p^I_n is the occupation operator of I atoms on n site, μ^I is the chemical potential of I atoms, and $\mathbf{H}$ the external magnetic field.

Using this Hamiltonian, Martínez-Herrera et al. [68] have calculated the phase diagram and the temperature dependence of the different (magnetic and chemical, short- and long-range) order parameters with in a Bethe approximation (the exactly-treated maximum cluster is the pairs). Figure 15c clearly puts into evidence the strong effect of the disappearance of the long range order on the magnetic moments and that of the Curie temperature on the chemical short range order parameters.

With such a Hamiltonian extended to the 4-atom clusters, the phase diagrams can be calculated within appropriate statistical approximations to describe the configurational entropy. For example this calculation was made in the frame of the cluster variation method with the tetrahedron as exactly-treated maximum cluster in the CoPt and NiPt systems [69–71] (Fig. 15). They found that the asymmetry of the phase diagram in concentration is due to both the stronger magnetism of one component and the concentration dependence of the interaction energies. The effect of the long range order on the Curie temperature is clearly evidenced: the calculation was made in the equilibrium state (empty triangles, dashed line) and in a metastable disordered quenched state (full triangles, dotted line). The Curie temperature is larger in the disordered state because the number of magnetic nearest neighbours of a magnetic atom (Co or Ni) is larger in average in this state (9 in M_3Pt, 6 in MPt and 3 in MPt_3) than in the ordered state (8 in M_3Pt, 4 in MPt and 0 in MPt_3). Moreover, the effect of the magnetic order on the chemical order can be seen at the crossing of the Curie temperature with the two-phase boundaries.

This interplay observed in bulk materials is also present in nanoalloys and still complicates the predictions of the thermodynamic and magnetic behaviour of these systems. The size, shape and internal structure present a more or less wide distribution and their effects on the magnetism or on the order state of a nanoparticle can only be understood when taking into account all phenomena.

7.3 Magnetic Anisotropy

In 1989, Bruno [73] has shown the close connection between magnetocrystalline anisotropy (MCA) and orbital moment in itinerant ferromagnets. In an alloy, if the chemical occupation of the lattice is random, the magnetocrystalline anisotropy is small. On the contrary in chemically ordered phases, the MCA can be very high, predominant over all other contributions (shape, strain, surface, and interface…). For example in $L1_0$ anisotropic alloys as MP with M = Co, Fe and P = Pt, Pd this anisotropy is much higher than in any pure metal: 0.2 (FePd) to 10 (FePt) MJ/m^3 [74]. This strong anisotropy is due to the spin–orbit coupling on the P atoms, combined with the large moment of the M atoms coupled to the induced moment on the P atoms. This makes these systems very interesting for their magnetic properties as the magnetic moments are much more stable (against temperature or parasitic fields) than in lower anisotropy systems. Many groups have calculated and measured the MCA in bulk, film [22, 23, 56, 75–78] and NP [26, 79] anisotropic alloys. Ab initio investigation of the XMCD and the magnetic properties have been performed for example for CoPt bulk [56] and for a Co or Fe monolayer on Pt(111) [77]. The different contributions (electronic, strain, interface, surface) were shown to be of the same order of magnitude in the latter case, giving rise to an oscillation of the MCA for the first deposited monolayers.

In partially ordered alloys, either stable or metastable [80, 81], and in disordered alloys with short range order, the anisotropy can be non negligible due to the anisotropic distribution of the pairs. In MP alloys, the anisotropic distribution of MP pairs induces a magnetic anisotropy, as observed combining EXAFS and magnetic measurements [82]. These results are in good agreement with the Néel model for magnetic anisotropy, which writes the magnetocrystalline energy, E_{MCA}, of a particle as a sum of pair interactions between nearest neighbors:

$$E_{MCA} = -\sum_{I,J,n,m} L_{IJ}(\mathbf{e}_{nm} \cdot \mathbf{M}/M) p_n^I p_m^J \tag{20}$$

where $\mathbf{e}_{nm}$ is the unit vector along the nm bond (same notations as before). The prefactor L_{IJ}, called the Néel anisotropy parameter, depends on the IJ pair nature. The extension of the empirical Néel model to the case of a bimetallic alloy has been recently exposed [83, 84]. The effect of chemical ordering on the MAE of CoPt and FePt particles has been determined: the evolutions of the anisotropy energy with the long range order and short range order parameters have been computed (Fig. 16).

In nanoalloys, the superparamagnetic regime can also be studied using ZFC–FC method, but the anisotropy constant K_{eff} itself can present a distribution for many reasons [78]: variation of partial order from NP to NP, a shape distribution (inducing a distribution in the *surface* contributions), concentration fluctuations, or different chemical configurations for a disordered alloy… This distribution induces a wider ZFC peak and a more spread out junction of both curves [84–86] (Fig. 17).

In such cases, the blocking temperature distribution is not only due to the volume distribution. In the simplest cases, the size and anisotropy distributions are independent. This has been found as adequate in size selected CoPt particles [86]: the activation energy distribution due to the volume spread enlarged by a Gaussian anisotropy constant distribution allows a perfect fit of the ZFC–FC curve opposite to a constant K_{eff} (Fig. 17). There are nevertheless many cases in which the quantitative analysis of ZFC–FC curves is more difficult—if not impossible—for

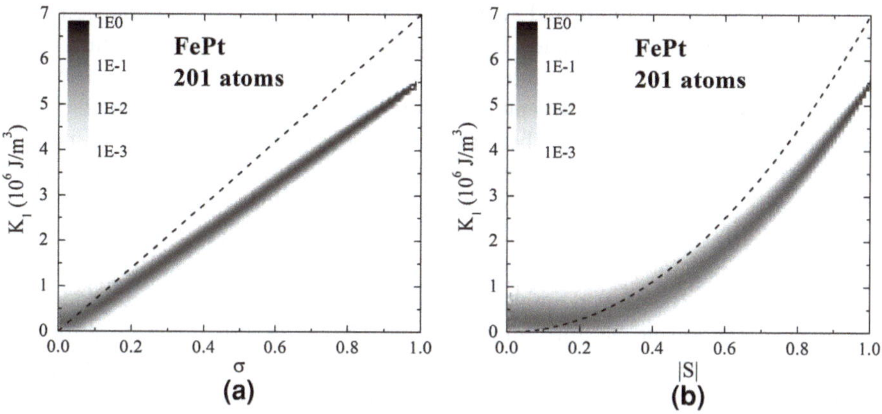

Fig. 16 Variation of the anisotropy constant K_1 versus the short range order parameter σ (**a**) or the long range order parameter S (**b**) for a 201 atoms FePt cluster. The color scale represents the density of probability for a cluster with a given chemical order parameter to have magnetic anisotropy constant K_1. The dashed line corresponds to the bulk behavior. From Ref. [83]

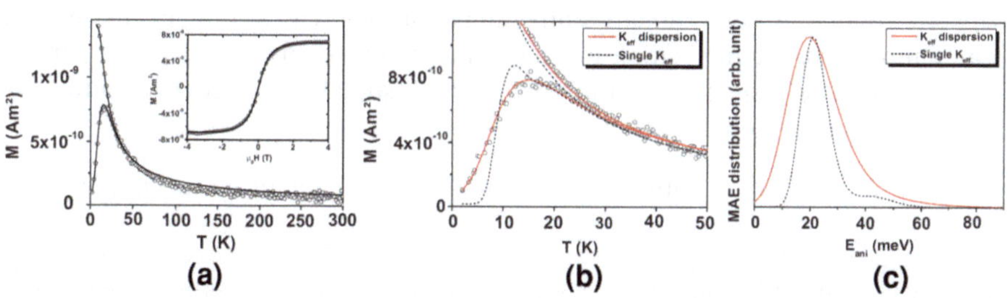

Fig. 17 **a** Experimental data points and fit for the ZFC/FC curves and superparamagnetic magnetization loop (in insert), for size-selected CoPt clusters. **b** Close-up around the ZFC peak where the fit with a K_{eff} dispersion is compared to the one with a single K_{eff} (*dashed curve*). The corresponding MAE distributions are displayed in (**c**). From Ref. [86], copyright (2010) by The American Physical Society

example if there is a correlation between the NP size and the long range order within the particle as observed in CoPt NPs [87].

8 Analysis of Magnetometry Measurements

We will give here a few indications on how experimental magnetometry measurements can be analyzed in order to characterize magnetic nanoparticles in general. The most common types of measurements will be described, without any pretention of exhaustiveness. A particular interest will be given to ZFC/FC curves analysis.

Since we are more interested in the intrinsic properties of nanoparticles, the samples are supposed to be made of diluted assemblies of nanoparticles far enough from each other for the interparticle interactions to be negligible. Such a condition may be in fact difficult to fulfil and can represent a synthesis challenge. Nevertheless, this assumption should be carefully verified (for instance by comparing measurements performed with different orientations of the applied magnetic field) before drawing any conclusion from experimental results, since interactions can strongly modify the response of a sample and make non-applicable the simple theoretical descriptions usually used. In addition, we will assume that nanoparticles in the sample behave as uniaxial macrospins, with their easy axis randomly oriented (*i.e.* there is no 'texture' in the sample). Note that if this last assumption is not fulfilled, the same kind of analysis may still be performed but by taking into account the 'texture' which can be complicated to determine.

8.1 Hysteresis Loops

A characteristic of ferromagnetic nanoparticles is that they are in the superparamagnetic regime at high enough temperature. In most cases, this means that the room temperature magnetization loops display no coercivity. For a single particle size the magnetic response is simply given by a so-called Langevin function (as Eq. (4) for atomic moments in Sect. 3):

$$m = N_{tot}\mu_P \times \mathcal{L}(x) \quad \text{with } \mathcal{L}(x) = \coth(x) - 1/x, \quad \text{with } x = \mu_0\mu_P H/k_B T, \tag{21}$$

where N_{tot} is the total number of particles and μ_P the magnetic moment of a particle, which can be written as $\mu_P = M_S V$ for a particle of volume V and saturation magnetization M_S.

In the case of a particle size distribution, $f(V)dV$ corresponds to the fraction of particles having a volume in the interval $[V, V + dV]$, and the total magnetic moment can be expressed as an integral of Langevin functions:

$$m(H) = N_{tot} M_S \int_0^\infty \left[\coth\left(\frac{\mu_0 M_S V H}{k_B T}\right) - \frac{k_B T}{\mu_0 M_S V H}\right] V f(V) dV \tag{22}$$

This equation can be used to fit experimental curves, in order to determine the magnetic size distribution (which can be different from the geometric size) or simply to assess the effect of the different parameters. However, it should be noted that it is quite easy to find a set of parameters providing a good agreement with an experimental curve: this fitting procedure alone is not very efficient for the discrimination of slightly different size distributions [89]. A correct agreement is a necessary condition but no a sufficient one to prove the validity of a given size distribution.

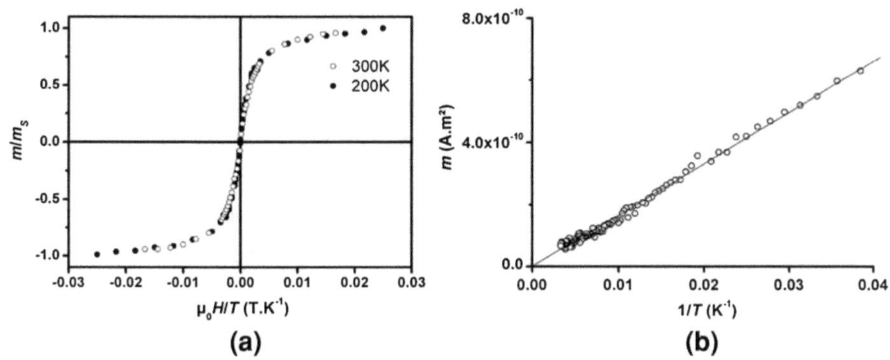

Fig. 18 **a** Magnetization loops for 3.1 nm CoPt particles at 300 and 200 K, plotted as a function of $\mu_0 H/T$. **b** Superparamagnetic part of the ZFC curve for 3.1 nm CoPt particles, plotted as a function of $1/T$. The *straight line* is a guide to the eyes. From Ref. [88], copyright (2011), with permission from Elsevier

As it can be seen from the theoretical expression $m(H)$ above, the magnetic moment appears to be a function of the parameter H/T. Therefore, as long as the system is in the superparamagnetic regime, all magnetization loops should give the same curve when plotted as a function of H/T (see Fig. 18). By checking if this property is satisfied one can detect if the anisotropy in fact still plays a role (*i.e.* the temperature is too low) or if there are interparticle interactions. Note that, in the regime where the magnetization loops show no coercivity but where the anisotropy still plays a role (typically at a temperature not too far from the T_{max} of the ZFC curve), a numerical analysis is still possible [90] but more delicate.

It is also standard to measure magnetization loops at low temperature (the lowest temperature possible, usually 2 K with commercial SQUID magnetometers), where the hysteresis loop displays a coercivity and a remanent magnetic moment. As explained before, within the Stoner–Wohlfarth model (see Sect. 4.2), the coercive field at 0 K is directly related to the anisotropy field H_A and is thus independent of the particle size (if K_{eff} and M_S are independent of the particle size). However, when the temperature increases, the coercivity decreases and the curves are modified differently for each cluster size: small particles may even reach the superparamagnetic regime (*i.e.* have a vanishing coercivity) while the largest particles may be almost unaffected by thermal agitation. For an assembly of size-distributed particles characterized at a non-zero temperature, the resulting hysteresis loop, and in particular the value of the coercive field, will then strongly depend on the size distribution. There is unfortunately no simple way to describe the curves and a fit is hardly possible.

Should the temperature be small enough so that it is reasonable to consider that the magnetic response is almost identical to that at zero temperature, it is then possible to numerically fit a hysteresis loop with a theoretical Stoner–Wohlfarth curve (see Fig. 19). The validity of the independent uniaxial macrospin

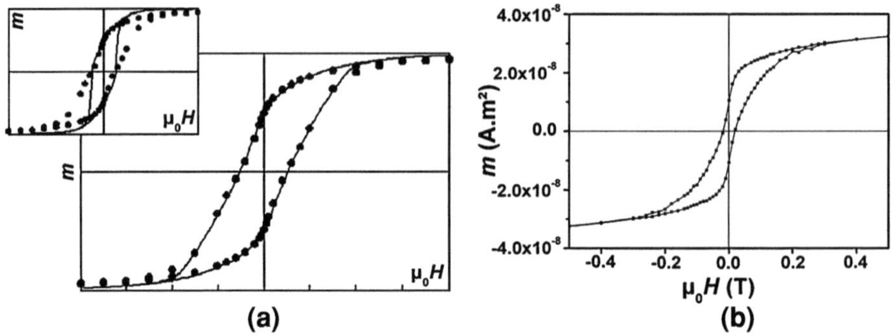

Fig. 19 **a** Low temperature (6 K) hysteresis loop of 4.5 nm Co particles embedded in Al_2O_3. The experimental curve (*dots*) is fitted using a Stoner–Wohlfarth loop with a distribution of anisotropy constants. In inset is shown the theoretical Stoner–Wohlfarth loop corresponding to a single magnetic anisotropy constant (from Ref. [94], copyright (2007), with permission from Elsevier). **b** Hysteresis loop at 16 K of 3.8 nm CoPt clusters embedded in amorphous carbon. The shape of the curve is typical of the superposition of blocked and superparamagnetic particles (adapted from Ref. [89])

approximation can be verified and the anisotropy constant (or anisotropy constant distribution) can subsequently be deduced from the measured H_{ani} (if M_S is known).

In the other cases, the only quantitative conclusion that can easily be drawn from the experimental hysteresis loop is a lower boundary for the value of the anisotropy field (and consequently K_{eff}): the coercive field H_C at low temperature is indeed always smaller than that at 0 K. The evolution of H_C with the temperature is sometimes used to extrapolate its theoretical value at 0 K and to determine the magnetic anisotropy constant (through the determined value of the blocking temperature T_B) from the analytical power law

$$H_C(T) = H_C(0)[1 - (T/T_B)^\alpha]. \tag{23}$$

This type of law can indeed be established in the case of an assembly of particles with a single size where we have $\alpha = 0.5$, for particle with their easy axis along the field direction, and $\alpha \sim 0.75$ for randomly oriented anisotropy axes [91–93]. However, such an analysis may prove highly unreliable because it does not take into account the particle size distribution, which deeply affect the value of $H_C(T)$.

Note also that hysteresis loops may present a characteristic shape with a 'wasp waist' (see right panel of Fig. 19), which is due to the coexistence of blocked and superparamagnetic particles in the sample: the curve is then the sum of an open magnetization loop and a S-shaped superparamagnetic curve. Such a hysteresis loop may sometimes be fitted using the combination of a 0 K Stoner–Wohlfarth loop and a superparamagnetic curve (integral of Langevin functions).

However, except if there are two really distinct particle populations in the sample, this approach is not physically sound since it completely dismisses the intermediate behaviours between a fully blocked regime and a fully superparamagnetic one. Qualitative information can also be inferred from a low temperature hysteresis loop. As explained earlier, within the Stoner–Wohlfarth model with randomly oriented particles, the remanence to saturation ratio m_R/m_S is equal to 0.5 at 0 K. This ratio then decreases when the temperature is increased since some particles become superparamagnetic. As a consequence, the experimentally measured m_R/m_S ratio should not be higher than 0.5, otherwise it means that there are interparticles interactions (most probable explanation), or a texture in the sample (partially oriented particles) or even that the particles are not purely uniaxial.

In conclusion, except in the superparamagnetic regime, magnetization loops measured on nanoparticle assemblies can often be only qualitatively analyzed and more demanding simulations [95], which cannot be routinely made, are necessary if a deeper insight is wanted. ZFC/FC susceptibility curves on the other hand are more suited for a quantitative analysis: they can provide an accurate particle size distribution and magnetic anisotropy determination (if the MCA is independent on the size).

8.2 ZFC/FC Susceptibility (dc and ac) and Curve Fitting

The ZFC/FC protocol has already been described (see Sect. 6.2): it is used to measure curves which display the signature of the blocked to superparamagnetic regime crossover of the particles in an assembly. For an assembly of N_{tot} particles of the same size, assuming that the applied magnetic field is small enough to be in the linear response regime, we can write the ZFC magnetic moment for the two extreme behaviours [45]:

$$m_b = N_{tot}\frac{\mu_0 H(M_S V)^2}{3K_{eff}V} \text{ in the blocked regime,} \tag{24}$$

$$m_{eq} = N_{tot}\frac{\mu_0 H(M_S V)^2}{3k_B T} \text{ in the superparamagnetic regime.} \tag{25}$$

Within the ‘abrupt change model’ or two states model originally introduced by Wohlfarth [96], the total ZFC magnetic moment of a particle assembly with a size distribution can be approximately written as the sum of two contributions (we have dropped the N_{tot} prefactor):

$$m_{ZFC} = \frac{\mu_0 H M_S^2}{3K_{eff}}\int_{V_b}^{\infty} Vf(V)dV + \frac{\mu_0 H M_S^2}{3k_B T}\int_0^{V_b} V^2 f(V)dV \tag{26}$$

The first term corresponds to the response of blocked particles, which are those having a volume higher than $V_b(T)$, and the second term is the response of superparamagnetic particles which have a volume smaller than $V_b(T)$. Since the blocking temperature T_B is a function of the MAE $K_{eff}V$, we can define for each temperature T the 'blocking' volume $V_b(T)$, used as the limit between the two regimes, by

$$T_B\left(K_{eff}V_b\right) = T \quad \text{which gives} \quad V_b = \varepsilon\, k_B T / K_{eff} \tag{27}$$

from the rule-of-thumb relation discussed earlier (see Sect. 6.1). Let us remind the reader that the value of the ε coefficient, often equal to 25, depends on the measuring time (Eq. (17)). Note that a more appropriate expression of $V_b(T)$ can be used in order to take into account the experimental temperature sweeping rate and in fact T_B is not the best choice as the crossover temperature (especially for FC) [44, 45].

By considering that for the FC process the blocked particles correspond to the superparamagnetic (equilibrium) magnetic moment at T_B, that is

$$m_b = m_{eq}(T_B) = \frac{\mu_0 H (M_S V)^2}{3k_B T_B} = \frac{\varepsilon\, \mu_0 H M_S^2 V}{3K_{eff}} \tag{28}$$

we can express the FC curve within the same 'abrupt change model' as

$$m_{FC} = \frac{\varepsilon\, \mu_0 H M_S^2}{3K_{eff}} \int_{V_b}^{\infty} V f(V) dV + \frac{\mu_0 H M_S^2}{3k_B T} \int_0^{V_b} V^2 f(V) dV \tag{29}$$

Once again, it should be noted that there exists a better choice than T_B (with the corresponding expression of V_b) for the blocked-superparamagnetic crossover. In particular, by using an improved model the correct limit of the FC is obtained for $T \to 0$ [45].

This model has been used in several investigations to fit ZFC/FC curves [97–100]. Let us emphasize that a significant number of studies have unfortunately been based on an erroneous formulation of the susceptibility curves [101–106] where a V term is missing in the integral of both superparamagnetic and blocked contributions.

A more elaborate model, called 'progressive crossover model' has been recently developed and provides a continuous analytical formula which can be used to describe the ZFC curve of a particle assembly [44]. We can write:

$$m_{ZFC} = m_b e^{-\nu \delta t} + m_{eq}(1 - e^{-\nu \delta t}) \tag{30}$$

for a single volume (single magnetic anisotropy energy), where, ν is the macrospin relaxation frequency which depends on the temperature (see Sect. 6) and δt is an effective waiting time which depends on several parameters (T, magnetic anisotropy energy, and experimental temperature sweeping rate). Thus, in the case of a size distribution, we simply have:

$$m_{ZFC} = N_{tot} \frac{\mu_0 H M_S^2}{3K_{eff}} \int_0^{\infty} \left[e^{-\nu \delta t} + \frac{K_{eff} V}{k_B T} \left(1 - e^{-\nu \delta t}\right) \right] V f(V) dV \tag{31}$$

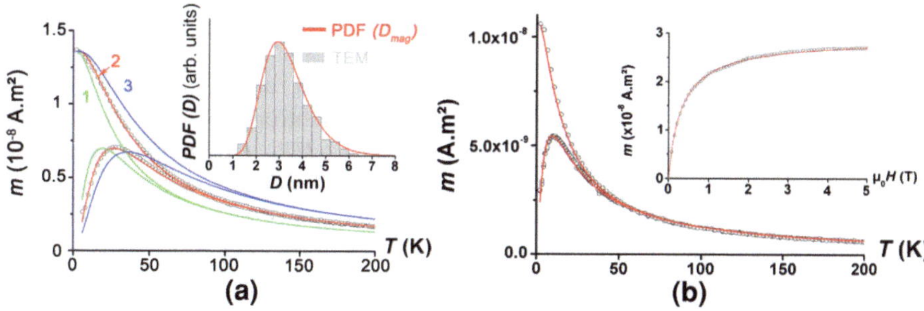

Fig. 20 **a** ZFC/FC magnetization curves (*dots*) of Co nanoparticles embedded in gold. The red lines (#2) correspond to the result of the 'triple fit' method. The other lines are constrained fits to the susceptibility curves based on different magnetic size distributions which are compatible with the room temperature superparamagnetic loop (not shown). The inset shows the size distributions as derived from the triple fit and from transmission electron microscopy (from Ref. [88], copyright (2011), with permission from Elsevier). **b** Experimental ZFC/FC curves and superparamagnetic loop (in inset) with the best fit (*Lines*) obtained using the 'triple fit' method for Co particles embedded in amorphous carbon (from Ref. [107], copyright (2010) by The American Physical Society)

A similar expression can be used for the FC, which may be viewed as a ZFC with a different starting point [44, 45].

In the end, the ZFC/FC curves and the room temperature (superparamagnetic) magnetization loop can be fitted using semi-analytical formulas, with only a limited number of parameters. Moreover, it is important to note that the curves share some common quantities, in particular the magnetic size distribution $f(V)$ and the total number of particles. A simultaneous fit of the three experimental curves, what is called the 'triple fit' procedure, is then subject to stringent constraints so that any fortuitous agreement is very unlikely: this means an improved accuracy of the inferred results [88]. The characterization of Co and CoPt nanoparticles [86, 88, 89, 107, 108] has proved the robustness of this technique (see Fig. 20). In addition, the failure of the triple fit can be a signature of inter-particle interactions, which makes the procedure a good way to test the usual underlying hypotheses (non-interacting macrospins) and to put into evidence subtle effects such as the anisotropy constant dispersion in nano-alloys [86], which is for example expected in ordering systems if there is a distribution of order parameter.

The theoretical description presented above can also be used to establish some scaling properties of the ZFC/FC curves [45]. This means that without any fit, susceptibility curves can provide interesting information especially when comparing two different samples: for instance, an anisotropy increase and the absence of coalescence upon annealing of CoPt particles can be confirmed by comparing normalized ZFC curves [109]. Another property that can be easily verified is the simple $1/T$ variation of the curves when the particles are in the superparamagnetic regime (when ZFC and FC have merged): the experimental curve plotted as a function of $1/T$ (or 1/m plotted as a function of T) should consist in a straight line passing through the origin (see Fig. 18b), otherwise there must be some interactions in the sample.

Let us also mention that the difference between a FC and ZFC curve, $\Delta m = m_{FC} - m_{ZFC}$ is related to the magnetic anisotropy energy distribution among the particles. More precisely, the quantity $(-1/T) \times (d\Delta m/dT)$ is proportional to the blocking temperature distribution $f(T_B)$. Therefore, once again without any fit, it is possible to have an idea of the distribution of blocking temperatures directly from the experimental curves [97, 105, 106, 110, 111]. Note however that because it involves a derivative and a $1/T$ term a significant noise can appear in the curve deduced from experimental data points.

ZFC susceptibility curves can also be measured in the *ac* mode. In this case, after having cooled down the sample without any applied field, a small alternative magnetic field *H* is applied at a pulsation ω and the subsequent magnetic moment oscillating at the same pulsation ω is measured, as a function of temperature. The magnetic response of the sample is made of an in-phase component and an out-of-phase one, which means that the complex amplitude of the induced magnetic moment can be written as: $m = m' - i\,m''$. Note that usually the experimental data are reported as a susceptibility (which is defined as $\chi = m/VH$, with *V* the sample volume), which is then decomposed into $\chi = \chi' - i\chi''$, but this can be made only if the total volume (or mass) of the magnetic particles is known.

A theoretical expression of the real and imaginary components can be established [44, 112–115] and we have, for a single volume *V*:

$$m' = m_b + \frac{m_{eq} - m_b}{1 + (\omega\tau)^2} \quad \text{and} \quad m'' = \omega\tau \frac{m_{eq} - m_b}{1 + (\omega\tau)^2} \tag{32}$$

where m_{eq} and m_b correspond respectively to the same expression of the equilibrium (superparamagnetic) and blocked magnetic moment as for a *dc* ZFC curve, and τ is the macrospin relaxation time which can be written using Eq. (15).

It can then be seen that at low temperature, since $\omega\tau \gg 1$, we have $m' \approx m_b$ and $m'' \approx 0$, while at high temperature, $\omega\tau \ll 1$ and we have $m' \approx m_{eq}$ and $m'' \approx 0$. This shows that the real part m' is similar to a *dc* ZFC, with a crossover around the temperature where $\omega\tau = 1$. This condition reads $T = T_B(\omega)$ where $T_B(\omega)$ is the blocking temperature corresponding to a measurement time $\tau_{meas} = 1/\omega$:

$$T_B(\omega) = \frac{K_{eff} V}{-k_B \ln(\omega\tau_0)} \tag{33}$$

On the other hand, the imaginary part m'' presents a peak situated (almost exactly) at $\omega\tau = 1$: it is non-zero only when the crossover between the blocked and superparamagnetic regimes occurs.

Experimental curves can be fitted with theoretical expressions similar to Eqs. (26–31), by performing a numerical integration in order to take into account the particle size distribution. It is also possible, as for *dc* curves, to fit the real part by using a two states or 'abrupt change' model which assumes that a particle is either fully blocked or fully superparamagnetic [105, 116]. As for a *dc* curve, the peak temperature T_{max} should not be confused with the mean or median blocking temperature: it strongly depends on the shape of the size distribution. We can also note that a relation between the imaginary part m'' and the blocking temperature distribution $f(T_B)$ has been used in the literature [105]. When the particle size distribution is known, it can then provide the value of the anisotropy constant K_{eff}

This type of measurements is interesting for several reasons: the imaginary part of the signal is insensitive to parasitic magnetic signals (in particular coming from the substrate), it can be easily interpreted, and by varying the pulsation ω we can have access to a quite wide range of timescales, which can be used to estimate the relaxation time τ_0 [106, 116]. The finding of an unphysical value for τ_0 can allow detecting the presence of significant interactions in a nanoparticle sample. Besides, it is also quite common to derive a single energy barrier value from the evolution of T_{max} with ω, using an Arrhenius-type plot [106, 116]. We can indeed write, for a single energy barrier (in which case $T_{max} \sim T_B$)

$$-\ln \omega = \ln \tau_0 + \frac{K_{eff} V}{k_B T_{\max}} \tag{34}$$

This means that a plot of $y = -\ln(\omega)$ as a function of $x = 1/T_{max}$ should consist in a straight line: its slope is directly related to the magnetic anisotropy energy and its crossing point with the y axis corresponds to $\ln(\tau_0)$.

8.3 Other Measurements

$m_R(T)$ remanence curves are obtained by measuring the magnetic moment as a function of increasing temperature, with no applied field, starting from the remanent state reached at low temperature after having saturated the sample. This provides another way to observe the progressive change from a blocked regime to a superparamagnetic regime where there is no more remanent magnetic moment. As for the difference between $\Delta m = m_{FC} - m_{ZFC}$, the curve is related to the distribution of magnetic anisotropy energy among the particles in the sample.

$m_S(T)$ saturation curves are also obtained by measuring the magnetic moment as a function of temperature, but usually by keeping the applied magnetic field to its maximum experimental value: it corresponds in fact to $m(H_{max})$ and not really to the saturation moment. Moreover, the signal coming from the nanoparticles may be altered by the magnetic response of impurities or of the substrate. Nevertheless this type of measurement can be used to detect a magnetic transition (determination of the Curie temperature for a ferromagnetic to paramagnetic transition for instance) and to verify the validity of the often made assumption (in particular in the fit of susceptibility curves) that m_S is almost constant on a given range of temperature.

Finally, let us mention the isothermal remanent magnetization (IRM) protocol and its counterpart the direct current demagnetization (DCD) protocol. IRM and DCD curves are measured at remanence, at a low enough temperature to have particles in the blocked regime, and reflect the irreversible switching of the particles. The initial configuration for the IRM is a demagnetized state obtained by cooling the sample with zero-field from a high temperature superparamagnetic state, whereas for the DCD, the initial configuration corresponds to a remanent state obtained after having saturated the sample under a $-H_{max}$ applied field. Then, by applying for a short moment a field H, which is increased step by step, before coming back to remanence, it is possible to plot $IRM(H)$ and $DCD(H)$ curves. Since each measurement is performed at a zero field, the curves are only sensitive to irreversible variations of the magnetic moment: this ensures that superparamagnetic particles, paramagnetic impurities, and the diamagnetic substrate (and matrix) do not contribute to the signal. The curve shape is linked to the switching field distribution among the particles, which is much less dependent on the size distribution than ZFC/FC curves. These measurements are interesting because they deal with a switching process different from the one involved in susceptibility measurements: here the barrier is removed by applying a given field, while in ZFC/FC measurements the anisotropy barrier is overcome by a thermal process. Moreover, it can be easily shown [117] that in the case of independent particles, IRM and DCD curves are linked by the theoretical relation:

$$m_R - DCD = 2\,IRM$$

If it happens that this relation is not fulfilled, then it is the signature of the presence of either magnetizing or demagnetizing interactions in the sample.

9 Other Experimental Techniques

We now review some other experimental techniques, well adapted to study the magnetic properties of binary alloy cluster assemblies and single NPs [118].

9.1 X-Ray Magnetic Circular Dichroism (XMCD)

The XMCD signal is a difference spectrum of two X-ray absorption spectra recorded under a magnetic field, one taken with left circularly polarized light, and the other with right circularly polarized X-ray light, both accessible on synchrotron radiation facilities [119]. XMCD techniques are particularly well adapted to the study of magnetic nanoalloy assemblies. Indeed the chemical selectivity is obtained by tuning the photon energy at the $L_{2,3}$ absorption edges of each component. Then, it is possible to separate the magnetic contribution on the orbital

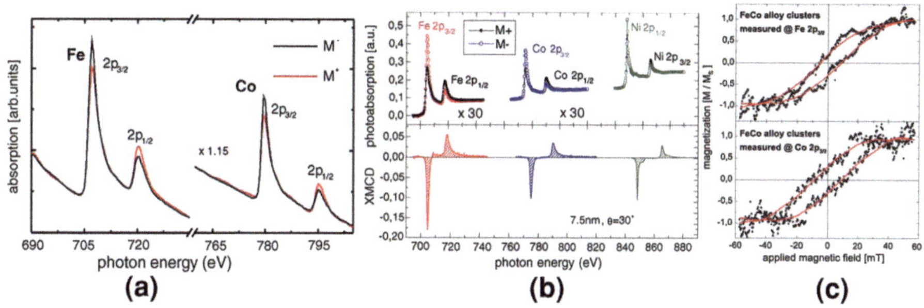

Fig. 21 XMCD spectra of 12 nm FeCo NPs on Si substrate (**a**) and 7.5 nm deposited on a Ni(1 1 1) film (**b**) recorded at the Fe and Co 2p3/2 core levels (for opposite magnetization directions M^+ and M^-). Element-specific hysteresis curves recorded of FeCo alloy nanoparticles deposited on Si at both absorption edges (**c**). From Refs. [120, 121], copyright (2004), with permission from Elsevier

(*L*) and spin (*S*) moment of both materials. Whereas the exchange interaction among electron spin is isotropic, the orbital term is connected, via the spin–orbit interaction, to the atomic structure of magnetic materials, hence giving rise to a magnetic anisotropy (MAE) [73]. Getzlaff et al. [120] have been able to put into evidence a strong magnetic coupling of Fe and Co in FeCo nanoclusters deposited on a Si substrate (see Fig. 21a, c) as one would expect from the enhanced total magnetic moment in bulk alloys. The same clusters deposited on a Ni surface apparently display an enhanced magnetic moment as compared to the bulk (see Fig. 21b) [121].

Theoretical calculations predict an increase of MAE in transition metal NPs related to a complicated, non-perturbative behavior as a function of cluster size, structure, bond length, and *d*-band filling [122].

One can list some experimental results to illustrate the difficulties to clearly understand the evolution from single atoms to finite-size NPs, which is complicated by nanoalloy, morphology and surrounding effects.

Gambardella et al. [123] have clearly put into evidence finite-sized effects from in situ XMCD measurements on Co-NPs up to 40 atoms deposited in ultrahigh vacuum by molecular epitaxity on a clean Pt(111) surface.

XMCD spectra performed at both the Fe and Pt $L_{3,2}$ edges on wet-chemically synthesized $Fe_{50}Pt_{50}$ particles (with mean diameter of 6.3 nm) after complete removal of the organic ligands (and the oxide shell) by soft hydrogen plasma, result in a pure metallic state [62]. After a thermal treatment, the authors have shown that the Fe orbital magnetic moment has increased by 330% while the Pt one is reduced by 30% and the effective spin moments have not changed.

In Ref. [60], chemically disordered alloyed Co_xPt_{1-x} nanoparticles prepared under UHV conditions by chemical synthesis have been studied. A linear increase of the orbital to spin moment ratio versus the Pt amount has been shown for the 2–8 nm size range and must be due to the effect of direct Co–Pt hybridization.

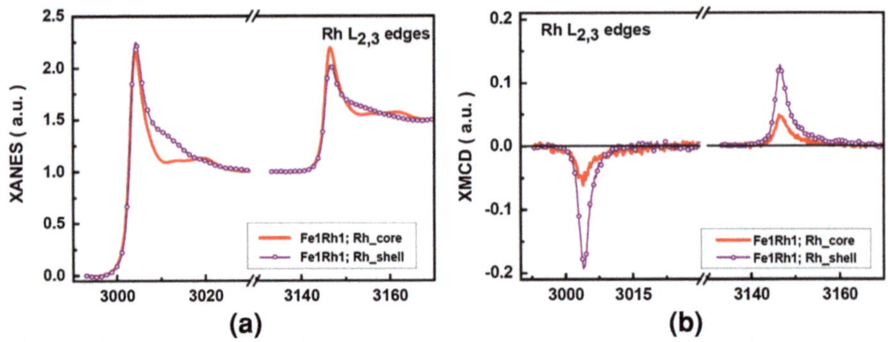

Fig. 22 XANES (**a**) and XMCD (**b**) spectra at the Rh L_3 and L_2 absorption edges for FeRh nanoparticles with a different core/shell order. From Ref. [124]

The magnetic signature of the $A1 \to L1_0$ chemical order transition has been put into evidence by XMCD on CoPt clusters embedded in an amorphous carbon matrix. Despite a striking change of the Co magnetic moment, the magnetic anisotropy of chemically ordered nanoparticles increases, with respect to the chemically disordered $A1$ phase, in much lower proportions than what is expected for the bulk [61].

Core–shell bimetallic MRh (M = Fe or Co) nanoparticles with mean diameter of 2 nm and either M@Rh or Rh@M core/shell structure have been investigated by XMCD experiments. At the same edges, it has been shown that 4*d* states of Rh atoms acquire an induced magnetic moment depending on the 3*d* transition metal and on the core/shell chemical order in the nanoparticle (see Fig. 22) [124].

In order to get a better understanding of the magnetic properties in nanoalloys and to avoid the complications due to distributions of particles sizes, orientations, etc., which are always present in assemblies of particles [89], single-particle measurement techniques have been developed such as magnetic microscopies (see Sects. 9.2 and 9.3). Another powerful experimental technique based on micro-Hall probes has been reported [125] and finally we briefly describe, in Sect. 9.4, micro-SQUID devices.

9.2 Magnetic Microscopies at Submicron Scale (5–100 nm)

"A remarkable number of methods for direct, real-space imaging in magnetic microscopy have been demonstrated over the past decade, and the pace of development shows no sign of slowing" wrote Freeman et al. [126] ten years ago in a review article. The last developments in spatial and temporal resolution of magnetic domain observation can be found in Table 1 and in the recent papers listed below (non exhaustive list):

Table 1 Current spatial resolution limit for different magnetic microscopy techniques

MOKE	PEEM	MFM	Lorentz	SEMPA	Holography
(~100 nm)	(~50 nm)	(~10 nm)	(~10 nm)	(~10 nm)	(~5 nm)

- The **Magneto-Optical Kerr Effect** (MOKE) microscopy is based on the rotation of the plane of polarization of linearly polarized light upon reflection from a magnetic surface (see for example [127]).
- The **PhotoEmission Electron Microscopy** (PEEM) is an imaging technique that uses the secondary electrons emitted from a sample surface upon absorption of photons. Magnetic sensitivity can be added using circularly polarized X-rays from synchrotron facilities, through the X-ray Magnetic Circular Dichoism (XMCD) effect (see for example [128]).
- The **Magnetic Force Microscopy** (MFM) is a particular type of atomic force microscopy, where a sharp magnetized tip scans a magnetic sample. The tip-sample magnetic interactions are detected and used to reconstruct the magnetic structure of the sample surface. Recently ultra-high-resolution MFM images have been obtained by using $Co_{90}Fe_{10}$-coated carbon nanotube probes [129].
- **Lorentz Scanning Transmission Electron Microscopy** (Lorentz STEM) is a method to detect, with high accuracy, the deflection of a focused electron beam caused by the Lorentz force at each point on the specimen, while the beam is scanned across it. Lorentz STEM thus gives access to microscopic distributions of magnetic induction as raster images in case of magnetic specimen [130].
- The **Scanning Electron Microscopy with Polarisation Analysis** (SEMPA) technique use the fact that when a beam of electrons strikes a magnetized surface, the secondary electrons emitted from this surface are spin polarized. The polarization of these electrons can be detected with a spin polarization detector such as the Mott detector. Images of the magnetic domain structure of the sample can be obtained [131].
- **Off-axis electron holography** in a transmission electron microscope (TEM) is capable of measuring the magnetic induction in a thin film quantitatively by illuminating magnetic sample coherently, with a spatial resolution that can approach 5 nm [132–135].

9.3 Magnetic Measurements on Single Nanoparticles with Atomic Spatial Resolution

To visualize spin mapping in the deep nanoscale, scanning probe microscopy techniques are required. The direct observation of spin structures of metallic and electrically insulating magnetic nanostructures, with atomic-scale resolution can be reached using low-temperature **Spin-Polarized Scanning-Tunneling Microscopy**

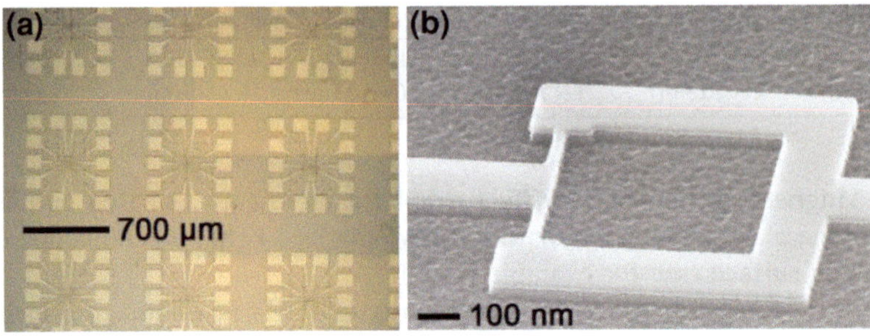

Fig. 23 SEM observation of lithographed chips containing 12 micro-SQUID (**a**) and micro-SQUID where the two micro-bridges act as Josephson junctions (**b**). From Ref. [27], copyright (2008) reprinted with permission from IEEE

(SP-STM) and **Magnetic Exchange Force Microscopy** (MExFM) which are described in the Wiesendanger's review article [136]. Discoveries of novel types of magnetic order at the nanoscale are presented as well as challenges for the future, including studies of local spin excitations based on spin-resolved inelastic tunneling spectroscopy and measurements of damping forces in MExFM experiments.

At room temperature, **Ballistic Electron Emission Microscopy** (BEEM) [137] is able to characterize both the magnetic properties and the shape of nanoparticles (for instance, in order to investigate the validity of the macrospin model for spherical or elongated particles). Briefly, this technique measures the transmission of hot electrons, emitted at a fixed energy over the sample Fermi level by the tip of a scanning tunnelling microscope, through a metallic film deposited on a semiconductor. The hot electrons are discriminated from the thermalized electrons at the metal/semiconductor interface, which acts as an energy filter thanks to the Schottky barrier [138, 139]. Due to spin dependent attenuation in ferromagnetic metals, the transmission is highly sensitive to magnetism so that it is possible to unravel magnetic structures and configurations in thin films and nanostructures [140].

9.4 Micro-SQUID

The micro-SQUID technique [141, 142] allows the detection of the magnetization reversal of an individual cluster made of only a few hundreds spins. This high sensitivity set-up is composed of a 20 nm-thick superconducting niobium film containing a low density of magnetic nanoparticles. Using an electron beam, a network of chips where each chip counts 12 micro-SQUIDs is lithographed (Fig. 23). The clusters are deposited near the Josephson junctions and their magnetic flux coupling is strong enough to produce a detectable signal induced by an applied magnetic field, using the micro-SQUID as a trigger. The angular dependence of the static switching field of a single nanoparticle can be measured at low temperature ($T = 35$ mK) in this way (see for example Fig. 9a).

10 Conclusions

The results in nanoalloys are today quite scarce because the subject is new and very difficult. The magnetism of nanoalloys will be surely developed widely during the next decade, due to their numerous applicative interests in different domains: information (magnetic storage media, spintronics…), energy (catalysis and electrocatalysis…), medecine [143] (hyperthermy, temporal and spatial site-specific drug delivery…) etc.

References

1. Nanomagnetism, Ultrathin Films, Multilayers and Nanostructures, ISBN 978-0-444-51680-0. In: Mills, D.L., Bland, J.A.C. (eds.) Contemporary Concepts of Condensed Matter Science, Elsevier (2006)
2. Guimarães, A.P.: Principles of Nanomagnetism. NanoScience and Technology. Springer, Berlin (2009). ISBN 978-3-642-01481-9, http://www.springer.com/materials/book/978-3-642-01481-9
3. Kittel, C.: Theory of the structure of ferromagnetic domains in films and small particles. Phys. Rev. **70**, 965 (1946)
4. Panissod, P., Drillon, M.: Magnetic ordering due to dipolar interaction in low dimensional materials. In: Miller, J.S., Drillon, M. (eds.): Magnetism: Molecules to Materials IV, Wiley VCH Verlag, Weinheim (2003)
5. Rave, W., Fabian, K., Hubert, A.: Magnetic states of small cubic particles with uniaxial anisotropy. J. Magn. Magn. Mat. **190**, 332 (1998)
6. Hubert, A., Schaefer, R., Magnetic domains, the analysis of microstructures. Springer, New York (2000). ISBN 3-540-64108-4
7. Hehn, M., et al.: Nanoscale magnetic domain structures in epitaxial cobalt films. Phys. Rev. B **54**, 3428 (1996)
8. Hehn, M., et al.: Nanoscale magnetic domains in mesoscopic magnets. Science **272**, 1782 (1996)
9. Kronmüller, H., Hertel, R.: Computational micromagnetism of magnetic structures and magnetisation processes in small particles. J. Magn. Magn. Mat **215–216**, 11 (2000)
10. Buda, L.D., et al.: Micromagnetic simulations of magnetisation in circular cobalt dots. Comp. Mat. Sci. **24**, 181 (2002)
11. Masseboeuf, A., et al.: Dimensionality crossover in magnetism: from domain walls (2D) to vortices (1D). Phys. Rev. Lett. **104**, 127204 (2010)
12. Ebels, U., et al.: Spin accumulation and domain wall magnetoresistance in 35 nm Co wires. Phys. Rev. Lett. **84**, 983 (2000)
13. Henry, Y., et al.: Magnetic anisotropy and domain patterns in electrodeposited cobalt nanowires. Eur. Phys. J. B **20**, 35 (2001)
14. Shinjo, T., Okuno, T., Hassdorf, R., Shigeto, K., Ono, T.: Magnetic vortex core observation in circular dots of permalloy. Science **289**, 930 (2000)
15. Natali, M., Prejbeanu, I.L., Lebib, A., Buda, L.D., Ounadjela, K., Chen, Y.: Correlated magnetic vortex chains in mesoscopic cobalt dot arrays. Phys. Rev. Lett. **88**, 157203 (2002)
16. Kechrakos, D., Trohidou, K.N.: Magnetic properties of dipolar interacting single-domain particles. Phys. Rev. B **58**, 12169 (1998)
17. Plumer, M.L., van Ek, J., Weller, D. (eds.): The Physics of Ultra-High-Density Magnetic Recording. Springer, Berlin (2001)

18. Plumer, M.L., et al.: Micromagnetic simulations of interacting dipoles on an fcc lattice: application to nanoparticle assemblies. J. Phys. Condens. Matter **22**, 296007 (2010)
19. Kittel, C.: Physique de l'état solide. Ed. Dunod (1995)
20. Bozorth, R. M.: Ferromagnetism. Van Nostrand, New York (1951)
21. Gradmann, U.: Magnetism in ultrathin transition metal films. In: Buschow, K.H.J. (ed.) Handbook of Magnetic Materials, vol. 7, Chap. 1. North-Holland, Amsterdam (1993)
22. Ersen, O., et al.: Growth conditions to optimise chemical order and magnetic properties in molecullar beam epitaxy grown CoPt/MgO(001) thin films. J. Appl. Phys. **93**, 2987 (2003)
23. Abes, M., et al.: Magnetic switching field distribution of patterned CoPt dots. J. Appl. Phys. **105**, 113916 (2009)
24. Christodoulides, J.A., et al.: Intrinsic and hysteresis properties of FePt nanoparticles. Phys. Rev. B **68**, 054428 (2003)
25. Stoner, E. Wohlfarth, E.: A Mechanism of Magnetic Hysteresis in Heterogeneous Alloys. Philos. Trans. R. Soc. London, Ser. A **240**, 599 (1948)
26. Thirion, C., et al.: Switching of magnetization by nonlinear resonance studied in single nanoparticles. Nat. Mater. **2**, 524 (2003)
27. Raufast, C., et al.: Microwave assisted magnetization reversal in individual isolated cluster of Co. IEEE Trans. Magn. **44**, 2812–2815 (2008)
28. Krause, S., et al.: Current-induced magnetization switching with a spin-polarized scanning tunneling microscope. Science **317**, 1537 (2007)
29. Osborn, J.A.: Demagnetizing factors of the general ellipsoid. Phys. Rev. **67**, 351 (1945)
30. Kneller, E.: Ferromagnetismus, Springer, Berlin (1962)
31. Mills, D.L.: The ferromagnetism of ultrathin films; from two to three dimensions. J. Magn. Magn. Mater. **100**, 515 (1991)
32. Bruno, P.: Physical origins and theoretical models of magnetic anisotropy. IFF-Ferienkurses, Jülich (1993)
33. Paige, D.M., Szpunar, B., Tanner, B.K.: The magnetocrystalline anisotropy of cobalt. J. Magn. Magn. Mater. **44**, 239 (1984)
34. Purcell, S.T., et al.: Structural and magnetic studies of ultrathin epitaxial Co films deposited on a Pd(111) single crystal. J. Magn. Magn. Mat. **93**, 25 (1991)
35. den Broeder, F.J.A., et al.: Magnetic anisotropy of multilayers. J. Magn. Magn. Mat. **93**, 562 (1991)
36. Rohart, S., et al.: Magnetic anisotropy of Co_xPt_{1-x} clusters embedded in a matrix: Influences of the cluster chemical composition and the matrix nature. Phys. Rev. B **74**, 104408 (2006)
37. Néel, L.: Anisotropie magnétique superficielle et surstructures d'orientation. J. Phys. Radium **15**, 225 (1954)
38. Jamet, M., et al.: Magnetic anisotropy of a single cobalt nano-cluster. Phys. Rev. Lett. **86**, 4676 (2001)
39. Jamet, M., et al.: Magnetic anisotropy in single clusters. Phys. Rev. B **69**, 024401 (2004)
40. Thiaville, A.: Coherent rotation of magnetization in three dimensions: a geometrical approach. Phys. Rev. B **61**, 12221 (2000)
41. Néel, L.: Théorie du traînage magnétique des ferromagnétiques en grains fins avec applications aux terres cuites. Ann. Geophys. **5**, 99 (1949)
42. Brown, W.F.: Thermal fluctuations of a single-domain particle. Phys. Rev. **130**, 1677 (1963)
43. Bean, C.P., Livingston, J.D.: Superparamagnetism. J. Appl. Phys. **30**, 120S (1959)
44. Tournus, F., Bonet, E.: Magnetic susceptibility curves of a nanoparticle assembly, I: Theoretical model and analytical expressions for a single magnetic anisotropy energy. J. Magn. Magn. Mater. **323**, 1109 (2011)
45. Tournus, F., Tamion, A.: Magnetic susceptibility curves of a nanoparticle assembly II. Simulation and analysis of ZFC/FC curves in the case of a magnetic anisotropy energy distribution. J. Magn. Magn. Mater. **323**, 1118 (2011)
46. Billas, I.M.L., et al.: Magnetic moments of iron clusters with 25 to 700 atoms and their dependence on temperature. Phys. Rev. Lett. **71**, 4067 (1993)

47. Billas, I.M.L., et al.: Magnetism of Fe, Co and Ni clusters in molecular beams. J. Magn. Magn. Mat. **168**, 64 (1997)
48. Li, C., Freeman, A.J.: Giant monolayer magnetization of Fe on MgO: a nearly ideal two-dimensional magnetic system. Phys. Rev. B **43**, 780 (1991)
49. Hjortstam, O., et al.: Calculated spin and orbital moments in the surfaces of the 3d metals Fe, Co, and Ni and their overlayers on Cu(001). Phys. Rev. B **53** (1996)
50. Guzmán-Ramírez, G., et al.: Stability, structural, and magnetic phase diagrams of ternary ferromagnetic 3d-transition-metal clusters with five and six atoms. J. Chem. Phys. **134**, 054101 (2011)
51. Apsel, S.E., et al.: Surface-enhanced magnetism in nickel clusters. Phys. Rev. Lett. **76**, 1441 (1996)
52. Vega, A., et al.: Topological antiferromagnetism at Cr surfaces and interfaces. Phys. Rev. B **49**, 12797 (1994)
53. Pizzini, S., et al.: Magnetic circular X-ray dichroism measurements of Fe-Co alloys and Fe/Co multilayers. Phys. Rev. B **50**, 3779 (1994)
54. Parette, G., Mirebeau, I.: Magnetic moment distribution in iron-cobalt alloys. Physica B **156–157**, 721 (1989)
55. Hamada, N.: Local environment effect on the atomic magnetic moment in ferromagnetic transition-metal alloys. J. Phys. Soc. Jpn **46**, 1759 (1979)
56. Grange, W., et al.: Experimental and theoretical X-ray magnetic-circular-dichroism study of the magnetic properties of CoPt thin films. Phys. Rev. B **62**, 1157 (2000)
57. Huttel, Y., et al.: Magnetic V embedded in copper evidenced by X-ray magnetic circular dichroism. Phys. Rev. B **67**, 052408 (2003)
58. Cox, A.J., et al.: Magnetism in 4d-transition metal clusters. Phys. Rev. B **49**, 12295 (1994)
59. Dreyssé, H., Demangeat, C.: Theoretical approaches of magnetism of transition-metal thin films and nanostructures on semi-infinite substrate. Surf. Sci. Rep. **28**, 65 (1997)
60. Imperia, P., et al.: XMCD studies of Co_xPt_{100-x} nanoparticles prepared by vapour deposition and chemical synthesis. Phys. Stat. Sol. (a) **205**, 1047 (2008)
61. Tournus, F., et al.: Evidence of $L1_0$ chemical order in CoPt nanoclusters: Direct observation and magnetic signature. Phys. Rev. B **77**, 144411 (2008)
62. Antoniak, C., et al.: Enhanced Orbital Magnetism in $Fe_{50}Pt_{50}$ Nanoparticles. Phys. Rev. Lett. **97**, 117201 (2006)
63. Lyubina, J., et al.: Influence of composition and order on the magnetism of Fe–Pt alloys: neutron powder diffraction and theory. Appl. Phys. Lett. **89**, 032506 (2006)
64. Aguilera-Granja, F., et al.: Theoretical investigation of free-standing CoPd nanoclusters as a function of cluster size and stoichiometry in the Pd-rich phase: geometry, chemical order, magnetism, and metallic behaviour. Phys. Rev. B **74**, 224405 (2006)
65. Kubachewski, O.: Iron Binary Phase Diagrams. Springer Verlag N.Y. (1982)
66. Hansen, M., Anderko, K.: Constitution of Binary Alloys. McGraw-Hill (1958)
67. Pierron-Bohnes, V., Cadeville, M.C., Gautier, F.: Magnetism and local order in dilute FeCo alloys. J. Phys. F: Met. Phys. **13**, 1689 (1983)
68. Martínez-Herrera, F.J., Mejia-Lira, F., Aguilera-Granja, F., Moran-Lopez, J.L.: Theory of phase equilibria in Co-Fe alloys. Phys. Rev. B **31**, 1686 (1985)
69. Leroux, C., et al.: Comparative investigation of structural and transport properties of $L1_0$ NiPt and CoPt phases; the role of magnetism. J. Phys. F: Met. Phys. **18**, 2033 (1988)
70. Sanchez, J.M., Moran-Lopez, J.L., Leroux, C., Cadeville, M.C.: Chemical and magnetic ordering in CoPt. J. Phys. C8, **49**, 107 (1988)
71. Sanchez, J.M., et al.: Magnetic properties and chemical ordering in Co-Pt. J. Phys.: Cond. Matter 1, 491 (1989)
72. Dahmani, C.E., et al.: NiPt phase diagram: experiment and theory. Phys. Rev. Lett. **55**, 1208 (1985)
73. Bruno, P.: Tight-binding approach to the orbital magnetic moment and magnetocrystalline anisotropy of transition-metal monolayers. Phys. Rev. B **39**, 865 (1989)

74. Farrow, R.F.C., et al.: Control of the axis of chemical ordering and magnetic anisotropy in epitaxial FePt films. J. Appl. Phys. **79**, 5967 (1996)
75. Montsouka, R.V.P., et al.: Preparation and magnetic properties of epitaxial $FeNiPt_2$ thin films on MgO (100). Phys. Rev. B **74**, 144409 (2006)
76. Gehanno, V., Marty, A., Gilles, B., Samson, Y.: Magnetic domains in epitaxial ordered FePd(001) thin films with perpendicular magnetic anisotropy. Phys. Rev. B **55**, 12552 (1997)
77. Lehnert, A., et al.: Magnetic anisotropy of Fe and Co ultrathin films deposited on Rh(111) and Pt(111) substrates: an experimental and first-principles investigation. Phys. Rev. B **82**, 094409 (2010)
78. Honolka, J., et al.: Magnetism of FePt surface alloys. Phys. Rev. Lett. **102**, 067207 (2009)
79. Liscio, F., et al.: Structural origin of perpendicular magnetic anisotropy in epitaxial $CoPt_3$ nanostructures grown on WSe_2(0001). Phys. Rev. B **81**, 125417 (2010)
80. Maret, M., et al.: Structural order related to the magnetic anisotropy in epitaxial (111) CoPt3 alloy films. J. Magn. Magn. Mater. **166**, 45 (1997)
81. Maret, M., et al.: Enhanced perpendicular magnetic anisotropy in chemically long-range ordered (0 0 0 1)Co_xPt_{1-x} films. J. Magn. Magn. Mater. **191**, 61 (1999)
82. Meneghini, C., et al.: Structural origin of magnetic anisotropy in Co-Pt alloy films probed by polarized XAFS. Eur. Phys. J. B **7**, 347 (1999)
83. Rohart, S., Tournus, F., Dupuis, V.: Chemical order and size effects on the magnetic anisotropy of FePt and CoPt nanoparticles. (http://fr.arxiv.org/abs/1105.6292)
84. Tournus, F., Rohart, S., Dupuis, V.: Magnetic anisotropy dispersion in CoPt nanoparticles: an evaluation using the Néel model. IEEE Trans. Magn. **44**, 3201 (2008)
85. Makarov, D., et al.: CoPt alloy grown on the WSe_2(0 0 0 1) van der Waals surface. J. Magn. Magn. Mater. **320**, 1858 (2008)
86. Tournus, F., et al.: Dispersion of magnetic anisotropy in size-selected CoPt clusters. Phys. Rev. B **81**, 220405 (2010)
87. Alloyeau, D., et al.: Size and shape effects on the order-disorder phase transition in CoPt nanoparticles. Nat. Mater. **8**, 940 (2009)
88. Tournus, F., Blanc, N., Tamion, A., Hillenkamp, M., Dupuis, V.: Synthesis and magnetic properties of size-selected CoPt nanoparticles. J. Magn. Magn. Mater. **323**, 1868 (2011)
89. Tamion, A., et al.: Accurate determination of the magnetic anisotropy in cluster-assembled nanostructures. Appl. Phys. Lett. **95**, 062503 (2009)
90. Respaud, M.: Magnetization process of noninteracting ferromagnetic cobalt nanoparticles in the superparamagnetic regime: deviation from Langevin law. J. Appl. Phys. **86**, 556 (1999)
91. Dorman, J. L., et al.: Magnetic Relaxation in Fine-particle systems. In: Advances in Chemical Physics, vol. **XCVIII**. Wiley, New Yok (1997)
92. Pfeiffer, H.: Determination of anisotropy field distribution in particle assemblies taking into account thermal fluctuations. Phys. Stat. Sol. A **118**, 295 (1990)
93. Garcia-Otero, J., Garcia-Bastida, A.J., Rivas, J.: Influence of temperature on the coercive field of non-interacting fine magnetic particles. J. Magn. Magn. Mater. **189**, 377 (1998)
94. Morel, R., Brenac, A., Portemont, C., Deutsch, T., Notin, L.: Magnetic anisotropy in icosahedral cobalt clusters. J. Magn. Magn. Mater. **308**, 296 (2007)
95. Chantrell, R.W., Weller, D., Klemmer, T.J., Sun, S., Fullerton, E.E.: Model of the magnetic properties of FePt granular media. J. Appl. Phys. **91**, 6866 (2002)
96. Wohlfarth, E.P.: The temperature dependence of the magnetic susceptibility of spin glasses. Phys. Lett. **70A**, 489 (1979)
97. Sappey, R., et al.: Nonmonotonic field dependence of the zero-field cooled magnetization peak in some systems of magnetic nanoparticles. Phys. Rev. B **56**, 14551 (1997)
98. Respaud, M., et al.: Surface effects on the magnetic properties of ultrafine cobalt particles. Phys. Rev. B **57**, 2925 (1998)
99. Antoniak, C., Lindner, J., Farle, M.: Magnetic anisotropy and its temperature dependence in iron-rich Fe_xPt_{1-x} nanoparticles. Europhys. Lett. **70**, 250 (2005)

100. Margeat, O., Tran, M., Spasova, M., Farle, M.: Magnetism and structure of chemically disordered $FePt_3$ nanocubes. Phys. Rev. B **75**, 134410 (2007)
101. Chantrell, R. W., El-Hilo, M., O'Grady, K.: IEEE Spin-glass behavior in a fine particle system. Trans. Magn. **27**, 3570 (1991)
102. El-Hilo, M., O'Grady, K., Chantrell, R.W.: Susceptibility phenomena in a fine particle system: I. concentration dependence of the peak. J. Magn. Magn. Mater. **114**, 295 (1992)
103. Pfeiffer, H., Chantrell, R.W.: Zero-field-cooled magnetization and initial susceptibility of magnetic particle systems. J. Magn. Magn. Mater. **120**, 203 (1993)
104. Yang, H.T., et al.: Achieving a noninteracting magnetic nanoparticle system through direct control of interparticle spacing. Appl. Phys. Lett. **94**, 013103 (2009)
105. Luis, F., et al.: Enhancement of the magnetic anisotropy of nanometer-sized Co clusters: influence of the surface and of interparticle interactions. Phys. Rev. B **65**, 094409 (2002)
106. Denardin, J.C., et al.: Thermoremanence and zero-field-cooled/field-cooled magnetization study of $Co_x(SiO_2)_{1-x}$ granular films. Phys. Rev. B **65**, 064422 (2002)
107. Tamion, A., et al.: Magnetic anisotropy of embedded Co nanoparticles: Influence of the surrounding matrix. Phys. Rev. B **81**, 144403 (2010)
108. Tournus, et al.: Signature of multimers on magnetic susceptibility curves for mass-selected Co particles. J. Appl. Phys. **109**, 07B502 (2011)
109. Tournus, F., Blanc, N., Tamion, A., Dupuis, V., Epicier, T.: Coalescence-free $L1_0$ ordering of embedded CoPt nanoparticles. J. Appl. Phys. **109**, 07B722 (2011)
110. Mamiya, H., et al.: Extraction of blocking temperature distribution from zero-field-cooled and field-cooled magnetization curves. IEEE Trans. Magn. **41**, 3394 (2005)
111. Larde, R., et al.: Evidence of Superparamagnetic Co clusters in pulsed laser deposition-grown $Zn_{0.9}Co_{0.1}O$ thin films using atom probe tomography. J. Am. Chem. Soc. **133**, 1451 (2011)
112. Shliomis, M.I., Stepanov, V.I.: Frequency dependence and long time relaxation of the susceptibility of the magnetic fluids. J. Magn. Magn. Mater. **122**, 176 (1993)
113. Garcia-Palacios, J.L.: On the statics and dynamics of magnetoanisotropic nanoparticles. In: Advances in Chemical Physics. I. Prigogine, S.A. Rice (eds.), Wiley & Sons, Inc. **112**, 1 (2000)
114. Raikher, Y.L., Stepanov, V.I.: Linear and cubic dynamic susceptibilities of superparamagnetic fine particles. Phys. Rev. B **55**, 15005 (1997)
115. Gittleman, J.I., Abeles, B., Bozowski, S.: Superparamagnetism and relaxation effects in granular $Ni\text{-}SiO_2$ and $Ni\text{-}Al_2O_3$ films. Phys. Rev. B **9**, 3891 (1974)
116. Dormann, J.L., et al.: Thermal variation of the relaxation time of the magnetic moment of $\gamma\text{-}Fe_2O_3$ nanoparticles with interparticle interactions of various strengths. Phys. Rev. B **53**, 14291 (1996)
117. Wohlfarth, E.P.: Relations between different modes of acquisition of the remanent magnetization of ferromagnetic particles. J. Appl. Phys. **29**, 595 (1958)
118. Bansmann, J., et al.: Magnetic and structural properties of isolated and assembled clusters. Surface Science Reports **56**, 189 (2005)
119. Mariot, J.M., Brouder, C.: Spectroscopy and magnetism: an introduction. In: Beaurepaire, E., Scheurer, F., Krill, G., Kappler, J.-P. (eds.): Magnetism and Synchrotron Radiation. Springer, Berlin. LNP 565, pp. 24–59 (2001)
120. Getzlaff, M., Kleibert, A., Methling, R.P., Bansmann, J., Meiwes-Broer, K.-H.: Mass-filtered ferromagnetic alloy clusters on surfaces. Surf. Sci. **566–568**, 332 (2004)
121. Bansmann, J., Kleibert, A.: Magnetism of mass-filtered nanoparticles on ferromagnetic supports. Appl. Phys. A Mater. Sci. Process. **80**, 957 (2005)
122. Pastor, G.M., Dorantes-Davila, J., Pick, S., Dreyssé, H.: Magnetic Anisotropy of 3d Transition-Metal Clusters. Phys. Rev. Lett. **75**, 326 (1995),
123. Gambardella, P., et al.: Giant magnetic anisotropy of single Co atoms and Nanoparticles. Science **300**, 1130 (2003)
124. Smekhova, A., et al.: XANES and XMCD Studies of FeRh and CoRh Nanoparticles. J. Phys.: Conf. Ser. **200**, 072091 (2010)

125. Theil, L., et al.: Magnetisation of isolated single crystalline Fe-nanoparticles measured by a ballistic Hall micro-magnetometer. Eur. Phys. J. D **10**, 259 (2000)
126. Freeman, M.R., Choi, B.C.: Advances in Magnetic Microscopy. Science **294**, 1484 (2001)
127. Cormier, M., et al.: High resolution polar Kerr magnetometer for nanomagnetism and nanospintronics. Review of Scientific Instruments **79**, 033706 (2008)
128. Kronast, F., et al.: Spin-resolved photoemission microscopy and magnetic imaging in applied magnetic fields. Surf. Interface Anal. **42**, 1532 (2010)
129. Choi, S.-J., et al.: Demonstration of ultra-high-resolution MFM images using $Co_{90}Fe_{10}$-coated CNT probes. J. Mag. Mag. Mat, **322**, 332 (2010)
130. Masseboeuf, A., et al.: The use of Lorentz microscopy for the determination of magnetic reversal mechanism of exchange-biased $Co_{30}Fe_{70}$/NiMn bilayer. J. Mag. Mag. Mat **321**, 3080 (2009)
131. Freeman, M.R., Choi, B.C.: Advances in Magnetic Microscopy. Science **294**, 1484 (2001)
132. Eisebitt, S., et al.: Lensless imaging of magnetic nanostructures by X-ray spectro-holography. Nature **432**, 885 (2004)
133. Kasama, T., Dunin-Borkowski, R.E., Eerenstein, W.: Off-axis electron holography observation of magnetic microstructure in a magnetite (001) thin film containing antiphase domains. Phys. Rev. B **73**, 104432 (2006)
134. Lichte, H., Lehmann, M.: Electron holography—basics and applications. Rep. Prog. Phys. **71**, 016102 (2008)
135. Snoeck, E., et al.: Magnetic Configurations of 30 nm Iron Nanocubes Studied by Electron Holography. Nano Lett. **8**, 4293 (2008)
136. Wiesendanger, R.: Spin mapping at the nanoscale and atomic scale. Rev. Mod. Phys. **81**, 1495 (2009)
137. Rippard, W.H., Buhrman, R.A.: Spin-dependent hot electron transport in Co/Cu thin films. Phys. Rev. Lett. **84**, 971 (2000)
138. Kaiser, W., Bell, D.: Direct investigation of subsurface interface electronic structure by ballistic-electron-emission microscopy. Phys. Rev. Lett **60**, 1406 (1988)
139. Lee, E.Y., et al.: Ballistic-electron-emission-microscopy investigation of hot-carrier transport in epitaxial $CoSi_2$ films on Si(100) and Si(111). Phys. Rev. B **52**, 1816 (1995)
140. Rippard, W.H., et al.: Observation of magnetization reversal of thin-film permalloy nanostructures using ballistic electron magnetic microscopy. Appl. Phys. Lett. **77**, 1357 (2000)
141. Wernsdorfer, W.: Thèse de l'université Joseph Fourier-Grenoble 1 (1996)
142. Wernsdorfer, W., et al.: High sensitivity magnetization measurements of nanoscale cobalt clusters. J. Appl. Phys. **78**, 7192 (1995)
143. Couvreur, P., Vauthier, C.: Nanotechnology: intelligent design to treat complex disease. Pharm. Research, **23**, 1417 (2006)

Optical, Structural and Magneto-Optical Properties of Metal Clusters and Nanoparticles

Emmanuel Cottancin, Natalia Del Fatti and Valérie Halté

Abstract Reduction of the size of a material to a nanometric scale leads to large modifications of its physical properties. This is in particular the case for the linear and nonlinear optical responses of metal nanoparticles with the appearance of a giant resonance, the so-called surface plasmon resonance. In magnetic systems, confinement can also strongly affects the static and dynamical magnetic properties, one consequence being for instance the appearance of superparamagnetism. In this chapter we will discuss some aspects of these properties and of the spectroscopy of nanoparticles formed by one or multiple metals.

1 Linear Optical Properties of Metal Clusters

1.1 Introduction

The peculiar optical properties of metal nanoparticles embedded in a solid or liquid matrix have fascinated mankind for a long time. The first example of "nanotechnology" goes back to the Roman era: it is the Lycurgus cup, which dates

E. Cottancin (✉)
Clusters and Nanostructures, LASIM, Université Lyon 1, CNRS,
43 Bd du 11 novembre, 69622 Villeurbanne cedex, France
e-mail: cottancin@lasim.univ-lyon1.fr

N. Del Fatti (✉)
FemtoNanoOptics, LASIM, Université Lyon 1, CNRS, 43 Bd du 11 novembre,
69622 Villeurbanne cedex, France
e-mail: delfatti@lasim.univ-lyon1.fr

V. Halté (✉)
FemtoMag, IPCMS-DON, Université de Strasbourg, CNRS, 23 rue du Loess, BP 43,
67034 Strasbourg cedex, France
e-mail: halte@ipcms.unistra.fr

D. Alloyeau et al. (eds.), *Nanoalloys*, Engineering Materials,
DOI: 10.1007/978-1-4471-4014-6_10, © Springer-Verlag London 2012

from the IV century AD, and can be admired at the British Museum in London. It shows a striking changing colour, from bright red to green, under different illumination conditions, i.e. lit from behind or in front. The origin of this unusual optical property is ascribed to the presence of noble metal nanoparticles in the glass. In the following we will describe from theoretical and experimental points of view the interaction of light with metal clusters leading to these effects.

1.1.1 Electromagnetic Interaction with Nanoparticles: Scattering, Absorption and Extinction of Light

The optical properties of metal particles in colloidal solutions were first experimentally investigated by M. Faraday, who was fascinated by the ruby colour of colloidal gold, in 1857 [1]. They were latter interpreted using the theory developed by G. Mie at the beginning of the twentieth century. This describes the optical response of a sphere of diameter *D* (radius *R*) and complex dielectric function ε, embedded in a transparent dielectric medium of refractive index $n_m = \varepsilon_m^{1/2}$ [2]. The incident electromagnetic wave is considered as a monochromatic plane wave of angular frequency ω (wavelength λ). This model is based on the solution of Maxwell equations, with boundary conditions at the surface of the spherical dielectric inclusion—representing the metal nanoparticle—in a dielectric infinite matrix (Fig. 1a) [3, 4]. Using an expansion of the incident electromagnetic plane wave in vector spherical harmonics (well adapted to a spherical geometry), the internal and scattered electromagnetic fields can be analytically determined for any arbitrary size *D*.

From an experimental point of view the incident light is partly absorbed and partly scattered by the particle. These interactions are described by the absorption (σ_a) and scattering (σ_s) cross-sections, defined as the ratio between the power which is absorbed and scattered by the particle, respectively, to the incident intensity of light. The extinction cross-section, σ_e, is the sum of the two contributions ($\sigma_e = \sigma_a + \sigma_s$). These optical cross-sections can be deduced from the scattered and internal fields obtained by the Mie theory. If the size of the spherical inclusion is small as compared to the optical wavelength, simplified approximate expressions can be obtained using a power series expansion as a function of the size parameter D/λ [3]. For very small nanoparticles (typically $D < \lambda/10$ [3, 5]), the absorption and scattering cross-sections can be computed at the lowest order of the development (dipolar approximation):

$$\sigma_a = \frac{18\pi V_{np}}{\lambda}\varepsilon_m^{3/2}\frac{\varepsilon_2}{|\varepsilon + 2\varepsilon_m|^2} \tag{1}$$

$$\sigma_s = \frac{24\pi^3 V_{np}^2}{\lambda^4}\varepsilon_m^2\left|\frac{\varepsilon - \varepsilon_m}{\varepsilon + 2\varepsilon_m}\right|^2 \tag{2}$$

where V_{np} is the nanoparticle volume and ε_2 is the imaginary part of the complex dielectric function of the nanoparticle ($\varepsilon = \varepsilon_1 + i\,\varepsilon_2$).

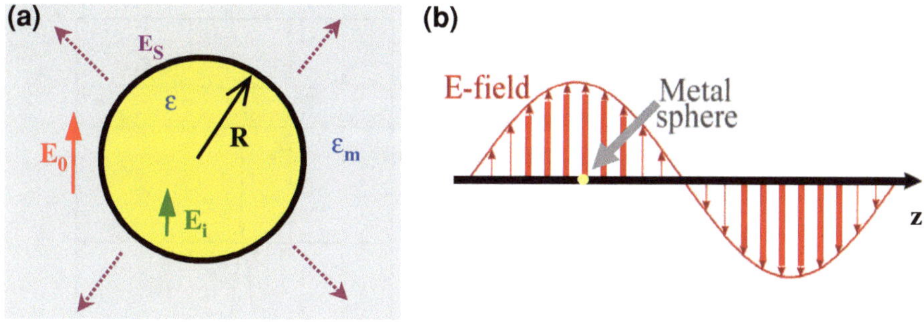

Fig. 1 **a** Interaction of a spherical dielectric inclusion of radius R and dielectric function ε with an incident electromagnetic field, E_O, in an homogeneous medium (of dielectric function ε_{m}). E_i and E_s are the internal and scattered electromagnetic field, respectively. **b** Validity of the dipolar approximation: $R \ll \lambda$

For these very small sizes $D \ll \lambda$, the optical extinction is dominated by absorption ($\sigma_e \approx \sigma_a$), the ratio between the scattering and absorption cross-sections being very small: $\sigma_s / \sigma_a \propto (D/\lambda)^3$. This is also known as the "quasi-static approximation", as it corresponds to a situation where the electromagnetic field can be considered uniform inside the sphere, as shown in Fig. 1b. In this case the light—matter interaction can be treated as a simple electrostatic problem. The internal field is proportional to the incident one, $\mathbf{E_i} = f(\omega)\mathbf{E_0}$, and the scattered field in the matrix correspond to the one radiated by an induced dipole at the centre of the sphere, of complex amplitude $\mathbf{p} = \varepsilon_0 V_{np}(\varepsilon - \varepsilon_m) f(\omega)\mathbf{E_0}$ [3]. Here ε_0 is the vacuum permittivity and the proportionality factor f, whose modulus also determines the optical cross-sections, is the so-called "dielectric confinement" (or "local field") factor:

$$f(\omega) = \frac{3\varepsilon_m}{\varepsilon + 2\varepsilon_m} \tag{3}$$

as it describes the local field enhancement in and around the particle due to the presence of the dielectric interface between two media (the nanoparticle and the external environment). For metal nanoparticles in a transparent medium, its modulus can be larger than one at some specific frequencies, as discussed in the following.

For larger sizes, the field retardation effects inside the particle must be taken into accounts and a multipolar treatment, using the complete Mie theory, is needed [2, 3].

1.1.2 Dielectric Function of Confined Metals

The wavelength or frequency dispersion of the dielectric function ε depends on the specific material constituting the nanoparticle. If we consider the simple case of noble metals (silver, gold and copper), from the schematic band structure shown in Fig. 2a it can be simply written as:

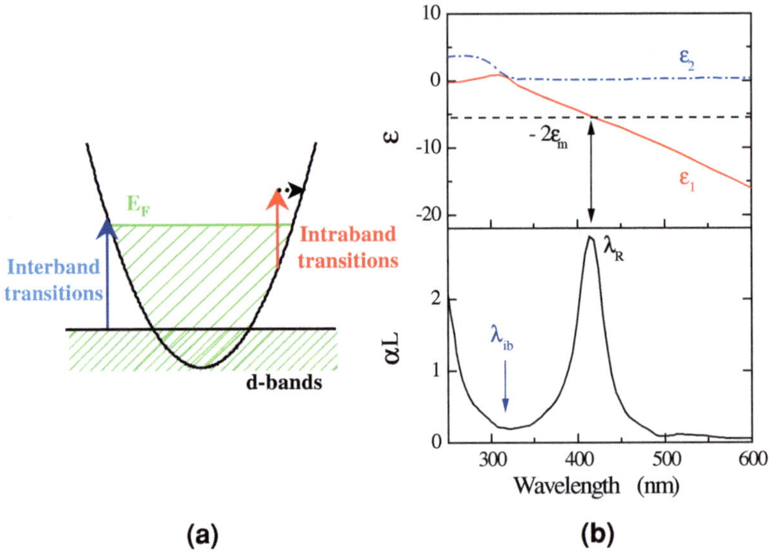

Fig. 2 **a** Schematic band structure (electronic energy vs. wavevector) for noble metals, showing the conduction band (E_F : Fermi energy) and d-bands. The two types of optical transitions are shown with *arrows*. **b** *Top panel*: real (ε_1) and imaginary (ε_2) part of the dielectric function of silver (results from [7]). *Bottom panel*: extinction spectra of silver nanoparticles embedded in a glass matrix ($D = 13$ nm, $\varepsilon_m = 2.75$). The SPR shows up at the wavelength λ_R satisfying the condition $\varepsilon_1 = -2\varepsilon_m$. Also indicated is the interband transition threshold in silver, $\lambda_{ib} \sim 320$ nm

$$\varepsilon(\omega) = \varepsilon^{ib}(\omega) - \omega_p^2 \Big/ \omega(\omega + i/\tau) \tag{4}$$

The first term is associated to interband transitions (from filled d-bands to conduction band). The second one is the Drude contribution from the quasi-free electrons in the conduction band. Here ω_p is the plasma frequency of the metal and τ is the mean optical collision time of the conduction electrons [6].

In confined metals, as nanoparticles, reduction of the size of the material leads to modification of their dielectric function. As a first approximation, a similar expression for ε remains valid, provided a modified size-dependent electron scattering rate is introduced in the Drude expression [4]. For spherical particles, this term thus writes $1/\tau(D) = 1/\tau_0 + 2gv_F/D,$ where the first contribution represent the scattering rate also present in bulk materials (dominated by electron–phonon interactions at room temperature). The second one is a consequence of electron quantum confinement in the metal. It can also be classically interpreted as an electron-surface scattering rate adding up to the other scattering processes (v_F is the electron Fermi velocity and g a proportionality factor of the order of unity) [4, 8, 9]. For very small sizes, additional quantum effects due to electronic confinement lead to further modifications of the dielectric function, and consequently of the optical response of metal clusters [10, 11]. At these sizes, interface effects (as the presence of ligands on the cluster surface) also strongly affect the electronic and optical properties of the systems, which then behave as molecular compounds [12, 13].

The dielectric function of the confined metal ε, together with the one of the matrix ε_m, determines the optical response of the embedded nanoparticle. The main effect of size reduction is the appearance of the dielectric confinement factor $f(\omega) = 3\varepsilon_m/(\varepsilon(\omega) + 2\varepsilon_m)$ introduced above. As shown in Fig. 2b, the real part of the dielectric function of the metal can be negative in the region where the Drude term is dominant, leading to large $f(\omega)$ for $|\varepsilon(\omega) + 2\varepsilon_m|$ minimum. The optical absorption, scattering and extinction of light by the nanoparticle are concomitantly enhanced at the frequency $\omega = \Omega_R$ (wavelength λ_R) minimizing $|\varepsilon(\omega) + 2\varepsilon_m|$. This condition corresponds to the "surface plasmon resonance" (SPR), i.e. an enhancement of all the linear and non-linear optical responses of the confined nanosphere. From a classical point of view, it can be seen as being associated to a collective oscillation of the electrons in the confined metal, resonantly driven by the external electromagnetic field.

For silver nanospheres embedded in glass, the SPR condition is fulfilled at a frequency far from the interband transitions (Fig. 2b): the resonance shows up as a quasi-Lorentzian line corresponding to an enhanced absorption by the conduction electrons [14]. Its frequency depends on both the interband and the matrix dielectric constants, and is given in the dipolar approximation by the simple expression $\Omega_R = \omega_P/\sqrt{\varepsilon_1^{ib} + 2\varepsilon_m}$ [4]. For other metals, as gold or copper, the SPR frequency lies close to the threshold for interband transitions (Fig. 3). The resonance then partly overlaps them, corresponding to an enhancement of both intraband and interband absorption by dielectric confinement effects.

In the quasi-static regime (typically $D < 30$ nm), the SPR spectral width is determined by the imaginary part of the confined metal dielectric function, $\varepsilon_2(\Omega_R)$. It depends on both the intraband Drude scattering rate $1/\tau(D)$ and the interband contribution $\varepsilon_2^{ib}(\Omega_R)$. Alkali clusters show a well marked resonance which is very weakly influenced by core electrons [15, 16]. For silver nanoparticles, the interband term is negligible and the resonance is also well defined (Fig. 2b). It broadens when reducing the size D, due to increase of the electron-surface scattering rate (see experimental Sect. 1.2.2). In copper or gold (and even more in other metals), the interband transitions in the SPR region considerably broaden the resonance, which may become difficult to observe (see for instance part 3-b). For larger sizes, multipolar effects also broaden the SPR, due to radiative damping [4].

1.1.3 Surface Plasmon Resonance: Shape Effects

The multipolar Mie theory has been generalized to spheroidal-shape objects, in an homogeneous environment and with a core–shell geometry [17, 18]. In the quasi-static approximation, simple analytical expressions are also available for prolate and oblate small nanoparticles [3].

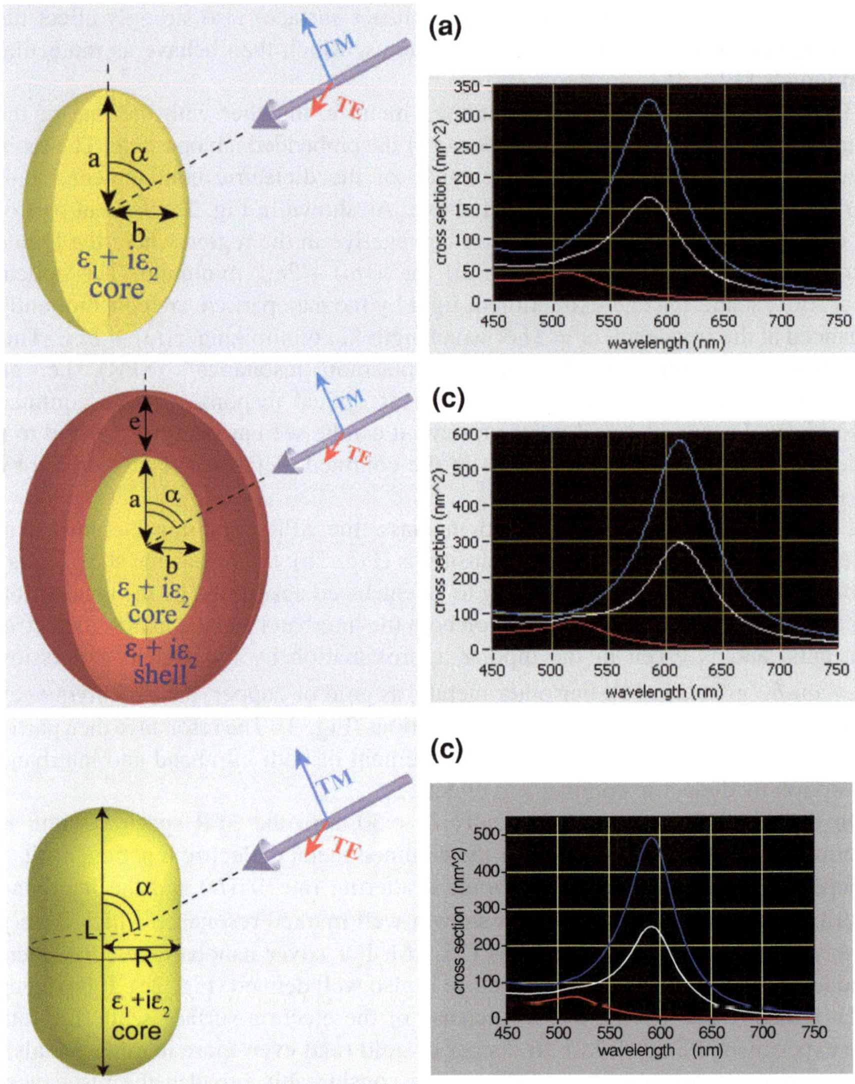

Fig. 3 Computed extinction cross-sections of a non-spherical gold nanoparticle in water ($\varepsilon_m = 1.69$, $g = 0.7$) for different geometries. The blue and red lines correspond to light polarization parallel to the long and short axis, respectively ($\alpha = 90°$), and the white line to non-polarized light. **a** Prolate shape with $2a = 20$ nm and $2b = 10$ nm. The SPR is computed with a multipolar treatment [17–19]. **b** Same as (a) in a core–shell structure, with a silica shell of $\varepsilon = 2.25$ and thickness $e = 5$ nm. **c** Cigar-like shape (cylinder capped by two hemispheres) with $L = 20$ nm and $2R = 10$ nm. The SPR is computed using FEM simulations [20]

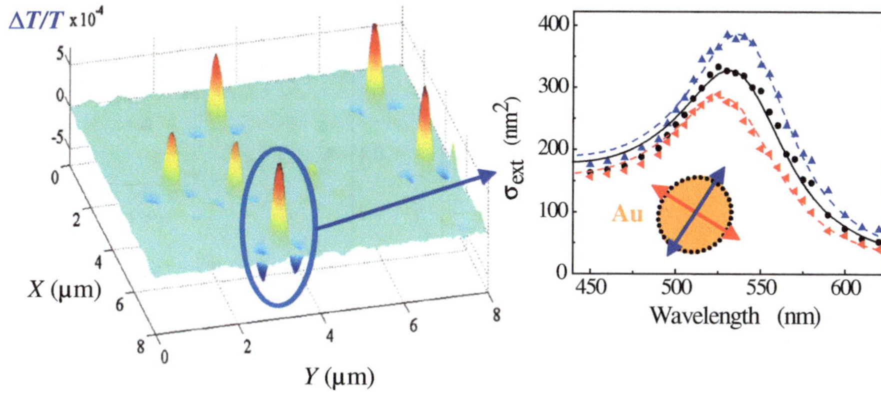

Fig. 4 Detection of single gold nanoparticles deposited on a glass substrate by spatial modulation microscopy (SMS). Each object gives rise to a three peak signal. The quantitative extinction cross-section of a selected particle is measured for two orthogonal polarizations (*red* and *blue dots*), and with non-polarized light (*black dots*). Data are quantitatively reproduced using the multipolar generalized Mie theory for ellipsoids ($2a = 19.5$ nm and $2b = 17.5$ nm). These sizes correspond to the ones measured by electron microscopy on the same nano-object. From Ref. [5]

In the more general case of an arbitrary shape, the optical response can be computed numerically by different methods, as Discret Dipole Approximation DDA [21, 22], or using finite element methods (FEM).

For non-spherical particles, the optical response generally depends on the light polarization direction. As an example, computations of the optical extinction cross-sections of a gold prolate nanoparticle in water, of a silica-coated nano-ellipsoid and of a cigar-like nanorod are shown in Fig. 3 [19, 20].

1.2 Monometallic Systems

1.2.1 Experimental Detection and Spectroscopy of a Single Nanoparticle

Nowadays light absorption or scattering by a single metal nanoparticle can be experimentally investigated using simple far-field optical methods [23–26]. Individual nano-object studies overcome the intrinsic limitations of ensemble optical measurements, subject to size, shape and environment fluctuations, thus permitting a direct comparison with theoretical models.

Figure 4 shows the optical response of quasi-spherical gold nanoparticles deposited on a glass substrate, detected by a spatial modulation technique (SMS) [26]. This non conventional far-field microscopy technique is based on the direct measurement of the extinction of light by a nano-object which absorbs or scatters a tiny amount of the

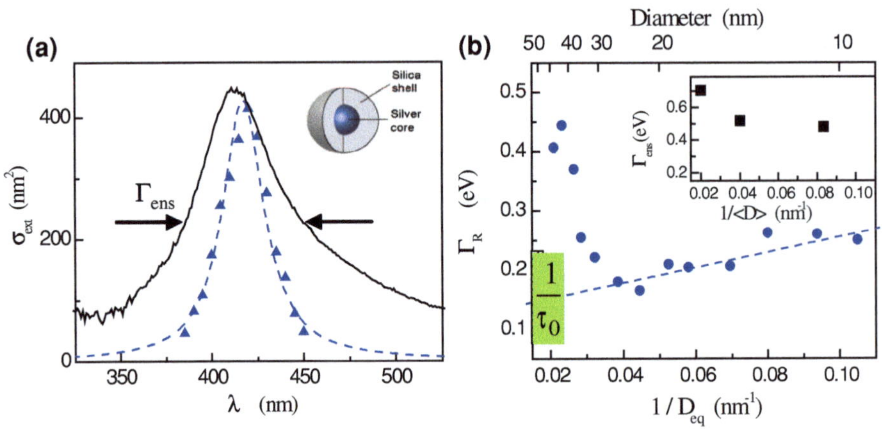

Fig. 5 **a** Experimental absorption spectrum of a colloidal solution of silica-coated D = 10 nm silver nanospheres (*black lines*), and extinction cross-section of a single particle (*blue dots*). **b** Measured SPR widths, Γ, of different single nanospheres as a function of their inverse diameter (*blue dots*). The inset shows the SPR widths of three ensembles of colloidal solutions with average diameters 10, 25 and 50 nm (*black dots* in inset). Single particle measurements shows quantum-size effects broadening of the SPR (*blue line*, corresponding to $g = 0.7$)

incident electromagnetic field. The presence of a nano-object under a strongly focused light beam is detected by the contrast in the transmitted light, $\Delta T/T$, when modulating the position of the nano-object *in* and *out* of the focal plane [27, 28]. The detected signal is proportional to the extinction cross-section of the illuminated nanoparticle. A spectroscopic measurement of its linear optical response is thus possible using a tunable light source (Fig. 4) [5, 29, 30].

As this technique provides the quantitative extinction cross-section, experimental results can be directly compared to theoretical models. Information on the size and shape anisotropy of the detected nanoparticle can thus be extracted using this comparison (Fig. 4): an "optical image" is realized, containing information close to the ones obtained by electron microscopy [31].

1.2.2 Confinement and Environment Effects

The capability of measuring the optical response at a single nanoparticle level opens the way to precise studies of the properties of confined systems as the electronic interactions in a confined metal. As stressed above (Sect. 1.1.2), quantum confinement is expected to induce a broadening of the SPR for sizes smaller than the bulk mean free path (typically $\sim$ 30 nm in noble metals) [4, 8]. Neglecting the interband contribution (e.g. for Ag particles), the SPR width is altered proportionally to the inverse diameter for nanospheres:

$$\Gamma \approx 1/\tau(D) = 1/\tau_0 + 2gv_F/D \tag{5}$$

Single particle experiments have provided a quantitative determination of these quantum effects. A clear evidence of the SPR broadening has been recently obtained in silica-coated single silver nanoparticles (Fig. 5) [32]. The silica shell provides a well controlled environment in comparison with surfactant-coated colloidal nanoparticles, whose optical response is strongly affected by chemical damping [33]. By optically measuring simultaneously the size of the individual nanoparticles and the width of their quasi-lorentzian silver SPR, a surface broadening parameter $g = 0.7$ has been determined [32, 34].

Single particle experiments also demonstrate the presence of an optimum size (close to $D = 25$ nm) minimizing Γ, i.e. maximizing the quality factor of the SPR: the SPR is broadened by quantum effects and by radiative damping for smaller or larger particles, respectively (Fig. 5).

1.3 Bimetallic Systems

Bimetallic nanoparticles (NPs) are of double interest as they may exhibit particular features of bulk alloys that usually do not simply derive from those of their constituents, and in addition, singular properties due to nanoscale confinement. In this respect, the study of bimetallic nanosystems is an emerging research field in numerous domains as such systems may adopt various chemical structures [35]. First of all, the atoms may arrange themselves to build an ordered nanoalloy or they may be randomly distributed. Moreover, new structures may exist only in the nanoscale range, like core/shell, nanoshell, multishell structures or Janus NPs [36] with separated phases. The internal structure of the clusters will be reflected in their optical response that may be used as a probe of it [37].

1.3.1 Theoretical Description of Bimetallic Systems

As described in details in the Chemical Order and Disorder in Alloys and Segregation and Phase Transitions in Reduced Dimension: From Bulk to Clusters via Surfaces, the structure of bimetallic clusters of two given components may be a priori guessed from thermodynamic considerations and from an insight into bulk properties of both components (Wigner–Seitz radius, melting temperature, cohesive and surface energies).

In the bulk phase, when two materials get mixed up there is a variation of enthalpy due to interactions between unlike atoms that do not exist in pure materials. At zero temperature, a positive enthalpy of mixing indicates that the homogeneous alloy is not stable: such mixed systems will separate into two phases. Nevertheless at finite temperature the entropic term will moderate this tendency lowering the Gibb's potential $\Delta G_{mixing} = \Delta H_{mixing} - T\Delta S_{mixing}$, this, all the more that the temperature is high. Therefore the miscibility will depend on temperature and on relative composition. On the other hand if the enthalpy of mixing is negative, the homogeneous alloy should be stable and the entropic term strengthens the miscibility with increasing temperature. Such considerations may partly explain the general pattern of phase diagrams which bring besides more information about the possible structure of the composite clusters. In the nanometer scale, the surface to volume ratio being increased, the surface energy differences between both constituents must be taken into account. As a rule of thumb, to minimize the total energy, the component with the smaller sublimation specific heat (or surface energy) is expected to accumulate preferentially at the surface. Nevertheless, the lattice mismatch between the two components may modify this tendency (the atoms with the higher lattice parameter are supposed to migrate onto the surface). Obviously, these thermodynamic laws have to be considered with caution for clusters produced by chemical ways because the surface energies are highly correlated to the surrounding medium and the surface adsorbed ions and the kinetics of reduction may drive the final structure. Therefore thermodynamical equilibrium conditions are not necessary reached and such particle architectures may be only metastable in this respect. Nevertheless, the main advantage of chemical ways is that various structures can be elaborated. To conclude on the structure of bimetallic nanosystems, one can underline that whatever the formation process (chemical or physical ways), the final size, shape and chemical arrangement are the result of a competition between thermodynamics and kinetics. Metastable compounds may be obtained without effective control of the chemical structure (order) except when thermodynamics rules the nucleation mechanism which can be modelled thanks to Monte Carlo simulations or molecular dynamics for very small systems.

In optics, the response of bimetallic NPs will be influenced by their internal structure and composition. The SPR can be spectrally tailored by playing on the relative concentration of both components. Conversely, we will see that the optical response may be viewed as a probe of their internal structure.

The main difficulty for "nanoalloys" is the knowledge of the dielectric function of the composite material A_xB_{1-x} to input in optical models. If the nanosystem adopts an ordered alloyed structure, similar to the bulk phase for which the dielectric function has been measured, this can be used in calculations. If it is not the case simplifications have to be made. The simplest way in current use is to take the weighted average of the dielectric function of each component to define the effective dielectric function of the nanoalloy $\left(\varepsilon_{eff}(\omega) = x\varepsilon_A(\omega) + (1-x)\varepsilon_B(\omega)\right)$. It is still a very crude hypothesis presupposing that both constituents are randomly distributed in pure nano-domains that can be described macroscopically by their own dielectric function. Another way to get an effective dielectric function comes from effective medium theories [38]. One can quote the Bruggeman or the Maxwell–Garnett models resulting from the Clausius-Mossoti equation:

$$\frac{\varepsilon_{eff}(\omega) - \varepsilon_o}{\varepsilon_{eff}(\omega) + 2\varepsilon_o} = \frac{1}{3\varepsilon_o}\sum_k N_k\alpha_k = \sum_k p_k \frac{\varepsilon_k(\omega) - \varepsilon_o}{\varepsilon_k(\omega) + 2\varepsilon_o} \tag{6}$$

This equation gives the effective dielectric function $\varepsilon_{eff}(\omega)$ of spherical inclusions of dielectric function $\varepsilon_k(\omega)$ randomly distributed in vacuum with a volume concentration p_k. Each inclusion is described as a dipole of size lower than the wavelength of excitation, thus submitted to a uniform electromagnetic field (quasi-static approximation). In the Bruggeman model, the dielectric function deduced from the Clausius-Mossoti equation is obtained by assuming that all the inclusions "see" an effective medium characterized by $\varepsilon_{eff}(\omega)$. It implies a total symmetry of both constituents and overestimates the dipolar interaction between inclusions. Results on the optical response of bimetallic clusters with the Bruggeman model, valid only for almost equal concentrations ($x \sim 0.5$), are not very different from those obtained with an averaged dielectric function of those of both components.

Conversely in the Maxwell-Garnett model, the interactions between inclusions are underestimated as one of the component is taken as a matrix in which are embedded the inclusions of the other components. Therefore this model is only valid for large interparticle distances, thus for volume concentrations of metal in the matrix lower than 5–10%. The only interest of this model is that it permits to deduce the metallic volumic concentration from absorbance measurements in samples in which the optical index of the matrix is well-known and the cluster size distribution relatively thin.

The only structure for which calculations are easily feasible is the segregated core/shell geometry [3] that may appear in a confined system. The problem can be exactly solved in the dipolar approximation for spherical or ellipsoidal shapes and may be generalized to multishell structures (as described in Sect. 1.1.3). By solving the Poisson's equation and applying the boundary conditions on each interface, one can deduce the dynamic polarizability and then the absorption cross-section. It takes the following form for a spherical core/shell NP, of core radius R_c and total radius R, composed of two materials with core and shell complex dielectric functions ε_c and ε_{sh} in a medium characterized by its dielectric function ε_m :

$$\sigma_{abs}(\omega) = \frac{3\omega\varepsilon_m^{1/2}}{c}\left(\frac{4}{3}\pi R^3\right)\Im\left(\frac{(\varepsilon_{sh}-\varepsilon_m)(\varepsilon_c+2\varepsilon_{sh})+f_v(\varepsilon_c-\varepsilon_{sh})(\varepsilon_m+2\varepsilon_{sh})}{(\varepsilon_{sh}+2\varepsilon_m)(\varepsilon_c+2\varepsilon_{sh})+2f_v(\varepsilon_{sh}-\varepsilon_m)(\varepsilon_c-\varepsilon_{sh})}\right) \tag{7}$$

where $f_v = \left(\frac{R_c}{R}\right)^3$ designs the volume ratio of the core.

Such calculations can be performed on various bimetallic clusters or in the case of metallic clusters with an oxide shell. Furthermore, the reduction of the mean free path in the core and in the shell may be also taken into account [39]. This calculation, made in the frame of the dipolar approximation, is only valid for relatively small particles but it may be extended to large NPs in the frame of the generalized Mie's theory by calculating scattering and extinction cross-sections [3, 40].

Concerning Janus particles of sufficiently large sizes, as they do not exhibit a simple symmetry, their optical response may be calculated thanks to numerical methods.

In all the simple models described above, each component is assumed to develop nano-domains in the NPs for which its corresponding bulk dielectric function can be defined. For "alloyed" structures at an atomic level and for very small clusters, it is not possible to define nano-domains and one has to resort to a full microscopic description to calculate the optical response.

Table 1 Wigner–Seitz radii and surface energies of several metals

	Ag	Au	Ni	Pt
Wigner-Seitz radius (Å)	1.598	1.592	1.376	1.532
Surface strain (dynes.cm^{-1})	925	1145	1796	1748

1.3.2 A Comparative Study of the AuAg, NiAg and AgPt Systems

The aim of this sub-section is not to give an overview of the optical studies on bimetallic clusters, but rather to discuss about the correlation between their internal structure and their optical response. Therefore, to see how the optical response may be used as a probe of the internal structure, we focus here on three different representative systems (i.e. AuAg, NiAg and AgPt) of the various structures that can be encountered in bimetallic systems. We will see that combined with other techniques of characterizations (electron microscopy, Low Energy Ion Scattering (LEIS) [41]), the optical properties of bimetallic clusters allow to get information about their internal structure.

From the analysis of bulk characteristics of the different species (Table 1) and of the phase diagrams [42], various tendencies can be intuited for the three systems. The phase diagram of the AuAg system shows that both metals are miscible whatever the proportion. Moreover there is no lattice mismatch and both metals have similar surface energies. One can thus expect a high miscibility without preferential migration onto the surface if thermodynamical laws dominate. Conversely the phase diagram of NiAg shows that Ni and Ag are immiscible over the entire composition range. Moreover as regards the lattice mismatch (14%) and the surface energies, silver is supposed to accumulate preferentially on the surface. Such a tendency is confirmed by molecular dynamics and Monte Carlo calculations [43, 44]. As for the AgPt system, the difference between surface energies are in favor of a migration of silver onto the surface, but the lattice mismatch is weaker (4%) and the phase diagram indicates that ordered alloys do exist.

For the three systems for which experimental results will be presented in the following, the samples consist of bimetallic clusters of a few nanometers in diameter embedded in a transparent matrix (alumina). Bi-metallic clusters have been generated by laser vaporization of an alloyed target, allowing to independently control the size and the composition of the produced clusters. Afterwards clusters are co-deposited on a substrate with the transparent matrix. The stoichiometry of the clusters, analyzed through Rutherford Back Scattering (RBS) and Energy dispersive x-ray measurements (EDX), corresponds to the one of the target.

The AuAg System

Among all studies on bimetallic clusters, AuAg clusters are probably the most investigated in optics. For alloyed systems the SPR appears between those of gold and silver and its precise spectral position depends on the relative proportion

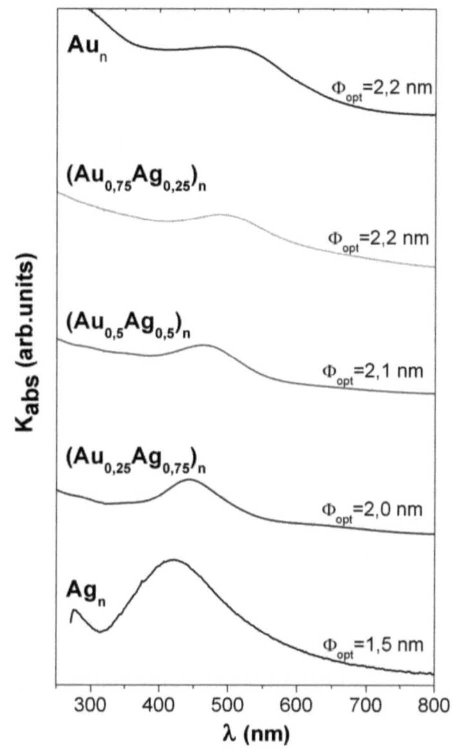

Fig. 6 Evolution of the optical absorption spectra of AuAg $(Au_xAg_{1-x})_n$ clusters embedded in alumina for various relative compositions x and almost same optical diameters (the optical diameter is defined as the following: $\phi_{opt} = \sqrt[3]{\langle\phi^3\rangle}$ where ϕ is the diameter). From Ref. [49], reproduced by permission of the royal society of chemistry

of both species [45, 46]. On the contrary in core/shell geometries (essentially obtained by chemical synthesis), the optical response is more complex and displays two resonance peaks for sufficiently large nanoparticles or an asymmetric resonance peak for smaller clusters of a few nanometers in diameter [47, 48].

Figure 6 shows the optical response of AuAg clusters embedded in alumina. Absorption spectra of same optical diameters and various compositions show a single resonance evolving regularly from the silver SPR to the gold one when the gold proportion increases. This observation thus rules out the possibility of a core/shell structure for which two resonances or at least a highly asymmetric resonance would be expected [48, 50]. Besides, in this very low size range (below 5 nm in diameter here), quantum effects may appear and semi-quantal calculations are required to correctly reproduce size and composition effects [50, 51]. Two models have been input to deduce the interband (IB) dielectric function of the bimetallic system. In the first one a simple weighted average of both IB dielectric functions of each component is used presupposing the existence of nano-domains in the clusters. A direct consequence is the appearance in the effective dielectric function of two IB thresholds due to silver and gold IB thresholds. In the second model the IB dielectric function reproduces the dielectric function measured in annealed alloy films. In this last case, there is only one IB threshold shifting from the one of silver to the one of gold with increasing gold proportion. In both cases a qualitative agreement may explain composition and size effects experimentally observed, but the shape of the spectra with the second model better reproduces the shape of the experimental ones. This suggests that bimetallic AuAg clusters produced by laser vaporization of an alloyed target adopt an alloyed structure at an atomic level, near from the ones of annealed alloys. This example illustrates the case of an alloyed structure whose dielectric function can be inferred from the corresponding one in bulk alloys. Let us underline that similar evolution of the optical response with a single resonance is observed in aqueous solutions of alloyed gold-silver nanoparticles of diameters ranging from 15 to 50 nm [52]. The Mie theory using the effective dielectric function deduced from the second model allows, here also, to better reproduce the composition evolution of the experimental spectra.

The NiAg and AgPt Systems

Works on NiAg clusters are very scarce, but studies on NiAg films show that silver and nickel weakly interact [53]. Experimental absorption spectra of samples consisting of $(Ni_xAg_{1-x})_n$ clusters in alumina for various composition x is depicted in Fig. 7. The optical response is intermediate between those of pure silver and pure nickel clusters, with a SPR broadened and blue-shifted compared to pure silver clusters [41]. This feature can be reproduced with the Mie theory in the dipolar approximation assuming segregation between both metals, either with a core/shell geometry or without any regular repartition (random dispersion of pure nano-domains in the cluster). In this last case the effective dielectric function of the mixed system is the simple weighted average of the ones of the pure metals. In the core/shell geometry, the reduction of the mean free path, inducing a resonance broadening is taken into account [4, 39].

As experimental results are well reproduced with both hypotheses, it is clear that the effective dielectric function is simply correlated with the ones of its constituents. Nevertheless, optical studies are not sufficient to discriminate between both structures. Consequently, other characterizations have to be performed to probe the cluster surface, as for instance, Energy Filtered Transmission Electron Microscopy (EFTEM) for which images of Cu/Ag clusters have shown that clusters adopt a core/shell structure [54]. Nevertheless, in this very low size range (a few nanometers in diameter) core/shell structures will not be easy to observe by EFTEM, and Low Energy Ion Scattering (LEIS) is an alternative allowing to probe the cluster surface. Such characterizations performed on $(Ni_{50}Ag_{50})_n$ clusters deposited under Ultra high vacuum give the evidence that the surface of the clusters is mainly covered with silver atoms [41]. Finally combining optics with LEIS measurements indicates that the core/shell geometry with silver on the surface is the most likely structure.

Concerning AgPt clusters in alumina [55], LEIS measurements lead also to the conclusion that the cluster surface is mainly composed of silver. On the other hand experimental absorption spectra displayed in Fig. 8 show a damped and broadened SPR whatever the composition, in total disagreement with theoretical calculations

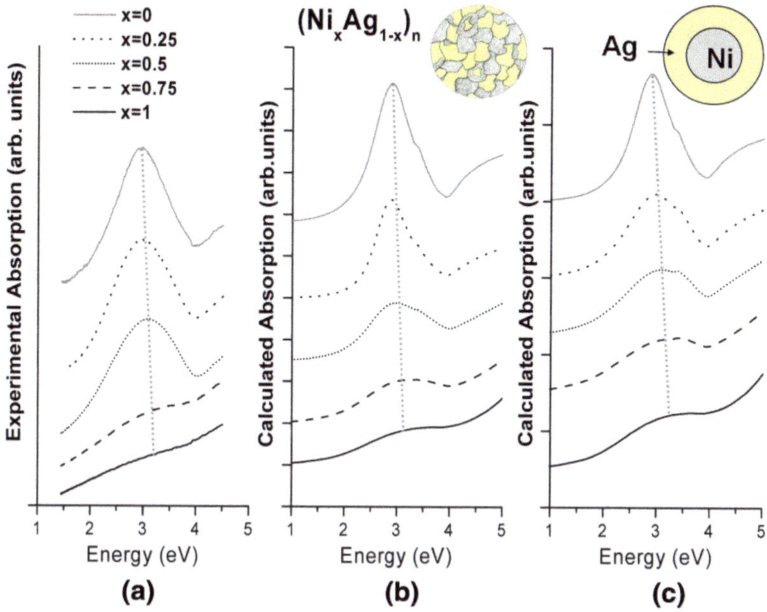

Fig. 7 **a** Absorption spectra of $(Ni_xAg_{1-x})_n$ clusters [of almost same optical diameters (2.6–2.7 nm)] in alumina for various composition x. **b** Theoretical absorption spectra in the dipolar approximation using an effective dielectric function defined as: $\varepsilon_{eff}(\omega) = x\varepsilon_{Ni}(\omega) + (1-x)\varepsilon_{Ag}(\omega)$. **c** Theoretical absorption spectra in the core/shell model by taking into account the reducing of the mean free path in the silver shell. From Ref. [49], reproduced by permission of the royal society of chemistry

for which a clear SPR similar to the case of NiAg is expected. In this example, it is thus obvious that the optical study allows to discriminate between a core/shell structure and an alloy at a atomic level. The results show that the dielectric function of the actual system is necessary completely different from those of both components. Indeed, even for silver proportions of 80%, the SPR is not manifest suggesting that the cores of the clusters are made of an alloy with its own dielectric function not correlated with the ones of silver and platinum. LEIS measurements giving evidence of a surface rich in silver, the system is probably composed of an alloy surrounded by an atomic shell of silver, the silver atoms on the surface permitting to lower the whole energy of the bimetallic system.

Monte Carlo simulations performed on small clusters of AuAg, NiAg and AgPt rich in silver (75% of silver atoms in clusters containing up to 309 atoms) lead to the same conclusions [56]. First, AuAg clusters are stable as a solid solution equilibrium phase below their melting point. At zero temperature segregation is expected in NiAg and AgPt systems with cluster surfaces entirely composed of silver atoms. When the temperature increases the NiAg system remains segregated. Besides, it displays a peculiar transformation to prolate shapes before the nickel core melts. Ag/Pt as for it, exhibits a continuous transition from core/shell to alloy core/pure Ag shell.

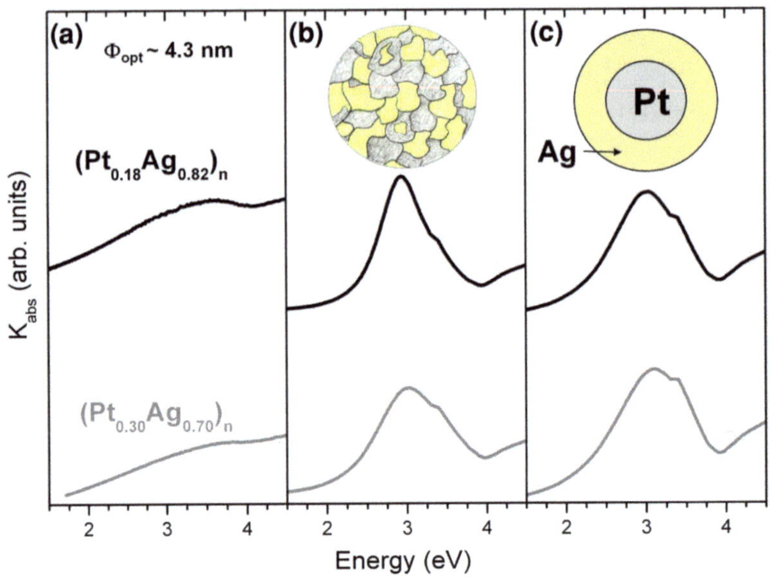

Fig. 8 **a** Absorption spectra of $(Pt_xAg_{1-x})_n$ clusters (of same optical diameters (4.3 nm)) embedded in alumina for various composition x. **b** Theoretical absorption spectra in the dipolar approximation using an effective dielectric function defined as: $\varepsilon_{eff}(\omega) = x\varepsilon_{Pt}(\omega) + (1 - x)\varepsilon_{Ag}(\omega)$. **c** Theoretical absorption spectra in the core/shell model by taking into account the reducing of the mean free path in the silver shell

To conclude, we have shown that the optical properties of bimetallic clusters may exhibit peculiar features as compared to their constituents. The relative composition can be used as a parameter to shape the Surface Plasmon Resonance. Moreover as the optical response reflects the dielectric function of the bimetallic system, it may reveal its internal structure if the other parameters (matrix index, shape and size dispersion, ...) influencing the optical response are well controlled. In this respect, the emergence of single NPs studies for which the size, the shape and the environment of the NP may be precisely known [31] will be surely promising.

In the examples depicted above the internal structure of bimetallic clusters was partly revealed by their optical response, but it has to be underlined that optics is generally not sufficient and other characteristic measurements have to be necessary performed (like LEIS or electron microscopy).

2 Ultrafast Optics and Femtomagnetism

Femtosecond optical spectroscopy is a powerful tool for analyzing the non-linear ultrafast optical properties of materials and the underlying electronic, vibrational and thermal kinetics and interaction processes. In this part of the chapter, after a brief introduction on the ultrafast excitation and electron relaxation in metals, we will focus on application of these techniques to study the magnetic properties of metal clusters.

2.1 *Ultrafast Electronic and Acoustic Response of Metal Nanoparticles*

Time-resolved spectroscopy with femtosecond lasers has been extensively used to investigate metallic materials in strongly non-equilibrium situations. In the usual "pump-probe" configuration, the electrons in the metal are first driven out of equilibrium by focusing a strong pump laser on the system. Their relaxation to equilibrium is subsequently followed by monitoring for instance transmission or reflection changes of a second pulse, the probe, delayed with respect to the pump. The time evolution of the optical properties of the materials is closely related to modification of the electron or lattice properties induced by the optical excitation. Depending on the pump and probe conditions (wavelength, polarization, pulse duration), the different processes in the metal relaxation can be analyzed, corresponding to energy redistribution among the electrons, towards the lattice, and to the environment [57].

These processes and the underlying interactions (electron–electron and electron–phonon scattering, metal—matrix coupling) are key parameters determining the properties of nanostructured materials. They have been extensively studied on ensembles of noble metal nanoparticles (see [14] and references there-in), and more recently in nickel and bimetallic Au–Ag and Ni–Ag clusters [49]. Extension of these studies to a single particle has been recently demonstrated, by combining the SMS setup (described in Sect. 1.2.1) with a high sensitivity two-color pump-probe technique [58].

Most of these investigations on single and ensembles of particles were performed for large sizes ($D > 3$ nm), for which a "small solid" modeling can be used. The characteristic internal thermalization time (typically a few hundred fs in bulk noble metals) has been found to remain almost unaffected by confinement effects for sizes larger than $D \sim 10$ nm, and to strongly decrease for smaller sizes [59]. A similar behavior has been observed for the electron-lattice energy exchange time [60]. This is illustrated Fig. 9 for Au, Ag and bimetallic AuAg clusters embedded in different matrices. These modifications have been ascribed to a reduction of the screening effects (i.e. increasing of the efficiency of electronic interactions) when reducing the size of the nanoparticle. Extension to small clusters ($D < 2$ nm) should be particularly promising to understand the transition between a small solid (nanoparticle) and a molecular-like behavior (in clusters).

The impact of size reduction on the acoustic response of a nano-object is also attracting considerable interest. This is motivated by fundamental issues, i.e., the impact of size on the elastic properties of a nano-object, and by technological ones, i.e., the development of high frequency THz nano-resonators and nano-electro-mechanical

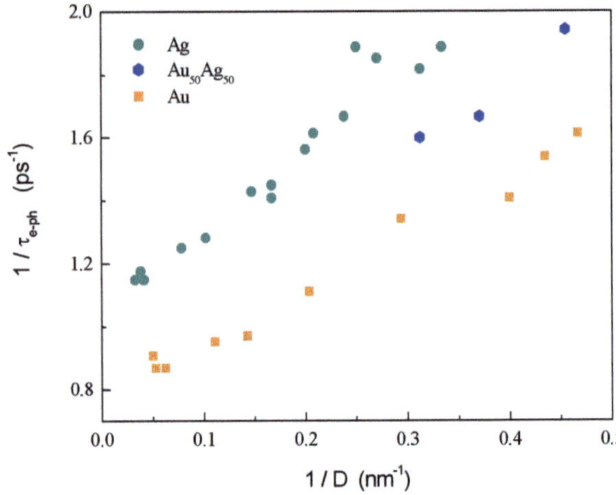

Fig. 9 Measured size evolution of the electron—lattice energy exchange rate for Au, Ag and $Au_{50}Ag_{50}$ clusters of diameters ranging from 30 to 2.2 nm. Results from Ref. [49, 60]

systems (NEMS) for application as high-sensitivity sensors or optomechanical transducers. In a pump-probe technique the ultrafast electron excitation and fast energy transfer to the lattice (faster than the typical mechanical oscillation period of the nanoparticle) impulsively launches the acoustic modes of the nano-objects. The coherent mechanical movement of all the excited particles subsequently leads to oscillations of the probe transmission change on a picosecond time scale, permitting its direct time-domain monitoring. This non-contact optical technique has emerged as a powerful tool for investigation of nanoparticles [61]. The impact of size, shape, structure and environment has been demonstrated in ensemble and single particle experiments, by measuring the vibrational mode frequencies and damping [62–69]. As the frequency of a resonator reflects its size, small clusters are expected to vibrate in the THz domain. At this scale, fundamental questions raise on the description of their acoustic response using continuum mechanics, and on the elastic properties of the constituting materials. Recent experiments have been performed on high-quality, surfactant-free quasi-spherical platinum nanoparticles with diameters between 3 and 1.3 nm (i.e., 940 to 75 atoms), vibrating at frequencies ranging from 1.1–2.6 THz [70]. Surprisingly, these values are in excellent quantitative agreement with the prediction of the macroscopic elastic model, using the continuum elastic model together with the bulk elastic constants of platinum. This demonstrates that, in contrast to the electronic response, which strongly modified by quantum effects, the mechanical vibration period of a metal cluster is very well described by a macroscopic approach, down to the one-nanometer size range (less than 100 atoms).

2.2 Ultrafast Optics and Magnetism: Femtomagnetism

In this section, we will describe an emerging and attractive research axis in magnetism: the so-called femtomagnetism. It consists in exploring the dynamics of magnetization of magnetic systems using femtosecond laser pulses. Here, we will focus on magnetic nanostructures where questions under debate are threefold and constitute the guideline of the three following subsections. First, what is the pathway of magnetization of magnetic nanoparticles under magnetization reversal induced by femtosecond pulses? Secondly, what is the role of interparticle interactions, known as an important parameter in static magnetic properties in the femtosecond regime? And finally, what brings spin photonics in ultrafast optics and nanophotonics?

2.2.1 State of the Art

In 1996, it has been demonstrated that one can induce an ultrafast modification of magnetization up to complete demagnetization in a ferromagnetic film using femtosecond laser pulses [71, 72]. This has been achieved by time resolved magneto-optical Kerr technique in a pump and probe configuration. Let us emphasize that these pioneering experiments led to many subsequent experimental works that corroborated their findings and initiated a new research field known as femtomagnetism [73–75]. Later, it has been shown that one can take advantage of the ultrafast demagnetization to launch a precession behavior around an effective field [76] or can induce a magnetic order phase transition by using femtosecond pulses [77]. Despite a large amount of works, little is known about the origin of the ultrafast demagnetization. Only a few theoretical attempts focused on the demagnetization process at a microscopic level [78]. Other models [79] have been proposed that seems to be rather controversial [80, 81]. More recently, using time-resolved X-ray magnetic circular dichroism, a first strong indication has been revealed about the role played by spin–orbit interaction in the ultrafast demagnetization induced by femtosecond laser pulses [82].

2.2.2 Time Resolved Magneto-Optical Techniques

If one associates magnetism to ultrafast optics, it is essential to describe time scale for magnetic mechanisms. In the millisecond to a few hundred of microsecond time range, a macroscopic effect takes place corresponding to the domain motion and to the propagation of their walls under applied magnetic field. On a shorter timescale (10^{-9} s), damping processes are involved that have been fruitfully modelled by different approaches such as Bloch [83], Landau-Lifshitz [84] or Gilbert [85] models. Then, in tens of picoseconds, precession behaviors around an effective field occur associated to time-dependent anisotropy. Finally, the sub-picoseconds time scale corresponds to the so-called femtomagnetism where many mechanisms may participate such as exchange interaction, spin–orbit interaction or energy relaxation from the electrons to the lattice. All these mechanisms are accessible by ultrafast spectroscopy techniques based on present femtosecond laser technology.

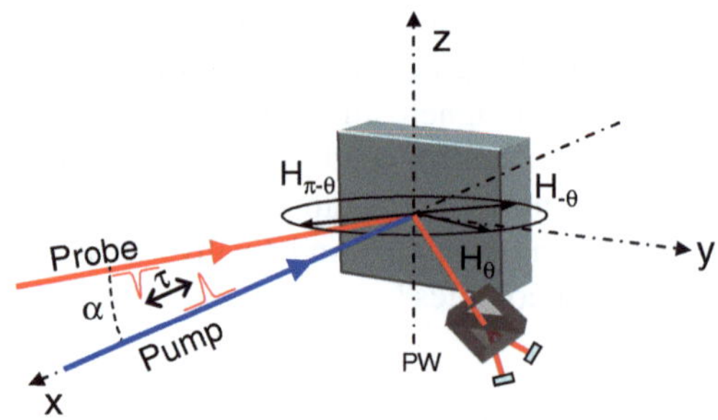

Fig. 10 Polarimetric setup used to extract the 3D trajectory of magnetization vector, presented here in case of polarization bridge detection using a prism Wollaston (PW). Copyright (2006), American Institute of Physics

Since the nineteenth century it is well known that crucial information on magnetization of a material are obtained by polarization changes analysis of a reflected (Kerr effect) or a transmitted (Faraday effect) light beam. Here, we will focus on their extension to the time domain using femtosecond pulses. Mainly, ultrafast dynamics of magnetic nanostructures have been obtained by the pump and probe technique as described earlier in this chapter. A permanent magnetic field is applied to the sample to measure magnetization dynamics. Generally, the static magnetization vector components can be retrieved for a given angle α of the impinging probe beam by analyzing polar, longitudinal and transverse components of the magneto-optical signals. We present here a dynamical version of this polarimetric technique used to follow in real time the variations of the three components of magnetization polar (Pol), longitudinal (Long) and transverse (Trans) of the magnetization **M** [76]. The three dimensional setup is shown on the Fig. 10 for the Kerr configuration. The sample is defined by y–z axis and the incident plane is xOy. Therefore x, y, z correspond to respectively polar, longitudinal and transverse components of the magnetization. The applied magnetic field **H** is kept perpendicular to the z-axis. The components of magnetization are obtained by linear combination of the Kerr signals for three complementary angles of **H** with respect to the normal of the sample: θ, $-\theta$, $\pi - \theta$ as follows:

$$Pol = \frac{S(\theta) - S(\pi - \theta)}{2} = \frac{S(\theta) + S(-\theta)}{2} \tag{8}$$

$$Long = \frac{S(\theta) - S(-\theta)}{2} = \frac{S(\theta) + S(\pi - \theta)}{2} \tag{9}$$

$$Trans = \frac{R(\theta) - R(\pi - \theta)}{2} \tag{10}$$

Let us emphasize that the transverse component is obtained without polarization analysis (R designs the reflectivity) but is only accessible for a p-polarized probe. The corresponding dynamics are obtained by applying these equations to the time dependent Kerr signals defined as $\Delta S(t)/S_0$ where $\Delta S(t) = S(\text{with pump})-S_0(\text{without pump})$. When it is possible, transmitted signals are also measured. The signals are measured by synchronous detection using a lock-in amplifier. Mainly, it exists two ways to analyze Kerr/Faraday rotation either by crossed polarizers or by a polarization bridge.

2.3 Non Interacting Magnetic Nanoparticles

The advancements of the last decade in the fabrication and characterization of magnetic materials at the nanoscale combined with the availability of new experimental techniques of investigation, as for example, ultrafast lasers or new generation synchrotrons sources, have made the study of magnetic nanostructures an important current topic in condensed matter physics. In recent years, this effervescence may be illustrated by some breakthroughs, as the discovery of the spin-torque transfer effect [86–88], the ultrafast demagnetization and coherent precession trigged by femtosecond laser pulses [71, 89, 90] or even by devices that have been developed since the 1990s based on the giant magnetoresistance effect [91]. For the magnetic storage industry one key point remains the production, understanding and control of the properties of magnetic nanoparticles, since one obvious way to increase the density of stored information in magnetic storage media is decreasing particle size. Obviously, as we increase the speed of manipulation of information in magnetic storage media, there is an increased interest in understanding the ultrafast magnetization dynamics in magnetic nanostructures. In this context, the present evolution of the density of data storage will quickly lead to a technological breakdown corresponding to the superparamagnetic limit. This limit is associated to the size of the magnetic nanoparticles used to store information where the decrease of volume anisotropy leads their magnetization to undergo thermal fluctuations resulting in an apparent zero magnetization on a time scale larger than the thermal fluctuations time τ. These aspects are treated in details in Magnetism of Low-Dimension Alloys. Clearly, to understand and eventually to control this dynamics it is important to know the pathway of magnetization vector during the magnetization reversal. In their early works, Néel and Brown have predicted that the pathway of magnetization in monodomain magnetic nanoparticles is similar to the one of a tiny gyroscope [92, 93]. We readdress here this long-standing problem taking advantage of the development of new femtosecond magneto-optical techniques that have proven to be a powerful tool to study magnetization dynamics in ferromagnetic films [71, 74, 89, 90, 94, 95].

2.3.1 Trajectory of Magnetization

We have studied the dynamics of magnetization of cobalt nanoparticles realized by ionic implantation of Co^+ at 160 keV either in SiO_2 or in sapphire matrices. The nanoparticles have been all implanted at 600°C with different fluxes F resulting in different sizes. Here, we focus on those made with $F = 10^{17}$ cm^{-2}. In case of SiO_2 matrix, the cobalt nanoparticles have a diameter of 10 nm and a ferromagnetic behavior at 300 K with a 105 Oe coercive field as shown on the

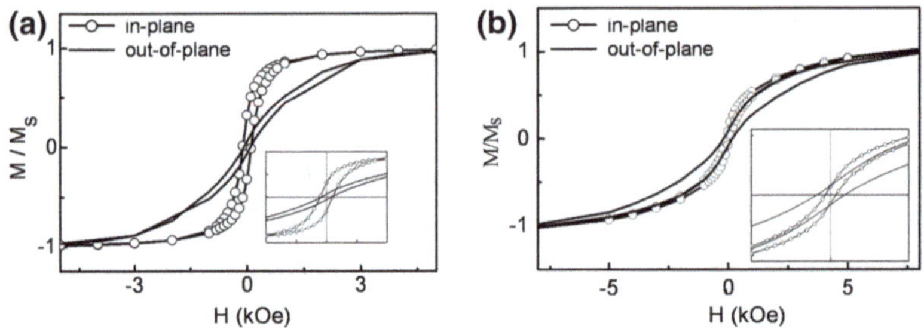

Fig. 11 Hysteresis loops of cobalt nanoparticles for in-plane and out-of-plane applied magnetic field **a** in SiO_2 at 300 K **b** in sapphire, at 30 K (*solid line* and *open circles*)

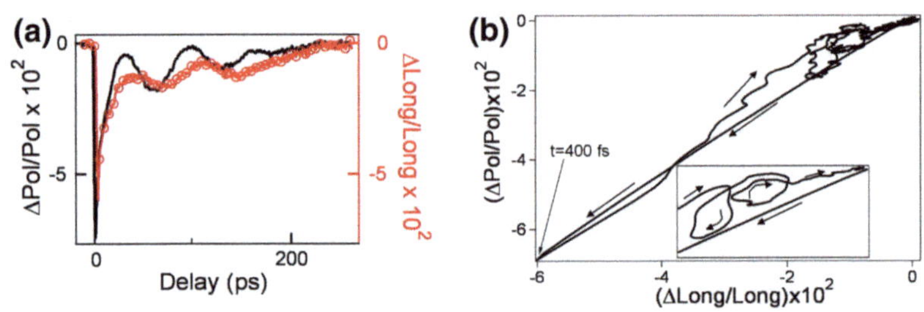

Fig. 12 **a** Dynamics of the polar (*black*) and longitudinal (*red*) components of magnetization of ferromagnetic nanoparticles. **b** Projection of the trajectory in the polar/longitudinal plane as a function of the delay. Detailed view of the precession motion (*down right*)

hysteresis loops of the Fig. 11a. In case of an Al_2O_3 substrate the coercive field is equal to zero at 300 K (Fig. 11b) which corresponds to a superparamagnetic phase. The nanoparticles have a 4 nm diameter which has been calculated thanks to their blocking temperature $T_B = 80$ K extracted from ZFC/FC (Zero Field Cooling/Field Cooling) measurements using the following formula [90]:

$$D = \sqrt[3]{\frac{150 k_B T_B}{\pi K}} \tag{11}$$

with K the anisotropy constant and k_B the Boltzmann constant.

The trajectory of magnetization in these cobalt nanoparticles samples have been measured using the polarimetric setup described in the Sect. 2.2 using laser pulses of 130 fs delivered by an amplified femtosecond laser system cadenced at 5 kHz. The probe beam is centred at 800 nm, the fundamental wavelength of the laser system and the intense pump beam is obtained by frequency doubling in a non linear BBO crystal.

The Fig. 12a shows the dynamics of the polar and longitudinal components of magnetization in ferromagnetic nanoparticles in silica matrix with an applied magnetic field of 2.8 kOe and a density of excitation of 0.3 mJ/cm^2. It shows an

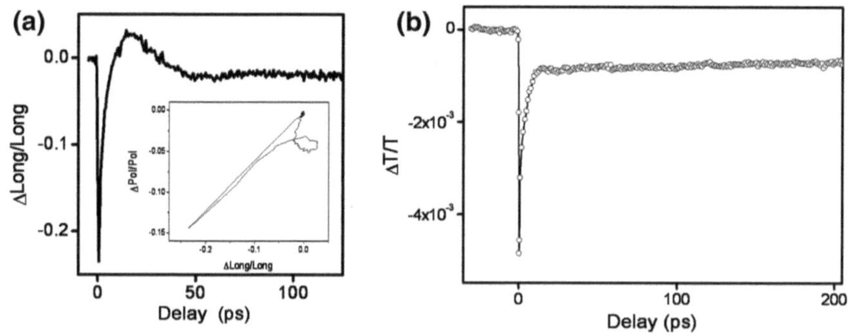

Fig. 13 **a** Dynamics of the longitudinal component for $I_p = 1.7$ mJ/cm^{-2} and $H_O = 2.8$ kOe, Inset: corresponding trajectory in polar/longitudinal plane. **b** Corresponding differential transmission dynamics. From Ref. [90], copyright (2006) by The American Physical Society

ultrafast demagnetization associated to the fast raise of the electronic temperature, followed by a partial re-magnetization as electrons cool down to the lattice through electrons (spins)-phonons interactions. At longer time delays, the dynamics exhibit oscillations with a period of 74 ps and a damping of 120 ps. These oscillations correspond to a motion of precession due to a modification of the effective field related to a temperature dependant anisotropy and exchange interaction. It corresponds to the analogous in time domain of the ferromagnetic resonance measured in the frequency domain induced by radio frequency magnetic field [96].

The correlation between charges and spin dynamics is displayed on the projection in the polar/longitudinal plane of the trajectory of magnetization (Fig. 12b). As one can see the maximum demagnetization is reached after 400 fs coinciding with the thermalization of electrons. Let us stress that the dynamics of electrons (not shown here) has the expected behavior in metals and is well described by the usual two temperatures model [97].

The question about the pathway of the magnetization vector under thermal fluctuations is of crucial importance. In the Fig. 13, we have represented the dynamics of the longitudinal component of the magnetization vector for superparamagnetic particles with $d = 4$ nm under an applied external magnetic field $H_0 = 2.8$ kOe and a density of excitation $I_p = 1.7$ mJ.cm^2. As can be observed, it exhibits an oscillatory behavior with a period $T = 50$ ps and a damping time $\eta = 90$ ps. The gyroscopic behavior is obvious in the inset of the Fig. 13a representing the trajectory of the magnetization vector in the plane polar/longitudinal. The first step is a fast demagnetization reaching its maximum at 400 fs corresponding to the thermalization time of electrons [90] followed by a partial re-magnetization. Finally precession behavior takes place which is rapidly damped as a single loop is observable before the magnetization comes back to its initial state. Let us notice how powerful our technique of investigation is. Indeed, our time resolution is orders of magnitude faster than the thermal fluctuations time which minimizes their influence. Moreover, we measure under a permanent

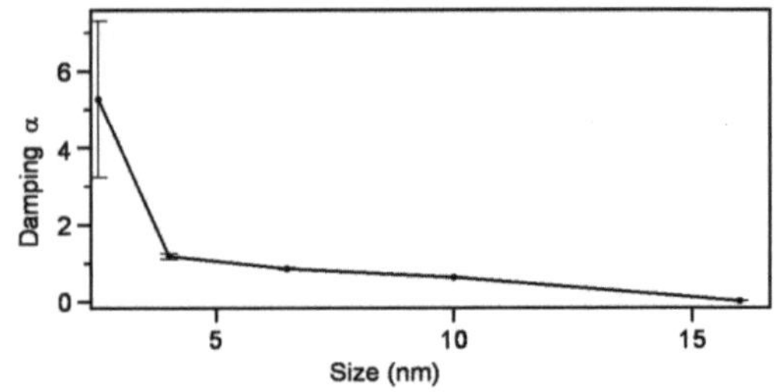

Fig. 14 Size effect of nanoparticles on the damping of Gilbert

magnetic field so that the nanoparticles have a magnetization close to the saturation value Ms. In this configuration we only explore an intra-well dynamics. Our methodology allows us to follow simultaneously the spin dynamics and the charge dynamics as shown in the Fig. 13b. As mentioned above, the correlation between magnetization and electrons dynamics is obvious. Initially, the electrons excited above the Fermi level are not thermalized due to Pauli principle. The process of thermalization is not instantaneous and lasts 400 fs. Then the thermalized electrons relax to the lattice via electron–phonon interaction $\tau_{e-l} = 1$ ps. As electrons and phonons are in equilibrium, the energy is transferred to the environment. Depending on the nature of the matrix, this process can occur in several steps. In our case where the nanoparticles are dispersed in sapphire, the heat transfer to the matrix lasts 7 ps due to the large surface/volume ratio [98, 99] then thermal diffusion out of the excited area occurs (730 ps).

We have also studied the evolution of the damping of precession as a function of the size of the particles. To lighten the role of size, we have compared the damping time η of different samples of nanoparticles and of a thin Co film. The thin film is a 16 nm thin Co film elaborated by MBE (Molecular Beam Epitaxy) on a sapphire substrate. The nanoparticles of cobalt are those described all along this paper. For each of these samples, we have calculated the Gilbert damping α which in small damping approximation is linked to the damping time η via the relation [99]:

$$\alpha = \frac{2}{M_s \eta \gamma_0} \tag{12}$$

where $\gamma_0 = 2.4 \times 10^5$ $A^{-1}.s^{-1}$ is the gyromagnetic factor of Co. The magnetization at saturation M_S and the damping time η are extracted from experimental measurements. The Fig. 14 represents the Gilbert damping α as a function of size. The reported value for the Co film is extracted from the Ref. [6]. It shows up that the damping continuously increases as the particle size decreases and is larger in nanoparticles than in bulk. Let us notice in case of nanoparticles with $d = 2.5$ nm the corresponding α value is probably overestimated. Indeed, in such case, the precession behavior is quickly damped and the oscillations are hardly observable. In these conditions, the approximation of small damping is also questionable. The mechanisms responsible for the increasing of damping are not quite clear. However, our results are consistent with previous ac susceptibility and

Mossbauer spectroscopy experiments [100]. We believe as sustained by Respaud et al. [101] that particle–matrix interface plays a crucial role on the magnetic disorder. Moreover, our results show that the particles interactions do not affect much the damping of precession. Indeed, when we compare the nanoparticles implanted in sapphire matrix, it appears that the smaller ones ($d = 2.5$ nm) however implanted with a lower dose of metal than the sample of nanoparticles with d = 4 nm, exhibit a larger damping. This result is in contradiction with Dormann's work [100] but pleads in favor of spin scattering at the surface of nanoparticles.

Finally, we have shown that the pathway of magnetization under magnetization reversal is probably not completely coherent for the smallest nanoparticles.

2.3.2 Role of the Anisotropy

We have also compared the magnetization dynamics of superparamagnetic nanoparticles in sapphire substrate ($d = 4$ nm) and ferromagnetic nanoparticles in silica matrix ($d = 10$ nm). Comparatively, the ferromagnetic nanoparticles demonstrate a planar anisotropy along the Oy axis. We think that if in cobalt films the anisotropy axis results from the cobalt itself, in ferromagnetic cobalt nanoparticles, there are probably others factors that affect the anisotropy orientation. The elaboration technique used for the Co nanoparticles tends to align preferentially the magnetization along the direction of implantation perpendicular to the sample. Indeed, the superparamagnetic nanoparticles exhibit a perpendicular anisotropy along Ox axis as shown by SQUID measurements associated to a small deformation of nanoparticles in the implantation direction and an absence of interactions at such sizes (Fig. 11b). However, the cobalt nanoparticles implanted in SiO_2 have in-plane anisotropy as shown in the Fig. 11a. It could be associated to dipolar interactions between nanoparticles. The initial trajectory of magnetization in the polar/longitudinal plane clearly depends on the anisotropy as shown in the Fig. 15 in the temporal range 0–30 ps with an applied field $H_0 = 2.8$ kOe and a density of pump excitation of $I_p = 2$ mJ.cm^{-2}. Both samples have an initial ultrafast demagnetization in the first hundred of femtoseconds. However, as the partial re-magnetization takes place the magnetization vector re-orientation follows a different pathway for each sample. In case of a planar anisotropy (Fig. 15a) the polar signals relaxes faster than the longitudinal component whereas for a perpendicular anisotropy (Fig. 15b), the magnetization direction rotates toward the Oy axis.

2.4 *Self-Organized Magnetic Nanostructures*

Self-organization is present in many natural phenomena such as the spontaneous magnetization, the crystallization at the atomic scale. This is ubiquitous at a larger scale ranging from nanometer to micrometer in living systems. This has inspired

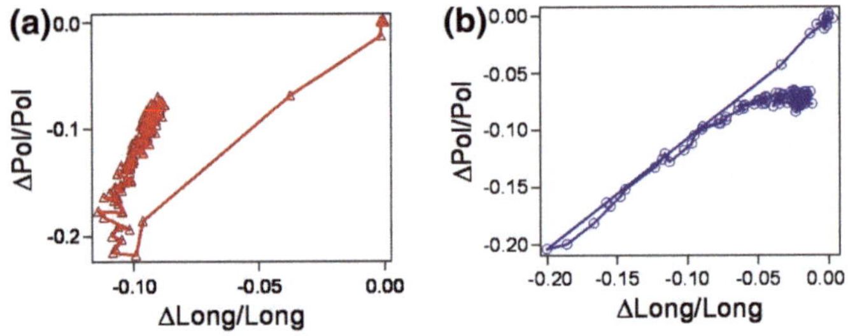

Fig. 15 Comparison between magnetization trajectories in **a** Co/SiO_2 and **b** Co/Al_2O_3 for $I_p = 2$ mJ.cm^{-2} and $H_0 = 2.8$ kOe

many research works that aim to mimic this building process, as in case of photonic crystals elaboration using silica microspheres [102]. At nanometer scale, 3D structures of metallic nanocrystals are interesting material because they exhibit new collective properties. In particular, magnetic self-organizations have attracted much interest in recent years associated to their potential technological applications mainly in high-density data storage [103]. As we mentioned earlier, our aim is to study the influence of interactions between nanoparticles on the optical and magneto-optical properties of such mesostructures using ultrafast spectroscopy combined to confocal microscopy. We will describe experimental details in the first sub-section of this paragraph. In the second one, dynamics of charges and spins in self-assemblies of cobalt nanoparticles will be discussed.

2.4.1 Time-Resolved Confocal Microscopy

In order to study the dynamics of electrons in self-organized structures, we have used a really ingenious instrument developed for the very first time at IPCMS in Strasbourg. It consists in a time-resolved confocal microscope in reflectivity which combines a sub micron spatial resolution to a sub-picosecond temporal resolution [104]. Let's stress that this device has been develop initially to achieve ultrafast measurements of magnetization dynamics in sub-micron ferromagnetic disks as we will show in the next section. However, we have exploited this smart device to study the dynamics of reflectivity in cobalt nanoparticles mesostructures.

The dynamical signals have been measured using the pump and probe technique. The pump and probe pulses are obtained from a titanium-sapphire amplified laser system cadenced at 5 kHz which delivers 150 fs pulses centered at 790 nm. The pump beam spectrum is centered at the fundamental wavelength and the probe beam ($\lambda = 395$ nm) is obtained by second harmonic generation in a BBO crystal. The two beams are focused on the sample collinearly through a microscope objective with a 0.65 numerical aperture (Fig. 16). The reflected probe beam is collected by a dichroic beam splitter and focused in a 20 μm diameter pinhole. Then, the signal is detected on a photomultiplier using a synchronous detection

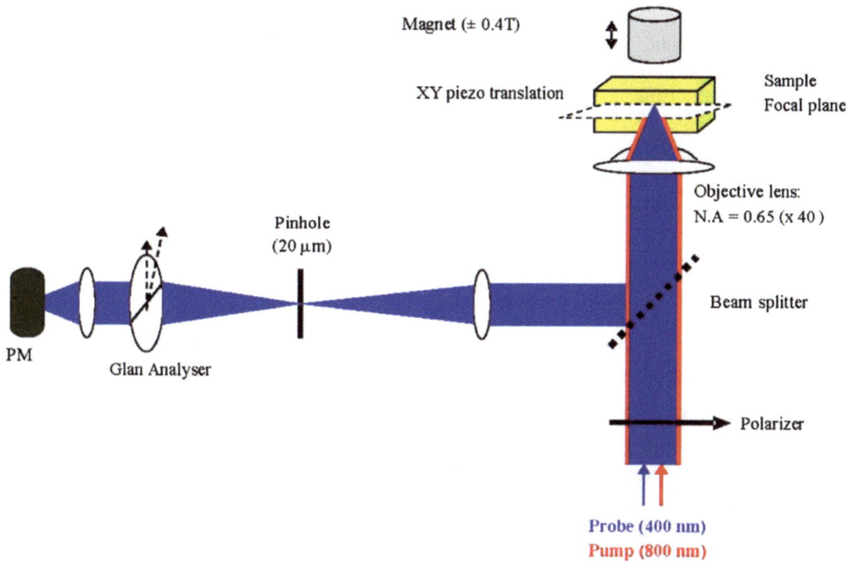

Fig. 16 Time resolved confocal Kerr microscopy setup. From Ref. [108], copyright (2007) by Springer Science + Business Media

scheme. The detection bench is also equipped by a Glan-Taylor polarizer which is used as an analyzer in magneto-optical Kerr measurements.

The two beams are focused at their diffraction limit which results in an overall spatial resolution of 500 nm.

In this sub-section, we investigate 2D and 3D well-organized suprastructures of cobalt nanoparticles. They have been elaborated by a soft chemistry technique [105]. Initially, they use micellar solutions of cobalt ions which are reduced by sodium borohydride to form cobalt nanocrystals. After being coated by dodecanoic chains, the nanocrystals are washed. The resulting black powder is then dispersed in hexane and centrifuged to remove the largest nanocrystals. This procedure allows obtaining a rather narrow size distribution. Our samples have a characteristic size distribution of 11%. Two sizes of particles have been studied here with either 6.5 or 7.5 nm of diameter. After synthesis, the colloidal solution of cobalt nanoparticles is evaporated on highly oriented pyrolitic graphite (HOPG) substrate. Depending on the conditions of evaporation different levels of organizations can be obtained. 2D assemblies are realized by drop wise deposition on HOPG. The resulting structures are films organized in a hexagonal network. In order to realize 3D arrangements, a HOPG substrate is immersed in the colloidal solution. In case of a well controlled slow evaporation process under nitrogen, highly ordered fcc supracrystals are obtained. These films are characterized by cracks and perfectly organized pavements with size upper than 10×10 μm. At contrary, amorphous structures are obtained by a fast evaporation of the colloidal solution in the air. For each batch of nanoparticles, 3D supracrystals and amorphous assemblies have been evaporated. For each kind

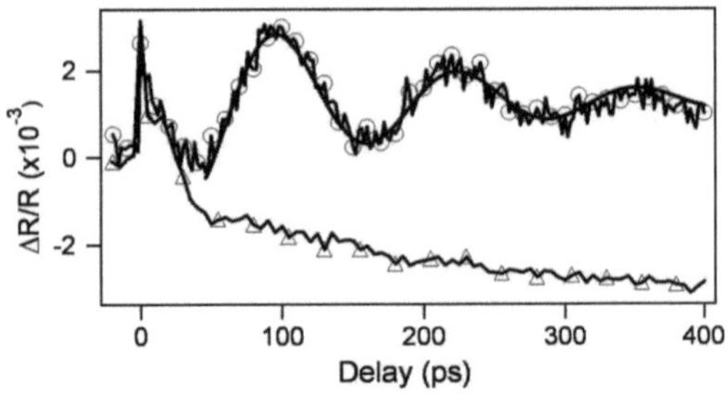

Fig. 17 Differential reflectivity dynamics of assemblies of 7.3 nm cobalt nanocrystals at long time range (*open triangles*: amorphous assembly; *open circles*: fcc supra-crystal)

of these native arrangements, annealing have been also performed at 350°C. This treatment does not affect the ordering of the film. However, it modifies significantly the crystallinity of the nanocrystals transforming poorly crystallized fcc nanoparticles to hcp nanocrystals.

The suprastructures reveal also interesting magnetic properties [106]. Indeed, under a critical size around 10 nm, individual cobalt nanoparticles behave like a macrospin and have a superparamagnetic behavior at room temperature. When self-organized, these nanoparticles have modified magnetic properties due to dipolar interactions. For instance, the blocking temperature T_B increases as the number of closest neighbors increases with the following values $T_B = 80$ K for isolated nanoparticles, 90 K in 2D arrangements and 110 K in 3D suprastructures. Another remarkable evolution in the magnetic properties has been observed as the nanoparticles are annealed. Indeed, it has been shown that the blocking temperature is drastically increased closed to room temperature ($T_B = 260$ K for 3D self-organizations). This is attributed to an increase of the anisotropy constant due to the structural change of individual nanoparticles from the polycrystalline fcc phase to pure hcp nanocrystals.

2.4.2 Vibration and Magnetization Dynamics

In this part, we investigate the ultrafast dynamics of electrons for fcc supracrystals and amorphous structures using the confocal microscope described in the previous section. The differential reflectivity is shown on the Fig. 17 for the ordered and disordered assemblies of cobalt nanoparticles with 7.3 nm diameter for a maximum density of excitation of 40 mJ/cm^2. The variation of the differential $\Delta R(t)/R$ signal reflects the variation of the electronic temperature following the pump excitation. As discussed above, the dynamics of electrons in metals is well described by the two temperature model. The femtosecond electron dynamics is characterized by an initial athermal distribution after the pump pulse excitation, which thermalizes in few hundred of femtoseconds in a hot electron distribution via electron–electron scatterings. Then, the electrons relax to the lattice via electron–phonon coupling with a characteristic time τ_{e-l} for both structures. This result is consistent as this short time mechanism is internal to the

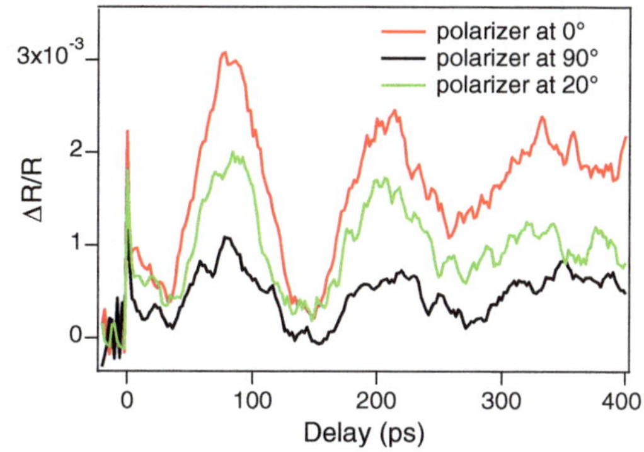

Fig. 18 Differential reflectivity of 3D supracrystal made of 7.3 nm nanoparticles for different polarizations of the probe

nanocrystals, which are issue of the same batch of nanocrystals for both ordered and disordered samples. At long time scale where the energy transfer to the environment takes place, the dynamics of both arrangements differ drastically. Indeed, in case of the amorphous structure, the decay is monotonous whereas one observes a periodic modulation with a characteristic time $T_{3D} = 131 \pm 5$ ps [107]. Let's remark that this period does not depend on the thickness of the explored region and that is the first observation in real time of coherent oscillation in self-organized assemblies of nanoparticles. Moreover, the spatial resolution of our microscope insures that we investigate one single well-organized pavement. One can observe also that these oscillations damp faster than the heat diffusion probably due to defects.

We have investigated the role of different parameters on the dynamics of reflectivity for such 3D supracrystals to understand the origin of these periodic features. We have measured the dynamics as a function of the incident polarization of light as shown on the Fig. 18. The polarizer sets at 0° corresponds to p-polarized probe. It proves that even if the contrast is modified, the period is not affected. So we can exclude a birefringence effect. In a recent study, it has been also demonstrated that the period of the oscillations decreases with the size of the nanoparticles [106]. It has been also shown that this coherent motion is consistent with a simple model of harmonic oscillator where nanoparticles assimilated to spheres are linked by forces due to the aliphatic chains acting as mechanical nanosprings.

We have also explored the dynamics of magnetization of these long-range ordered mesostructures. This has been achieved using a time-resolved magneto-optical Kerr setup similar to the one described in the Sect. 2.3.1. As can be seen on the Fig. 19 in case of a 2D structure, the dynamics of the polar component have been performed at room temperature and under the blocking temperature ($T_B = 90$ K) for a density of pump of 1.5 mJ/cm^2. The low temperature measurements are realized in a cryostat where a directional permanent magnet of 0.4 T insures a well defined magnetic initial state. After the fast demagnetization, the re-magnetization takes place as electrons cool down to the lattice for both temperatures. However, as we cross the blocking temperature, the dynamics shows a drastic different behavior with a maximum at 40 ps. We have also studied 3D

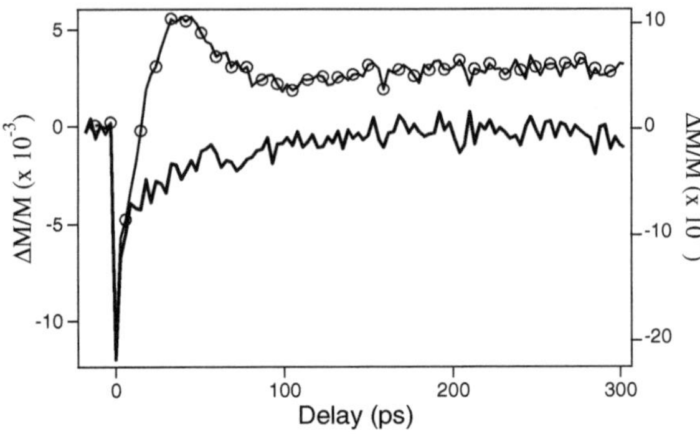

Fig. 19 Magnetization dynamics of 2D self-organizations for T = 300 K (*solid line, right axis*) and T = 71 K (*open circles, left axis*)

supra crystals made of natives nanoparticles. It shows up that the maximum at 40 ps appears also in the ferromagnetic phase. Moreover, it is also present, even less pronounced, in case of annealed 3D supracrystals at room temperature. This behavior may be associated to a precession behavior drastically damped or more likely, we think it corresponds to a preferential alignment of magnetic moments along the direction of the applied magnetic field as the dipolar interactions are temporally destroyed.

In this section, we show that dipolar interactions have also large consequences on the dynamics of charges and spins of nanoparticles assemblies.

2.5 Spin Photonics

This last section will treat about a new promising way for magnetic data storage which consists in modifying magnetization with photons called spin photonics. Indeed, in this technological context, even spin electronics has made important progresses; there is still some space for this other research direction. As the frequencies and the densities to reach correspond to the physical relevant scales in space: 100 to 10 nm and in time 100 ps to 100 fs, it is obvious to think at combining ultrafast optics and magnetism.

In the following, we have studied the magnetization dynamics of individual ferromagnetic disks with sub-micron size.

2.5.1 Magnetization Dynamics and Sub-micron Spatial Resolution

The measurements have been realized on the confocal microscope described earlier. A permanent magnet of ± 0.4 T magnetizes the sample out of plane. The magneto-optical signals obtained by the pump-probe technique are detected using either crossed polarizers or polarization bridge. The sample is mounted on a piezoelectric stage so that we can scan the sample.

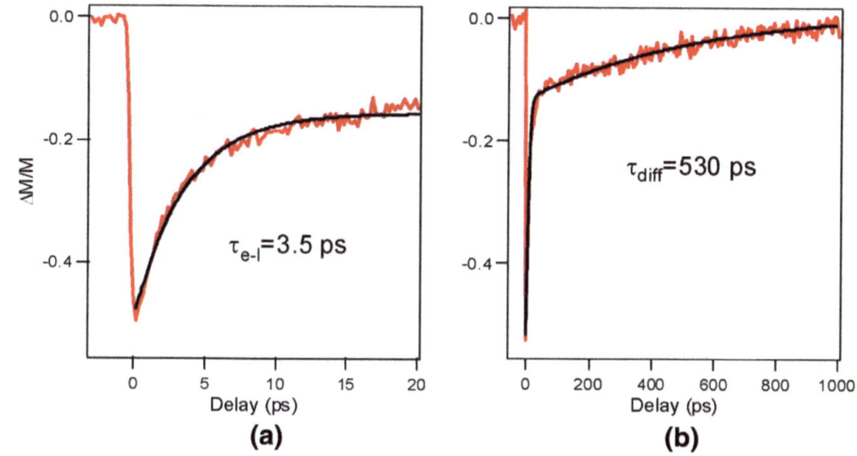

Fig. 20 Magnetization dynamics of a $CoPt_3$ disk with 1 μm of diameter at short **a** and long **b** delays for a density of excitation of 4 mJ/cm^2 and an applied magnetic field of 4 kOe. From Ref. [109], copyright (2007), American Institute of Physics

The samples have been elaborated by electronic lithography. They consist in 15 nm thick $CoPt_3$ dots on a Pt buffer layer deposited on a 500 μm oriented sapphire crystal (0001). The diameter d of the dots varies from 0.2 to 1 μm. These ferromagnetic dots have a high perpendicular magneto-crystalline anisotropy and a large coercive field. We present here the dynamics of magnetization for a $CoPt_3$ disk with d = 1 μm. At short delays (Fig. 20a), we observe the ultrafast demagnetization during the pump excitation, followed by the re-magnetization as electrons cool down to the lattice in a time $\tau_{e-l} = 3.2$ ps. At longer delay (Fig. 20b) as equilibrium between electrons and phonons is reached, the heat diffusion to the environment takes place and lasts 530 ps [108, 109].

We have also demonstrated that using this confocal Kerr microscope, we are able to follow a precession behavior on magnetization dynamics of permalloy squares [109]. Indeed, in this case, the anisotropy is in-plane and coercive field is weaker so, by applying a magnetic field perpendicularly to the film, we temporally align the magnetization out-of-plane and take advantage of the ultrafast demagnetization to launch the motion of precession around the effective field. In particular, it has been shown that the precession damping time is of the same order than the diffusion time, which is dissimilar to the case of cobalt thin films [110].

2.5.2 Spatio-Temporal Imaging of Magnetization

Using the piezoelectric motor stage, we have scanned the ferromagnetic disk for different pump-probe delays so that we can follow the spatial distribution as a function of time [111]. As shown on the Fig. 21, at 500 fs there is 50% demagnetization on the whole surface of the disk, followed by a partial re-magnetization

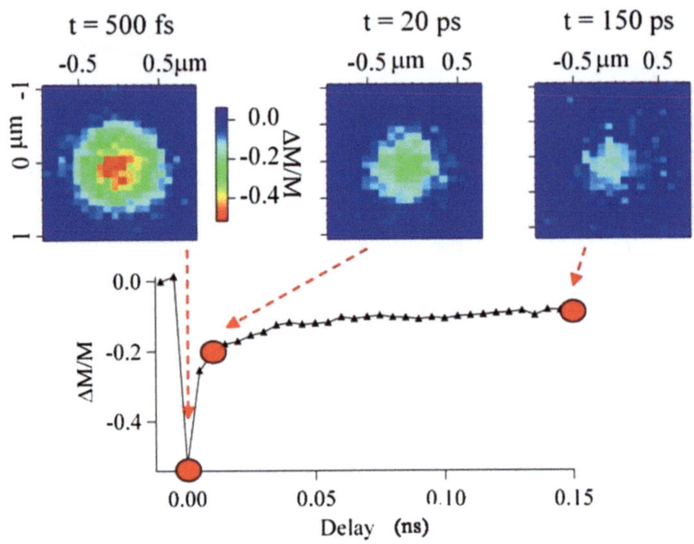

Fig. 21 Spatio-temporal evolution of magnetization of an individual $CoPt_3$ disk (d = 1 μm) for 4 mJ/cm^2 of density of excitation

at 20 ps and finally the quasi-total re-magnetization after 150 ps where the signal has almost disappeared as shown on the corresponding low contrast image. Let us mention that the contrast on each image has been adjusted to the maximum at 500 fs. So we can't see any spatial expansion in time. However, by taking the cross section for each image, we have observed an expansion of the diameter of the demagnetized area as pump-probe delay increases. We attribute this result to the heat diffusion along the lateral direction of the disk. It is also possible by using this approach to read, in small perturbations regime, magnetic domains written by intense laser pulses in ferromagnetic films or disks.

In this last section of the chapter, we have demonstrated that one can advantageously use ultrafast lasers pulses to study ultrafast dynamics of magnetization of magnetic systems even in superparamagnetic phase. We have also shown how powerful our technique is, based on the spatial resolution of a confocal microscope and the temporal resolution of femtosecond lasers. It has allowed us to study the dynamics of electrons in long-range ordered assemblies of nanoparticles and the spatio-temporal imaging of magnetization of individual sub-micron disks.

3 Conclusion

The new properties of mono-metallic nano-objects and the fundamental mechanisms at their origin have been extensively investigated during the last two decades. Though many questions are still open, these properties can now be largely tailored by playing with the object size, shape or environment. Their control, using physical or chemical synthesis techniques, has led to the development of many applications such as in nano-optics or plasmonics, nano-magnetism and catalysis.

More complex nano-objects, formed by two or more metallic components in the same particle, either alloys or segregated, can now be fabricated. This opens-up many new possibilities for adapting their linear and non-linear optical and magnetic responses. However, their detailed investigation and understanding, as a function of the nano-object composition and morphology for instance, are still very challenging, requiring a high control of the particle synthesis, multiple characterization tools and adapted theoretical models.

Acknowledgments E. Cottancin would like to warmly thank all the persons who took part in the works on bimetallic clusters, especially M. Broyer, F. Calvo, M. Gaudry, J. Lermé, M. Pellarin, B. Prével, J.-L. Rousset and J.-L. Vialle. N. Del Fatti thanks all the persons who contributed to this work and in particular F. Vallée, D. Christofilos, P. Langot, O. Muskens, A. Crut, P. Maioli, H. Baida. The Institut Universitaire de France (IUF) is also acknowledged. V. Halté would like to sincerely thank Jean-Yves Bigot for his precious and constant help to obtain the results presented here and for fruitful discussions about the concerned physics. She would like also to thank her colleagues: L. Guidoni, M. Vomir, M. Albrecht, A. Derory. Many thanks to A. Laraoui that realized most of the measurements shown here during his PhD. Finally, she is also grateful to C. Petit for his active participation to ultrafast dynamics measurements on self-organized assemblies. *Contacts* Natalia Del Fatti, *Linear and ultrafast optical properties of metal nanoparticles:* delfatti@lasim.univ-lyon1.fr; Emmanuel Cottancin, *Bimetallic systems*: cottancin@lasim.univ-lyon1.fr; Valérie Halté, *Ultrafast optics and magnetism - Femtomagnetism*: halte@ipcms.unistra.fr.

References

1. Faraday, M.: The Bakerian lecture: experimental relations of gold (and Other metals) to light. Philos. Trans. R. Soc. Lond. **147**, 145–181 (1857)
2. Mie, G.: Beiträge zur Optik trüber Medien, speziell kolloidaler Metallösungen. Ann. Phys. **25**, 377 (1908)
3. Bohren, C.F., Huffman, D.P.: Absorption and Scattering of Light by Small Particles. Wiley, New York (1983)
4. Kreibig, U., Vollmer, M.: Optical Properties of Metal Clusters. Springer, Berlin (1995)
5. Muskens, O., Christofilos, D., Del Fatti, N., Vallee, F.: Optical response of a single noble metal nanoparticle. J. Opt. a-Pure Appl. Opt. **8**, 264–272 (2006)
6. Ashcroft, N.W., Mermin, N.D.: Solid State Physics, International Edition. Saunders College, Philadelphia (1976)
7. Johnson, P.B., Christy, R.W.: Optical constants of the noble metals. Phys. Rev. B **6**, 4370–4379 (1972)
8. Kawabata, A., Kubo, R.: Electronic Properties of Fine Metallic Particles. II. Plasma Resonance Absorption. J. Phys. Soc. Jpn. **21**, 1765 (1966)
9. Perenboom, J.A.A.J., Wyder, P., Meier, F.: Electronic-properties of small metallic particles. Phys. Rep. **78**, 173–292 (1981)
10. Cottancin, E., Celep, G., Lermé, J., Pellarin, M., Huntzinger, J.-R., Vialle, J.-L., Broyer, M.: Optical properties of noble metal clusters as a function of the size : comparison between experiments and a semi-quantal theory. Theor. Chem. Acc. **116**, 514 (2006)

11. Kreibig, U.: Anomalous frequency and temperature dependence of the optical absorption of small gold particles. J. Phys. **38**, 97 (1977)
12. Wyrwas, R.B., Alvarez, M.M., Khoury, J.T., Price, R.C., Schaaff, T.G., Whetten, R.L.: The colours of nanometric gold—optical response functions of selected gold-cluster thiolates. Eur. Phys. J. D, **43**, 91–95 (2007)
13. Walter, M., Akola, J., Lopez-Acevedo, O., Jadzinsky, P.D., Calero, G., Ackerson, C.J., Whetten, R.L., Grönbeck, H., Häkkinen, H.: A unified view of ligand-protected gold clusters as superatom complexes. Proc. Nat. Acad. Sci. U.S.A. **105**, 9157–9162 (2008)
14. Voisin, C., Del Fatti, N., Christofilos, D., Vallee, F.: Ultrafast electron dynamics and optical nonlinearities in metal nanoparticles. J. Phys. Chem. B **105**, 2264–2280 (2001)
15. Bréchignac, C., Cahuzac, P., Kebaïli, N., Leygnier, J., Sarfati, A.: Collective resonance in large free potassium cluster ions. Phys. Rev. Lett. **68**, 3916 (1992)
16. Bréchignac, C., Cahuzac, P., Leygnier, J., Sarfati, A.: Optical response of large lithium clusters: evolution toward the bulk. Phys. Rev. Lett. **70**, 2036–2039 (1993)
17. Voshchinnikov, N.V., Farafonov, V.G.: Optical-properties of spheroidal particles. Astrophys. Space Sci. **204**, 19–86 (1993)
18. Farafonov, V.G., Voshchinnikov, N.V., Somsikov, V.V.: Light scattering by a core-mantle spheroidal particle. Appl. Opt. **35**, 5412–5426 (1996)
19. computed with SigmaMie, a code developed in FemtoNanoOptics group under LabVIEW
20. computed with a code developed in FemtoNanoOptics group under COMSOL Multiphysics
21. Purcell, E.M., Pennypacker, C.R.: Scattering and absorption of light by nonspherical dielectric grains. Astrophys. J. **186**, 705–714 (1973)
22. Draine, B.T., Flatau, P.J.: Discrete-dipole approximation for scattering calculations. J. Opt. Soc. Am. A **11**, 1491–1499 (1994)
23. Sönnischen, C., Franzl, T., Wilk, T., Plessen, G.v., Feldmann, J.: Plasmon resonances in large noble-metal clusters. New J. Phys. **4**, 93.91–93.98 (2002)
24. Boyer, D., Tamarat, P., Maali, A., Lounis, B., Orrit, M.: Photothermal imaging of nanometer-sized metal particles among scatterers. Science **297**, 1160 (2002)
25. Lindfors, K., Kalkbrenner, T., Stoller, P., Sandoghdar, V.: Detection and spectroscopy of gold nanoparticles using supercontinuum white light confocal microscopy. Phys. Rev. Lett. **93**, 037401 (2004)
26. Arbouet, A., Christofilos, D., Del Fatti, N., Vallée, F., Huntzinger, J.R., Arnaud, L., Billaud, P., Broyer, M.: Direct measurement of the single-metal-cluster optical absorption. Phys. Rev. Lett. **93**, 127401 (2004)
27. Muskens, O.L., Billaud, P., Broyer, M., Del Fatti, N., Vallée, F.: Optical extinction spectrum of a single metal nanoparticle: quantitative characterization of a particle and of its local environment. Phys. Rev. B **78**, 205410 (2008)
28. Billaud, P., Marhaba, S., Grillet, N., Cottancin, E., Bonnet, C., Lermé, J., Vialle, J.-L., Broyer, M., Pellarin, M.: Absolute optical extinction measurements of single nano-objects by spatial modulation spectroscopy using a white lamp. Rev. Sci. Instrum. 81, 043101 (2010)
29. Muskens, O.L., Del Fatti, N., Vallee, F., Huntzinger, J.R., Billaud, P., Broyer, M.: Single metal nanoparticle absorption spectroscopy and optical characterization. Appl. Phys. Lett. **88**, 063109 (2006)
30. Muskens, O.L., Bachelier, G., Del Fatti, N., Vallee, F., Brioude, A., Jiang, X.C., Pileni, M.P.: Quantitative absorption spectroscopy of a single gold nanorod. J. Phys. Chem. C **112**, 8917–8921 (2008)
31. Billaud, P., Marhaba, S.: Cottancin, E., Arnaud, L., Bachelier, G., Bonnet, C., Del Fatti, N., Lermé, J., Vallée, F., Vialle, J.-L., Broyer, M., Pellarin, M.: Correlation between the extinction spectrum of a single metal nanoparticle ant its electron microscopy image. J. Phys. Chem. C **112**, 978–982 (2008)
32. Baida, H., Billaud, P., Marhaba, S., Christofilos, D., Cottancin, E., Crut, A., Lermé, J., Maioli, P., Pellarin, M., Broyer, M., Del Fatti, N., Vallée, F., Sanchez-Iglesias, A., Pastoriza-Santos, I., Liz-Marzan, L.M.: Quantitative determination of the size Dependence

of surface plasmon resonance damping in single Ag@SiO_2 nanoparticles. Nano Lett. **9**, 3463–3469 (2009)
33. Hovel, H., Fritz, S., Hilger, A., Kreibig, U., Vollmer, M.: Width of cluster plasmon resonances—bulk dielectric functions and chemical interface damping. Phys. Rev. B **48**, 18178–18188 (1993)
34. Lerme, J., Baida, H., Bonnet, C., Broyer, M., Cottancin, E., Crut, A., Maioli, P., Del Fatti, N., Vallee, F., Pellarin, M.: Size dependence of the surface plasmon resonance damping in metal nanospheres. J. Phys. Chem. Lett. **1**, 2922–2928 (2010)
35. Ferrando, R., Jellinek, J., Johnston, R.L.: Nanoalloys: from theory to applications of alloys clusters and nanoparticles. Chem. Rev. **108**, 845 (2008)
36. Glaser, N., Adams, D.J., Böker, A., Krausch, G.: Janus particles at liquid–liquid interfaces. Langmuir **22**, 5227–5229 (2006)
37. Hubenthal, F., Ziegler, T., Hendrich, C., Alschinger, M., Träger, F.: Tuning the surface plasmon resonance by preparation of gold-core/silver-shell and alloy nanoparticles. Eur. Phys. J. D **34**, 165–168 (2005)
38. Berthier, S.: Optique des milieux composites. Polytechnica, Paris (1993)
39. Granqvist, C., Hunderi, O.: Optical absorption of ultrafine metal spheres with dielectric cores. Zeitschrift für Physik B Condens. Matter **30**, 47–51 (1978)
40. Cottancin, E., Broyer, M., Lermé, J., Pellarin, M.: Chapter 24: optical properties of metal clusters and nanoparticles. In: Sattler, K.D., (ed.) Handbook of Nanophysics,Vol. Nanoelectronics and Nanophotonics, pp. 1–25, Taylor & Francis CRC Press, Boca Raton (2010)
41. Gaudry, M., Cottancin, E., Pellarin, M., Lermé, J., Arnaud, L., Huntzinger, J.-R., Vialle, J.-L., Broyer, M.: Size and composition dependence in the optical properties of mixed (transition metal/noble metal) embedded clusters. Phys. Rev. B **67**, 155409 (2003)
42. Massalski, T.B., Murray, J.L., Bernett, L.H., Baker, H.: Binary Alloy Phase Diagrams, vol 1. Amercian Society for Metals, Metals Park, Ohio (1986)
43. Baletto, F., Mottet, C., Rapallo, A., Rossi, G., Ferrando, R.: Growth and energetic stability of AgNi core-shell clusters. Surf. Sci. **556–568**, 192–196 (2004)
44. Rossi, G., Rapallo, A., Mottet, C., Fortunelli, A., Baletto, F., Ferrando, R.: Magic Polyicosahedral core-shell clusters. Phys. Rev. Lett. **93**, 105503 (2004)
45. Sinzig, J., Radtke, U., Quinten, M., Kreibig, U.: Binary clusters: homogeneous alloys and nucleus-shell structure. Z. Phys. D **26**, 242–245 (1993)
46. Link, S., Wang, Z.L., El-Sayed, M.A.: Alloy formation of Gold-Silver nanoparticles and the dependence on their absorption. J. Chem. Phys. **103**, 3529–3533 (1999)
47. Aden, A.L., Kerker, M.: Scattering of electromagnetic waves from two concentric spheres. J. Appl. Phys. **22**, 1242 (1951)
48. Morriss, R.H., Collins, L.F.: Optical properties of multilayer colloids. J. Chem. Phys. **41**, 3357–3363 (1964)
49. Broyer, M., Cottancin, E., Lermé, J., Pellarin, M., fatti, N. D., Vallée, F., Burgin, J., Guillon, C., Langot, P.: Optical properties and relaxation processes at femtosecond scale of bimetallic clusters. Faraday Discussion, **138**, 137 (2008)
50. Cottancin, E., Lermé, J., Gaudry, M., Pellarin, M., Vialle, J.L., Broyer, M., Prével, B., Treilleux, M., Mélinon, P.: Optical properties of AunAgn clusters. Phys. Rev. B **62**, 5179 (2000)
51. Gaudry, M., Lermé, J., Cottancin, E., Pellarin, M., Vialle, J., Broyer, M., Prével, B., Treilleux, M., Mélinon, P.: Optical properties of (AuxAg1-x)n clusters embedded in alumina : evolution with size and stoichiometry. Phys. Rev. B **64**, 085407 (2001)
52. Russier-Antoine, I., Bachelier, G., Sablonière, V., Duboisset, J., Bénichou, E., Jonin, C., Bertorelle, F., Brevet, P.-F.: Surface heterogeneity in Au-Ag nanoparticles probed by hyper-Rayleigh scattering. Phys. Rev. B **78**, 35436 (2008)
53. Proux, O., Mimault, J., Regnard, J., Revenant, C., Mevel, B., Dieny, B.: Structural coherence between phases in Ni0.35Ag0.65 thin films. J. Phys. Condens. Matter **12**, 3939 (2000)

54. Langlois, C., Alloyeau, D., Bouar, Y.L., Loiseau, A., Oikawa, T., Mottet, C., Ricolleau, C.: Growth and structural properties of CuAg and CoPt bimetallic nanoparticles. Faraday Discuss. **138**, 375 (2008)
55. Cottancin, E., Gaudry, M., Pellarin, M., Lermé, J., Arnaud, L., Huntzinger, J.-R., Vialle, J.-L., Treilleux, M., Mélinon, P., Rousset, J.-L., Broyer, M.: Optical properties of mixed clusters: comparative study of NiAg and AgPt clusters. Eur. Phys. J. D **24**, 111–114 (2003)
56. Calvo, F., Cottancin, E., Broyer, M.: Segregation, core alloying, and shape transitions in bimetallic nanoclusters: Monte Carlo simulations. Phys. Rev. B **77**, 121406(R) (2008)
57. Del Fatti, N., Voisin, C., Achermann, M., Tzortzakis, S., Christofilos, D., Vallee, F.: Nonequilibrium electron dynamics in noble metals. Phys. Rev. B **61**, 16956–16966 (2000)
58. Muskens, O.L., Del Fatti, N., Vallee, F.: Femtosecond response of a single metal nanoparticle. Nano Lett. **6**, 552–556 (2006)
59. Voisin, C., Christofilos, D., Del Fatti, N., Vallée, F., Prével, B., Cottancin, E., Lermé, J., Pellarin, M., Broyer, M.: Size-dependent electron–electron interactions in metal nanoparticles. Phys. Rev. Lett. **85**, 2200 (2000)
60. Arbouet, A., Voisin, C., Christofilos, D., Del Fatti, N., Vallée, F., Lermé, J., Celep, G., Cottancin, E., Gaudry, M., Pellarin, M., Broyer, M., Maillard, M., Pileni, M.P., Treguer, M.: Electron-phonon scattering in metal clusters. Phys. Rev. Lett. **90**, 177401 (2003)
61. Hartland, G.V.: Coherent excitation of vibrational modes in metallic nanoparticles. Annu. Rev. Phys. Chem. **57**, 403–430 (2006)
62. Del Fatti, N., Voisin, C., Chevy, F., Vallee, F., Flytzanis, C.: Coherent acoustic mode oscillation and damping in silver nanoparticles. J. Chem. Phys. **110**, 11484–11487 (1999)
63. Del Fatti, N., Voisin, C., Christofilos, D., Vallee, F., Flytzanis, C.: Acoustic vibration of metal films and nanoparticles. J. Phys. Chem. A **104**, 4321–4326 (2000)
64. Voisin, C., Christofilos, D., Del Fatti, N., Vallee, F.: Environment effect on the acoustic vibration of metal nanoparticles. Phys. B-Condens. Matter **316**, 89–94 (2002)
65. Guillon, C., Langot, P., Del Fatti, N., Vallee, F., Kirakosyan, A.S., Shahbazyan, T.V., Cardinal, T., Treguer, M.: Coherent acoustic vibration of metal nanoshells. Nano Lett. **7**, 138–142 (2007)
66. Burgin, J., Langot, P., Del Fatti, N., Vallee, F., Huang, W., El-Sayed, M.A.: Time-resolved investigation of the acoustic vibration of a single gold nanoprism pair. J. Phys. Chem. C **112**, 11231–11235 (2008)
67. Zijlstra, P., Tchebotareva, A.L., Chon, J.W.M., Gu, M., Orrit, M.: Acoustic oscillations and elastic moduli of single gold nanorods. Nano Lett. **8**, 3493–3497 (2008)
68. Staleva, H., Hartland, G.V.: Vibrational dynamics of silver nanocubes and nanowires studied by single-particle transient absorption spectroscopy. Adv. Funct. Mater. **18**, 3809–3817 (2008)
69. Petrova, H., Lin, C.H., Hu, M., Chen, J.Y., Siekkinen, A.R., Xia, Y.N., Sader, J.E., Hartland, G.V.: Vibrational response of Au–Ag nanoboxes and nanocages to ultrafast laser-induced heating. Nano Lett. **7**, 1059–1063 (2007)
70. Juve, V., Crut, A., Maioli, P., Pellarin, M., Broyer, M., Del Fatti, N., Vallee, F.: Probing elasticity at the nanoscale: terahertz acoustic vibration of small metal nanoparticles. Nano Lett. **10**, 1853–1858 (2010)
71. Beaurepaire, E., Merle, J.C., Daunois, A., Bigot, J.Y.: Ultrafast spin dynamics in ferromagnetic nickel. Phys. Rev. Lett. **76**, 4250–4253 (1996)
72. Beaurepaire, E., Maret, M., Halté, V., Merle, J.C., Daunois, A., Bigot, J.Y.: Spin dynamics in $CoPt_3$ alloy films: a magnetic phase transition in the femtosecond time scale. Phys. Rev. B **58**, 12134–12137 (1998)
73. Hohlfeld, J., Matthias, E., Knorren, R., Bennemann, K.H.: Nonequilibrium magnetization dynamics of nickel. Phys. Rev. Lett. **78**, 4861–4864 (1997)
74. Ju, G.P., Nurmikko, A.V., Farrow, R.F.C., Marks, R.F., Carey, M.J., Gurney, B.A.: Ultrafast time resolved photoinduced magnetization rotation in a ferromagnetic/antiferromagnetic exchange coupled system. Phys. Rev. Lett. **82**, 3705–3708 (1999)

75. Rhie, H. S., Durr, H. A., Eberhardt, W.: Femtosecond electron and spin dynamics in Ni/W(110) films. Physical Review Letters, **90**, 247201 (2003)
76. Vomir, M., Andrade, L.H. F., Guidoni, L., Beaurepaire, E., Bigot, J.Y.: Real space trajectory of the ultrafast magnetization dynamics in ferromagnetic metals. Phys. Rev. Lett. **94**, 237601 (2005)
77. Thiele, J.U., Buess, M., Back, C.H.: Spin dynamics of the antiferromagnetic-to-ferromagnetic phase transition in FeRh on a sub-picosecond time scale. Appl. Phys. Lett. **85**, 2857–2859 (2004)
78. Hübner, W., Zhang, G.P.: Ultrafast spin dynamics in nickel. Phys. Rev. B **58**, R5920–R5923 (1998)
79. Koopmans, B., Ruigrok, J.J.M., Longa, F.D., de Jonge, W.J.M.: Unifying ultrafast magnetization dynamics. Phys. Rev. Lett. **95**, (2005)
80. Radu, I., Woltersdorf, G., Kiessling, M., Melnikov, A., Bovensiepen, U., Thiele, J.U., Back, C.H.: Laser-induced magnetization dynamics of lanthanide-doped permalloy thin films. Phys. Rev. Lett. **102**, 117201 (2009)
81. Bigot, J.Y., Vomir, M., Beaurepaire, E.: Coherent ultrafast magnetism induced by femtosecond laser pulses. Nat. Phys. **5**, 515–520 (2009)
82. Boeglin, C., Beaurepaire, E., Halte, V., Lopez-Flores, V., Stamm, C., Pontius, N., Durr, H.A., Bigot, J.Y.: Distinguishing the ultrafast dynamics of spin and orbital moments in solids. Nature **465**, 458–461 (2010)
83. Bloch, F.: Nuclear Induction. Phys. Rev. **70**, 460–474 (1946)
84. Landau, L., Lifshitz, E.: On the theory of the dispersion of magnetic permeability in ferromagnetic bodies. Phys. Z. Sov. Union **8**, 153 (1935)
85. Gilbert, T.L.: A lagrangian formulation of the gyromagnetic equation of the magnetization field. Phys. Rev. **100**, 1243 (1955)
86. Berger, L.: Emission of spin waves by a magnetic multilayer traversed by a current. Phys.Rev. B **54**, 9353–9358 (1996)
87. Slonczewski, J.C.: Current-driven excitation of magnetic multilayers. J. Magn. Magn. Mater. **159**, L1–L7 (1996)
88. Katine, J.A., Albert, F.J., Buhrman, R.A., Myers, E.B., Ralph, D.C.: Current-driven magnetization reversal and spin-wave excitations in Co/Cu/Co pillars. Phys. Rev. Lett. **84**, 3149–3152 (2000)
89. van Kampen, M., Jozsa, C., Kohlhepp, J.T., LeClair, P., Lagae, L., de Jonge, W.J.M., Koopmans, B.: All-optical probe of coherent spin waves. Phys. Rev. Lett. **88**, 227201 (2002)
90. Andrade, L.H.F., Laraoui, A., Vomir, M., Muller, D., Stoquert, J.P., Estournes, C., Beaurepaire, E., Bigot, J.Y.: Damped precession of the magnetization vector of superparamagnetic nanoparticles excited by femtosecond optical pulses. Phys. Rev. Lett. **97**, 127401 (2006)
91. Thompson, S.M.: The discovery, development and future of GMR: the Nobel Prize 2007. J. Phys. D Appl. Phys. **41**, 093001 (2008)
92. Néel, L.: Théorie du traînage magnétique des ferromagnétiques en grains fins avec applications aux terres cuites. Ann. Geophys. **5**, 99 (1949)
93. Brown, W.F.: Thermal fluctuations of a single-domain particle. Phys. Rev. **130**, 1677 (1963)
94. Acremann, Y., Back, C.H., Buess, M., Portmann, O., Vaterlaus, A., Pescia, D., Melchior, H.: Imaging precessional motion of the magnetization vector. Science **290**, 492–495 (2000)
95. Gerrits, T., van den Berg, H.A.M., Hohlfeld, J., Bar, L., Rasing, T.: Ultrafast precessional magnetization reversal by picosecond magnetic field pulse shaping. Nature **418**, 509–512 (2002)
96. Griffiths, J.H.E.: Anomalous high-frequency reistance of ferromagnetic metals. Nature **158**, 670–671 (1946)
97. Halté, V., Guille, J., Merle, J.C., Perakis, I., Bigot, J.Y.: Electron dynamics in silver nanoparticles: comparison between thin films and glass embedded nanoparticles. Phys. Rev. B **60**, 11738–11746 (1999)

98. Nisoli, M., Stagira, S., DeSilvestri, S., Stella, A., Tognini, P., Cheyssac, P., Kofman, R.: Ultrafast electronic dynamics in solid and liquid gallium nanoparticles. Phys. Rev. Lett. **78**, 3575–3578 (1997)
99. Halté, V., Bigot, J.Y., Palpant, B., Broyer, M., Prevel, B., Perez, A.: Size dependence of the energy relaxation in silver nanoparticles embedded in dielectric matrices. Appl. Phys. Lett. **75**, 3799–3801 (1999)
100. Dormann, J.L., DOrazio, F., Lucari, F., Tronc, E., Prene, P., Jolivet, J.P., Fiorani, D., Cherkaoui, R., Nogues, M.: Thermal variation of the relaxation time of the magnetic moment of gamma-Fe2O3 nanoparticles with interparticle interactions of various strengths. Phys. Rev. B, **53**, 14291–14297 (1996)
101. Respaud, M., Goiran, M., Broto, J.M., Yang, F.H., Ould-Ely, T., Amiens, C., Chaudret, B.: High-frequency ferromagnetic resonance on ultrafine cobalt particles. Phys. Rev. B **59**, R3934–R3937 (1999)
102. Lopez, C.: Materials aspects of photonic crystals. Adv. Mater. **15**, 1679–1704 (2003)
103. Terris, B.D., Thomson, T.: Nanofabricated and self-assembled magnetic structures as data storage media. J. Phys. D Appl. Phys. **38**, R199–R222 (2005)
104. Laraoui, A., Albrecht, M., Bigot, J.Y.: Femtosecond magneto-optical Kerr microscopy. Opt. Lett. **32**, 936–938 (2007)
105. Lisiecki, I., Albouy, P.A., Pileni, M.P.: “Supra” crystal: Control of the ordering of self-organization of cobalt nanocrystals at the mesoscopic scale. J. Phys. Chem. B **108**, 20050–20055 (2004)
106. Parker, D., Lisiecki, I., Salzemann, C., Pileni, M.P.: Emergence of new collective properties of cobalt nanocrystals ordered in fcc supracrystals: II, magnetic investigation. J. Phys. Chem. C **111**, 12632–12638 (2007)
107. Lisiecki, I., Halte, V., Petit, C., Pileni, M.P., Bigot, J.Y.: Vibration dynamics of supra-crystals of cobalt nanocrystals studied with femtosecond laser pulses. Adv. Mater. **20**, 4176 (2008)
108. Laraoui, A., Halté, V., Vomir, M., Venuat, J., Albrecht, M., Beaurepaire, E., Bigot, J.Y.: Ultrafast spin dynamics of an individual $CoPt_3$ ferromagnetic dot. Eur. Phys. J. D **43**, 251–253 (2007)
109. Laraoui, A., Venuat, J., Halte, V., Albrecht, M., Beaurepaire, E., Bigot, J.Y.: Study of individual ferromagnetic disks with femtosecond optical pulses. J. Appl. Phys. **101**, 09C105 (2007)
110. Bigot, J.Y., Vomir, M., Andrade, L.H.F., Beaurepaire, E.: Ultrafast magnetization dynamics in ferromagnetic cobalt: the role of the anisotropy. Chem. Phys. **318**, 137–146 (2005)
111. Bigot, J.Y., Laraoui, A., Vomir, M., Albrecht, M. : Conference on lasers and electro-optics & quantum electronics and laser science conference in IEEE lasers and electro-optics society (LEOS) annual meeting, New-York, NY, USA, San Jose, 1093–1094 (2008)

Surface Studies of Catalysis by Metals: Nanosize and Alloying Effects

Laurent Piccolo

Abstract Supported metallic nanoparticles have long been employed to catalyze a number of industrially-relevant chemical reactions. In many cases, metal addition has allowed one to increase the activity, selectivity and/or stability of single-metal catalysts. However, a detailed understanding of catalysis by metal nanoparticles, including nanoalloys, requires the use of model catalysts such as single-crystal surfaces and well-defined supported nanoparticles. In this chapter, after a brief presentation of its basic concepts, the structural aspects of heterogeneous catalysis by metals and alloys will be illustrated by several examples from (mainly) surface science. The so-called "size" and "alloying" effects, which have been classically described in terms of geometric and electronic effects, might have more subtle origins (morphology, support, etc.) and be interrelated. In turn, the structure of supported nanoparticles is highly sensitive to the reaction conditions, as illustrated by examples of adsorption-induced surface restructuring and segregation. In spite of this complexity, it will be shown that the recent advances in *operando* experimentation and computer simulation open the way to a "rational design" of bimetallic catalysts.

1 Introduction

About 80–90% of manufactured products include catalytic processes in the course of their fabrication, showing the outstanding importance of catalysis in the chemical industry. The recent Nobel Prizes in chemistry attributed to Chauvin,

L. Piccolo (✉)
Institut de recherches sur la catalyse et l'environnement de Lyon (IRCELYON), UMR 5256 CNRS and Université Lyon 1, 2 avenue Albert Einstein, 69626 Villeurbanne, France
e-mail: laurent.piccolo@ircelyon.univ-lyon1.fr

D. Alloyeau et al. (eds.), *Nanoalloys*, Engineering Materials,
DOI: 10.1007/978-1-4471-4014-6_11,

Grubbs and Schrock in 2005, Ertl in 2007, and Heck, Negishi and Suzuki in 2010 celebrate catalysis as a major science. While the first descriptions of reaction rates (i.e., kinetics) were reported around 1805, catalysis really expanded at the beginning of the twentieth century with the production of fertilizers from nitrogen via the catalytic synthesis of ammonia [1].

Catalysis consists in the acceleration of reactions through the use of substances called *catalysts*, which should themselves not be consumed by the reactions. Catalysis considerably decreases the energetic cost of thermodynamically-feasible reactions. In *homogeneous catalysis*, the reactants and the catalysts belong to the same phase (typically, liquid), while in *heterogeneous catalysis* the catalysts are solid and the reactants are liquid or gaseous.

Heterogeneous catalysis by metals, which is the subject of this chapter, constitutes an important part of the catalysis field. Metals are used in important processes like the reforming of naphtha to gasoline (over Pt-based catalysts), the epoxidation of ethylene to ethylene oxide (Ag), the steam reforming of methane and the methanation (Ni), the abovementioned ammonia synthesis (Fe), among many others [1]. Another well-known application of catalysis is the automotive pollution control using Pt, Pd and Rh. Let us point out that the catalytically efficient metals are mostly transition metals from 8 and 1B groups of the periodic table, since they bind molecules with intermediate strength (referred to as the Sabatier's Principle, see Sect. 2). If the adsorption is too weak, the reactants cannot be stabilized at the catalyst's surface. If it is too strong, the energy barrier to the reaction is too high.

Given that heterogeneous catalysis is an interface phenomenon, precious metals have to be finely dispersed over suitable supports in order to improve the efficiency and reduce the cost of the catalysts. This is where nanoscience meets catalysis: typical heterogeneous catalysts are supported metal particles with sizes below 10 nm. This leads to the so-called *particle size and support effects* on catalytic properties. The performances of a catalyst are generally described in terms of activity, selectivity, and stability. The use of metal combinations is a way to maximize the catalyst performances by changing the atomic and electronic structure of the nanoparticles surface through the so-called *alloying effects*. As a matter of fact, as we will see, there are numerous examples where multimetallic catalysts perform better than their single-metallic counterparts. Although the first works on bimetallic catalysts were reported in the late 1940 s, the first industrially important bimetallic catalysts were the Pt–Re and Pt–Ir systems, patented in 1968 and 1976 by Chevron [2] and Exxon [3], respectively. These catalysts are used for the abovementioned reforming of naphtha, a complex mixture of C_6–C_{10} hydrocarbons, into high-quality gasoline [4].

It is important to keep in mind that "bimetallic" catalysts do not necessarily contain bimetallic particles! As the homogeneous preparation and the atomic-scale characterization of such particles remain a difficult task, the question of knowing the exact structure of bimetallic particles is still relevant nowadays, and somewhat neglected in many papers dealing with catalysis. The structure of nanoalloys depends on several factors, such as the catalyst preparation method, the nature of the support, and the nature of the physicochemical environment (reactants, pressure, and temperature).

The present chapter aims at providing to non-specialist readers (although already familiar with the basic solid-state and surface-science concepts) an overview of fundamental heterogeneous catalysis by metals. The basics of adsorption and catalysis on metals are introduced in Sect. 2, along with notions of theoretical modeling. The core of this chapter addresses the question of structural effects in catalysis, including nanosize and alloying effects. To this aim, in addition to some examples of well-characterized powder-supported catalytic nanoalloys, we use the surface science approach of planar model catalysts (i.e., single-crystal surfaces and nanoparticles supported on planar substrates), often combined with theoretical modeling. Indeed, nothing better than surface science can access the intimate atomic-scale processes of catalysis (the corresponding methods are described in Sect. 3). In order to introduce the concepts gradually, we will first focus on "nano" effects for single-metallic supported particles (Sect. 4), then on "alloy" effects for extended surfaces and well-defined nanoalloys (Sect. 5). Relevant references for more "technical" supported bimetallic catalysts may be found in various books and reviews [5–12].

2 Basic Concepts of Heterogeneous Catalysis

In this section, only the terms and concepts needed for the understanding of the subsequent sections are introduced. Among other references, the excellent book by Chorkendorff and Niemantsverdriet [1] has inspired this overview.

2.1 Kinetics and Mechanisms

Let us briefly introduce some important definitions. The role of catalysts is to increase the *rates* of thermodynamically possible reactions, which else would not occur within a reasonable time, at reasonable temperature and pressure. The catalyst *activity* corresponds to the rate at which the reactants are consumed or the products are formed at the catalyst surface. Reaction *kinetics* consists in the measurement of reaction rates, or, more generally, relates to the reaction rates themselves. By selecting suitable catalysts, the reactants can be converted into the desired products. For a given set of reactants, the catalytic *selectivity* to the product P_i is defined by the rate of P_i formation divided by the total conversion rate. Although the catalyst is not involved in the overall reaction scheme, it forms bonds with the reactants and products, and so may be affected by the catalytic process, e.g., lose part of its activity and/or selectivity in the course of this process. The catalyst *stability,* i.e., the conservation of the activity and the selectivity during "time-on-stream", is thus an important parameter for characterizing the catalyst performances.

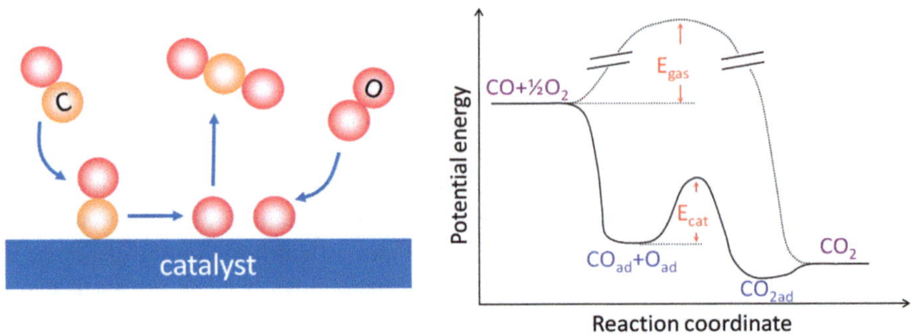

Fig. 1 Reaction scheme (*left*) and potential energy diagram (*right*) for the oxidation of CO by O_2 on Pt-group metals

Heterogeneous catalysis involves a number of elementary steps, such as *adsorption* of the reactants, *surface diffusion*, elementary *reaction* between the reactants to form the products, and *desorption* (Fig. 1). Depending on the system, the adsorption may be *associative* (or "molecular") or *dissociative*. In addition, depending on the strength and the nature of the adsorbate–surface interaction, the adsorption may be physical (*physisorption*, essentially a weak Van der Waals interaction) or chemical (*chemisorption*). Catalysis by metals generally involves chemisorption steps, which consist in the formation of true chemical bonds, i.e., with sharing of electrons between the adsorbates and the surface. This corresponds to a relatively high absolute value of the *heat of adsorption* (or "adsorption energy"), i.e., the energy difference between the adsorbed state and the gaseous state (typically -100 kJ mol^{-1}, *versus* -20 kJ mol^{-1} for physisorption, both processes being exothermic). Associative chemisorption is in general preceded by a physisorption state and may be followed by dissociation if the interaction with the surface is strong and the corresponding *activation energy barrier* is overcome.

A catalytic reaction can be described by a *potential energy diagram*, which represents the potential energy *versus* the reaction coordinate, i.e., the timescale relevant to the sequence of reaction steps forming the *reaction pathway*. Obviously, the knowledge of the *reaction mechanism* is required to plot the associated energy diagram. Figure 1 presents the simple example of carbon monoxide oxidation ($CO + ½O_2 \rightarrow CO_2$), an important reaction for car-exhaust pollution control. In the non-catalyzed case, the energy (E_{gas}) required to perform the $CO + O_2$ reaction is ca. 500 kJ mol^{-1}. This is the energy needed to break the O–O bond prior to O–CO combination, leading to CO_2. As the corresponding rate is proportional to $\exp(-E_{gas}/kT)$, the reaction in the gas phase would require a very high temperature. However, O_2 easily dissociates on Pt-group metals, so that the overall activation energy of the catalyzed process (E_{cat}) is in fact the energy needed for the elementary reaction between adsorbed CO and O to form adsorbed CO_2 (all three species are *reaction intermediates*). E_{cat} (50–100 kJ/mol) is much lower than E_{gas}. This step is called the *rate-determining step* since it governs the

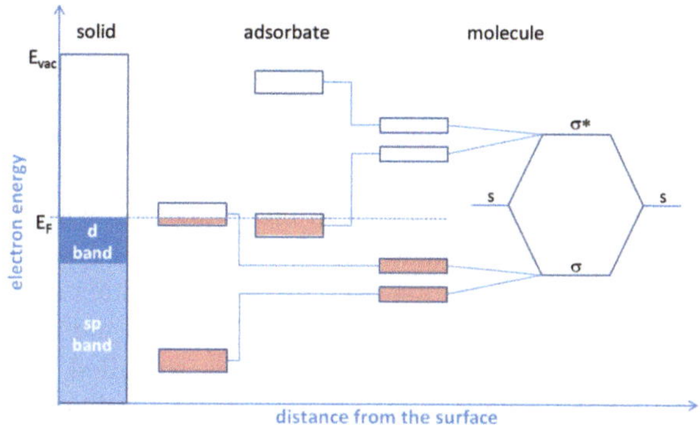

Fig. 2 Schematic representation of the interaction between a two-orbital molecule and a transition metal

rate of the whole reaction, and the molecular state at the top of the barrier is the *transition state*. The final step of the *catalytic cycle*, at the end of which the catalytic system recovers its initial state, is the (fast) desorption of adsorbed CO_2. Let us notice that for CO oxidation, as for various other reactions, the beneficial action of the catalyst consists in the breaking of a strong bond.

2.2 Chemisorption at Metal Surfaces

In order to understand how chemical bonding works, some elements of quantum chemistry and solid-state physics must be introduced here. Simply, when two atoms of the same element are brought together to form a molecule, the atomic energy levels, or "states", split into *bonding* (low energy) and *antibonding* (high energy) states (Fig. 2). Molecular orbitals are filled with electrons, starting from the lowest-energy bonding orbital. The more the antibonding orbital is filled with electrons, the less the molecule is stable. If the antibonding orbital is full, as in the case of a virtual He_2 molecule (two electrons in the 1σ bonding orbital and two electrons in the 2σ antibonding orbital), the molecule cannot form.

Besides, the electronic structure of solids is described using the "band theory" (see Chap. 5). In the case of metals, outer s and p electrons are delocalized and form broad sp bands (the limit case is the "free electron gas"), which coexist with a narrower but more intense d band corresponding to more localized atomic d orbitals. For transition metals, which are the most interesting metals for catalysis, going to the left or downwards in the periodic table, the d band broadens, i.e., the overlap between the (less localized) d orbitals increases. The bands are filled with valence electrons up to the Fermi level. Similarly to the case of molecular orbitals, the more the upper "antibonding" energy levels of the band are filled with electrons, the lower the metal cohesion. The highest cohesive energy is found for 5d transition metals with a half-filled d band. In fact, most of the variations in physical and chemical properties of solids depend on the *degree of filling of the d band*. The surface electronic structure governs the reactivity of the solid. At the surface, the bands are narrower since the atoms have less nearest-neighbors than in the bulk, i.e., there is a smaller overlap between the atomic orbitals.

The chemisorption of a simple diatomic molecule like H_2 or CO on a transition metal surface is depicted in Fig. 2. Upon approach, the strong interaction between the molecular states and the metal sp band leads to a broadening and a downshift of the states. Moreover, from the interaction with the metal d band, which can be seen as a large orbital, each of the molecular states splits into bonding and antibonding chemisorption orbitals. These orbitals are filled up to the Fermi level of the metal. Importantly, the partial filling of the initially antibonding orbital of the molecule by electrons from the metal (so-called "back donation") implies that this orbital contributes to the chemisorption bond, and that the intramolecular bond in the adsorbate is weaker than in the original molecule. Above a certain degree of back donation, the adsorbate may dissociate.

Computational chemistry, within the *Density Functional Theory* (DFT) approach, is nowadays an indispensable tool for catalysis [13–19]. In recent years, calculations based on the DFT have allowed accessing, from first principles (i.e., "ab initio"), the electronic structure and energetics of adsorbates and transition states, as well as the intimate mechanisms of various surface reactions. Briefly, the DFT relies on an averaging approach considering that the valence electrons (i.e., those which are involved in chemical bonds) "move" in a periodic potential generated by the core electrons. Instead of treating each electron of a system individually, the theory considers that, in the ground state, the energy of the system is a unique functional of the electron density (see Chap. 5). Solving the Shrödinger equation using the DFT principle combined with the so-called *Generalized Gradient Approximation* (GGA) then allows the computing of the electron density. Periodic slabs of three to five atomic layers, each containing a few tenths of atoms, are commonly used to model the solid with a sufficient accuracy in a reasonable computing time. The metal slabs are separated by "vacuum slabs" to simulate the surface. DFT-GGA calculations, often using the *Vienna* Ab initio *Simulation Package* (VASP) code [20], are frequently combined or compared with experiments and enable the prediction of trends in surface adsorption and catalysis, as will be reported in Sect. 2.3.

As an application of the scheme in Fig. 2, Fig. 3 shows the DFT-calculated electronic structure of CO, a classical probe of surface reactivity, adsorbed on Al ("free electron" metal) and Pt (transition metal) surfaces [13]. Upon adsorption, the sharp 5σ and 2π states of CO broaden and downshift through interaction with the sp bands of the metals.[1] In the case of Pt, the downshifted orbitals also split into bonding and antibonding orbitals (see dotted lines) through interaction with the d band. Looking at the respective positions of the Pt(111) d band, the adsorbate states, and the Fermi level ($\varepsilon = 0$), it appears that the main contribution to chemisorption comes from the 2π orbital interaction with the d band. This

---

[1] The additional structure at the bottom of the downshifted 5σ orbital is only due to the interaction with the 4σ orbital of CO.

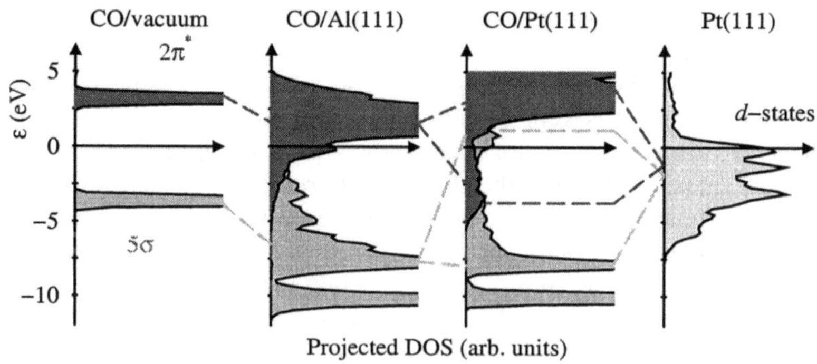

Fig. 3 Self-consistent calculated DOS projected onto the bonding 5σ and antibonding 2π orbitals of CO in vacuum and on Al(111) and Pt(111) surfaces. Also shown is the d band of Pt(111). From Ref. [13], copyright (2000), with permission from Elsevier

contribution is attractive, since before adsorption the 2π orbital was empty, and upon adsorption only the bonding part of the orbital is occupied, i.e., the corresponding projected density of states (DOS, number of energy states per energy unit at a given energy) in Fig. 3 is below the Fermi level.

Having introduced the important concepts of the chemisorption phenomenon (bonding and antibonding orbitals, metal d band…) and described the energetics of a particular system (CO/Pt(111)), we will now examine how this evolves along the periodic table.

2.3 Trends in Surface Reactivity

For the chemisorption of atomic and diatomic molecules on late transition metals, the hybridization energy is the sum of an attractive term and a repulsive term. In the case of CO, this can be approximated as follows [1]:

$$\Delta E_{d-hyb} \approx -2f\frac{\beta_{2\pi}^2}{\varepsilon_{2\pi} - \varepsilon_d} + 2f\,\gamma_{2\pi}\beta_{2\pi}^2 \tag{1}$$

The surface-metal bond strength is thus governed by three factors: (i) the degree of filling of the d band (f); (ii) the overlap between the molecular states and the metal d states, which is proportional to the corresponding interaction matrix element $\beta_{2\pi}$ and correlated to the d band width; (iii) the energy difference between the original molecular state and the center of the d band ($\varepsilon_{2\pi}$–ε_d). Note that $\gamma_{2\pi}$ is a proportionality constant.

An important consequence of Eq. (1) is that the chemisorption strength increases when moving to the left in the periodic table [13]. In the case of molecules like CO, this is mainly due to the upshift of the d band center (ε_d). Figure 3 shows that this corresponds to a greater filling of the antibonding part of the adsorbate orbital, i.e., a stronger adsorbate–surface interaction at the expense of the intramolecular bond.

Equation (1) also shows that $\Delta E_{d\text{-}hyb}$ increases (i.e., decreases in absolute value) as ε_d decreases when f is kept constant. This implies the possibility to tune the hybridization energy by growing a pseudomorphic layer of metal A onto the surface of a metal B. Indeed, the dilatation (*strain*) or compression (*stress*, or "compressive strain") in the A layer (depending on the lattice constants) lead to narrower or broader d bands shifted up or down in energy, respectively, to maintain the filling degree. In turn, the interaction with the 2π orbital of adsorbed CO, and thus the chemisorption bond, increases or decreases, respectively. These trends have been demonstrated experimentally [21] and theoretically [22] for a number of pseudomorphic overlayers.

In terms of local electronic structure, strain is analogous to a decrease in atom coordination. The smaller overlap between the d orbitals within the surface layer induces a narrowing and upshift of the d band, leading to stronger chemisorption. This explains why CO and similar molecules adsorb much more strongly at under-coordinated sites, such as steps on surfaces and edges on nanoparticles. "Morphology effects" and "strain effects" in catalysis will be illustrated in Sects. 4 and 5, respectively.

We have previously seen that an important role of a catalyst is to promote the dissociation of at least one reactant. In general, this step is activated, as will be shown later for N_2 dissociation prior to ammonia synthesis. DFT modeling of this type of reaction requires the structural determination of the transition state as well as the initial and final chemisorption states. The energies associated to these states have to be minimized in order to find the most probable (and fastest) reaction pathway. In the case of N_2 adsorption on Ru(0001), the transition state corresponds to the stretched N_2 molecule metal-bonded parallel to the surface [23]. However, the dissociation probability depends not only on the activation energy but also on the energy of the final state (see the BEP relation herein below). In the example of CO, the molecule does not dissociate on metals for which the dissociated adsorbates (C_{ad} and O_{ad}) are poorly stable with respect to non-dissociated CO_{ad}. As a consequence, for transition metals, dissociative chemisorption is favored when moving towards the left of the periodic table, and associative chemisorption is favored towards the right. For example, CO adsorption is dissociative on Mo and associative on Pd and Ag [13].

Till now, only the case of diatomic molecules has been examined. In Sects. 4 and 5, reactions involving more complex molecules, such as hydrocarbons, will be considered. Besides, in contrast to dissociative chemisorption, the actual *reaction* intermediates are not necessarily the most strongly bonded ones, especially at high adsorbate coverages. In the example of the Pd-catalyzed ethylene ($H_2C{=}CH_2$) hydrogenation, di-σ bonding, in which the C atoms bind to separate metal atoms, is the most stable chemisorption state for ethylene [24]. However, it has been demonstrated that the high-coverage mechanism involves weakly π-bonded ethylene, in which the C=C double bond is above a single metal atom [25]. This allows the formation of the complex depicted in Fig. 4b, consisting in the adsorbed

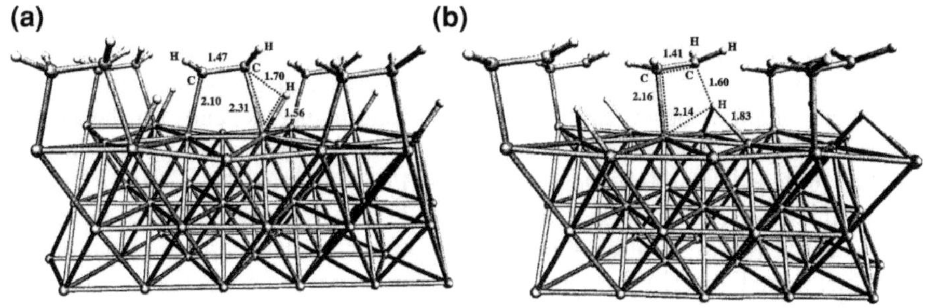

Fig. 4 Structure of the transition state for ethylene hydrogenation on Pd(111) at low coverage **a** and high coverage **b**. From Ref. [24], copyright (2000) American Chemical Society

H atom (issuing from dissociated H_2) and the ethylene molecule bonded to distinct metal atoms [24]. At low coverage, the formation of the most favorable transition state is more energetic since it requires the bonding of H_{ad} and a C atom of ethylene on the same Pd atom (Fig. 4a). In conclusion, the nature of the transition state is as important as that of the chemisorbed species for the reaction kinetics, and weakly bonded adsorbates may react faster than more stable ones.

Now we can understand an important qualitative concept of catalysis, the *Sabatier's Principle*. The reactants must be stable on the surface, but not too much in order to make the reaction possible. As a consequence of the trends mentioned above, the rate of a given catalytic reaction follows a volcano curve along the rows of the periodic table. For example, in the case of ammonia synthesis ($N_2 + 3H_2 \rightarrow 2NH_3$), which is rate-determined by N_2 dissociation, the best catalysts are Fe, Ru and Os. On their left in the periodic table, the metals dissociate N_2 even faster but form excessively stable N adatoms, while the metals to the right are unable to dissociate N_2.

For this type of reaction involving a rate-determining dissociative chemisorption step, a more quantitative picture of the Sabatier's Principle is given by the so-called *Brønsted-Evans-Polanyi* (BEP) *relationships*. They postulate that the activation energy for dissociation (of, e.g., N_2) is proportional to the heat of adsorption (negative value for exothermic processes) of the dissociation products (e.g., N_{ad} and H_{ad}). This means that the dissociation rate increases with the stability of the resulting adsorbates on the surface. Nørskov et al. have demonstrated the validity of these relationships for a number of d metals and diatomic molecules (N_2, CO, NO, and O_2) [26]. It appears that the same linear relation is valid for all these reactants and for metals with close-packed surface structure (face-centered cubic (111), body-centered cubic (110), and hexagonal close-packed (0001)). For stepped surfaces, the activation energy decreases but the slope remains the same. Such a universal relation facilitates the discovery of improved catalysts basing on a simple parameter like the interaction strength between the relevant intermediate and the surface, as we will see in Sect. 5.3. The universality of the BEP relation originates from the fact that, for similar surface geometries (e.g., flat close-packed surfaces), the transition state structure is essentially independent of the type of diatomic molecule and d metal. Indeed, within the dissociation transition state the two atoms having already separated from each other to a large extent and lost their molecular identity, the transition state geometry mainly depends on the heat of adsorption of the dissociated atoms. Through a comprehensive theoretical study of acrolein (O=CH–CH=CH_2) hydrogenation on Pt(111), Loffreda et al. have recently shown that the classical BEP relation could be extended to other energetic parameters [19]. As a matter of fact, for unsaturated aldehyde hydrogenation, the abovementioned linearity does not hold but there appears a novel linearity between the energy of the transition state and that of the precursor state. The latter corresponds to a metastable complex formed by the adsorbed reactants just before the transition.

From the previous results, one can find several ways to adjust the reactivity of metal surfaces by strengthening or weakening the adsorbate–surface bond. This might be done via: (i) the choice of the metal; (ii) the tuning of the atom coordination (surface orientation, steps, particle size and shape, etc.); (iii) alloying (strain effects, etc.). Various examples of experimental and theoretical investigations of these effects will be given in Sect. 5.

Most of the theoretical works evoked above rely on structural models and energetic data issuing from experimental studies, especially surface science studies. The surface science methods relevant to heterogeneous catalysis are now introduced.

3 Surface Science Methods

For about 50 years, surface science has provided a detailed knowledge of solid surface structures and molecule-surface interaction processes, among which chemisorption and catalysis. This was made possible by the development of various techniques of spectroscopy, diffraction, and microscopy dedicated to the analysis of the first atomic layers of solid surfaces, and was performed essentially under *ultrahigh vacuum* (UHV, i.e., pressure lower than 10^{-9} Torr [1 Torr = 133 Pa]). UHV allows one to carry out contaminant-free experiments and minimize the interaction of the particles used to probe the surfaces (photons, electrons, ions, atoms) with the ambient atmosphere. UHV setups make use of a number of specific tools for low-pressure production and measurement, sample transfer, heating, etc. [27].

3.1 Model Catalysts

Two main types of metal-based *model catalysts* may be distinguished in surface-science studies: extended metal single-crystal surfaces and metal nanoparticles supported on planar substrates (also referred to as "supported model catalysts" [28]).

Extended single-crystal surfaces can be either low-Miller-index ("nominal") surfaces, such as fcc (111), (100) and (110) orientations, or higher-index "vicinal" ones exhibiting, e.g., an alternation of terraces and steps. Vicinal surfaces have been extensively compared to their nominal counterparts in order to analyze the structure-sensitivity of catalytic properties [29]. Bimetallic single-crystal surfaces can be either bulk alloy surfaces or metal A/metal B deposits. The latter are generally prepared under UHV by atomic deposition of a metal A onto a metal B single-crystal surface, followed by the suitable annealing. In all cases, metal single-crystal surfaces, which are already finely polished when purchased, must be cleaned under UHV by repeated cycles of sputtering (generally with Ar^+ ions) and annealing. Some examples of single-crystal surface imaging at the atomic scale are provided in Chap. 2 of the present book.

Supported model catalysts, as the one depicted in Fig. 5 [30], can be prepared in several ways. The substrates are generally bulk oxide single-crystals (e.g., MgO(100)) or oxide thin films epitaxially grown on a metallic substrate (e.g., Al_2O_3 over Ni_3Al(111) or NiAl(110)). The most usual metal deposition method is the atomic deposition under UHV by condensation of a metallic vapor onto the substrate (*physical vapor deposition*). Several reviews address this type of preparation and the characterization of such systems [28, 31, 32]. Clusters may also be produced before deposition on the substrate by, e.g., laser vaporization of a polycrystalline (multi)metallic rod under vacuum [33, 34], or colloidal methods [35] (see also Chap. 1). In the latter case, it might be necessary to remove the ligands surrounding the metal nanoparticles using, e.g., thermal treatments or oxygen plasmas. In the case of epitaxial metal-on-oxide growth, the choice of the support influences the morphology of the nanoparticles, and, as a consequence, their reactivity [36]. The growth of bimetallic nanoparticles is described in details in Chaps. 1 and 2.

3.2 Surface Techniques

The objective of this section is not to describe all the surface-science techniques. The reader interested in specific methods may consult Chap. 2 for scanning probe microscopies, Chap. 3 for X-ray scattering and absorption methods, and dedicated books [29, 37]. Although not specific to surface science, such bulk analysis techniques as transmission electron microscopy (and related methods like, e.g., HAADF-STEM and EDX, see Chap. 4), are of obvious importance. Table 1 gives an overview of the most useful and used techniques for the investigation of catalytic metal and alloy surfaces. Selected examples of their application will be given in the following sections. Note that some of these techniques or their counterparts in adapted configurations (XRD, XPS, LEIS, TPD, etc.) are also used to analyze powder-type catalysts. Ideally, a comprehensive approach of

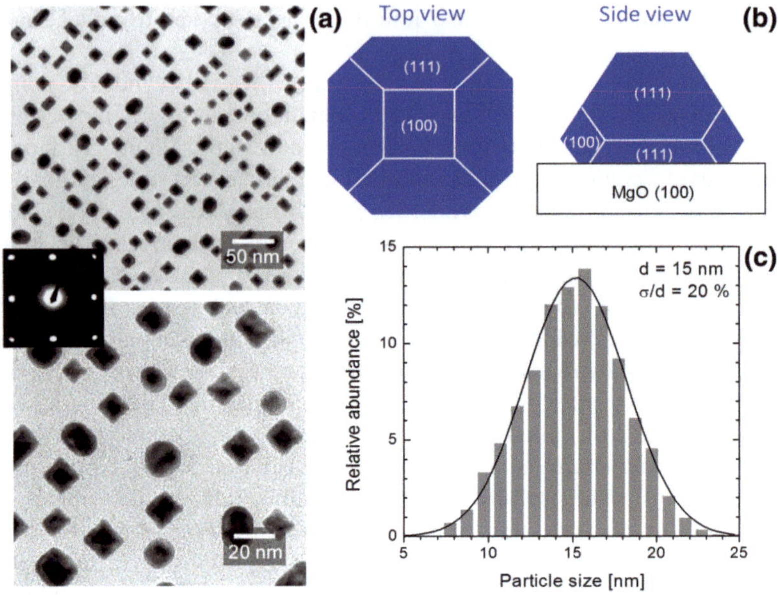

Fig. 5 **a** Example of model supported catalyst observed by TEM: Pd/MgO(100). The diffractogram in insert proves the epitaxial relationship between the particles and the substrate along the (100) orientation. **b** The drawing shows the main particle morphology (truncated octahedron further truncated by the support). **c** Size histogram. From Ref. [30]

heterogeneous catalysis should combine in situ characterization techniques of "realistic" catalysts with experimental and theoretical surface science methods [38].

A typical *catalytic surface-science setup* combines tools for routine sample preparation/characterization (typically, argon ion gun for surface cleaning, LEED, and XPS or AES devices) with one or several more specific techniques to investigate surface reactivity. The latter may be UHV methods like TPR, MBRS, or HREELS (see Table 1). However, a proper evaluation of catalytic properties requires the use of near-atmospheric-pressure reaction cells, combined with gas-phase product detection techniques (mass spectrometry, MS, and/or gas chromatography, GC) [39]. MS is faster than GC but the latter is indispensable for products identification in complex mixtures. Liquid-phase catalysis uses specific methods which are not addressed here. Due to the small surface area of the investigated samples (typically 1 cm^2), the reactor volume should be small enough to get a sufficient sensitivity of the detection method towards the surface catalytic activity. Figure 6 shows an example of catalytic surface-science apparatus combining sample preparation and LEED/AES characterization facilities with a static ("batch") low-volume reaction cell coupled to MS [40]. For on-line product detection by GC, some batch reactors use a gas

Table 1 Surface-science techniques commonly used for the study of model catalysts

Diffraction techniques		
LEED	Low energy electron diffraction	Surface atomic structure under UHV
SXRD	Surface X-ray diffraction	Surface atomic structure. Requires intense beams produced in synchrotrons
Scanning probe microscopies		
STM	Scanning tunneling microscopy	Surface structure and topography (electronic states imaging). Conducting surfaces only. Spatial resolution ca. 0.1 nm
AFM	Atomic force microscopy	Surface structure and topography. Atomic-scale imaging
Electron and ion spectroscopies		
AES	Auger electron spectroscopy	Chemical composition of the first atomic layers under UHV
XPS (or ESCA)	X-ray photoelectron spectroscopy	Chemical composition of the first atomic layers. Oxidation states. Recent in situ XPS allows experiments under gas pressure
UPS	Ultraviolet photoelectron spectroscopy	Structure of the surface electronic valence band under UHV
LEIS	Low energy ion scattering	Chemical composition of the topmost atomic layer under UHV
Vibrational spectroscopies		
HREELS	High resolution electron energy loss spectroscopy	Identification of adsorbates, determination of reaction mechanisms under UHV
IRAS (or IRRAS or RAIRS)	Infrared reflection absorption spectroscopy	Identification of adsorbates, determination of reaction mechanisms. Polarization modulation allows for increased surface sensitivity in the presence of a gas phase
SFG	Sum frequency generation	Identification of adsorbates, determination of reaction mechanisms. Highly surface sensitive
Mass spectrometries		
TDS (or TPD)	Thermal desorption spectroscopy	Evaluation of adsorption energy, quantification of adsorbates under UHV
TPR	Temperature-programmed reaction	Evaluation of reaction temperatures and mechanisms under UHV
MBRS (or MBMS)	Molecular beam relaxation spectroscopy	Measurement of adsorption energy, determination of reaction mechanisms under UHV. Molecular beams can be either thermal Knudsen beams or supersonic jets (higher intensity)

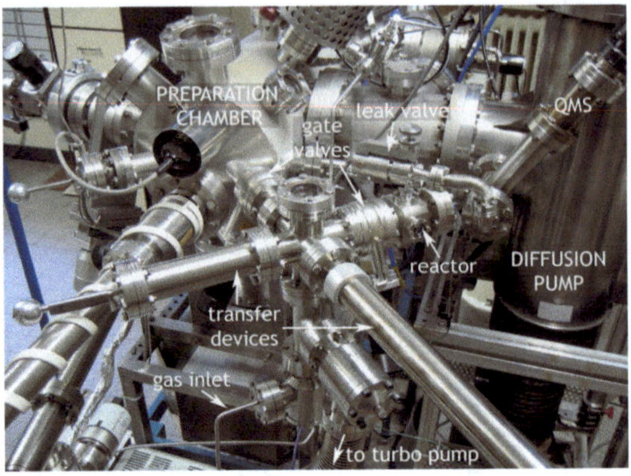

Fig. 6 Device for the preparation and catalytic testing of model catalysts. From Ref. [40], copyright (2005), with permission from Elsevier

recirculation system [41, 42]. Other types of reactors enable continuous gas flowing [43, 44].

Although UHV characterization techniques remain indispensable today (such as LEIS for alloy surface composition analysis [40, 45]), modern surface science of catalysis aims at characterizing the model catalysts *operando* under near-ambient pressure. This implies the simultaneous measurement of the catalytic activity on the one hand, and the in situ identification of adsorbates (by, e.g., IRAS [46–48] or SFG [48]), determination of the surface structure (STM [44, 47, 49, 50] or SXRD [51, 52]), or analysis of the composition/oxidation state (XPS [53, 54]) on the other hand. Several examples of the use of characterization techniques in relation to catalysis will be shown in the following sections.

Now, before reviewing alloying effects in Sect. 5, we provide an overview of the structural effects in catalysis by considering only single-metallic catalysts.

4 Structure Sensitivity: Size, Morphology and Support Effects

The effect of surface structure on chemisorption and catalysis has long been investigated using extended metal poly-crystal or single-crystal surfaces. In the example of ammonia synthesis on iron, Somorjai and coworkers have shown the superior activities of the (111) and (211) orientations over the (100), (210), and (110) ones [55]. This result has been ascribed to the presence of C7 *active sites* (Fe atoms surrounded by 7 nearest neighbors) only on the (111) and (211) surfaces. As already evoked in Sect. 2, this shows the critical influence of surface structure on catalytic properties in the case of the so-called *structure-sensitive reactions.*

More recently, several research groups have attempted to "bridge the material gap" between surface science and practical catalysis by synthesizing model planar supported catalysts. Although less simple and well-defined than extended metal surfaces, these metal/oxide systems are more similar to real catalysts. First, they allow one to account for possible nanoparticle size and support effects. Second, the metallic sites present at the nanoparticle surface, like edges, corners and particle/

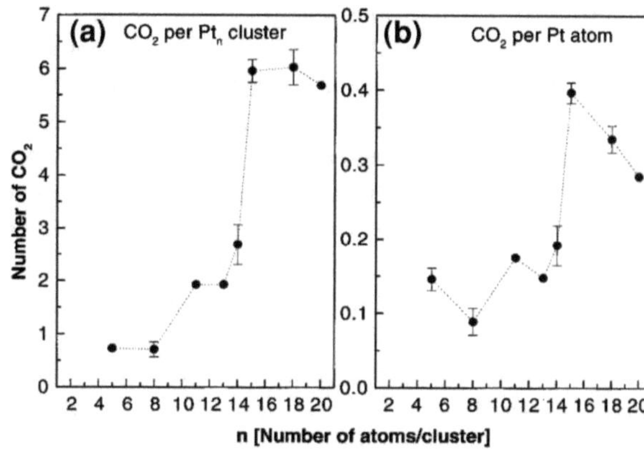

Fig. 7 Number of CO_2 molecules produced per Pt cluster **a** or per Pt atom **b** as a function of cluster size. From Ref. [60], copyright (1999) American Chemical Society

support interfacial sites, can be structurally different from those present at single-crystal surfaces. Several comprehensive reviews on model supported catalysts are available in the literature [28, 32, 56]. Here we illustrate the topic of structure sensitivity in catalysis by a few examples.

4.1 Size and Morphology Effects

The occurrence of size effects in catalysis is well-known [57–59]. Depending on the catalytic system, the reaction rate may increase (e.g., in the case of CO hydrogenation on Ru/Al_2O_3), decrease (e.g., ethane hydrogenolysis on Pt/SiO_2) or remain constant with the metal particle size (e.g., benzene hydrogenation on Pt/SiO_2) [57].

In fact, the dependence of surface reactivity on particle size depends on the size range. For clusters smaller than ca. 1 nm, all (or most) atoms are in contact with the reactants. Their electronic structure is strongly different from that of their bulk counterpart. They typically react as molecules, and their reactivity can be directly related to their orbital structure, which varies strongly with the number of atoms [59]. In this extreme case, size-reactivity relationships are called *quantum size effects*. Heiz and coworkers have reported several examples of atom-by-atom dependence of the catalytic properties of size-controlled metal clusters [33]. In their experimental setup, the clusters were formed by laser evaporation, supersonic expansion, size selection and low-energy deposition onto oxide substrates. The catalytic properties were assessed by temperature-programmed reaction (TPR) after low-temperature coadsorption of the reactants on the model catalyst. The number of product molecules formed in each experiment was deduced from the area of the TPR signal. Figure 7 shows the results obtained in the case of CO oxidation over Pt clusters deposited onto a thin MgO(100) film [60]. It is seen that the reactivity is highly dependent on the number of atoms in the cluster. A maximum activity is obtained for the Pt_{15} clusters. These results were discussed in terms of: (i) morphologic transition from 2D to 3D configuration when the number of atoms per Pt cluster exceeds 13; (ii) interaction of the O_2 antibonding orbitals with the clusters d band, the mean energy of which is size-dependent.

However, in most cases, the particle size is greater than 2 nm. Using the simple model developed by Van Hardeveld and Hartog for unsupported polyhedral clusters [61], it can be noticed that the number of edge atoms strongly increases over the total number of surface atoms when the size decreases below ca. 5 nm. If, for a given reaction, the catalytic reactivity of metal sites (e.g., atoms located at facets, edges, or corners) changes with their coordination number (see Sect. 2.3), a *morphologic size effect* is then expected. This effect, of topmost importance in heterogeneous catalysis, includes the case where specific particle shapes, with preferential facet orientations, have enhanced catalytic activities [62–65]. It is mostly an electronic phenomenon related to the coordination numbers of the exposed atoms, in close similarity to the abovementioned orientation-dependent reactivity of single-crystal surfaces. Note that the coordination-dependent sites present at extended surfaces (terraces *vs.* steps and kinks) have been often used as surface-science models to mimic the reactivity of their equivalents on nanoparticles (facets *vs.* edges and corners). However, the actual coordination numbers may be different between, e.g., the surface steps and the particles edges considered. Moreover, in some cases, the internal structure (crystallographic phase) and the external structure (relaxation, reconstruction) of nanoparticles may differ from those of their bulk material counterparts [66], which may lead to reactivity differences.

Finally, if the reactants require a particular geometry for the adsorption site (i.e., an ensemble of metal atoms), this geometric size effect is called *ensemble size effect*. This is particularly important in the case of alloys, as discussed in Sect. 5. The separation between electronic and geometric effects is in general difficult, since the electronic and atomic structures are obviously interrelated. However, a theoretical tool to decompose structure sensitivity into independent geometric and electronic effects has been recently proposed [67].

Fe and Ru-catalyzed ammonia synthesis is again a good illustration of how surface science and theoretical modeling can unravel the mechanisms and the relevant active sites for an important catalytic process. The pioneering work of Ertl and coworkers has enabled the determination of the elementary steps of this reaction over iron surfaces, as shown by Fig. 8 [68, 69].

It was also pointed out that the mechanistic results established under UHV could be extrapolated to realistic elevated-pressure conditions [55, 70]. Later on, the work of Nørskov and coworkers on ruthenium surfaces demonstrated the unique role of step edges in the dissociation of N_2. As mentioned in Sect. 2 and shown by the diagram of Fig. 8 for Fe, this step determines the rate of the whole reaction [23]. At ca. 230°C, the measured rate of dissociation on the Ru(0001) steps is at least 9 orders of magnitude higher than on the terraces. The corresponding DFT-calculated difference in activation energy is 1.5 eV. Recently, from DFT calculations on stepped Ru(0001), the authors succeeded in linking the experimental activity of a supported Ru-based catalyst to the total number of active sites present at the catalyst surface [71]. The only experimental input was

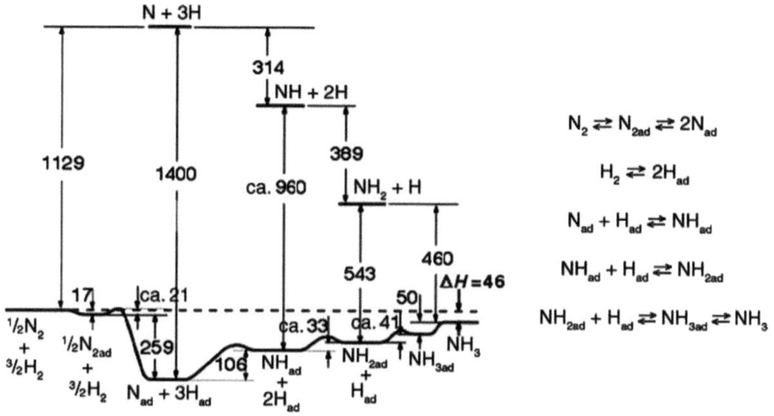

Fig. 8 Potential energy diagram (*left*) and mechanistic reaction scheme (*right*) of ammonia synthesis catalyzed by an iron surface. From Ref. [69]

the particle size distribution determined by TEM. These so-called B5 sites are step-like sites resulting from the reconstruction of Ru nanoparticles larger than 2 nm. Interestingly, the B5 sites on Ru(0001) are very similar to the C7 sites on Fe(111), which were also shown to be very active in the NH_3 synthesis, as mentioned above [55].

Van Santen has recently proposed a molecular theory rationalizing the different classes of structure sensitivity [59]. The key role of step edges in dissociative adsorption (and correlatively in the reactions governed by this elementary step), turns out to be general for the cleavage of π-bonds in diatomic molecules such as N_2, O_2, NO and CO. If the proper step-like structure is not present on very small metal particles, the reactivity may vanish below a given size, as in the previous example [71]. Similarly, the rate of cleavage of σ-bonds such as C–C in alkane hydrogenolysis and C–H in methane activation, which requires a single metal atom, generally increases with the coordinative unsaturation of the metal active site, i.e., with the presence of steps or kinks, or when the particle size decreases. Conversely, hydrogenation reactions essentially show structure-insensitivity.

4.2 Support Effects

The support of the nanoparticles may affect their catalytic properties in several manners. At the synthesis stage, the nanoparticle adhesion and shape depend on the nature of the support. In the case of epitaxial metal-on-oxide growth, the extent of particle truncation by the support is governed by the interfacial energy [28]. In chemical impregnation methods, the nature and amount of functional groups (e.g., hydroxyls) present at the support surface during the preparation play an important role in the resulting catalyst structure [8]. At the catalytic reaction stage,

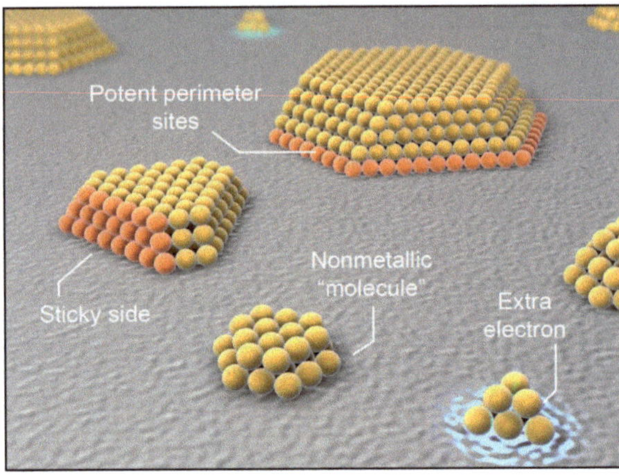

Fig. 9 Schematic representation of possible active sites on supported gold nanoparticles. "Sticky side" and "nonmetallic molecule" nanoparticles correspond to morphologic and quantum size effects (see Sect. 4.1), respectively. "Potent perimeter sites" and "extra electron" clusters relate to support effects. From Ref. [75], reprinted with permission from AAAS

which is the central scope of the present chapter, the nature of the support may also have a crucial importance. Depending on the catalytic system, several types of *support effects* have been observed, as we discuss now.

The exciting field of catalysis by gold illustrates the possible influence of the support on the catalytic properties of nanoparticles. Although this topic has emerged more than 30 years ago with the findings of Haruta and coworkers [72], it is still nowadays a subject of intense research. Small Au nanoparticles supported on suitable supports are able to catalyze numerous reactions, often at temperatures lower than those needed with conventional catalysts [73, 74]. The most studied reaction over gold catalysts is the oxidation of CO. The rate of this reaction essentially increases as gold particle size decreases. Even for this simple process, the reaction mechanism is still a question of debate. Figure 9 illustrates the possible active sites present on gold nanoparticles [75].

Whereas Nørskov and coworkers have proposed that size effects are mainly assignable to low-coordination gold atoms [76], Chen and Goodman have insisted on the specific reactivity of Au bilayer structures [77]. However, several authors have suggested that the rate-determining oxygen "activation" proceeds at the particle-support interface [74]. This would explain why the catalytic activity depends on the support nature [78]. DFT calculations have predicted efficient reaction pathways for the direct reaction between CO and molecular O_2 at energetically favorable interfacial sites, the dissociation of O_2 on Au being highly activated [79].

Moreover, in the case of small gold clusters, the substrate can have a direct influence on the electronic structure of gold. For example, Heiz and coworkers have shown that the oxygen vacancies ("F-centers") present on a defective

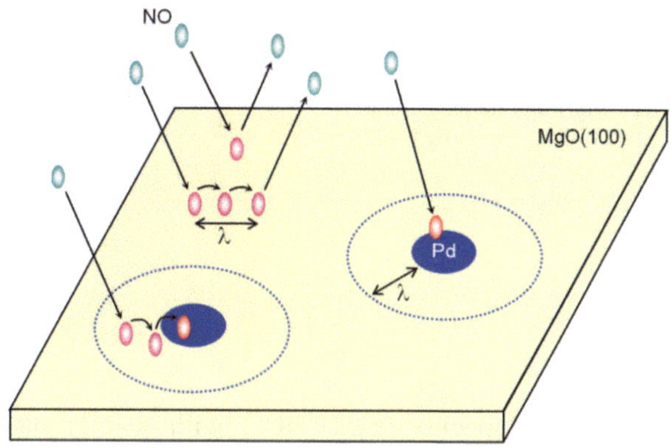

Fig. 10 Illustration of the "capture zone" model for CO or NO adsorption on Pd/MgO(100). λ is the mean free path of the CO molecules on MgO. It is proportional to $\exp(\Delta E/2kT)$, where ΔE is the difference between desorption and diffusion activation energies

MgO(100) surface strongly bind Au_8 clusters and increase their electronic charge. This would provide a low-energy reaction pathway for CO oxidation [80]. As shown from recent STM and IRAS experiments by Freund and coworkers on CO adsorption over Au/MgO(100), even for larger gold nanoparticles the specific oxidation state of the perimeter Au atoms affects the overall reactivity [81].

The support may also influence the catalytic properties of metal nanoparticles via cooperative effects [36]. In the case of *bifunctional catalysts* made of metal nanoparticles supported on an acidic support, the reactivity is governed by the metal–acid site balance and, consequently, depends on the metal particle size [82].

Another cooperative role of the support is the so-called *spillover* phenomenon, which is quite well-known in applied catalysis [83]. It relates to reactants which adsorb on the metal nanoparticles and afterwards diffuse onto the support. In the classical case of H_2, the dissociative adsorption of the molecule on the metal nanoparticles followed by spillover provides a low-energy pathway to the formation of H atoms on the oxide support. The reverse phenomenon, so-called "reverse spillover", has been analyzed in details in the case of the interaction of CO with a Pd/MgO(100) model catalyst [84] using a molecular beam technique [85]. The impinging CO molecules which physisorb on the MgO substrate diffuse on the surface, and can be "trapped" by the Pd particles through chemisorption. In the presence of another reactant such as O_2, this reverse spillover increases the nanoparticle activity, provided that the reaction rate is normalized over the number of surface metal sites (the corresponding rate is called "turnover frequency", TOF). However, taking quantitatively this effect into account through a capture-zone model (Fig. 10) has allowed the extraction of the intrinsic size- or morphology-dependent catalytic activity of Pd particles in the cases of the $CO + O_2$ [86] and CO + NO [62] reactions. This concept has also been used to explain the variations of the hydrogenation TOF with the Pd particle surface density on high-surface-area alumina supports [87].

Finally, in the case of noble metals supported on reducible oxides (e.g., the prototypical Pt/TiO_2 system), a so-called *strong metal-support interaction* (SMSI) effect has been evidenced in the late seventies [88]. It relates to the migration of

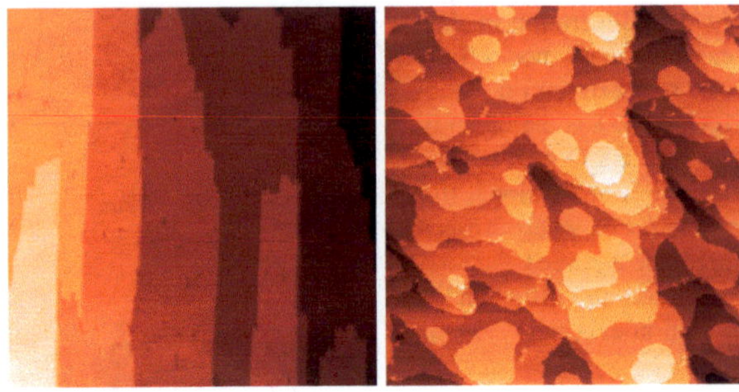

Fig. 11 STM images (350 × 350 nm) of the Au(110) stepped surface in UHV (*left*) and under 20 Torr of CO (*right*). From Ref. [47]

"suboxide" species (e.g., TiO_x, with $x < 2$) onto the metal particles under reductive conditions (typically in the presence of H_2 or under UHV at high temperature). In general, the resulting encapsulation thin film suppresses the nanoparticle activity [89–93]. Although not fully understood, SMSI is thought to be driven by the formation of strong metal-suboxide bonds. From the nanoparticle side, this phenomenon can be viewed as a combined effect of the support and the catalytic environment.

4.3 Adsorption-Induced Restructuring

Depending on the reactive atmosphere and the type of support, metal particles may be subjected to surface oxide formation (including SMSI), morphological restructuring, sintering (agglomeration or coalescence), redispersion (the opposite of sintering), etc. [92, 94, 95]. Recent advances in surface science have changed our ancient vision of static catalytic surfaces. In fact, under gaseous or liquid environments, the adsorbates may create their own adsorption sites by changing the surface structure. In turn, this possibly generates a change of the catalytic reactivity with respect to vacuum conditions.

For example, even for such a simple process as the room-temperature CO adsorption on "noble" Au or Pt surfaces, the CO pressure induces a spectacular roughening (Fig. 11) [47] or clustering [50, 96]. Similarly, single-crystal surfaces may exhibit adsorption-induced reconstruction [97, 98] or "nanofaceting" [99] in order to minimize their energy.

Getting closer to catalysis, Hendriksen and Frenken have investigated the oxidation of CO on Pt(110) using a STM combined with a small-volume flow reactor [44]. Under oxygen-rich conditions, the Pt surface appears to roughen and become more active, which has been ascribed to the formation of a surface Pt oxide. In addition, several studies have shown that the active phases during the Pd-catalyzed selective hydrogenation of alkenes or alkynes are actually surface Pd carbides and/or hydrides [40, 100].

In conclusion of this section, the use of model catalysts and the combination of surface-science techniques and computer simulations have enabled a detailed understanding of structural effects in catalysis. Those can be caused by the reduced dimensionality of metal clusters, the morphology of nanoparticles and/or the support contribution. The catalyst structure is, in turn, dependent on the reactive environment. As will be shown below, the combination of several metals within a catalyst brings additional possibilities to finely tune the catalytic efficiency.

5 Alloy Surfaces and Nanoalloys

In the 1960s, using surface-science methods, Sachtler and coworkers investigated the structure, composition and reactivity of UHV-evaporated Cu–Ni films. They also discussed the nature of the bond between an adsorbate and the alloyed atoms [101, 102]. Since these pioneering studies, a huge number of articles dealing with the catalytic properties of metal/metal deposits and bulk alloy surfaces have been published [103–106]. In this section, the main alloying effects are illustrated using examples from the recent literature.

Alloying a metal with another one might be seen as another way to modify the surface properties of nanoparticles. The objective is obviously to enhance the catalytic performances of pure metals, and possibly to decrease the amount of noble metals. In some cases, no *synergistic effect* is observed upon alloying. The catalytic activity of the alloy particles may only result from the addition of the intrinsic activities of the pure metals, taking into account the corresponding surface concentrations [107].

Classical interpretations of alloying effects in catalysis make a distinction between the *geometric* (or *ensemble*) effects and the *electronic* (or *ligand*) effects [6, 108, 109]. In the first (idealized) case, if (i) some atoms of an inert metal B replace atoms in a metal A surface layer (or cover them) without any electronic modification, and (ii) the reactant adsorption requires several contiguous A atoms ("ensemble"), then A–B alloying simply lowers the overall adsorption capacity of the surface. In the case of electronic effects, if B atoms are located in the first or subsurface layer and induce a modification of the electronic structure of the A atoms within the surface layer, then the reactivity of the A–B alloy surface is different from those of the separate A and B surfaces. In practice, geometric and electronic effects are often intermixed.

5.1 Alloying Effects

An illustration of the ensemble effect has been provided by Maroun et al. for CO and H_2 adsorptions on Pd–Au(111) electrodes in a sulfate solution [110]. Surface alloys were obtained by electrodeposition of Pd over Au(111). In situ STM, infrared spectroscopy and electrochemical measurements have shown that single Pd atoms are the smallest "ensembles" needed for CO adsorption and oxidation, whereas hydrogen adsorption requires at least Pd dimers. Ensemble effects have also been invoked in the industrially-relevant gas-phase acetoxylation of ethylene to vinyl acetate on submonolayer deposits of Pd over Au(111) and Au(100) surfaces [111]. In this case, the critical ensemble consists of two noncontiguous Pd atoms, suitably separated within the Au "matrix".

Besides, the selective blocking of some specific surface sites, such as the low-coordinated ones, with a poorly reactive metal like Au or Ag, may increase the selectivity of catalytic reactions and/or improve the catalyst resistance towards poisoning. For example, Vang et al. exposed a Ni(111) surface to a silver vapor at room temperature, before annealing at ca. 530°C [112]. A decoration of all the steps of the Ni surface with Ag was observed. Only in the case of the Ag-free Ni surface the exposure to ethylene led to a "brim" structure observed by STM and ascribed to decomposed ethylene. These results show that, for this system, the active sites for C–C bond breaking are mostly located at the step edges. DFT calculations further indicate that C–H bond breaking is less sensitive to the presence of steps. Thus, alloying Ni with a small amount of Ag may favor hydrocarbon dehydrogenation over C–C bond breaking. Similarly, the decoration of the Rh(553) surface with Ni atoms inhibits the step-promoted dissociation of CO [113], and the addition of Au to Ni improves the catalyst resistance to graphite formation during the "steam reforming" of butane [114]. In these cases, Ag, Ni, and Au may be viewed as *additives* or *promoters* of the catalytic reactivity.

In analogy with step decoration and orientation-dependent segregation [115] at extended surfaces, site-dependent surface segregation can be observed in the case of nanoalloys. For example, Yudanov and Neyman have shown through DFT calculations that Au atoms are stabilized at the edges of truncated octahedral Pd–Au nanoparticles. This is due to the easier ability of edges to allow the relaxation of the strains induced by the larger size of Au atoms [116]. Using TPD, SFG, IRAS and DFT methods, and CO as a probe molecule, Abbott et al. have confirmed the surface segregation of Au and its preference for edges in the case of Pd–Au nanoalloys supported on $Fe_3O_4(111)$, MgO(100), and $CeO_2(111)$ [117]. In these cases, the nature of the support does not affect significantly the structure of the nanoparticles.

Another example from electrocatalysis now illustrates the electronic effect. The oxygen reduction reaction (ORR, $\frac{1}{2}O_2 + 2H^+ + 2e^- \rightarrow H_2O$) in polymer electrolyte membrane fuel cells (PEMFC) is slow and limits their application in the automotive industry. Recently, Stamenkovic et al. have demonstrated that the $Pt_3Ni(111)$ surface is tenfold more active than the Pt(111) surface in the ORR [118]. The difference has been ascribed to the compositional oscillations in the near-surface layers of the Pt_3Ni surface, which lead to a downshift of the d band center (as determined from UPS). This weakens the adsorption of nonreactive OH species on surface Pt atoms, at the benefit of reactive O_2 species. A correlation has been observed between the position of the d band center and the activity in the ORR for various Pt_3M surfaces (M = Ni, Co, Fe, Ti, V) [119]. Pt_3Co appears to be the most efficient combination. The volcano-type behavior of the catalytic activity is governed by the respective adsorption strengths of reactive intermediates and spectator (blocking) species. Recent results have suggested that these concepts, formulated from surface science experiments, are also applicable to supported Pt_3Ni catalysts [120].

Strasser et al. have synthesized Pt–Cu nanoparticles supported on a carbon powder using impregnation methods [121]. An electrochemical "dealloying" protocol in which Cu was removed preferentially from the precursor Pt–Cu particles, has allowed them to obtain Pt–Cu@Pt core–shell particles (overall size ca. 4 nm, shell thickness ca. 1 nm), as evidenced from various microscopic and spectroscopic analyses. They have shown that the ORR activity of this catalyst increases with the Cu content in the alloyed cores. This result has been ascribed to an increasing compressive strain within the Pt shell, which induces a modification of the d band structure and a weakening of the Pt–O bond (see Sect. 2.3). Contrasting with the previous example of "Pt-skin" systems due to the large shell thickness, this *strain effect* [122–124] can be here discriminated from a ligand effect induced by the Cu-containing core. In addition to the lattice mismatch, the nanometer size of the particles can induce a compressive strain ("surface stress") in the exposed facets. The compression is again thought to be responsible of the ORR activity enhancement observed for Pd@Pt, Pd_3Co@Pt [125] and Pt_3Co@Pt [126] nanoalloys with respect to pure Pt nanoparticles. In the previous cases, the strain effect appears as a particular electronic effect but it may also be seen as a geometric one if the adsorption of a reactant, or the reaction itself, is facilitated by a specific spacing between the surface atoms, similarly to the above case of vinyl acetate.

The selective hydrogenation of 1,3-butadiene is an interesting reaction both from practical and fundamental points of view. In order to purify 1-butene before polymerization, butadiene (CH_2=CH–CH=CH_2) impurities have to be partially hydrogenated to 1-butene (CH_2=CH–CH_2–CH_3), while avoiding the total hydrogenation to undesired butane (CH_3–CH_2–CH_2–CH_3). Although Pd is the best catalytic metal for the selective hydrogenation, the addition of a second metal such as Au can further increase the selectivity. It has been shown by LEIS that the $Pd_{70}Au_{30}$ (111) and (110) topmost surfaces contain ca. 80% of Au under UHV (Fig. 12), while Pd segregates to the surface under ca. 5 Torr of the hydrocarbon–hydrogen mixture (1:10 ratio) at room temperature. However, the Pd-Au surfaces still contain enough Au to be even more selective towards butene formation than the pure Pd surfaces (Fig. 12) [40].

The enhanced selectivity of Pd–Au has been ascribed to an Au-induced weakening of the butene-Pd bond. Once formed, butene molecules desorb from the Pd–Au surface and cannot be further hydrogenated to butane. More "realistic" supported Pd–Au and Pd–Ag catalysts also exhibit high selectivity to butenes [127, 128]. Similar results have been obtained on the Pt_3Sn(111) [129] and Sn/Pt(111) [130] surfaces, Sn being in this case a "promoter" of Pt reactivity, as Au and Ag for Pd in the previous case. Detailed DFT calculations have shown the dual role of tin: (i) site blocking that forces unselective pathways to adopt distorted, high-energy transition states and (ii) ligand effect that decreases the energy barriers for the selective pathway to butenes [131]. Both geometric and electronic alloying effects are thus involved.

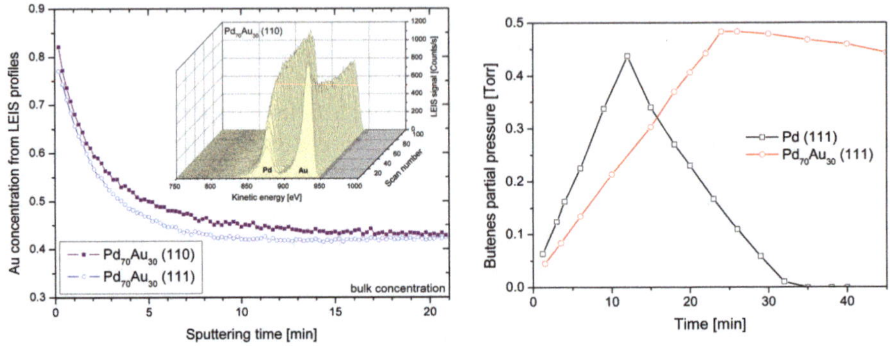

Fig. 12 *Left*: LEIS characterization of Pd–Au surfaces showing surface segregation of Au. *Right*: Evolution of the butenes partial pressure during butadiene hydrogenation over Pd and $Pd_{70}Au_{30}$ (111) after introduction of 5 Torr of H_2 and 0.5 Torr of butadiene in the static reactor depicted in Fig. 6. Butadiene is first converted to butenes, which are then converted to butane (butadiene and butane pressures are not shown for sake of clarity). The graph shows the slower butenes consumption over the Pd–Au surface. From Ref. [40], copyright (2005), with permission from Elsevier

Other DFT calculations by González et al. on Pd–Ag(111) surfaces have further suggested that reverse surface segregation induced by hydrogen adsorption (Pd migrates to the surface layer, Ag to the subsurface layer) suppresses the formation of subsurface interstitial H species [132]. As a result, the possible full hydrogenation pathway might be inhibited. In fact, hydrogen can diffuse into the bulk of Pd-rich compounds and form Pd hydrides. This can have a detrimental effect on the surface reaction rate [133]. A way to prevent the hydride formation while maximizing the hydrogenation rate is to alloy Pd with a metal in which hydrogen cannot be dissolved, and onto which Pd may segregate, such as Ni or Cu. The Pd_8Ni_{92}(110) surface, which exhibits a full Pd overlayer, has been shown to be more active in the selective hydrogenation of butadiene than Pd(110) [106], especially during the initial transient period when hydrogen absorption competes with the hydrogenation reaction [134].[2] The superior activity of Pd/Ni surfaces has been additionally attributed to their complex Pd overlayer structure resulting from the compressive strain relaxation [52, 135, 136].

5.2 Adsorption-Induced Restructuring

In general, adsorption on a bimetallic A–B surface leads to the segregation of the most reactive element (say B) to the surface, provided that the temperature is sufficient to overcome the kinetic barriers. If the surface is A-rich under UHV, this

[2] These examples show again that the actual catalytic phase is not necessarily metallic, but may also be in hydride, carbide, or oxide -like forms, depending on the reaction conditions.

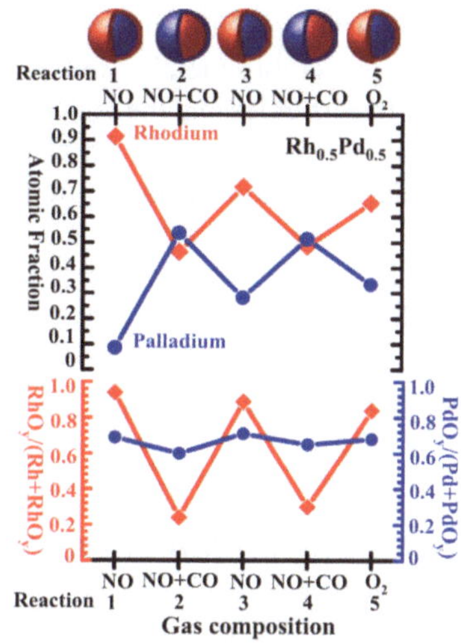

Fig. 13 *Top*: Evolution of Rh ($Rh^0 + Rh^{2y+}$) and Pd ($Pd^0 + Pd^{2y+}$) atomic fractions in the 3–4 first surface layers of $Rh_{0.5}Pd_{0.5}$ nanoparticles as determined by in situ XPS at 300°C under oxidizing conditions (0.1 Torr of NO or O_2) and catalytic conditions (0.1 Torr NO + 0.1 Torr CO). *Bottom*: Corresponding evolution of the fraction of the oxidized Rh (*left y axis*) and Pd (*right y axis*) atoms. From Ref. [54]. Reprinted with permission from AAAS

phenomenon is called *surface segregation reversal* (or "segregation inversion" or "resegregation"), as mentioned above for Pd-Au and Pd–Ag. Through this process, the energy gain brought about by the strong bonding of the adsorbate on the resulting surface is greater than the energy cost of the restructuring in the solid.

For example, Tao et al. have followed by in situ XPS the surface composition of Rh@Pd and Pt@Pd core–shell colloidal nanoparticles (ca. 15 nm) supported on Si wafers [54]. Depending on the atmosphere (UHV or CO, NO, O_2, H_2, or CO + NO, with 0.1 Torr of each gas at 300°C), reversible changes in the surface composition were observed, especially in the case of Rh–Pd (Fig. 13). The Rh oxide being more stable than the Pd oxide, the former is the most abundant surface phase upon NO or O_2 exposure, while metallic Pd and Rh are equally present at the surface under reducing conditions.

However, adsorption-induced surface segregation may be counter-intuitive in some cases. Andersson et al. have investigated the adsorption of CO on Cu/Pt(111) using various surface-science techniques and DFT calculations [137]. Under UHV, a near-surface Cu–Pt alloy is formed, with only Pt atoms in the top layer. Under ca. 2 Torr of CO at 200–450°C, Cu segregates to the surface, forming an ordered bidimensional alloy with Pt, although for pure metal surfaces the Pt–CO bond is stronger than the Cu-CO one. The driving force has been proposed to be the very strong bonding between the CO molecules and the Pt atoms surrounded by "naked" Cu atoms. In particular, this configuration would minimize the repulsive dipolar interactions between the CO molecules.

Combined with thermal annealing at 200°C, CO-induced Pt segregation has been used to obtain carbon-supported Pt–Co@Pt core–shell nanoparticles highly active in the abovementioned ORR [138]. However, under potential cycling in alkaline electrolyte at room temperature, Co segregates to the surface and is gradually dissolved in the liquid phase (*leaching*), causing a *dealloying* of the nanoparticles. This suggests a probable degradation of PEM fuel cell cathodes if such non-noble metals are combined with Pt [139]. Another possible effect of the reactive species on the nanoalloy structure is the formation of stable *oxides* (or carbides, sulfides, etc.) of, at least, the less noble metal component, leading to an oxide shell. For instance, it has been shown that Co–Pt nanoparticles exposed to oxygen exhibit a wurtzite-CoO oxide shell [140]. The nanometer size appears to stabilize this phase, which is metastable in the bulk. The interaction of nanoalloys with reactive gases may even cause *phase separation*. For example, the exposure of silica-supported Ni–Au nanoparticles to H_2 at 600°C, then to O_2 at 300°C, leads to Au/NiO aggregates [141].

5.3 Computational Design of Bimetallic Catalysts

An important objective for catalysis is the ability to predict which catalyst would be the most efficient one for a given reaction. Bimetallic alloys are versatile materials enabling a computer-assisted "rational design" of catalysts, since, in principle, the surface electronic structure of single-crystals or nanoparticles can be tuned by combining the appropriate elements of the periodic table [14, 17, 142].

As already mentioned, correlations have been evidenced from DFT calculations between, e.g., CO or H_2 adsorption energies and the surface d band center, which is sensitive to ligand and strain effects [22, 122]. However, the molecules involved in industrial processes can be much bigger than these diatomic ones. It is thus desirable to identify so-called *descriptors* of the catalytic activity (as in the BEP relations, see Sect. 2.3) in order to make possible a computational screening of bimetallic alloys suitable for specific catalytic reactions [14].

We first illustrate this approach with the partial hydrogenation of acetylene ($CH{\equiv}CH$) to ethylene ($CH_2{=}CH_2$). This reaction is used to remove the acetylene impurities present in ethylene, without forming ethane (CH_3–CH_3), before ethylene polymerization. From DFT calculations, it has been shown that acetylene and ethylene adsorption energies are proportional to the methyl group (CH_3) adsorption energy (the descriptor), whatever the (bi)metallic surface (Fig. 14a) [14, 143]. Similarly to the case of butadiene selective hydrogenation, a good catalyst should adsorb acetylene sufficiently strongly, to favor the activity, but not ethylene, to favor the selectivity. As acetylene and ethylene adsorption energies are linked, suitable catalysts must lie in the window defined by the dotted lines in the diagram of Fig. 14a. This turns out to be the case for the industrially used Pd–Ag system, as well as for Ni–Zn, which is a much cheaper material (Fig. 14b). A Ni–Zn supported catalyst has been tested in a flow reactor and has appeared to be as efficient as the Pd–Ag one. This non-noble metal combination is thus a potential alternative to the expensive industrial catalyst.

Similarly, using the chemisorption energy of nitrogen atoms as an activity descriptor of the ammonia decomposition reaction ($2NH_3 \rightarrow N_2 + 3H_2$), Hansgen et al. have performed microkinetic modeling combined with DFT [142]. They

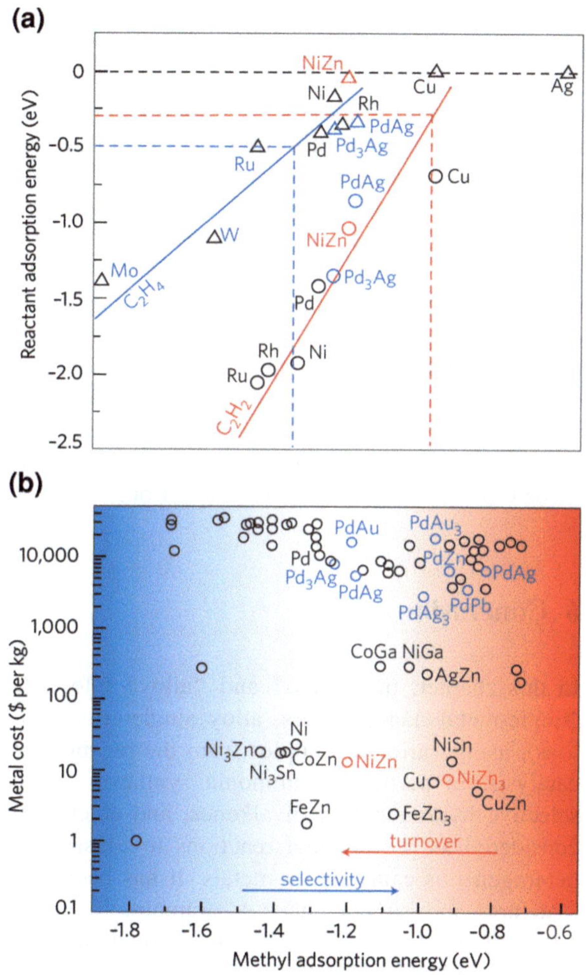

Fig. 14 **a** Calculated adsorption energy of acetylene and ethylene as a function of methyl adsorption energy over a number of metals and alloys. **b** Constituent cost of 70 binary intermetallic compounds as a function of their calculated methyl adsorption energies. From Ref. [143], reprinted with permission from AAAS

have shown that monolayer Ni over Pt could be a promising alternative to expensive Ru. According to the authors, this bimetallic system maximizes NH_x dehydrogenation and N_2 desorption rates, which are the kinetically significant steps of the overall reaction.

Besides, focusing on ethylene epoxidation, Linic et al. have calculated from DFT the difference in formation energy barrier between the undesired product (acetaldehyde, which readily transforms into combustion products) and the desired one (ethylene oxide) for a number of Ag-based alloys [144]. They have found that the Cu/Ag system maximizes this descriptor and exhibits the highest selectivity to ethylene oxide, as confirmed by experiments performed on supported catalysts. Recent DFT calculations and in situ XPS experiments suggest that the actual surface phase consists of a thin copper oxide [145].

Let us close this section by an example involving nanoalloys. Nilekar et al. have reported on the preferential oxidation of CO in the presence of H_2 [146]. The so-called PROX reaction, during which the oxidation of H_2 to water should be avoided, is useful to remove CO impurities from "reformate" H_2 fuel, in order to avoid the poisoning of PEM fuel cell electrodes. The authors have calculated from DFT the adsorbate chemisorption energies and the activation energies of the elementary PROX reaction steps for Pt(111) and various other metals covered with a Pt overlayer. To do so, they have used a reaction mechanism, similar to that proposed for Au catalysts [147], involving CO, O, H, OH, and OOH adsorbed intermediates. From computational screening, the Pt/Ru system appears to be the best one, which has been confirmed experimentally using alumina-supported metastable metal@Pt core–shell nanoparticles prepared by colloidal methods. Interestingly, the Pt-Ru nanoalloys which exhibit both Pt and Ru atoms at their surface (i.e., without a Pt shell) are far less active [148]. These results are again ascribed to strain and ligand effects, which limit the CO adsorption strength and favor CO oxidation over water formation.

6 Conclusion

In this chapter, the "nano" and "alloy" effects have been analyzed separately (single-metal nanoparticles, alloy single-crystal surfaces) and together (nanoalloys), using various examples from the recent literature in model heterogeneous catalysis. In particular, ammonia synthesis and decomposition, CO oxidation, selective hydrogenation of alkenes, and electro-reduction of oxygen have been considered as prototypical reactions to illustrate the main structural aspects of heterogeneous catalysis by metals. It has been shown that the nanoparticle size, structure, and support have a dramatic influence on the catalytic properties. Alloying effects can be understood through such simple concepts as "ensemble", "strain" and "ligand" effects, although these effects are often combined.

Recent advances in DFT-based computer simulations have enabled a better understanding of the physical origin of the structural effects and opened a new way towards the "rational design" of bimetallic catalysts. These methods have allowed the optimization of catalytic performances, essentially in the cases of extended alloy surfaces and rather simple reactions. However, due to the inherent complexity of catalytic materials and their sensitivity to the reaction environment (surface restructuring, segregation, oxidation, etc.), progresses in theoretical models and *operando* experimental methods are needed.

The major difficulty in catalysis by nanoalloys lies in the synthesis of well-defined and stable supported particles. Semi-model systems like nanoalloys synthesized from colloidal methods have recently attracted much attention since they enable a relative control of the particle size, shape and composition. However, the stability of such objects under operating conditions, and the need to remove the surface ligands, are still open questions. This explains why many studies using colloids focus on as-prepared nanoparticles in liquid-phase reactions performed at moderate temperature. Conversely, gas-phase studies of colloidally or physically prepared supported nanoalloys are scarce. As far as physical methods are concerned, the UHV-growth of nanoalloys homogeneous in size and composition is not straightforward and requires the use of nanostructured substrates (see Chaps. 1 and 2).

Studies of bimetallic catalysts prepared from more conventional chemical methods (e.g., impregnation of high-surface-area supports) were not the subject of the present chapter. However, these methods are the most industrially used ones and have long been successfully applied to the synthesis of efficient catalysts. Since, also in this case, the nanoparticle composition is difficult to control, future improvements in the preparation and characterization of this type of multimetallic catalysts are expected.

Acknowledgments I greatly acknowledge my colleagues Dr. Claude Descorme and Dr. Christophe Geantet for critical reading of the manuscript and fruitful discussions.

References

1. Chorkendorff, I., Niemantsverdriet, J.W.: Concepts of Modern Catalysis and Kinetics. Wiley-VCH, Weinheim (2007)
2. Kluksdahl, H.E.: Reforming a sulfur-free naphtha with a platinum-rhenium catalyst. Patent US 3,415,737, 1968
3. Sinfelt, J.H.: Combination reforming process. Patent US 3,791,961, 1974
4. Day, M.: Supported metal catalysts in reforming. In: Anderson, J.A., Fernandez Garcia, M. (eds.) Supported Metals in Catalysis. Imperial College Press, London (2005)
5. Sinfelt, J.H.: Bimetallic Catalysts: Discoveries, Concepts, and Applications. Wiley, New York (1983)
6. Ponec, V., Bond, G.C.: Catalysis by Metals and Alloys. Elsevier, Amsterdam (1995)
7. Epron, F., Especel, C., Lafaye, G., Marécot, P.: Multimetallic nanoparticles prepared by redox processes applied in catalysis. In: Astruc, D. (ed.) Nanoparticles and Catalysis. Wiley-VCH, Weinheim (2008)
8. Ertl, G., Knozinger, H., Weitkamp, J. (eds.): Handbook of Heterogeneous Catalysis. Wiley-VCH, Weinheim (2008)
9. Sinfelt, J.H.: Catalysis by alloys and bimetallic clusters. Acc. Chem. Res. **10**, 15–20 (1977)
10. Coq, B., Figueras, F.: Bimetallic palladium catalysts: influence of the co-metal on the catalyst performance. J. Mol. Catal. A. **173**, 117–134 (2001)
11. Guczi, L.: Bimetallic nano-particles: featuring structure and reactivity. Catal. Today **101**, 53–64 (2005)
12. Ferrando, R., Jellinek, J., Johnston, R.L.: Nanoalloys: from theory to applications of alloy clusters and nanoparticles. Chem. Rev. **108**, 845–910 (2008)
13. Hammer, B., Nørskov, J.K.: Theoretical surface science and catalysis—calculations and concepts. Adv. Catal. **45**, 71–129 (2000)
14. Nørskov, J.K., Bligaard, T., Rossmeisl, J., Christensen, C.H.: Towards the computational design of solid catalysts. Nat. Chem. **1**, 37–46 (2009)
15. Neurock, M.: Perspectives on the first principles elucidation and the design of active sites. J. Catal. **216**, 73–88 (2003)

16. Reuter, K., Frenkel, D., Scheffler, M.: The steady state of heterogeneous catalysis studied by first-principles statistical mechanics. Phys. Rev. Lett. **93**, 116105-1-4 (2004)
17. Greeley, J., Mavrikakis, M.: Alloy catalysts designed from first principles. Nat. Mater. **3**, 810–815 (2004)
18. Digne, M., Sautet, P., Raybaud, P., Euzen, P., Toulhoat, H.: Use of DFT to achieve a rational understanding of acid-basic properties of γ-alumina surfaces. J. Catal. **226**, 54–68 (2004)
19. Loffreda, D., Delbecq, F., Vigné, F., Sautet, P.: Fast prediction of selectivity in heterogeneous catalysis from extended Bronsted-Evans-Polanyi relations: a theoretical insight. Angew. Chem. Int. Ed. **48**, 8978–8980 (2009)
20. Kresse, G., Furthmüller, J.: Efficient iterative schemes for ab initio total-energy calculations using a plane-wave basis set. Phys. Rev. B. **54**, 11169–11186 (1996)
21. Rodriguez, J.A., Goodman, D.W.: The nature of the metal–metal bond in bimetallic surfaces. Science **257**, 897–903 (1992)
22. Hammer, B., Morikawa, Y., Nørskov, J.K.: CO chemisorption at metal surfaces and overlayers. Phys. Rev. Lett. **76**, 2141–2144 (1996)
23. Dahl, S., Logadottir, A., Egeberg, R.C., Larsen, J.H., Chorkendorff, I., Törnqvist, E., Nørskov, J.K.: Role of steps in N2 activation on Ru(0001). Phys. Rev. Lett. **83**, 1814–1817 (1999)
24. Neurock, M., Pallassana, V., van Santen, R.A.: The importance of transient states at higher coverages in catalytic reactions. J. Am. Chem. Soc. **122**, 1150–1153 (2000)
25. Cremer, P.S., Su, X., Shen, Y.R., Somorjai, G.A.: Ethylene hydrogenation on Pt(111) monitored in situ at high pressures using sum frequency generation. J. Am. Chem. Soc. **118**, 2942–2949 (1996)
26. Nørskov, J.K., Bligaard, T., Logadottir, A., Bahn, S., Hansen, L.B., Bollinger, M., Bengaard, H., Hammer, B., Sljivancanin, Z., Mavrikakis, M., Xu, Y., Dahl, S., Jacobsen, C.J.H.: Universality in heterogeneous catalysis. J. Catal. **209**, 275–278 (2002)
27. Harris, N.: Modern Vacuum Practice. Nigel Harris, England (2007)
28. Henry, C.R.: Surface studies of supported model catalysts. Surf. Sci. Rep. **31**, 231–325 (1998)
29. Somorjai, G.A.: Introduction to Surface Chemistry and Catalysis. Wiley, New York (1994)
30. Piccolo, L.: Réactivité d'agrégats de palladium supportés sur MgO(100): adsorption du NO, réaction CO + NO et CO + O. Thesis, University of Marseille (1999)
31. Baümer, M., Freund, H.J.: Metal deposits on well-ordered oxide films. Prog. Surf. Sci. **61**, 127–198 (1999)
32. Santra, A.K., Goodman, D.W.: Oxide-supported metal clusters: models for heterogeneous catalysts. J. Phys. Condens. Mat. **15**, R31–R62 (2003)
33. Heiz, U., Schneider, W.D.: Nanoassembled model catalysts. J. Phys. Condens. Mat. **33**, R85–R102 (2000)
34. Rousset, J.L., Renouprez, A.J., Cadrot, A.M.: Ion-scattering study and Monte Carlo simulations of surface segregation in Pd–Pt nanoclusters obtained by laser vaporization of bulk alloys. Phys. Rev. B. **58**, 2150–2156 (1998)
35. Jürgens, B., Borchert, H., Ahrenstorf, K., Sonström, P., Pretorius, A., Schowalter, M., Gries, K., Zielasek, V., Rosenauer, A., Weller, H., Bäumer, M.: Colloidally prepared nanoparticles for the synthesis of structurally well-defined and highly active heterogeneous catalysts. Angew. Chem. Int. Ed. **47**, 8946–8949 (2008)
36. Uzio, D., Berhault, G.: Factors governing the catalytic reactivity of metallic nanoparticles. Catal. Rev. Sci. Eng. **52**, 106–131 (2010)
37. Kolasinski, K.W.: Surface science: foundations of catalysis and nanoscience. Wiley, Chichester (2008)
38. Molenbroek, A.M., Helveg, S., Topsøe, H., Clausen, B.S.: Nano-particles in heterogeneous catalysis. Top. Catal. **52**, 1303–1311 (2009)
39. Rodriguez, J.A., Goodman, D.W.: High-pressure catalytic reactions over single-crystal metal surfaces. Surf. Sci. Rep. **14**, 1–107 (1991)

40. Piccolo, L., Piednoir, A., Bertolini, J.: Pd–Au single-crystal surfaces: segregation properties and catalytic activity in the selective hydrogenation of 1,3-butadiene. Surf. Sci. **592**, 169–181 (2005)
41. Rucker, T.G., Franck, K., Colomb, D., Logan, M.A., Somorjai, G.A.: Novel high-pressure isolation cell capable of reaching 120 atm mounted in an ultrahigh-vacuum chamber. Rev. Sci. Instrum. **58**, 2292–2294 (1987)
42. Wang, L., Tysoe, W.T.: An investigation of ethylene hydrogenation catalyzed by metallic molybdenum using an isolatable high-pressure reactor: Identification of the reaction site and the role of carbonaceous deposits. J. Catal. **128**, 320–336 (1991)
43. Zhao, Z., Diemant, T., Häring, T., Rauscher, H., Behm, R.J.: Small-volume, ultrahigh-vacuum-compatible high-pressure reaction cell for combined kinetic and in situ IR spectroscopic measurements on planar model catalysts. Rev. Sci. Instrum. **76**, 123903-1-8 (2005)
44. Hendriksen, B.L.M., Frenken, J.W.M.: CO oxidation on Pt(110): scanning tunneling microscopy inside a high-pressure flow reactor. Phys. Rev. Lett. **89**, 046101-1-4 (2002)
45. Brongersma, H., Draxler, M., Deridder, M., Bauer, P.: Surface composition analysis by low-energy ion scattering. Surf. Sci. Rep. **62**, 63–109 (2007)
46. Endo, M., Matsumoto, T., Kubota, J., Domen, K., Hirose, C.: Oxidation of methanol by molecularly adsorbed oxygen on Pt(111) under vacuum and ambient pressure conditions studied by infrared reflection absorption spectroscopy: identification of formate intermediate. Surf. Sci. **441**, L931–L937 (1999)
47. Jugnet, Y., Aires, F.J.C.S., Deranlot, C., Piccolo, L., Bertolini, J.C.: CO chemisorption on Au(1 1 0) investigated under elevated pressures by polarized reflection absorption infrared spectroscopy and scanning tunneling microscopy. Surf. Sci. **521**, L639–L644 (2002)
48. Rupprechter, G.: Sum Frequency generation and polarization-modulation infrared reflection absorption spectroscopy of functioning model catalysts from ultrahigh vacuum to ambient pressure. Adv. Catal. **51**, 133–263 (2007)
49. Vang, R.T., Lauritsen, J.V., Lægsgaard, E., Besenbacher, F.: Scanning tunneling microscopy as a tool to study catalytically relevant model systems. Chem. Soc. Rev. **37**, 2191–2203 (2008)
50. Tao, F., Dag, S., Wang, L.-W., Liu, Z., Butcher, D.R., Bluhm, H., Salmeron, M., Somorjai, G.A.: Break-up of stepped platinum catalyst surfaces by high CO coverage. Science **327**, 850–853 (2010)
51. Ackermann, M., Pedersen, T., Hendriksen, B., Robach, O., Bobaru, S., Popa, I., Quiros, C., Kim, H., Hammer, B., Ferrer, S., Frenken, J.: Structure and reactivity of surface oxides on Pt(110) during catalytic CO oxidation. Phys. Rev. Lett. **95**, 255505-1-4 (2005)
52. Saint-Lager, M.C., Jugnet, Y., Dolle, P., Piccolo, L., Baudoing-Savois, R., Bertolini, J.C., Bailly, A., Robach, O., Walker, C., Ferrer, S.: Pd8Ni92(110) surface structure from surface X-ray diffraction. Surface evolution under hydrogen and butadiene reactants at elevated pressure. Surf. Sci. **587**, 229–235 (2005)
53. Salmeron, M., Schlögl, R.: Ambient pressure photoelectron spectroscopy: a new tool for surface science and nanotechnology. Surf. Sci. Rep. **63**, 169–199 (2008)
54. Tao, F., Grass, M.E., Zhang, Y., Butcher, D.R., Renzas, J.R., Liu, Z., Chung, J.Y., Mun, B.S., Salmeron, M., Somorjai, G.A.: Reaction-driven restructuring of Rh–Pd and Pt–Pd core-shell nanoparticles. Science **322**, 932–934 (2008)
55. Strongin, D.R., Carrazza, J., Bare, S.R., Somorjai, G.A.: The importance of C7 sites and surface roughness in the ammonia synthesis reaction over iron. J. Catal. **103**, 213–215 (1987)
56. Freund, H.J., Baümer, M., Kuhlenbeck, H.: Catalysis and surface science: what do we learn from studies of oxide-supported cluster model systems? Adv. Catal. **45**, 333–384 (2000)
57. Che, M., Bennett, C.O.: The influence of particle size on the catalytic properties of supported metals. Adv. Catal. **36**, 55–172 (1989)
58. Henry, C.R., Chapon, C., Giorgio, S., Goyhenex, C.: Size effect in heterogeneous catalysis: a surface science approach. In: Lambert, R.M., Pacchioni, G. (eds.) Chemisorption and

reactivity on supported clusters and thin films. Kluwer Academic Publishers, Dordrecht (1997)
59. Van Santen, R.A.: Complementary structure sensitive and insensitive catalytic relationships. Acc. Chem. Res. **42**, 57–66 (2009)
60. Heiz, U., Sanchez, A., Abbet, S., Schneider, W.D.: Catalytic oxidation of carbon monoxide on monodispersed platinum clusters: each atom counts. J. Am. Chem. Soc. **121**, 3214–3217 (1999)
61. Van Hardeveld, R., Hartog, F.: The statistics of surface atoms and surface sites on metal crystals. Surf. Sci. **15**, 189–230 (1969)
62. Piccolo, L., Henry, C.R.: NO-CO reaction kinetics on Pd/MgO model catalysts: morphology and support effects. J. Mol. Catal. A. **167**, 181–190 (2001)
63. Freund, H.J., Libuda, J., Baümer, M., Risse, T., Carlsson, A.: Cluster, facets, and edges: Site-dependent selective chemistry on model catalysts. Chem. Rec. **3**, 181–200 (2003)
64. Piccolo, L., Valcarcel, A., Bausach, M., Thomazeau, C., Uzio, D., Berhault, G.: Tuning the shape of nanoparticles to control their catalytic properties: selective hydrogenation of 1,3-butadiene on Pd/Al_2O_3. Phys. Chem. Chem. Phys. **10**, 5504–5506 (2008)
65. Li, Y., Liu, Q., Shen, W.: Morphology-dependent nanocatalysis: metal particles. Dalton Trans. **40**, 5811 (2011)
66. Kitakami, O., Sato, H., Shimada, Y., Sato, F., Tanaka, M.: Size effect on the crystal phase of cobalt fine particles. Phys. Rev. B. **56**, 13849–13854 (1997)
67. Nørskov, J.K., Bligaard, T., Hvolbæk, B., Abild-Pedersen, F., Chorkendorff, I., Christensen, C.H.: The nature of the active site in heterogeneous metal catalysis. Chem. Soc. Rev. **37**, 2163–2171 (2008)
68. Ertl, G.: Reactions at surfaces: from atoms to complexity (Nobel lecture). Angew. Chem. Int. Ed. **47**, 3524–3535 (2008)
69. Ertl, G.: Surface science and catalysis—studies on the mechanism of ammonia synthesis: the P. H. Emmett Award address. Catal. Rev. Sci. Eng. **21**, 201–223 (1980)
70. Stoltze, P., Nørskov, J.K.: Bridging the "pressure gap" between ultrahigh-vacuum surface physics and high-pressure catalysis. Phys. Rev. Lett. **55**, 2502–2505 (1985)
71. Honkala, K., Hellman, A., Remediakis, I.N., Logadottir, A., Carlsson, A., Dahl, S., Christensen, C.H., Nørskov, J.K.: Ammonia synthesis from first-principles calculations. Science **307**, 555–558 (2005)
72. Haruta, M., Yamada, N., Kobayashi, T., Iijima, S.: Gold catalysts prepared by coprecipitation for low-temperature oxidation of hydrogen and of carbon monoxide. J. Catal. **115**, 301–309 (1989)
73. Haruta, M.: Gold as a novel catalyst in the 21st century: preparation, working mechanism and applications. Gold. Bull. **37**, 27–36 (2004)
74. Bond, G.C., Louis, C., Thompson, D.T.: Catalysis by Gold. Imperial College Press, London (2006)
75. Cho, A.: Connecting the dots to custom catalysts. Science **299**, 1684–1685 (2003)
76. Hvolbæk, B., Janssens, T.V.W., Clausen, B.S., Falsig, H., Christensen, C.H., Nørskov, J.K.: Catalytic activity of Au nanoparticles. Nano Today **2**, 14–18 (2007)
77. Chen, M.S., Goodman, D.W.: The Structure of catalytically active gold on titania. Science **306**, 252–255 (2004)
78. Arrii, S., Morfin, F., Renouprez, A.J., Rousset, J.L.: Oxidation of CO on gold supported catalysts prepared by laser vaporization: direct evidence of support contribution. J. Am. Chem. Soc. **126**, 1199–1205 (2004)
79. Molina, L.M., Hammer, B.: Some recent theoretical advances in the understanding of the catalytic activity of Au. Appl. Catal. A **291**, 21–31 (2005)
80. Yoon, B., Häkkinen, H., Landman, U., Wörz, A.S., Antonietti, J.-M., Abbet, S., Judai, K., Heiz, U.: Charging effects on bonding and catalyzed oxidation of CO on Au8 clusters on MgO. Science **307**, 403–407 (2005)
81. Lin, X., Yang, B., Benia, H.-M., Myrach, P., Yulikov, M., Aumer, A., Brown, M.A., Sterrer, M., Bondarchuk, O., Kieseritzky, E., Rocker, J., Risse, T., Gao, H.-J., Nilius, N., Freund, H.-J.:

Charge-mediated adsorption behavior of CO on MgO-supported Au clusters. J. Am. Chem. Soc. **132**, 7745–7749 (2010)
82. Nassreddine, S., Massin, L., Aouine, M., Geantet, C., Piccolo, L.: Thiotolerant Ir/SiO2-Al2O3 bifunctional catalysts: Effect of metal-acid site balance on tetralin hydroconversion. J. Catal. **278**, 253–265 (2011)
83. Pajonk, G.M.: Contribution of spillover effects to heterogeneous catalysis. Appl. Catal. A **202**, 157–169 (2000)
84. Henry, C.R., Chapon, C., Duriez, C.: Precursor state in the chemisorption of CO on supported palladium clusters. J. Chem. Phys. **95**, 700–705 (1991)
85. Libuda, J., Freund, H.J.: Molecular beam experiments on model catalysts. Surf. Sci. Rep. **57**, 157–298 (2005)
86. Röttgen, M.A., Abbet, S., Judai, K., Antonietti, J.-M., Wörz, A.S., Arenz, M., Henry, C.R., Heiz, U.: Cluster chemistry: size-dependent reactivity induced by reverse spill-over. J. Am. Chem. Soc. **129**, 9635–9639 (2007)
87. Benkhaled, M., Descorme, C., Duprez, D., Morin, S., Thomazeau, C., Uzio, D.: Study of hydrogen surface mobility and hydrogenation reactions over alumina-supported palladium catalysts. Appl. Catal. A **346**, 36–43 (2008)
88. Tauster, S.J., Fung, S.C., Garten, R.L.: Strong metal-support interactions. Group 8 noble metals supported on titanium dioxide. J. Am. Chem. Soc. **100**, 170–175 (1978)
89. Pesty, F., Steinrück, H.-P., Madey, T.E.: Thermal stability of Pt films on TiO_2(110): evidence for encapsulation. Surf. Sci. **339**, 83–95 (1995)
90. Dulub, O., Hebenstreit, W., Diebold, U.: Imaging cluster surfaces with atomic resolution: the strong metal-support interaction state of Pt supported on TiO_2(110). Phys. Rev. Lett. **84**, 3646 (2000)
91. Haller, G.: New catalytic concepts from new materials: understanding catalysis from a fundamental perspective, past, present, and future. J. Catal. **216**, 12–22 (2003)
92. Fu, Q., Wagner, T.: Interaction of nanostructured metal overlayers with oxide surfaces. Surf. Sci. Rep. **62**, 431–498 (2007)
93. Qin, Z.-H., Lewandowski, M., Sun, Y.-N., Shaikhutdinov, S., Freund, H.-J.: Encapsulation of Pt nanoparticles as a result of strong metal−support interaction with Fe_3O_4 (111). J. Phys. Chem. C **112**, 10209–10213 (2008)
94. Graoui, H., Giorgio, S., Henry, C.R.: Shape variations of Pd particles under oxygen adsorption. Surf. Sci. **417**, 350–360 (1998)
95. Yang, F., Chen, M.S., Goodman, D.W.: Sintering of Au particles supported on TiO_2(110) during CO oxidation. J. Phys. Chem. C **113**, 254–260 (2009)
96. Tao, F., Dag, S., Wang, L.-W., Liu, Z., Butcher, D.R., Salmeron, M., Somorjai, G.A.: Restructuring of hex-Pt(100) under CO gas environments: formation of 2-D nanoclusters. Nano Lett. **9**, 2167–2171 (2009)
97. Peters, K.F., Steadman, P., Isern, H., Alvarez, J., Ferrer, S.: Elevated-pressure chemical reactivity of carbon monoxide over Au(111). Surf. Sci. **467**, 10–22 (2000)
98. Loffreda, D., Piccolo, L., Sautet, P.: Surface restructuring under gas pressure from first principles: a mechanism for CO-induced removal of the Au(110)-(1×2) reconstruction. Phys. Rev. B. **71**, 113414-1-4 (2005)
99. Madey, T.E., Chen, W., Wang, H., Kaghazchi, P., Jacob, T.: Nanoscale surface chemistry over faceted substrates: structure, reactivity and nanotemplates. Chem. Soc. Rev. **37**, 2310–2327 (2008)
100. Teschner, D., Borsodi, J., Wootsch, A., Revay, Z., Havecker, M., Knop-Gericke, A., Jackson, S.D., Schlogl, R.: The roles of subsurface carbon and hydrogen in palladium-catalyzed alkyne hydrogenation. Science **320**, 86–89 (2008)
101. Sachtler, W.M.H., Dorgelo, G.J.H., Jongepier, R.: Phase composition and work function of vacuum-deposited copper-nickel alloy films. J. Catal. **4**, 100–102 (1965)
102. Sachtler, W.M.H., Van Der Plank, P.: The role of individual surface atoms in chemisorption and catalysis by nickel-copper alloys. Surf. Sci. **18**, 62–79 (1969)
103. Campbell, C.T.: Bimetallic surface chemistry. Annu. Rev. Phys. Chem. **41**, 775–837 (1990)

104. Rodriguez, J.A.: Physical and chemical properties of bimetallic surfaces. Surf. Sci. Rep. **24**, 223–287 (1996)
105. Chen, J.G., Menning, C.A., Zellner, M.B.: Monolayer bimetallic surfaces: Experimental and theoretical studies of trends in electronic and chemical properties. Surf. Sci. Rep. **63**, 201–254 (2008)
106. Bertolini, J.-C.: Model catalysis by metals and alloys: from single-crystal surfaces to well-defined nano-particles. Catal. Today **138**, 84–96 (2008)
107. Rousset, J.L., Stievano, L., Aires, F.J.C.S., Geantet, C., Renouprez, A.J., Pellarin, M.: Hydrogenation of toluene over γ-Al_2O_3-supported Pt, Pd, and Pd–Pt model catalysts obtained by laser vaporization of bulk metals. J. Catal. **197**, 335–343 (2001)
108. Ormerod, R.M., Baddeley, C.J., Lambert, R.M.: Geometrical and electronic effects in the conversion of acetylene to benzene on Au(111)/Pd and Au/Pd surface alloys. Surf. Sci. **259**, L709–L713 (1991)
109. Ponec, V.: Alloy catalysts: the concepts. Appl. Catal. A **222**, 31–45 (2001)
110. Maroun, F., Ozanam, F., Magnussen, O.M., Behm, R.J.: The role of atomic ensembles in the reactivity of bimetallic electrocatalysts. Science **293**, 1811–1814 (2001)
111. Chen, M., Kumar, D., Yi, C.W., Goodman, D.W.: The promotional effect of gold in catalysis by palladium-gold. Science **310**, 291–293 (2005)
112. Vang, R.T., Honkala, K., Dahl, S., Vestergaard, E.K., Schnadt, J., Laegsgaard, E., Clausen, B.S., Nørskov, J.K., Besenbacher, F.: Controlling the catalytic bond-breaking selectivity of Ni surfaces by step blocking. Nat. Mater. **4**, 160–162 (2005)
113. Stroppa, A., Mittendorfer, F., Andersen, J.N., Parteder, G., Allegretti, F., Surnev, S., Netzer, F.P.: Adsorption and dissociation of CO on bare and Ni-decorated stepped Rh(553) surfaces. J. Phys. Chem. C **113**, 942–949 (2009)
114. Besenbacher, F., Chorkendorff, I., Clausen, B.S., Hammer, B., Molenbroek, A.M., Nørskov, J.K., Stensgaard, I.: Design of a surface alloy catalyst for steam reforming. Science **279**, 1913–1915 (1998)
115. Weigand, P., Jelinek, B., Hofer, W., Varga, P.: Preferential sputtering and segregation reversal: (100) and (110) surfaces of Pt25Ni75 single crystal alloys. Surf. Sci. **301**, 306–312 (1994)
116. Yudanov, I.V., Neyman, K.M.: Stabilization of Au at edges of bimetallic PdAu nanocrystallites. Phys. Chem. Chem. Phys. **12**, 5094–5100 (2010)
117. Abbott, H.L., Aumer, A., Lei, Y., Asokan, C., Meyer, R.J., Sterrer, M., Shaikhutdinov, S., Freund, H.-J.: CO adsorption on monometallic and bimetallic Au–Pd nanoparticles supported on oxide thin films. J. Phys. Chem. C **114**, 17099–17104 (2010)
118. Stamenkovic, V.R., Fowler, B., Mun, B.S., Wang, G., Ross, P.N., Lucas, C.A., Marković, N.M.: Improved oxygen reduction activity on Pt3Ni(111) via increased surface site availability. Science **315**, 493–497 (2007)
119. Stamenkovic, V.R., Mun, B.S., Arenz, M., Mayrhofer, K.J.J., Lucas, C.A., Wang, G., Ross, P.N., Markovic, N.M.: Trends in electrocatalysis on extended and nanoscale Pt-bimetallic alloy surfaces. Nat. Mater. **6**, 241–247 (2007)
120. Wu, J., Zhang, J., Peng, Z., Yang, S., Wagner, F.T., Yang, H.: Truncated octahedral Pt3Ni oxygen reduction reaction electrocatalysts. J. Am. Chem. Soc. **132**, 4984–4985 (2010)
121. Strasser, P., Koh, S., Anniyev, T., Greeley, J., More, K., Yu, C., Liu, Z., Kaya, S., Nordlund, D., Ogasawara, H., Toney, M.F., Nilsson, A.: Lattice-strain control of the activity in dealloyed core–shell fuel cell catalysts. Nat. Chem. **2**, 454–460 (2010)
122. Kitchin, J.R., Nrskov, J.K., Barteau, M.A., Chen, J.G.: Role of strain and ligand effects in the modification of the electronic and chemical properties of bimetallic surfaces. Phys. Rev. Lett. **93**, 156801-1-4 (2004)
123. Schlapka, A., Lischka, M., Groß, A., Käsberger, U., Jakob, P.: Surface strain versus substrate interaction in heteroepitaxial metal layers: Pt on Ru(0001). Phys. Rev. Lett. **91**, 016101-1-4 (2003)
124. Kibler, L.A., El-Aziz, A.M., Hoyer, R., Kolb, D.M.: Tuning reaction rates by lateral strain in a palladium monolayer. Angew. Chem. Int. Ed. **44**, 2080–2084 (2005)

125. Wang, J.X., Inada, H., Wu, L., Zhu, Y., Choi, Y., Liu, P., Zhou, W.-P., Adzic, R.R.: Oxygen reduction on well-defined core−shell nanocatalysts: particle size, facet, and Pt shell thickness effects. J. Am. Chem. Soc. **131**, 17298–17302 (2009)
126. Chen, S., Ferreira, P.J., Sheng, W., Yabuuchi, N., Allard, L.F., Shao-Horn, Y.: Enhanced activity for oxygen reduction reaction on "Pt3Co" nanoparticles: direct evidence of percolated and sandwich-segregation structures. J. Am. Chem. Soc. **130**, 13818–13819 (2008)
127. Miura, H.: Preparation of supported bimetallic catalysts by means of selective deposition using mobile metal compounds as precursors. Catal. Today **28**, 215–221 (1996)
128. Hugon, A., Delannoy, L., Krafft, J.-M., Louis, C.: Selective hydrogenation of 1,3-butadiene in the presence of an excess of alkenes over supported bimetallic gold−palladium catalysts. J. Phys. Chem. C **114**, 10823–10835 (2010)
129. Jugnet, Y., Sedrati, R., Bertolini, J.: Selective hydrogenation of 1,3-butadiene on Pt3Sn(111) alloys: comparison to Pt(111). J. Catal. **229**, 252–258 (2005)
130. Zhao, H., Koel, B.E.: Hydrogenation of 1,3-butadiene on two ordered Sn/Pt(111) surface alloys. J. Catal. **234**, 24–32 (2005)
131. Vigné, F., Haubrich, J., Loffreda, D., Sautet, P., Delbecq, F.: Highly selective hydrogenation of butadiene on Pt/Sn alloy elucidated by first-principles calculations. J. Catal. **275**, 129–139 (2010)
132. Gonzalez, S., Neyman, K.M., Shaikhutdinov, S., Freund, H.-J., Illas, F.: On the promoting role of Ag in selective hydrogenation reactions over Pd–Ag bimetallic catalysts: a theoretical study. J. Phys. Chem. C **111**, 6852–6856 (2007)
133. Piccolo, L., Piednoir, A., Bertolini, J.: Absorption and oxidation of hydrogen at Pd and Pd–Au (111) surfaces. Surf. Sci. **600**, 4211–4215 (2006)
134. Valcarcel, A., Morfin, F., Piccolo, L.: Alkene hydrogenation on metal surfaces: Why and when are Pd overlayers more efficient catalysts than bulk Pd? J. Catal. **263**, 315–320 (2009)
135. Filhol, J.-S., Simon, D., Sautet, P.: Understanding the high activity of a nanostructured catalyst obtained by a deposit of Pd on Ni: first principle calculations. J. Am. Chem. Soc. **126**, 3228–3233 (2004)
136. Valcarcel, A., Loffreda, D., Delbecq, F., Piccolo, L.: Structure of the Pd8Ni92(110) catalytic surface from first principles. Phys. Rev. B. **76**, 125406-1-7 (2007)
137. Andersson, K.J., Calle-Vallejo, F., Rossmeisl, J., Chorkendorff, I.: Adsorption-driven surface segregation of the less reactive alloy component. J. Am. Chem. Soc. **131**, 2404–2407 (2009)
138. Mayrhofer, K.J.J., Juhart, V., Hartl, K., Hanzlik, M., Arenz, M.: Adsorbate-induced surface segregation for core-shell nanocatalysts. Angew. Chem. Int. Ed. **48**, 3529–3531 (2009)
139. Mayrhofer, K.J.J., Hartl, K., Juhart, V., Arenz, M.: Degradation of carbon-supported Pt bimetallic nanoparticles by surface segregation. J. Am. Chem. Soc. **131**, 16348–16349 (2009)
140. Papaefthimiou, V., Dintzer, T., Dupuis, V., Tamion, A., Tournus, F., Teschner, D., Hävecker, M., Knop-Gericke, A., Schlögl, R., Zafeiratos, S.: When a metastable oxide stabilizes at the nanoscale: wurtzite CoO formation upon dealloying of PtCo nanoparticles. J. Phys. Chem. Lett. **2**, 900–904 (2011)
141. Zhou, S., Yin, H., Schwartz, V., Wu, Z., Mullins, D., Eichhorn, B., Overbury, S.H., Dai, S.: In situ phase separation of NiAu alloy nanoparticles for preparing highly active Au/NiO CO oxidation catalysts. ChemPhysChem **9**, 2475–2479 (2008)
142. Hansgen, D.A., Vlachos, D.G., Chen, J.G.: Using first principles to predict bimetallic catalysts for the ammonia decomposition reaction. Nat. Chem. **2**, 484–489 (2010)
143. Studt, F., Abild-Pedersen, F., Bligaard, T., Sorensen, R.Z., Christensen, C.H., Nørskov, J.K.: Identification of non-precious metal alloy catalysts for selective hydrogenation of acetylene. Science **320**, 1320–1322 (2008)
144. Linic, S., Jankowiak, J., Barteau, M.A.: Selectivity driven design of bimetallic ethylene epoxidation catalysts from first principles. J. Catal. **224**, 489–493 (2004)

145. Piccinin, S., Zafeiratos, S., Stampfl, C., Hansen, T.W., Hävecker, M., Teschner, D., Bukhtiyarov, V.I., Girgsdies, F., Knop-Gericke, A., Schlögl, R., Scheffler, M.: Alloy catalyst in a reactive environment: the example of Ag–Cu particles for ethylene epoxidation. Phys. Rev. Lett. **104**, 035503 (2010)
146. Nilekar, A.U., Alayoglu, S., Eichhorn, B., Mavrikakis, M.: Preferential CO oxidation in hydrogen: reactivity of core−shell nanoparticles. J. Am. Chem. Soc. **132**, 7418–7428 (2010)
147. Quinet, E., Piccolo, L., Morfin, F., Avenier, P., Diehl, F., Caps, V., Rousset, J.L.: On the mechanism of hydrogen-promoted gold-catalyzed CO oxidation. J. Catal. **268**, 384–389 (2009)
148. Alayoglu, S., Nilekar, A.U., Mavrikakis, M., Eichhorn, B.: Ru–Pt core-shell nanoparticles for preferential oxidation of carbon monoxide in hydrogen. Nat. Mater. **7**, 333–338 (2008)

Index

D. Alloyeau et al. (eds.), *Nanoalloys*, Engineering Materials
DOI: 10.1007/978-1-4471-4014-6,

D

E

F

K

L

M

N

T

MIX
Papier aus verantwortungsvollen Quellen
Paper from responsible sources
FSC® C105338

If you have any concerns about our products,
you can contact us on
ProductSafety@springernature.com

In case Publisher is established outside the EU,
the EU authorized representative is:
Springer Nature Customer Service Center GmbH
Europaplatz 3, 69115 Heidelberg, Germany

Printed by Libri Plureos GmbH
in Hamburg, Germany